Zoology in Early Modern Culture

Intersections

Interdisciplinary Studies in Early Modern Culture

VOLUME 32 – 2014

The titles published in this series are listed at *brill.com/inte*

Zoology in Early Modern Culture

Intersections of Science, Theology, Philology, and Political and Religious Education

Edited by

Karl A.E. Enenkel and Paul J. Smith

BRILL

LEIDEN | BOSTON

Cover illustration: Excerpts from the images in the chapters by Karl A.E. Enenkel and Amanda Herrin.

Library of Congress Cataloging-in-Publication Data

Zoology in early modern culture : intersections of science, theology, philology, and political and religious education / edited by Karl A.E. Enenkel and Paul J. Smith.
pages cm. — (Intersections, ISSN 1568-1181 ; volume 32)
Includes bibliographical references and index.
ISBN 978-90-04-26823-4 (hardback : alk. paper) — ISBN 978-90-04-27917-9 (e-book) 1. Zoology—History. I. Enenkel, K. A. E., editor. II. Smith, P. J. (Paul J.) editor.

QL15.E25 2014
590—dc23

2014023662

This publication has been typeset in the multilingual 'Brill' typeface. With over 5,100 characters covering Latin, IPA, Greek, and Cyrillic, this typeface is especially suitable for use in the humanities.
For more information, please see brill.com/brill-typeface.

ISSN 1568-1181
ISBN 978-90-04-26823-4 (hardback)
ISBN 978-90-04-27917-9 (e-book)

This book is printed on acid-free paper.

Contents

Physiology and Political Ideology

Acknowledgements

The majority of the contributions published in the present volume were selected from papers delivered at the conference *Natural History and the Arts from the Perspective of Religion and Politics, 15th–18th Centuries*, held at the Westfälische Wilhelms-Universität Münster in May 2012, and they appear now in much revised and extended forms. The conference *Natural History and the Arts* was organized by the *Seminar für Lateinische Philologie des Mittelalters und der Neuzeit*, directed by Karl Enenkel, in collaboration with the research project *Cultural Representations of Living Nature: Dynamics of Intermedial Recording in Text and Image* (*ca.* 1550–1670) of Leiden University, directed by Paul Smith and funded by the Netherlands Organisation for Scientific Research (NWO). The contribution "The Species and Beyond [...]" was written as part of Enenkel's research programme *The New Management of Knowledge in the Early Modern Period*, funded by the NWO. The above-mentioned conference would not have been possible without generous grants from the Westfälische Wilhelms-Universität, the *Seminar für Lateinische Philologie des Mittelalters und der Neuzeit*, and the *Cluster of Excellence "Religion und Politik"*. For assistance with the organisation of the conference we are grateful to Christian Peters.

List of Illustrations

Figures 2.1–2.38 (accompanying the article by Karl Enenkel, "The Species and Beyond [...]"):

Figures 3.1a–3.11 (accompanying the article by Sophia Hendrikx):

Figures 8.1–8.9 (accompanying the article by Paul Smith):

Figures 9.1–9.22 (accompanying the article by Amanda Herrin):

Figures 10.1–10.11 (accompanying the article by Marrigje Rikken):

Figures 11.1–11.5 (accompanying the article by Sabine Kalff):

Notes on the Editors

Karl Enenkel

is Professor of Medieval Latin and Neo-Latin at the University of Münster (Germany). Previously he was Professor of Neo-Latin at Leiden University (Netherlands). He is a member of the *Royal Netherlands Academy of Arts and Sciences*. He has published widely on international Humanism, early modern culture, paratexts, literary genres 1300–1600, Neo-Latin emblems, word and image relationships, and the history of scholarship and science. Among his major book publications are *Francesco Petrarca: De vita solitaria, Buch 1.* (1991), *Die Erfindung des Menschen. Die Autobiographik des frühneuzeitlichen Humanismus von Petrarca bis Lipsius* (2008), and *Die Stiftung von Autorschaft in der neulateinischen Literatur (ca. 1350–ca. 1650). Zur autorisierenden und wissensvermittelnden Funktion von Widmungen, Vorworttexten, Autorporträts und Dedikationsbildern* (2014). He has (co)edited and co-authored some 25 volumes, among others, *Modelling the Individual. Biography and Portrait in the Renaissance* (1998), *Recreating Ancient History* (2001), *Mundus Emblematicus. Studies in Neo-Latin Emblem Books* (2003), *Cognition and the Book* (2004), *Petrarch and his Readers* (2006), *Early Modern Zoology* (2007), *The Sense of Suffering. Constructions of Physical Pain in Early Modern Culture* (2009), *The Neo-Latin Epigram* (2009), *Meditatio—Refashioning the Self. Theory and Practice in Late Medieval and Early Modern Intellectual Culture* (2011), *Portuguese Humanism* (2011), *The Authority of the Word* (2011), *Discourses of Power. Ideology and Politics in Neo-Latin Literature* (2012), *The Reception of Erasmus* (2013), *Transformation of the Classics* (2013), *Die* Vita *als Vermittlerin von Wissenschaft und Werk* (2013), and *Neo-Latin Commentaries and the Management of Knowledge* (2013). He has founded the international series *Intersections* (Brill); *Proteus. Studies in Early Modern Identity Formation*; *Speculum Sanitatis: Studies in Medieval and Early Modern Medical Culture (500–1800)* (both Brepols), and *Scientia universalis. Studien und Texteditionen zur Wissensgeschichte der Vormoderne* (LIT-Verlag). He is member of the board of, among others, *Humanistica Lovaniensia*, and the *Conseil international pour l'édition des oeuvres complètes d'Erasme.*

Paul J. Smith

is Professor of French literature at Leiden University. He has widely published on 16th, 17th, and 20th century French literature, its reception in the Netherlands, French and Dutch fable and emblem books, literary rhetoric, intermediality, and animal symbolism and early modern zoology, and its presence in art and literature. His main book publications include *Voyage et écriture. Etude sur le* Quart Livre *de Rabelais* (1987), *Het schouwtoneel der dieren. Embleemfabels in de Nederlanden (1567–ca. 1670)* (2006), *Dispositio. Problematic Ordering in French Renaissance Literature* (2007), and *Réécrire la Renaissance, de Marcel Proust à Michel Tournier. Exercices de lecture rapprochée* (2009) He is co-author of *Francis Ponge: lectures et méthodes* (2004), editor of *Editer et traduire Rabelais à travers les âges* (1997), *Translating Montaigne* (2011) and co-editor of *Lectures de René Char* (1990), *Fabuleux La Fontaine* (1996), *Le paradoxe en linguistique et en littérature* (1996), *Montaigne and the Low Countries (1580–1700)* (2007) and *Early Modern Zoology. The Construction of Animals in Science, Literature and the Visual Arts* (2007). He is member of the editorial board of *Intersections, Neophilologus, Montaigne Studies*, and *Renaissance and Reformation.*

Notes on the Contributors

Tamás Demeter
is Senior Research Fellow at the Institute of Philosophy, *Hungarian Academy of Sciences*, and teaches at the University of Pécs, Hungary. He received his PhD from the University of Cambridge. He has held various fellowships at the Universities of Bristol, Cambridge, and Edinburgh; the National University of Singapore; the *Netherlands Institute for Advanced Study*; and the *Max Planck Institute for the History of Science* in Berlin. He has published widely on early modern natural and moral philosophiy and is a member of the editorial board of *Studies in East European Thought.*

Sophia Hendrikx
took her MA in history and book history at the University of Oxford and the University of Amsterdam (UvA, Netherlands). She is a rare books specialist at *Rare Fish Books Antiquarian Booksellers* (Amsterdam). Currently she is working on a research project on Conrad Gessner's ichthyological publications.

Amanda K. Herrin
took her PhD at the *Institute of Fine Arts*, New York University. She specializes in sixteenth- and seventeenth-century Netherlandish, Dutch, and Flemish art, as well as Pre-Columbian and Colonial Latin American art. Research interests include sixteenth- and seventeenth-century printmaking as visual exegesis; the intersections between art and science in the early modern period; issues of originality and imitation; and the trans-Atlantic art market.

Eric Jorink
works as a Senior Researcher at the *Huygens Institute for the History of the Netherlands* (The Hague), which is part of the *Royal Netherlands Academy of Arts and Sciences* (KNAW), and holds a special chair at the University of Leiden on behalf of the Teylers Foundation, dedicated to Enlightenment and religion from historical, social, and cultural perspectives. He has studied history at the University of Groningen and the *École des Hautes Études en Sciences Sociales* in Paris. He has published widely on scientific culture in early modern Europe, and his works include *Reading the Book of Nature in the Dutch Golden Age, 1575–1715* (2010). He is especially interested in the relationship between art, religion, and scholarship. In 2012–2013 he was

Andrew W. Mellon Visiting Professor at the *Courtauld Institute of Art*, University of London. Currently he is working on a biography of Johannes Swammerdam.

Sabine Kalff
is a research fellow in the Department of German Literature at the Humboldt University in Berlin. She studied German literature, cultural history, and American studies at Humboldt University and at La Sapienza in Rome. She completed her doctoral thesis on political medicine in the early modern period in 2011. She has published on early modern medicine and natural philosophy, and she recently worked on the interrelation between genres of life-writing since the early modern period and cultural techniques, such as the calculation of birth charts or list-making. She co-edited (with Nikola Roßbach, Thomas Stäcker, Flemming Schock, Constanze Baum, and Imke Harjes) *Welt und Wissen auf der Bühne. Die Theatrum-Literatur der Frühen Neuzeit. Repertorium* (2011).

Alexander Loose
is a lecturer in medieval Latin and Neo-Latin at the Martin-Luther-Universität in Halle/Wittenberg. He studied theology at the Humboldt-University in Berlin and Latin at the Martin-Luther-Universität in Halle/Wittenberg, where he took his PhD. He is the author of a commented edition of Johannes Dubravius, *Theriobulia* (2011).

Brian W. Ogilvie
is an associate professor of history at the University of Massachusetts–Amherst, where he specializes in the history of science in Europe from the Renaissance to the Enlightenment. He is the author of *The Science of Describing: Natural History in Renaissance Europe* (2006).

Marrigje Rikken
is a PhD candidate at Leiden University and a university lecturer in art history at the University of Amsterdam. She worked at the Rijksmuseum Amsterdam from 2006 to 2008 as assistant curator of Dutch 17th-century paintings. She specializes in sixteenth- and seventeenth-century Dutch and Flemish paintings, prints, and drawings. She is the author of a book on *Melchior d'Hondecoeter: Bird Painter* (2008). Her dissertation discusses the relation of Flemish artists to natural historical knowledge, between 1550 and 1650.

Bernd Roling
is professor for classical and medieval Latin at the *Institute for Greek and Latin Philology* of the Freie Universität in Berlin. His research interests include high medieval and early modern Latin poetry; medieval and early modern philosophy, especially the philosophy of language; the history of early modern science; university history, with a special focus on Scandinavia; and early modern esoteric traditions. He is the author of, among others, *Christliche Kabbalah und aristotelische Naturphilosophie im Werk des Paulus Ritius* (2007); *Locutio angelica. Die Diskussion der Engelsprache im Mittelalter und der Frühen Neuzeit als Antizipation einer Sprechakttheorie* (2008); *Drachen und Sirenen: Die Aufarbeitung und Abwicklung der Mythologie an den europäischen Universitäten* (2010); and *Physica Sacra: Wunder, Naturwissenschaft und historischer Schriftsinn zwischen Mittelalter und Früher Neuzeit* (2013). Together with Iolanda Ventura and Baudoin van den Abeele he edited Bartholomaeus Anglicus, *De proprietatibus rerum*, vol. 1 (2007). He is currently preparing a book on the Swedish polymath Olaus Rudbeck and his reception in eighteenth-century Northern Europe.

INTRODUCTION

Intersections of Science, Theology, Philology, and Political and Religious Education

Karl A.E. Enenkel

The present volume is a follow-up project of *Early Modern Zoology. The Construction of Animals in Science, Literature and the Visual Arts*, published some seven years ago in the same series.[1] In the present volume, it is our aim to further explore the fascinating alterity of early modern zoology by focusing on *theology* and *philology*, and their interplay with other methods of early modern knowledge production. Theology and philology certainly belong to the most important alterity aspects of early modern zoology. If one looks at the basic principles, methods, tools, and research practices of the modern science of zoology (in the 20th and 21st centuries), it is clear that there is in fact no place for religion or theology, and hardly any for philology; and there is only a very limited amount of attention devoted to what may be called literary tradition. The very foundation of modern zoology—evolutionary theory—excludes all religious worldviews that are based on the assumption of a single *creatio ex nihilo* of the species by a godly creator. Evolutionary theory demonstrates that the species presently living have evolved during an evolutionary process of hundreds of millions of years, which has caused all kinds of adaptations; life once started from amoebas, and it has slowly developed into the most complex organisms, the mammals, with man as the highest developed species at the top of the pyramid of evolution. According to *Genesis*, however, man, plants, and animals (i.e. *all* known species) were created by God at approximately the same time in Paradise.

Moreover, in Christian theology man was completely set apart from the animals: he was not part of animal life, and nobody regarded him as the offspring of animal ancestors. In the modern zoological system of species, as a matter of course man (*Homo sapiens*) partakes in evolution: he is regarded as belonging to the category of the *Primates*, and he has the same early pedigree as chimpanzees, orangutans, gorillas, and gibbons, but also the *Lemures*, makis and bushbabies (*Galagonidae*). The more direct ancestors of *Homo sapiens*, the family

1 Enenkel K.A.E. – Smith P.J. (eds.), *Early Modern Zoology. The Construction of Animals in Science, Literature and the Visual Arts*, Intersections 7 (Leiden – Boston: 2007), 2 vols.

 | DOI 10.1163/9789004279179_002

Hominidae or great apes (consisting of seven surviving species), diverged from the family of the *Cercopithecidae* (guenons) some 24 million to 30 million years ago, and from that of the gibbons or *Hylobatidae* some 12 million to 18 million years ago; *Homo* (not yet *Homo sapiens*) diverged from other *Homidae* or great ape species, such as chimpanzees, orangutans, and gorillas, some 3 million to 4 million years ago. In comparison, according to the Bible the creation of both man and all animal species took place only some 6,000 years ago. Modern zoology does not attribute to man (*Homo sapiens*), the only surviving subspecies of *Homo*, an ontological status that would differ from that of other *mammalia* species. Ethological research and behavioural biology treat man and other animal species along the same lines, and the same goes for comparative morphology and physiology. The academic discipline of theology is in fact irrelevant to the science of zoology, at least if one talks about the methods and practices of zoological research. The understanding of nature is no longer intertwined with theological knowledge. The modern understanding of the science of zoology is characterised by sharp chronological distinctions: zoology before and after Linnaeus, before and after Darwin.

While modern zoology hardly leaves any room for theology, there are a number of intersections with political ideology. These pertain to the status of animals in the 20th and 21st centuries, animal rights, regulations of professional and commercial animal keeping, food politics, environmental questions in the broadest sense, and the preservation of wild living species. With respect to the use of animals and their status in the modern world, a consensus seems to have developed that some kind of preservation of wild living species is necessary in order to prevent the extinction of species. Nowadays, animals can be regarded as endangered beings. Many political parties in the Western world include some sort of environmental issues in their ideological programme, while some countries even have political parties in defence of animals.

In the early modern period, however, almost nobody felt the need to preserve wild living species. Especially the European *Homo sapiens* was about to conquer and "civilise" more and more of the areas of "wild" nature. Animals were predominantly seen either as useful, as harmful, or as a mirror of God's wisdom, and they were treated accordingly. On the one hand, Christian belief legitimised the use of animals through the authority of the Bible; on the other hand, animals were regarded as an important part of God's creation, and therefore as a manifestation of his unlimited prudence, wisdom, goodness, and beauty. It is exactly this fact that made them an excellent object of study. The understanding of nature was somehow equated with the understanding of God. God was the author of the "Book of Nature", and it was man's task to read and interpret it, and to contemplate on it. And it is clear this could be done in many different ways.

The contributions of the present volume try to map out the different ways in which the reading of the "Book of Nature" by early modern naturalists took shape. The "Book of Nature" comprises, among other things, the careful description of animal species, using certain literary, scholarly, or pictorial traditions, as well as empirical observations, vivisection, and eyewitness accounts (contributions by Enenkel on "The Species and Beyond", Hendrikx, Jorink, and Ogilvie); furthermore, it also involves the production of zoological illustrations in woodcuts and engravings (Hendrikx, Enenkel on "The Species and Beyond", Jorink, and Herrin), and the "translation" of zoological species into visual art for different purposes, such as religious devotion and prayer, scientific and philosophical contemplation, or scholarly curiosity (contributions by Jorink, Smith, Herrin, and Rikken). It includes theoretical, philosophical, and theological thinking regarding God's creation, the Flood, and the generation and procreation of animals (contributions by Enenkel on "The Species and Beyond", Roling on Bartholin, Roling on palingenesis, Jorink, and Ogilvie); new attempts with respect to animal nomenclature and taxonomy (contributions by Hendrikx, Enenkel on the "Species and Beyond", and Ogilvie); biblical exegesis in word and image (contributions by Smith, Herrin, and Enenkel); philological comment upon classical authors (contributions by Enenkel on "The Species and Beyond" and Smith); the application of etymology, proverbial wisdom, and Erasmus's *Adagia* (contributions by Smith, Loose, and Kalff); translations, either of Greek sources into Latin or classical sources into vernacular languages; impressive collections of either literary or pictorial sources (contributions by Hendrikx and Herrin), of natural objects in early modern *Wunderkammern* (contributions by Roling on Bartholin, Rikken, and Herrin), or of animals in (princely) menageries (Rikken and Herrin). It also includes engagement in practical issues and questions, e.g. the problems and methods of animal breeding (contribution by Enenkel on "The Species and Beyond") and maintenance; cooking; medicinal recipes with regard to animal substances (contribution by Roling on Bartholin); and analysis of the anatomy and physiology of animals by vivisection and experiments (contributions by Jorink and Roling on Bartholin). At the same time it also includes—even on a large scale—the production and discussion of symbolic meanings ascribed to animals, with respect to ethics, religion, politics, and social hierarchy; and, in more general terms, the use of animals as transmitters of various kinds of applicable knowledge in different literary and scholarly contexts, such as emblematics, fable literature, books of memory, satire, political and religious pamphlets, etc. (contributions by Kalff, Loose, Herrin, and Smith). In many cases, these different approaches were interconnected and were applied in various combinations: in early modern zoology rationalisation, analysis, and empirical observation were always intertwined with religious devotion and the search for admirable or miraculous aspects.

Indeed, most of the above-mentioned approaches, fields of interest, and methods of research were in one way or another connected with a distinct search for the admirable or wondrous (*mirabile*). This goes back to the major classical sources of early modern zoology—Aristotle, Pliny, and the Bible—and to medieval zoology as well, for example Albertus Magnus. As is demonstrated in the contribution by Karl Enenkel on "Die antike Vorgeschichte der Verankerung der Naturgeschichte in Politik und Religion", Pliny preferred not to present his animal descriptions in the framework of a clear and plausible system that would explain the common morphological and physiological features of certain families or classes, but he always deliberately stressed the *uniqueness* of the various species. In his zoology, Pliny generally emphasised the miraculous character of God's creation, and nature's inherent intelligence, wisdom, and creative power. As Enenkel shows, Pliny constructed his zoology as a lemmatical collection of *mirabilia naturae*. Aristotle, although his zoological works were of a different kind, nevertheless did not refrain from including strange and astonishing aspects.

Early modern zoology proceeded to work along those lines, albeit in part approaching the topics from different perspectives and using different means. Its different perspectives were, of course, formed by the mainstream features of Christian theology, including scholastic debates; from the middle of the 16th century on, the theological framework of zoology was shaped by Protestantism and the Counter-Reformation, and especially Jesuit scholarship and science. Early modern zoology's different means and methods refer to a new and distinct interest in animal anatomy based on vivisection, and furthermore to the invention of the microscope; the increase in travelling; the discovery of the New World; the invention of the printed book, especially animal illustration in woodcuts and engravings; and the new role scientific exchange played, for example in terms of scholarly correspondence.

The present volume aims to present in-depth case studies that shed light on the fascinating amalgam of intellectual pursuits that shaped early modern zoology. In fact, all of the contributions are dedicated to the various intersections and combinations of the above-mentioned approaches, interests, and methods: the intersections of new "scientific" methods and refined theological argumentation (contributions by Ogilvie, Jorink, Enenkel, Roling on Bartholin, and Roling on palingenesis); of empirical observations and literary traditions (the majority of the contributions); of attempts to construct an integrative systematic taxonomy and of the belief in wondrous and "phantastic" animals, or singular "monsters" (contribution by Enenkel on "The Species and Beyond"); of empirical observation and the rise of a new "teratology" (contributions by Enenkel on "The Species and Beyond" and Roling on Bartholin); of various constructions of nomenclature, both new and traditional (contribution by Hendrikx); the intersections of religious or mythological animal painting and of 16th-century zoology (contribution by Rikken); of graphical representations of animals (woodcuts, copper etchings, drawings) and philology, including etymology and proverbial wisdom (contribution by Smith); of animal painting as a virtuoso piece of art and *Wunderkammer* collector's item, and microscopic zoological research (contribution by Jorink); of representations of animals in the visual arts and of biblical exegesis (contributions by Smith, Herrin, and Rikken); and of printed representations of animals (woodcuts and engravings) as scientific illustrations (contributions by Hendrikx, and Enenkel on "The Species and Beyond") and as emblematic, symbolic images (contributions by Herrin and Kalff).

The second section ("The Order of Nature") is especially dedicated to early modern views on animal classification and generation. Enenkel's contribution on "The Species and Beyond" focuses on one of the most striking paradigms of early modern zoology: hybridisation (cross-breeding) and hybrids. He discusses the methods and patterns of argumentation of animal classification and description in zoological treatises and manuals of the 16th and 17th centuries, especially those by Edward Wotton, Conrad Gessner, Wolfgang Franzius, Giovanni Battista della Porta, Juan Eusebio Nieremberg, John Jonston, Athanasius Kircher, and Caspar Schott. Enenkel elaborates on the intersections of philology, theology, and empirical observation. He shows that it would be misleading to assume that there was a linear "progress" in zoology from the middle of the 16th century to the end of the 17th century, in terms of either method or results. The belief in hybrids is evident in the 16th century, and it still flourished in the second half of the 17th century. The various arguments developed by the above-mentioned zoologists are determined and influenced by their different interests, purposes, intellectual contexts, and theological affiliations.

The contributions by Sophia Hendrikx and Karl Enenkel on "The Species and Beyond" demonstrate that a single zoologist (Gessner) was able to work with different methods and principles. Whereas in the first edition of his work on the quadrupeds (*mammalia* and part of the *reptilia*) Gessner deliberately refrained from presenting the species in an (integrative) taxonomical system, he used such a system in the *Icones* and, in an even more refined and elaborate way, in his works on fish species. Hendrikx shows in her case study on the herring species (*Clupeidae*) that Gessner made a systematic attempt to construct a taxonomical family of species based on empirical study, eyewitness accounts, and scientific exchange; and that the methods he applied led to results which come surprisingly close to the later ichthyological classifications based on the Linnaean binary nomenclature.

Bernd Roling in his contribution on Bartholin describes the breath-taking, but complex and ambiguous career of the *unicorn* in the zoology of the 17th century, from a quadruped to a fish, the narwhal (*Monodon monoceros*, a sea mammal); he especially sheds light on the pivotal role the Danish naturalist, medical doctor, and university professor Thomas Bartholin and his family (his father, Caspar, and his brother-in-law Ole Worm) played in this process. Although at a certain point it became clear that the spectacular horns of unicorns that were shown in collections and *Wunderkammern* all over Europe did not belong to the famous but mysterious even-toed quadruped (the existence of which, however, was legitimised by an immeasurable amount of literary *auctoritates*), but to a whale species, the narwhal, Bartholin and other scholars, such as the Dutch medical doctor Albert Kyper and the German scholar Paul Ludwig Sachse, still insisted on the (supposed) medicinal value of the so-called "horn"—which was in fact the tooth of the narwhal—as an antidote to poisons. Interestingly, the medieval belief in the almost magical power of the unicorn goes closely together with a "modern" interest in the animal for medical experiments. A number of cats and dogs were the poor victims of these experiments.

Erik Jorink elaborates on the intriguing intersections of advanced empirical zoology (based on systematic microscopic observations) and technically brilliant *Wunderkammer* painting or virtuoso artistry in the 17th-century Dutch Republic, against the background of contemporary physico-theological debates. Jorink focuses on the intellectual and artistic exchange between the remarkable, but thus far little-known animal painter Otto Marseus van Schrieck (born ca. 1620) and the pharmacist, naturalist, and pioneer in entomology and microscopic research Johannes Swammerdam (who authored the ground-breaking *Historia generalis insectorum*, 1669), in the Amsterdam circle of learned *curiosi*. Both Marseus and Swammerdam engaged in comparative anatomy, and both were fascinated by the problem of the mysterious "spontaneous generation". Through his advanced microscopic research Swammerdam was able to observe and describe the genitals of insects, which of course made spontaneous generation superfluous, at least with respect to the species researched by the Dutch naturalist. Swammerdam's advanced empirical research, however, did not exclude theological thinking. Through microscopic research, God's creation turned out to be even more ingenious and perfect.

The same is true for John Ray's zoological studies on insects, to which Brian Ogilvie dedicates an in-depth study. Ogilvie tackles the question of how seriously physico-theologians took the discoveries of natural history, and to what degree they included them in their devotional and apologetic argumentations. With respect to John Ray's works, Ogilvie presents a positive answer: the enormous number of insect species testified to the 'magnitude of God's creative power', and their remarkable anatomy to the 'immense subtlety of divine craftsmanship'. Ogilvie elaborates on the precision and scientific devotion with which Ray described and classified insect species, and demonstrates that in the succeeding editions of his works, Ray was eager to improve, correct, or complete his findings. In this way God manifests himself in the smallest details of zoological research. Each little step toward more precision is an important one toward the knowledge of God.

In his second contribution Bernd Roling deals with another important theory of generation, *palingenesis*. He shows that 'already in the 17th century, scholars such as Caspar Posner had recognised that the Scotist hypothesis of a *resurrectio naturalis* could open a clear path to accepting palingenesis. If the connection of form and matter could operate by natural means at the moment of resurrection, then they were both perhaps continuously together even after death. Why should the form of the body not simply be re-activated as the life-principle in its remains after death? For Paracelsus, the form stayed in the remains of a creature and merely waited for its revivification'. Roling demonstrates that many naturalists at the end of the 16th and the early 17th century adopted Paracelsus' view and attempted to prove it by experiments.

The third section ("Images of *Genesis*"), containing three contributions, is dedicated to the representations of animals in the early modern visual arts (painting, woodcut, engraving/copper etching, and drawing), and their intersections with theology, philology, and the 'modern' zoological knowledge displayed in the important manuals of the 16th century. All three contributions deal with biblical topics or scenes from *Genesis*: *The Fall of Man* (Smith); *The Creation of the Animals in Paradise*; *Adam Naming the Animals*; *The Animals Boarding the Ark of Noah*, and *The Animals Leaving the Ark* (all Herrin and Rikken). The second, fourth and fifth topic also played an important part with respect to the problem of hybridisation and hybrid speciation, especially in zoological discussions of the 17th century—such as those by Wolfgang Franzius, Juan Eusebio Nieremberg, Athansius Kircher, and Caspar Schott (contribution by Enenkel, on "The Species and Beyond")—and to the question of spontaneous generation treated by Jan Swammerdam and Otto Marseus (contribution by Jorink).

Paul Smith analyses Albrecht Dürer's copper etching of *The Fall of Man* (1504), and his woodcut (1510) and drawing (again from 1510) on the same topic. In each of these three pieces of art Adam and Eve are accompanied by different animals: on the 1504 print by the serpent, elk, cat, mouse, hare, ox (or cow), parrot, and ibex (or he-goat); on the 1510 woodcut by badger and bison; and on the 1510 drawing by serpent, lion, and stag. In these devotional pieces of art, Dürer's animal representations are on the one hand very realistic, but on the other highly symbolic. Dürer's inventions are based on a profound interest in philology, etymology, and proverbial expressions, inspired by the advanced humanism of the early 16th century, especially Erasmus's *Adagia*. Paul Smith convincingly elaborates on these intersections of painting and humanist philology. Dürer has constructed the representation of the animals as intellectual enigmas, and he expects the viewer to solve them in a process of intellectual discovery. Probably only highly educated viewers were able to fully understand Dürer's symbolic inventions.

Amanda Herrin deals with the intersections of the visual arts at the end of the 16th century and important zoological works that appeared in the second half of the 16th century, especially Gessner's *Historia animalium* (1st edition, 1551–1558). She focuses on a series of engravings invented by the Antwerp artist Maarten de Vos, *Imago Bonitatis* (ca. 1587), and executed by Jan de Sadeler, which she convincingly explains in the context of the zoological interest and knowledge of the circles of artists in Antwerp, and later in Prague. Among the artists inspired by Gessner's animal illustrations were Marcus Gheeraerts, Jan Collaert, and Joris Hoefnagel.

Marrigje Rikken further explores this specific interest of artists in the Low Countries and in Prague, by focusing on Jan Brueghel the Elder and Roeland Savery. Rikken demonstrates that the inventions of Breughel and Saverij are to a spectacular degree dominated by their fascination with exotic or rare species. Both try to depict as many exotic species as possible, and often they hide other elements of the biblical narrative (e.g. the Ark of Noah) in the background or do not depict them at all. It looks as if painting scenes from *Genesis* has developed in a genre of animal painting *pur sang*.

The fourth section focuses on the symbolic use of animals in political education. Sabine Kalff discusses the political meaning of the crane as an emblematic animal. Alexander Loose explains in what ways the Bohemian humanist Johannes Dubravius used animal allegory as a mirror of princes. The last section deals with physiology and early modern political ideology.

Tamás Demeter shows that in the first half of the eighteenth century, the representation of human functioning in physiology and moral philosophy went through a significant transformation. The ideal of a mechanical description give way to a vitalistic approach more sensitive to qualitative differences than mathematical formulae. This theoretical transformation coincided with a large-scale social and political change in Scotland after the Glorious Revolution and the Union of 1707. Demeter argues that the mechanical image of human functioning served the purposes of Scottish Jacobite political apology. In contrast, the vitalistic image of man, developed in physiology and moral philosophy at the same time, could be put to ideological use serving the social and political aims of the post-Union political and intellectual elites.

The editors hope that the contributions in this volume may inspire others to further research the intriguing and complex field of early modern zoology.

Legite Feliciter! Monasterii, Kalendis Martiis A.D. MMXIIIIo

Selective Bibliography

Ackermann J.M., "Early Renaissance 'naturalism' and scientific illustration", in Ellenius A. (ed.), *The Natural Sciences and the Arts* (Uppsala: 1985) 1–17.

Allan D.E., *The Naturalists in Britain. A Social History* (London: 1976).

Asúa M. De – French R., *A New World of Animals. Early Modern Europeans on the Creatures of Iberian America* (Aldershot: 2005).

Bäumer A., *Geschichte der Biologie*, 3 vols. (Frankfurt a.M. – Bern – New York – Paris: 1991).

Bono J., *The Word of God and the Language of Men: Interpreting Nature in Early Modern Science and Medicine*, vol. 1 (Madison: 1995).

Céard J., "Tératologie et Tératomancie au XVIe siècle", in Jones-Davies M.T. (ed.), *Monstres et prodiges au temps de la Renaissance* (Paris: 1980) 5–15.

Céard J., "Analogie et zoologie chez les naturalistes de la Renaissance", in Christian L.C. (ed.), *Theatrum mundi. The History of an Idea* (New York: 1987).

Cherry J. (ed.), *Fabeltiere. Von Drachen, Einhörnern und anderen mythischen Wesen* (Stuttgart: 1997).

Cunningham A., *The Anatomical Renaissance* (Aldershot: 1997).

Davies M., "Making Sense of Pliny in the Quattrocento", *Renaissance Studies* 9 (1995) 240–257.

Delauny P., *La Zoologie au XVIe siècle*, Histoire de la pensée 7 (Paris: 1962; reprint, Paris: 1997).

Doody A., *Pliny's Encyclopedia: The Reception of the Natural History* (Cambridge: 2010).

Enenkel K.A.E., "Zur Konstituierung der Zoologie als Wissenschaft in der Frühen Neuzeit: Diskursanalyse zweier Großprojekte (Wotton, Gesner)", in idem – Smith P.J. (eds.), *Early Modern Zoology. The Construction of Animals in Science, Literature and the Visual Arts*, Intersections 7 (Leiden – Boston: 2007), 2 vols., vol. I, 15–74.

Enenkel K.A.E. – Smith P.J. (eds.), *Early Modern Zoology. The Construction of Animals in Science, Literature and the Visual Arts*, Intersections 7 (Leiden – Boston: 2007), 2 vols.

Faber Kolb A., *Jan Brueghel the Elder: The Entry of the Animals into Noah's Ark* (Los Angeles: 2005).

Felfe R., *Naturgeschichte als kunstvolle Synthese. Physikotheologie und Bildpraxis bei Johann Jakob Scheuchzer* (Berlin: 2003).

Findlen P., *Possessing Nature. Museums, Collecting and Scientific Culture in Early Modern Italy* (Berkeley: 1999).

Findlen P. (ed.), *Athanasius Kircher: The Last Man Who Knew Everything* (New York: 2004).

French R.K., *Ancient Natural History: Histories of Nature* (London: 1994).

French R.K., *Dissection and Vivisection in the European Renaissance* (Aldershot: 1999).

French R.K. – Greenaway F. (eds.), *Science in the Roman Empire: Pliny the Elder, his Sources, his Influence* (London: 1986).

French R.K. – Wear A. (eds.), *Medical Revolution in the 17th Century* (Cambridge: 1989).

George W., "Sources and Background to Discoveries of New Animals in the Sixteenth and Seventeenth Centuries", *History of Science* 18 (1980) 70–104.

George W., "Alive or Dead: Zoological Collections in Seventeenth Century", in Impey O. – MacGregor A. (eds.), *The Origins of Museums. The Cabinet of Curiosities in Sixteenth- and Seventeenth-Century Europe* (Oxford: 1986) 179–187.

Gmelig-Nijboer C.A., *Conrad Gesner's "Historia Animalium": An Inventory of Renaissance Zoology* (Diss. Utrecht: 1977).

Guerrini A., *Experimenting with Humans and Animals. From Galen to Animal Rights* (Baltimore – London: 2003).

Harrison P., *The Bible, Protestantism, and the Rise of Natural Science* (Cambridge: 1998).

Heller J.L., "Classical Mythology in the Systema Naturae", in idem, *Studies in Linnaean Method and Nomenclature* (Berlin – Bern: 1983) 1–30.

Heller J.L., "The Early History of Binomial Nomenclature", in idem, *Studies in Linnaean Method and Nomenclature* (Berlin – Bern: 1983) 41–75.

Helmcke J.G., "Der Humanist Conrad Gesner auf der Wende von mittelalterlicher Tierkunde zur neuzeitlichen Zoologie", *Physis* 12 (1970) 339–346.

Heuvelmans B., *Les bêtes humaines d'Afrique* (Paris: 1980).

Heuvelmans B., "How many animal species remain to be discovered", *Cryptozoology* 2 (1983) 1–24.

Heuvelmans B., "The Metamorphosis of Unknown Animals into Fabulous Beasts and of Fabulous Beasts into Known Animals", *Cryptozoology* 9 (1990) 1–12.
Heuvelmans B., *On the Track of Unknown Animals* (London – New York: 1995).
Heuvelmans B., *The Natural History of Hidden Animals*, ed. by Hopkins P.G. (London: 2007).
Hoeniger F.D. – Hoeniger J.F.M., *The Development of Natural History in Tudor England* (Charlottesville: 1969).
Hünemörder Ch., "Aims and Intentions of Botanical and Zoological Classification in the Middle Ages and in the Renaissance", *History and Philosophy of the Life Science* 5 (1983) 53–67.
Jahn I., *Grundzüge der Biologiegeschichte* (Jena: 1990).
Jahn I. (ed.), *Geschichte der Biologie* (Hamburg: 2004; 3rd edition, Jena: 1998).
Jardine N. – Secord J.A. – Spary E.C. (eds.), *Cultures of Natural History* (Cambridge: 1996).
Kaufmann T.D., *The Mastery of Nature. Aspects of Art, Science and Humanism in the Renaissance* (Princeton: 1993).
Knight D., *Zoological Illustration. An Essay towards a History of Printed Zoological Pictures* (Folkestone: 1977).
Koreny F., *Albrecht Dürer and the Animal and Plant Studies* (Boston: 1988).
Leu U.B., "Theologia naturalis", in idem, *Conrad Gesner als Theologe. Ein Beitrag zur Zürcher Geistesgeschichte des 16. Jahrhunderts*, Zürcher Beiträge zur Reformationsgeschichte 14 (Bern – Frankfurt a.M. – New York – Paris: 1990) 31–100.
Lloyd J.E.B., *African Animals in Renaissance Literature and Art* (Oxford: 1971; New York: 1972).
Lugli A., *Naturalia et Mirabilia. Il collezionismo nelle Wunderkammern d'Europa* (Mailand: 1983).
Mason P., *Infelicities. Representations of Exotic* (Baltimore – London: 1998).
Mason P., *Before Disenchantment: Images of Exotic Plants and Animals in the Early Modern World* (London: 2009).
Mayr E., *Systematics and the Origin of Species from the View Point of a Zoologist* (Cambridge, MA – London: 1999).
Mendelsohn E.E., "Philosophical Biology versus Experimental Biology: Spontaneous Generation in the Seventeenth Century", in Grene M. – Everett E.E. (eds.), *Topics in the Philosophy of Biology* (Dordrecht: 1976) 37–65.
Mesnard P., "L'horizon zoologique de la Renaissance. Les animaux anciens et les animaux modernes", in *Sciences de la Renaissance. VIIe congrès international de Tours* (Paris: 1973).
Münch P., "Verwandtschaft oder Differenz? Zur Theorie des Mensch/Tier-Verhältnisses im 17. Jahrhundert", in Lehmann H. – Trepp A.-Ch. (eds.), *Im Zeichen der Krise. Religiösität im Europa des 17. Jahrhunderts* (Göttingen: 1999) 517–535.

Nauert Ch., "Humanists, Scientists, and Pliny: Changing Approaches to a Classical Author", *American Historical Review* 84.1 (1974) 72–85.

Newton M., *Encyclopedia of Cryptozoology. A Global Guide* (Jefferson – London: 2005).

Nissen C., *Die zoologische Buchillustration, ihre Bibliographie und Geschichte* (Stuttgart: 1971).

Ogilvie B., *The Science of Describing. Natural History in Renaissance Europe* (Chicago – London: 2006).

Petit G. – Theodorides J., *Histoire de la zoologie des origines à Linné* (Paris: 1962).

Raven Ch. E., *English Naturalists from Neckham to Ray: A Study of the Making of the Modern World* (Cambridge: 1947).

Roling B., *Drachen und Sirenen. Die Rationalisierung und Abwicklung der Mythologie an den europäischen Universitäten*, Mittellateinische Studien und Texte 42 (Leiden – Boston: 2010).

Rowland B., *Animals with Human Faces. A Guide to Animal Symbolism* (Knoxville: 1975).

Ruestow E.G., "Leeuwenhoeck and the Campaign against Spontaneous Generation", *Journal of the History of Biology* 17 (1984) 225–248.

Smith J.E.H., "Degeneration and Hybridism in Early Modern Species Debate: Towards the Philosophical Roots of the Creation-Evolution Controversy", in Wolfe Ch.T. (ed.), *Monsters and Philosophy* (London: 2005) 109–130.

Smith J.E.H., *The Problem of Animal Generation in Early Modern Philosophy* (Cambridge: 2006).

Strick J.E., *Sparks of Life. Darwinism and the Victorian Debates over Spontaneous Generation* (Cambridge, MA: 2000).

Thomas K., *Man and the Natural World: Changing Attitudes in England 1500–1800* (London: 1983).

Verner L., *The Epistemology of the Monstrous in the Middle Ages* (New York – London: 2005).

Intersections of Zoology, Religion and Politics in Antiquity

CHAPTER 1

Die antike Vorgeschichte der Verankerung der Naturgeschichte in Politik und Religion: Plinius' Zoologie und der römische Imperialismus *With an English Summary*

Karl A.E. Enenkel

Plinius' Zoologie in seiner *Naturalis historia* ist eine jener Wissenssammlungen, die für die frühneuzeitliche Zoologie bis ins 18. Jahrhundert hinein grundlegend waren.[1] Dabei erscheint das Werk neueren Lesern, v.a. klassischen Philologen und Wissenschaftshistorikern, als ein sperriger und qualitätsarmer Text, dem man die Stempel unsystematischer Präsentationsweise, anekdotenhaften Erzählens, eklatanten Originalitätsmangels, kritikloser Übernahme überlieferten Wissens (besonders des von Aristoteles gesammelten), ja sogar der Unwissenschaftlichkeit aufgeprägt hat. Etwa in Ilse Jahns in mehreren Auflagen erschienenen *Geschichte der Biologie* (1982–2004) werfen Georg Harig und Jutta Kollesch Plinius vor, Unwissenschaftliches und Wissenschaftliches durcheinander gewürfelt zu haben[2] und seiner Zoologie kein systematisches Prinzip zugrunde gelegt zu haben;[3] dabei fällt er das vernichtende Urteil, Plinius' Zoologie hätte gegenüber Aristoteles so gut wie keinen

1 Vgl. u.a. Doody A., *Pliny's Encyclopedia: The Reception of the Natural History* (Cambridge: 2010); Borst A., *Das Buch der Naturgeschichte: Plinius und seine Leser im Zeitalter des Pergaments* (Heidelberg: 1994); Ogilvie B.W., *The Science of Describing: Natural History in Renaissance* Europe (Chicago: 2006); Nauert Ch., „Humanists, Scientists, and Pliny: Changing Approaches to a Classical Author", *American Historical Review* 84.1 (1974) 72–85; ders., „Caius Plinius Secundus", in Cranz F.E., Kristeller P.O. et alii (Hrsg.), *Catalogus translationum et commentariorum: Mediaeval and Renaissance Latin Translations and Commentaries. Annotated Lists and Guides* (Washington: 1980 ff.), Bd. IV, 297–422; Chibnall M., „Pliny's Natural History and the Middle Ages", in Dorey T.A. (Hrsg.), *Empire and Aftermath: Silver Latin II* (London – Boston: 1975) 57–78; Gudger E.W., „Pliny's *Historia Naturalis*: The Most Popular Natural History Ever Published", *Isis* 6 (1924) 269–281.

2 „Die enzyklopädische Literatur", in Jahn I. (Hrsg.), *Geschichte der Biologie* (Hamburg: 2004; hier zitiert nach der dritten Auflage Jena: 1998) 79: ‚Das Normale [bei Plinius] sind dagegen Berichte, in denen wissenschaftliche und unwissenschaftliche Nachrichten vermischt sind'.

3 Ebd. 79–80: ‚Dieser Einteilung in Landtiere, Wassertiere, fliegende Tiere und Insekten liegt, kein systematisches Prinzip zugrunde [...]'. Eine Ausnahme scheinen Harig und Kollesch

 | DOI 10.1163/9789004279179_003

wissenschaftlichen Fortschritt gebracht;[4] ein Urteil, das teleologisch orientierte Wissenschaftsgeschichte kennzeichnet. Es ist kein Zufall, dass man in Änne Bäumers *Geschichte der Biologie*, Bd. 1, ähnliche Aussagen antrifft.[5] Vordem hatte Olaf Gigon Plinius' Werk gar mit dem ‚Zerfall der antiken Naturwissenschaft' gleichgesetzt.[6] In den letzten 20–30 Jahren hat zwar nicht spezifisch Plinius' Zoologie, jedoch seine *Naturalis historia* insgesamt verstärkte Aufmerksamkeit erhalten, unter anderen in den Monographien von Mary Beagon, Trevor Murphy, Roger Kenneth French, Valérie Naas und Sorcha Carey.[7] Besonders Trevor Murphy hat in seiner Studie aus dem Jahr 2004, *Pliny the Elder's Natural History. The Empire in the Encyclopedia* einige Ansätze geliefert, die mir vielversprechend erscheinen. Murphy geht es darum, Plinius' Enzyklopädie als Kulturprodukt (‚artefact') zu betrachten.[8] Dabei bezieht er die ideologischen Bedingungen mit ein, unter denen das Werk zustande

---

diesbezüglich nur im Hinblick auf die Schwämme und Aktinien zu machen, die Plinius an der Grenze zwischen Tier- und Pflanzenreich ansiedelte.

4 Ebd. 80 (mehrfach).

5 Bäumer A., *Die Biologie von der Antike bis zur Renaissance*, in Dies., *Geschichte der Biologie*, Bd. 1 (Frankfurt a.M. – Bern – New York – Paris: 1991).

6 Gigon O., „Plinius und der Zerfall der antiken Naturwissenschaft", in ders., *Studien zur antiken Philosophie* (Berlin: 1972), 370–390; urspr. in *Arktos* 4 (1966) 23–45.

7 Monographien und Sammelbände: Murphy T., *Pliny the Elder's* Natural History. *The Empire in Encyclopedia* (Oxford: 2004); Beagon M., *Roman Nature: the Thought of Pliny the Elder* (Oxford: 1992); Carey S., *Pliny's Catalogue of Culture: Art and Empire in the Natural History* (Oxford: 2003); French R.K., *Ancient Natural History: Histories of Nature* (London: 1994); Ders. – Greenanway F. (Hrsg.), *Science in the Roman Empire: Pliny the Elder, his Sources, his Influence* (London: 1986); Naas V., *Le projet encyclopédique de Pline l'Ancien* (Rom: 2002); Citroni Marchetti S., *Plinio il Vecchio e la tradizione del moralismo Romano* (Pisa: 1991); Detlefsen D., *Untersuchungen über die Zusammensetzung der Naturgeschichte des Plinius* (Berlin: 1899). Für Einzelbeiträge vgl. u.a. Naas V., „Réflexions sur la méthode de travail de Pline l'Ancien", *Revue de Philologie* 70 (1996) 305–322; Baldwin B., „The Composition of Pliny's Natural History", *Symbolae Osloenses: Norwegian Journal of Greek and Latin Studies* 70 (1995) 72–81; Locher A., „The Structure of Pliny the Elder's *Natural History*", in French – Greenaway (Hrsg.), *Science in the Roman Empire* 20–29; Conte G.B., „The Inventory of the World: Form of Nature and Encyclopedic Project in the Work of Pliny the Elder", in ders., *Genres and Readers: Lucretius, Love Elegy, Pliny's Encyclopedia*, trans. G.W. Most (Baltimore: 1994); Howe N.P., „In Defense of the Encyclopedic Mode: On Pliny's Preface of the Natural History", *Latomus* 44 (1985) 561–576. Spezifisch zur Zoologie vgl. Bodson L., „Le témoignage de Pline l'Ancien sur la conception romaine de l'animal", in Cassin B. – Labbarière J.-L. (Hrsg.), *L'animal dans l'Antiquité* (Paris: 1997) 325–354; Seeck G.A., „Plinius und Aristoteles als Naturwissenschaftler", *Gymnasium* 92 (1985) 419–434; Vegetti M., „Zoologia et antropologia in Plinio", in *Plinio il Vecchio sotto il profile storico e letterario: Atti del Convegno di Como 1979* (Como: 1982) 117–131.

8 Vgl. „Introduction: Encyclopedia as Artefact", in Murphy, *Pliny the Elder's* Natural History 1–3.

gekommen sei. Indem er die *Naturalis historia* mit der *Encyclopedia Brittanica*, ein Produkt des British Empire, vergleicht, versucht er zu zeigen, wie ein etabliertes Weltreich sozusagen sich selbst beschreibt und verherrlicht.[9] Der plinianischen Anekdote[10] schreibt Murphy den Rang eines *triumphalistischen* Darstellungsmittels zu. Sie wird seines Dafürhaltens von Plinius eingesetzt, um das, was sich dem konkreten Wissen entzieht, enzyklopädisch zu vereinnahmen. D.h. Plinius vermag damit die Grenzen des bis dato bekannten Wissens zu überschreiten. Murphy fokussiert dabei, wie seine Perspektivierung nahelegen mag, auf die Geographie und die Ethnographie. Plinius' Anekdoten sollen dabei die Ränder der Erde erklären, die Gebiete und Völker, die noch nie jemand besucht hat. Die Zoologie klammerte Murphy jedoch aus—vielleicht weil ihm das Material zu sperrig und zu wenig einschlägig erschien. Wie der vorliegende einleitende Beitrag, in dem ich das 8. Buch der *Naturalis historia*, die Zoologie der sog. „Landtiere" (zumeist Säugetiere),[11] betrachte, zeigen wird, erachte ich Murphys Ansatz für inspirierend, stimme mit ihm jedoch, u.a., was die Zielrichtung des Darstellungsmittels der Anekdote betrifft, nicht überein. Auch glaube ich, dass der Imperialismus in Plinius' Zoologie nicht nur eher metaphorisch und sekundär zum Tragen kommt, insofern er die Motivation bildet, Weltwissen umfassend darzustellen, sondern dass *imperialistische Gedanken* sowohl die Grundzüge der Wissensorganisation selbst als auch die *inventio* und die Selektion des Materials mit bedingten. Zoologie ist für Plinius wesentlich mit politischen und religiösen Gedanken verwoben.[12]

---

9 Vgl. ebd. 2: ‚In this book I shall argue for a reading of Pliny's Encyclopedia as a political document, a cultural artefact of the Roman Empire just as much as the *Encyclopedia Britannica* was an artefact of the British Empire [...]'; 5: ‚Roman power has united the world and opened it up to be looked at [...]. This triumphalism is fundamental to the *Natural History*'.

10 Lloyd G.E.R., *Science, Folklore and Ideology* (Cambridge: 1983) 56, hat Plinius' als oberflächlichen Anekdotensammler hingestellt, wie sie in Rom zu dutzenden zu finden waren. Beagon, *Roman Nature* 129, widersetzte sich dieser Betrachtungsweise. Zur Geschichte der Anekdote vgl. Fineman J., „The History of the Anecdote: Fiction and Fiction", in Aram Veeser H. (Hrsg.), *The New Historicism* (New York: 1989) 49–76.

11 Zum Text siehe Gaius Plinius Secundus der Ältere, *Naturkunde. Lateinisch—deutsch. Buch VIII. Zoologie: Landtiere.* Herausgegeben und übersetzt von Roderich König in Zusammenarbeit mit Gerhard Winkler (Kempten: 1976; Tusculum-Reihe), im Folgenden zitiert mit *Naturalis historia / Naturkunde*.

12 Für die religiösen Gedanken, die der plinianischen Enzyklopädie zugrunde liegen, vgl. Beagon, *Roman Nature*, chapter I „Divina Natura: the Roots of Pliny's Thoughts" (26–54). Für Beagon ist das Lob der Natur bzw. der (stoische) *natura-artifex*-Gedanke ein ‚overall theme' der *Naturalis historia* (*Roman Nature* 131 und *passim*). Für Plinius' Haltung zu und Verarbeitung der römischen Religion in der *Naturalis historia* vgl. Köves-Zulauf Th.,

Zur näheren Bestimmung von Plinius' Zoologie, scheint es mir heuristisch sinnvoll, einige Merkmale der modernen Zoologie kurz zu skizzieren.[13] Moderne Zoologie ist ihrem Selbstverständnis nach eine *systematische* Wissenschaft, deren Grundlage die *Taxonomie* im Verein mit der *Evolutionsbiologie* bildet. Die Taxonomie ordnet die *Species*[14] (Tierarten) innerhalb eines feinmaschigen Klassifizierungssystems ein, in welchem sie die Verwandtschaftsverhältnisse der verschiedenen Arten festlegt. Dabei geht es nicht nur um eine Klassifizierung der noch lebenden Arten, sondern zugleich um die Einordnung derselben in einem „Stammbaum der Lebewesen", der durch die Evolutionslehre erklärt wird. Ein Wissenschaftszweig, der die Klassifizierung unterstützt und speist, ist die Paläontologie, welche die Tierwelt der weit zurückliegenden Vergangenheit untersucht und die in der Gegenwart lebenden Arten auf diese zurückführt. So ergibt sich aus paläontologischen Forschungen, dass die Mehrzahl der heute lebenden Mammalia (insgesamt etwa 4250–4700 Arten)[15] aus dem Zweig (Teilklasse) der Eutheria (Plazentatiere) hervorgeht, von denen sich vor ca. 200 Millionen die eierlegenden Protheria (Kloakentiere) und vor etwa 120 Millionen Jahren die Metatiere (Beuteltiere) abgespalten hatten. Vor etwa 100–90 Millionen Jahren gliederten sich die Plazentatiere in drei Hauptzweige auf: Glires (2 Ordnungen) und Euarchonta (4 Ordnungen); Laurasiatheria (6 Ordnungen); Afrotheria (6 Ordnungen). Mit der Taxonomie sind verschiedene Forschungsmethoden und—zweige verbunden (u.a. Vergleichende Anatomie, Physiologie, Biochemie, Genetik, Verhaltensbiologie), die ebenfalls Erkenntnisse zur Klassifizierung beitragen können, wie sich nicht zuletzt in der modernen Definition

„Plinius der Ältere und die Römische Religion", in *Aufstieg und Niedergang der Römischen Welt* [...] II, 16.1 (Berlin – New York: 1978) 187–288.

13 Für diese Skizze vgl. die Arbeit des Verf., „Zur Konstituierung der Zoologie als Wissenschaft in der Frühen Neuzeit: Diskursanalyse zweier Grossprojekte (Wotton, Gesner)", in: Enenkel K.A.E. – Smith P.J. (Hrsg.), *Early Modern Zoology. The Construction of Animals in Science, Literature and the Visual Arts*, Brill, Leiden – Boston 2007, 2 vols., vol. I, (15–74); Macdonald D. (Hrsg.), *Enzyklopädie der Säugetiere* (Königswinter: 2003; urspr. englisch, Oxford: 2001), Vorwort. In den meisten neueren Studien zu Plinius' *Naturalis historia* fehlt eine diesbezügliche Klärung.

14 An dieser Stelle und im Folgenden wird der *Species*-Begriff im Sinn der modernen Taxonomie verwendet. In der Antike wurde das Wort bekanntlich auf flexible und mehrdeutige Weise benutzt; vgl. Hünemörder Ch., „Aims and Intentions of Botanical and Zoological Classification in the Middle Ages and in the Renaissance", *History and Philosophy of the Life Science* 5 (1983) 53 ff.

15 Diese und die nachfolgenden Angaben wurden Macdonald D. (Hrsg.), *Enzyklopädie der Säugetiere* Xvi entnommen.

der *Säugetiere* zeigt: als solche werden *warmblütige Wirbeltiere bezeichnet, die ihre Jungen mit Milch säugen sowie einen behaarten Körper und ein einteiliges Kiefergelenk besitzen*.[16] In dieser Definition treffen sich Morphologie (Wirbeltiere, einteiliges Kiefergelenk), Physiologie (Warmblütigkeit, Thermoregulation u.a. durch behaarten Körper), Biochemie (Endothermie durch Oxidation von Nahrung im Körper) und Verhaltensbiologie (Säugen).

Moderne Zoologie insgesamt und Taxonomie im Besonderen werden zum einem Gutteil durch funktionsanalytische Gedankenformationen bestimmt. Z.B. wird bei den Säugetieren das Körpermerkmal der Behaarung nicht nur als morphologisches Fakt, sondern als ein in bezug auf seine physiologische Funktionalität—die Thermoregulation—wesentliches Definitionskriterium aufgefasst: Für die Säuger ist grundlegend, dass sie ihre Körpertemperatur von innen heraus (Endothermie), durch Oxidation von Nahrung im Körper, regeln. Dies hat den Vorteil einer relativen Unabhängigkeit von stets wechselnden Umweltbedingungen und sorgt für eine möglichst große biochemische Stabilität des Organismus, hat jedoch den Nachteil, dass dieses Energiesystem relativ kostspielig ist. Die Maßnahmen, die erforderlich sind, um das System rentabel zu gestalten, werden evolutionsbiologisch als *Anpassungen* gedeutet: Solche Anpassungen der Mammalia stellen Behaarung, Talgdrüsen und Schweißdrüsen dar. Ebenso fasst man die Reduktion auf einen Unterkieferknochen nicht einfach als morphologisches Fakt auf, sondern deutet sie als Anpassungsschritt zur Entwicklung höherer Lebensformen: Die Reduktion auf einen Unterkieferknochen ermöglicht, dass sich die Knochen, die das Gehirn umgeben, verstärken konnten—dies macht eine Vergrößerung des Gehirns und eine Ausdehnung des Nervensystems möglich, und bildet somit die Voraussetzung zur Entwicklung höherer Intelligenz.

Die Herangehens- und Arbeitsweise der modernen Zoologie wird weiter von einer Fokussierung auf genaue Messdaten, Zählungen und andere auf empirischem Weg eruierte Daten gekennzeichnet. Bei Beschreibungen von Species werden u.a. Körperlänge, Schulterhöhe, Schwanzlänge, das Körpergewicht, in der Regel differenziert nach Geschlecht und Alter, mit möglichst präzisen Durchschnittswerten (ev. auch mittels Eckdaten) angegeben; weiter wird die Nahrung bestimmt und gewogen, ebenfalls differenziert nach Geschlecht und Alter. Die Geschlechtsreife wird in Monaten, die Tragzeit in Tagen, die durchschnittliche Lebenserwartung in Jahren angegeben. Territorien werden vermessen, Verbreitungsgebiete kartographisch beschrieben, die Verbreitungsdichte durch Zählungen bestimmt. Die Verbreitungsdichte besitzt insofern besondere Relevanz, als zahlreiche Species

16 Vgl. ebd.

von Wildtieren insbesondere seit dem 19. Jahrhundert vom Menschen stark zurückgedrängt wurden und werden. Es ist klar, dass unter einer bestimmten Verbreitungsdichte das Fortbestehen einer Art nicht mehr gesichert ist bzw. unwahrscheinlich wird, wobei Gegenmaßnahmen weder einfach zu nehmen sind noch leicht greifen.

Schon aufgrund dieser notgedrungen viel zu kurzen und rudimentären Skizze dürfte klar sein, dass in der modernen Zoologie Religion so gut wie keine Rolle spielt. Die Art der wissenschaftlichen Methodik, die grundlegend mit der Evolutionstheorie verbunden ist, schließt vor allem eines aus: das Denkmodell eines *Schöpfergottes*, der die Welt und die in ihr lebenden Wesen/Tierarten in einem einmaligen Schöpfungsakt hervorgebracht hat. Die moderne Zoologie versucht zu demonstrieren, dass sich die in der Gegenwart lebenden Arten erst im Laufe eines hunderte Millionen von Jahren dauernden Evolutionsprozesses durch immer neue Anpassungen herauskristallisiert haben. Sie waren nicht vom Anfang der Welt an da. Wesentlich ist weiter, dass von der modernen Zoologie eine scharfe, ontologische Trennung zwischen Mensch und Tier—anders als in den meisten Religionen—abgelehnt wird. In der modernen Zoologie figuriert der Mensch in der Ordnung *Primates* (Herrentiere), d.h. er ist Bestandteil des Tierreiches und Teilhaber an der Evolution der Species. Über die Ethologie ist die Verhaltenslehre des Menschen weiter mit der Verhaltenslehre der Tiere verbunden. Die Wissenschaft der Theologie ist für die Forschungsmethoden und -praktiken der modernen Zoologie im Grunde irrelevant. Man verbindet die Erkenntnis der Natur nicht mehr grundlegend mit der Erkenntnis Gottes. Das Wissenschaftsverständnis ist von scharfen Zäsuren gekennzeichnet: Zoologie vor und nach Linnäus, vor und nach Darwin. Ebenso dürfte klar sein, dass in der modernen Zoologie „Überlieferungswissen", besonders das in Heiligen Büchern überlieferte Wissen, keine besondere Relevanz besitzt: Dieses wurde von empirisch eruierten Daten völlig verdrängt; Glauben und Wissenschaft schließen sich aus, jedenfalls, wenn es um Forschungsmethoden und -praktiken geht.

Während für Religion und Glauben in der modernen Zoologie im Grunde kein Platz mehr ist, lassen sich zahlreiche Berührungspunkte zwischen Zoologie und Politik feststellen. Diese betreffen die Rolle des Tiers in der modernen Welt/ Gesellschaft, die Modalitäten und Regelungen der Nutztierhaltung, Ernährungspolitik, Umweltfragen im weitesten Sinn, räumliche Ordnung, Flächenwidmungspläne, Landschaftsschutz, Tierschutz im allgemeinen, den Artenschutz wild lebender Arten im besonderen, die Errichtung etwaiger Schutzzonen, Naturparks oder Reservate, Jagdordnungen/ Maßnahmen bei Überpopulation oder bei Populationsrückgang usw. Was die Nutzung von

Tieren betrifft, scheint sich in westlichen Gesellschaften—übrigens von unterschiedlichen Gesichtspunkten her—immer mehr die Überzeugung durchzusetzen, dass Tiere vom Menschen nicht ohne Beschränkungen und Regeln genutzt werden dürfen. In Bezug auf wild lebende Arten ist klar, dass deren Lebensraum generell immer mehr schmilzt und dass viele Arten ohne eingehende Schutzmaßnahmen nicht überleben werden können. Während ganz unterschiedliche Belange im Spiel sind, scheint sich, jedenfalls in westlichen Gesellschaften, ein Consens herauszubilden, der besagt, dass Artenschutz in irgendeiner Form erforderlich ist. Die in der „Natur" lebenden Wildtiere werden zurecht immer mehr als bedrohte Species betrachtet; die „Natur" bildet demnach nicht eine Größe, die die Existenz des Menschen grundsätzlich gefährdet—vielmehr ist es der Mensch, der die Natur bedroht und verdrängt. Es gibt seit einiger Zeit nicht nur politische Parteien, die den Natur-, Arten-, Umwelt- und Tierschutz in ihre Programme aufgenommen haben, sondern es wurden verschiedentlich sogar eigene Parteien für Tiere bzw. Umweltschutz errichtet. Parteien, die sich hauptsächlich dem Tier widmen, sind zwar klein, jedoch ist für die gegenwärtige Situation bezeichnend, dass es sie gibt.

Während im Zuge der Globalisierung die Naturräume immer kleiner werden und der Mensch dem Tier überall auf der Welt immer mehr „auf den Pelz rückt", bewirkt die Globalisierung über die elektronischen Medien, dass die wildlebenden Species auf den jeweiligen anderen Kontinenten immer mehr den Reiz des Seltsamen bzw. Exotischen verlieren. Denn digital sind sie immer und überall gegenwärtig. Sogar die entlegensten, am schwersten zugänglichen Gebiete der Erde rücken uns immer näher: Die Tiefsee, Arktis und Antarktis, der Regenwald des Amazonasgebietes (immerhin der Ort der größten Artenvielfalt), die Regenwälder und Gebirgszonen Afrikas, verschiedene Wüstengebiete, Sibirien usw.—dies alles öffnet sich auf Abruf modernen Augen. In der Natur lebende Wildtiere werden zwar als bedroht betrachtet, jedoch erwecken sie viel weniger als noch vor fünfzig, hundert oder hundertfünfzig Jahren unser Erstaunen und unsere Verwunderung.

Plinius verfasste seine *Naturalis historia* in einer völlig anders gelagerten wissenschaftlichen, intellektuellen, politischen und lebenswirklichen Umwelt. Zoologie hatte, trotz der Arbeiten und Bemühungen des Aristoteles, nicht den Status einer eigenständigen, systematischen Wissenschaft. Es herrschte ein großer Mangel an gesicherten Messdaten, Quantitätsangaben, überhaupt an Daten aller Art. Die Parameter und die Art des Wissens waren nicht ganz klar. Plinius verfügte aufgrund seiner Ämter und seiner Reisen sicherlich über ein beachtliches empirisches Wissen. Dennoch ist sein empirisches Wissen weder die alleinige noch die Hauptgrundlage seiner Zoologie. Ein Gutteil seiner Zoologie gründete sich auf Überlieferungswissen.[17] Sein Werk richtete sich an die römische Oberschicht, insbesondere an die Intellektuellen bzw. die im weitesten Sinn an den *studia humanitatis* Interessierten. Er sammelte, selektierte und ordnete seine Daten so, dass dies optimal den Bedürfnissen und Interessen seiner Leserschaft entsprach. Diese Leserschaft verlangte nicht nach einer möglichst großen Ansammlung von Messdaten, schon deshalb, weil sie nicht gewusst hätte, wie es diese hätte anwenden und verwerten können; gleiches gilt für eine möglichst vollständige Beschreibung aller damals lebenden Arten. Es gab dutzende damals lebender Tierarten, die kaum interessierten und die deshalb nicht behandelt wurden. Dies zeigt sich insbesondere bei den kleineren Tierarten. Z.B. werden in der Ordnung der Nagetiere (Rodentia) von modernen Enzyklopädien ca. 2000 Species in ca. 430 Gattungen erfasst. In Plinius' Zoologie werden lediglich der Biber (*Castor fiber*, VIII, 109),[18] das Eichhörnchen (VIII, 138, *Sciurus vulgaris*) und das Murmeltier (VIII, 132)—übrigens beide im Verein mit dem artenfernen Dachs (einem Raubtier), die Maus (VIII, 221–223), die Spitzmaus und der Siebenschläfer (beide VIII, 224) behandelt.

Wie die verschiedenen Stellen, an denen Plinius diese Arten beschrieb, zeigen, betrachtete er die *Rodentia*-Species nicht alle als verwandt. Der Biber ähnelt in Plinius' Beschreibung eher einem Raubtier: Das Gebiss des Bibers soll so scharf sein, dass es, ‚wenn es einen Menschen packt', ihm ‚die Knochen krachend zerbricht'.[19] Die Bemerkung über das ungeheuer starke Raubtiergebiss des Bibers hat Plinius der *Historia animalium* des Aristoteles

17 Vgl. French R.K. – Greenaway F., *Science in the Roman Empire: Pliny the Elder, his Sources, his Influence* (London: 1986). Kollesch und Haring bemerken zurecht, dass „eigene Bobachtung bei ihm [Plinius] nur gelegentlich zu Worte" kämen; vgl. Jahn, *Geschichte der Biologie* 79. Für die *Naturalis historia* insgesamt behauptet Locher Plinius' Augenzeugenschaft für ‚a considerable number of passages' („The Structure of Pliny the Elder's Natural History" 22). Über genaue Quantitätsverhältnisse lassen sich in der augenblicklichen Lage schwerlich Aussagen treffen; jedoch ist jedenfalls in der Zoologie, Buch 8 (Landtiere), der Anteil des empirischen Wissens sehr gering. Dies steht im Zusammenhang mit einer bestimmten Diskursausrichtung und mit bestimmten Hauptkategorien des zoologischen Sammelns (*mirabilia*), vgl. unten.

18 Die Species *Castor fiber* bezeichnet den eurasiatischen Biber, im Gegensatz zum Kanadabiber (*Castor canadensis*). Vgl. dazu Macdonald, *Enzyklopädie der Säugetiere* 590–595.

19 *Naturalis historia*/ *Naturkunde* VIII, 109: ‚hominis parte comprehensa non ante quam fracta concrepuerint ossa morsus resolvit'. Davon, dass der Biber mit den Nagetieren verwandt sei, berichtet Plinius nicht.

entnommen.[20] Aristoteles hatte sie jedoch auf den Fischotter, nicht auf den Biber, gemünzt, also in der Tat auf ein Raubtier.[21] Außerdem hatte Aristoteles die Relevanz der „Beobachtung" stark eingeschränkt, indem er sie der Kategorie des Hörensagens zuschrieb. Er stützt die Angabe weder auf empirische Beobachtungen noch bestätigt er sie.[22] Die Art, wie Plinius hier mit seiner Quelle verfährt, ist bezeichnend für seine Fokussierung auf Ausgefallenes, Erstaunliches und Seltsames. Es wäre in Plinius' Darstellung viel weniger effektiv gewesen, wenn er festgestellt hätte, dass der Biber ein harmloses Nagetier ist, das dem Menschen nichts zuleide tut. Der römische Zoologe betrachtete es offensichtlich als sinnvoll, den Biber als „gefährliches" Tier darzustellen.

Es geht Plinius, wie nicht zuletzt dieses Beispiel andeutet, *nicht* darum, möglichst klar und plausibel die Verwandtschaftsverhältnisse der einzelnen Species zueinander zu eruieren. Er betrachtet die *Species* am liebsten *als Einzelphänomene*, die bestimmte möglichst seltsame und auffällige Eigenschaften haben. Dass Biber so scharfe Zähne haben, dass sie Menschenknochen brechen, ist ein solches auffälliges Sonderphänomen. Es ist übrigens fraglich, ob man ein derartiges Beschreibungsmerkmal als Anekdote auffassen sollte. Für Anekdoten gilt, dass sie bestimmte individuelle einmalige Ereignisse wiedergeben. Bei dem „Biberbiss" handelt es sich jedoch um etwas, das—jedenfalls aus der Perspektive von Plinius' Darstellung—schon oft aufgetreten ist und von daher die Eigenschaft einer Species mitbestimmen soll. Gleiches gilt für eine andere wundersame Eigenschaft, die Plinius dem Biber zuschreibt: Er soll sich bei Gefahr die Geschlechtsteile abbeißen (VIII, 109). In diesem Fall kommt hinzu, dass Plinius das bizarre „Factum" bestätigt und begründet: Die Biber tun dies, weil sie wissen, dass man sie wegen des begehrten Bibergeils jagt. Das Bibergeil lokalisierte Plinius—übrigens zu Unrecht—in den Geschlechtsteilen.[23] Ein anderes Nagetier, das Murmeltier, bringt Plinius ebenfalls in die Nähe eines Raubtieres (Dachs): Das Murmeltier (von Plinius als ‚Mus alpinum',[24] ‚Alpenmaus', bezeichnet, VIII, 132) soll die

20 *Historia animalium* VIII, 5.

21 Ebd., engl. Übers. von D'Arcy Wentworth Thompson: ‚The otter will bite a man, and it is said that whenever it bites it will never let go until it hears a bone crack'.

22 Dies zurecht; es ist nicht anzunehmen, dass Fischotter Knochen eines Menschen, z.B. Armknochen, brechen können.

23 In Wirklichkeit handelt es sich bei dem „Bibergeil" um ein Sekret, das in der Bauchhöhle produziert wird. Die Geschlechtsteile sind in dieser Beziehung irrelevant.

24 Helmut Leitner hat eine Identifizierung der in Plinius' Zoologie genannten Arten mit den von der modernen Taxonomie erfassten Species versucht: *Zoologische Terminologie beim älteren Plinius* (Hildesheim: 1972). Ebd. 175 s.v. „Mus 1" identifiziert er „mus Alpinus" ebenfalls mit dem Alpenmurmeltier (*Marmota marmota*).

Größe eines Dachses besitzen. Obwohl das Alpenmurmeltier (*Marmota marmota*) die größte Hörnchenart darstellt (53–73 cm Körperlänge, Gewicht 4–8 kg), sind Dachse in der Regel doch um einiges grösser und schwerer (80 cm-1 m, Gewicht bis etwa 12 kg).[25] Aufgrund dieser Abweichung vermuteten König und Winkler, dass Plinius mit dem ‚Mus alpinum' nicht das Murmeltier, sondern ‚eine Marderart' gemeint habe.[26] Mir erscheint der „Dachsvergleich" eher als eine weitere Übertreibung, die dazu dienen soll, eine unbedeutende Species eindrucksvoller zu gestalten. Im Übrigen werden die Dachse selbst von Plinius wundersam „aufgeblasen": Sie sollen die Eigenschaft besitzen, dass sie, wenn Gefahr droht, ihre Haut von innen mit Luft aufblähen, damit gewissermaßen Luftkissen bzw. Airbags schaffen, die sie vor den Bissen von Hunden oder den Schlägen von Menschen schützen.[27] Kein anderes Säugetier verfügt über eine vergleichbare, ingeniöse Methode, seine Feinde abzuwehren. Auch dieses Beschreibungsmerkmal zielt auf einen sorgfältig konstruierten *mirabile*-Effekt ab. Bei Aggression stellen Dachse ihre Rückenhaare auf und machen sich dadurch bedrohlicher: Dieses Verhalten hat Plinius als „Aufblähen mit Luft" gedeutet.

Plinius' Aufgliederung des Tierreiches in klassifikatorisch bzw. taxonomisch nicht mit einander zusammenhängende Einzeltierarten hängt mit seinen Darstellungszielen und dem von ihm angenommenen Leserinteresse zusammen. Er will seinen Lesern die Natur als *wunderbare Schöpfung* präsentieren: als wundersame Schöpfung Gottes, der im Großen wie im Kleinen den Lebewesen erstaunliche Klugheit, Intelligenz und Kraft verliehen hat.[28] Es lässt sich nachvollziehen, dass er zu diesem Zweck stets jene Aspekte des Tierreichs, die Bewunderung und Staunen erwecken, hervorhebt. Es geht ihm um den *mirabile-Effekt,*[29] und dieser ist stets auch mit Plinius' theologischer

25 Vgl. Macdonalds, *Enzyklopädie der Säugetiere* 117 und 599.

26 Plinius, *Naturalis historia/ Naturkunde* 222.

27 *Naturalis historia/ Naturkunde* VIII, 138: ‚Alia sollertia in metu melibus: sufflatae cutis distentu ictus hominum et morsus canum arcent'.

28 Vgl. dazu das Kapitel „Divina Natura: the Roots of Pliny's Thought", in Beagon, *Roman Nature* 26–54. Locher, „The Structure of Pliny the Elder's *Natural History*" 29, bemerkte: ‚Love of nature, which in Pliny is undoubtedly of a religious nature [...]'. Für Plinius' Bewunderung der kleinsten Lebewesen, welche für ihn die Insekten sind, vgl. *Naturalis historia* XII, 2–4: ‚[...] in his tam parvis atque tam nullis quae ratio, quanta vis, quam inextricabilis perfectio!' (‚Welche Klugheit, welche Kraft, welch kaum nachvollziehbare Perfektion findet man bei diesen kleinen und nichtigen Lebewesen!').

29 Für Plinius' Betonung des Wunderbaren vgl. u.a. Beagon, *Roman Nature* 129 (1992) und Locher, „The Structure of Pliny the Elder's *Natural History*" 23 ff. (1986). Locher hat überzeugend dargelegt, dass das von ihm als *curiosum* und *mirabile* bezeichnete eines der

Naturbetrachtung verbunden. Das *mirabile* ist eine der Hauptkategorien, nach denen Plinius die Bausteine seiner *Naturalis historia* insgesamt,[30] besonders aber der Zoologie, zusammengetragen hat. Das achte Buch über die Landtiere etwa liest sich wie eine lange lemmatische Sammlung von eigens dafür ausgewählten *mirabilia naturae*. Aus dieser Diskursausrichtung wird verständlich, warum sich Plinius kaum um möglichst genaue Messdaten, Zählungen und andere auf empirischem Weg eruierte Daten kümmert. Derartiges würde den *mirabile-Effekt* ansehnlich verringern und insofern Plinius' Demonstrationsziel entgegenwirken. Dasselbe gilt für physiologische und anatomische Zusammenhänge, die das „gewöhnliche" Funktionieren tierischer Organismen erklären würden. Plinius' Selektionsprinzip ist ausschließlich auf das ausgerichtet, was dazu dient, die wunderbare Macht und Kraft der Natur *ad hoc* und kasuistisch hervorzukehren. Für Anatomie und Physiologie interessiert sich Plinius in der Beschreibung der Tierarten (=*Naturalis historia*, Buch VIII, 1–XI, 120) folgerichtig vor allem, wenn etwas merkwürdiges, ausgefallenes, ja einzigartiges zu verzeichnen ist. Z.B. vermerkt er das—vermeintliche—Fehlen der Halswirbel beim Löwen, beim Wolf und bei der Hyäne,[31] sowie

---

strukturbestimmenden Bestandteile der *Naturalis historia* ist und dass dieses sicherlich zu den Kriterien gehört, nach denen Plinius seine Informationen sammelte. Dabei hat er das „Wunderbare" (*mirabile*) zur Gänze mit dem Anekdotischen (*curiosum*) verquickt. Vgl. a.a.O. 23: ‚A further structural unit seems to me to be represented by the *curiosum* or *mirabile*, the amazing or grotesque incident or paradox, which, however, can in part coincide with *historiae*'. Der Unterschied liegt in der Tatsache, dass das *mirabile* der Natur eine religiöse bzw. theologische Komponente besitzt, das bloß Anekdotische der *Historia* hingegen nicht. Beagon wendet sich a.a.O. daher zu Recht gegen die Engführung des Wunderbaren zum Anekdotischen hin, wie sie auch Lloyd, *Science, Folklore and Ideology* 56, dem Plinius zuschreibt. Stattdessen betrachtet Beagon das Wunderbare als positive Kategorie, die Plinius' Hochschätzung und Lob der Natur charakterisiert. Für den Aspekt des Wunderbaren bei Plinius vgl. weiter Naas V., „Extraordinaire et merveilleux dans l'Historia Naturalis: Réflexions sur la définition de ‚l'encyclopédie' plinienne", in *Colloque international, Paris-X-Nanterre, 11–13 janvier 1996* (Paris: 1997) 139–168.

30 Dies hat Locher, „The Structure of Pliny the Elder's *Natural History*" 23 ff., gezeigt.

31 Vgl. *Naturalis historia*/ *Naturkunde* XI, 177: ‚[cervix] leoni tantum et lupo et hyaenae singulis rectisque ossibus rigens' (‚Der Löwe, der Wolf und die Hyäne besitzen einen einzigen starren Nackenknochen'); VIII, 105 (Hyäne): ‚[Hyaenae] collum et iuba unitate spinae porrigitur flectique nisi circumactu totius corporis non quit' (‚Der mähnentragende Hals und das Rückgrat [der Hyäne] werden durch einen Knochen aus einem Stück gebildet, und das Tier kann sich nicht wenden, ohne sich mit dem ganzen Körper umzudrehen'; König übersetzte ‚Hals wie Mähne laufen mit dem Rückgrat in einem Stück [...]'. Für das Fehlen der Halswirbel beim Löwen vgl. Enenkel, „Zur Konstituierung der Zoologie als Wissenschaft in der Frühen Neuzeit" 37–39.

das—ebenso vermeintliche—Fehlen der Zunge beim Krokodil.[32] Diese *mirabilia* gehen in beiden Fällen auf Aristoteles zurück. Aristoteles vermerkte in seiner *Historia animalium*, dass alle eierlegenden und blutführenden Vierfüßer die herkömmlichen Sinnesorgane besäßen, mit Ausnahme des Nilkrokodils, dem die Zunge fehle.[33]

Weiter vermeldet Plinius in der Regel nicht, welches Alter, welche Körperlänge und Schulterhöhe die Tiere durchschnittlich erreichen, auch dann nicht, wenn dies im Prinzip bekannt war, wie z.B. beim Haushund oder beim Pferd. Beim Krokodil, womit er das Nilkrokodil (*Crocodylus Niloticus*) meint, verzeichnet Plinius jedoch, dass es ‚häufig' länger als 18 Ellen, (d.h. 18 mal 44 cm, also 7.92 Meter) werde.[34] Das ist abermals ein ausgefallenes und merkwürdiges „Fakt", welches Plinius eben nur aus jenem Grunde anführt. Klar ist jedenfalls, dass er Nilkrokodile nicht vermessen ließ; seine Längenangabe stimmt nicht mit der Realität überein. Erwachsene Krokodilweibchen haben eine Länge zwischen 2.4 und 3.8 Meter, erwachsene Männchen zwischen 3.5 und 5 Meter. Plinius hat eine etwa doppelt so große Normlänge angegeben, wohl um das Staunen seiner Leser zu vergrößern. Außerdem setzt er eine weitere Unikats-Information hinzu: dass ‚nach manchen' das Krokodil das einzige Tier sei, das niemals einen stabilen Erwachsenenstatus erreiche, sondern ‚sein ganzes Leben lang wachse'.[35] Diese vermeintliche Eigenschaft gewinnt noch an Spektakularität angesichts der Tatsache, dass das Tier sehr alt wird'.[36] Das bedeutet nichts weniger als dass das Krokodil das Potential besitzt, geradezu ins Unermessliche zu wachsen.

---

32 *Naturalis historia / Naturkunde* VIII, 89: ‚Unum hoc animal terrestre (sc. Crocodilum) linguae usu caret [...]'.

33 *Historia animalium* II, 10, engl. Übers. von D'Arcy Wentworth Thompson: ‚Oviparous and blooded quadrupeds—and, by the way, no terrestrial blooded animal is oviparous unless it is quadrupedal or is devoid of feet altogether—are furnished with a head, a neck, a back, upper and under parts, the front legs and hind legs, and the part analogous to the chest, all as in the case of viviparous quadrupeds, and with a tail, usually large, in exceptional cases small. And all these creatures are many-toed, and the several toes are cloven apart. Furthermore, they all have the ordinary organs of sensation, including a tongue, with the exception of the Egyptian crocodile'.

34 *Naturalis historia / Naturkunde* VIII, 89: ‚Magnitudine excedit plerumque duodeviginti cubita'.

35 Ebd. VIII, 94: ‚quidam hoc unum [sc. animal], quamdiu vivat, crescere arbitrantur'.

36 Diese (wenngleich vage) Angabe entspricht der Wirklichkeit: Nilkrokodile können 70–100 Jahre alt werden.

Ähnlich verfährt Plinius mit einem anderen Tier Ägyptens, dem Nilpferd oder Flusspferd (*Hippopotamus amphibius*, von Plinius ‚Hippopotamius'[37] genannt). Es soll nach Plinius in der Regel sogar über 8 Meter lang werden,[38] d.h. noch länger als das schon wunderbar lange Krokodil. In Wirklichkeit werden erwachsene Flusspferde zwischen 3.3 und 5.2 Meter lang. In der Beschreibung des Elefanten trifft man die Angabe an, dass er in der Regel 200–300 Jahre alt wird und seine Reifezeit erst ab dem 60. Jahr anfängt.[39] Tatsächlich werden Elefanten durchschnittlich bis zu 60 Jahre alt, während die Reifezeit für Bullen mit 5, für Kühe mit etwa 10 Jahren anfängt. Es geht hier nicht lediglich um einen an sich vielleicht bedauernswerten Wissensmangel im Hinblick auf einige exotische Species, sondern um ein *Darstellungsprinzip*. Das lässt sich nicht zuletzt aus folgendem ersehen: Das Durchschnittsalter von Pferden war wohlbekannt. Es ist jedoch bezeichnend, dass Plinius auch hier auf Ausnahmefälle fokussiert: In Opus soll es Hengste gegeben haben, die noch mit 40 Jahren zeugungsfähig waren—jedoch habe man ihnen dann helfen müssen, im Klartext, man habe die altersschwachen Hengste auf die Stuten geschoben oder vielmehr gehoben (VIII, 164). Auch gibt Plinius das merkwürdig hohe Alter eines Pferdes an, das 75 Jahre alt geworden sei.[40] Übertriebene Angaben zu Körpergröße und Alter haben die Aufgabe, sowohl die Species als auch ihre Bemeisterung durch den Menschen eindrucksvoller gestalten.

Aus dieser vom *mirabile*-Effekt bestimmten Diskursorientierung erklärt sich, dass für Plinius in der Beschreibung der Tierarten (*Naturalis historia*, Buch VIII, 1–XI, 120) weder die Anatomie noch die Physiologie eine reizvolle Klassifizierungsgrundlage abgab. Dabei muss man berücksichtigen, dass auch dies nicht ausschließlich auf einen Wissensnotstand zurückzuführen ist. Denn Aristoteles, auf dessen Werke sich Plinius beruft, hatte sich um eine möglichst schlüssige Einteilung des Tierreichs bemüht, die er im Zuge seiner Studien nicht mehr vornehmlich auf die Logik, sondern auf vergleichende anatomisch-morphologische und physiologische Beobachtungen gründete

37 Wohl in der Nachfolge des Aristoteles (‚hippos ho hippopotamios'); vgl. Leitner, *Zoologische Terminologie beim älteren Plinius* 138. An anderer Stelle hat Plinius Aristoteles' ‚hippos ho hippopotamios' ins Lateinische übersetzt, mit ‚equs fluviatilis' (vgl. *Naturalis historia* VIII, 73).

38 *Naturalis historia / Naturkunde* VIII, 95. Plinius hat seine Angaben aus Herodot II, 7 und Aristoteles, *Historia animalium* II, 7, 502 A 9–15, bezogen; vgl. Leitner, *Zoologische Terminologie beim älteren Plinius* 138.

39 *Naturalis historia* VIII, 28.

40 Ebd. VIII, 164.

(auch wenn diese nicht immer stichhaltig waren).[41] Zwar hat Aristoteles keine integrale klassifizierende Beschreibung des Tierreichs hinterlassen, jedoch hat er in seiner *Historia animalium* und in seinem Traktat *De partibus animalium* zu einer solchen brauchbare Grundlagen geschaffen. Aristoteles entwarf u.a. die Kategorien oder Klassen der blutführenden im Gegensatz zu den blutlosen Tieren (also Insekten, Testacea oder Schalentiere, Crustacea oder Krusten- bzw. Krebstiere und Cephalopodia oder Tintenfische);[42] unter den blutführenden Tieren die Klassen der Fische, Vögel und Vierfüßer. Die Vierfüßer unterteilte er in lebendgebärende (zoiotoka) und eierlegende (oiotoka) Arten. Natürlich sind die systematischen Kategorien und Klassen des Aristoteles nicht mit jenen der modernen Taxonomie identisch. Jedoch hat er immerhin Kategorien entworfen, nach denen sich eine sinnvolle Einteilung des Tierreichs vornehmen hätte lassen. Dass dies der Fall ist, hat viel später—in der Mitte des 16. Jahrhunderts—der englische Arzt Edward Wotton mit seiner integralen Beschreibung des Tierreichs in *De differentiis animalium* gezeigt.[43] Z.B. ist Aristoteles' Kategorie der *lebendgebärenden Vierfüßer* im Großen und Ganzen deckungsgleich mit der modernen taxonomischen Klasse der Mammalia.[44]

Vielleicht, weil sie zu erkennen meinen, dass Plinius Aristoteles „schlecht verstanden" habe, sind manche Forscher geneigt, ihm abzusprechen, dass er Aristoteles' zoologische Schriften im Original gekannt oder gründlich gelesen hat. Etwa Kollesch-Harig bemerken: ‚Der Vergleich lässt aber auch erkennen, dass Plinius den Aristoteles nicht im Original gelesen, sondern

41 Zu Aristoteles' Klassifizierung des Tierreiches vgl. Pellegrin P., *La classification des animaux chez Aristote* (Paris: 1982); Lloyd G.E.R., „The Development of Aristotle's Theory of the Classification of Animals", *Phronesis* 6 (1961) 59–81; Heck L., *Die Hauptgruppen des Thiersystems bei Aristoteles und seinen Nachfolgern, ein Beitrag zur Geschichte der zoologischen Systematik* (Diss. Leipzig: 1885); Christian Hünemörder, „Aims and Intentions of Botanical and Zoological Classification in the Middle Ages and in the Renaissance", *History and Philosophy of the Life Science* 5 (1983) (53–67) 53–56. Lloyd (1961) zeigte, dass sich Aristoteles' Einteilung des Tierreichs in seinen früheren philosophischen Werken auf die Logik gründete; später verwendete er andere Kriterien. Hünemörder, „Aims and Intentions" 54, veranschlagt Aristoteles' Leistung bei der Klassifizierung von Tierarten nicht sehr hoch: ‚he was not very succesful in classifying living beings'. Dabei scheint er davon auszugehen, dass sich Aristoteles nur auf die Logik berufen hätte.

42 Vgl. Hünemörder, „Aims and Intentions" 54–55.

43 Siehe dazu Enenkel, „Zur Konstituierung der Zoologie als Wissenschaft in der Frühen Neuzeit", bsd. 26–51, sowie, in vorliegendem Band, ders., „The Species and Beyond [...]", bsd. 3 ff.

44 Aufgrund dieser Tatsache könnte man, in einer teleologischen Herangehensweise, Aristoteles' Zoologie für fortschrittlicher halten als Plinius' Zoologie.

nur Zusammenstellungen seiner Schriften gekannt hat, so dass bei ihm viele Tierarten, die von Aristoteles beschrieben worden sind, fehlen—ein Umstand, der den Niedergang der zoologischen Forschung nach Aristoteles [...] deutlich vor Augen führt'.[45] Demgegenüber hatte Plinius in einer programmatischen Aussage im achten Buch der *Naturalis historia* explizite festgestellt, dass seine Zoologie auf Aristoteles' ‚50 zoologische Bücher' aufbaue: Er habe das von Aristoteles dargebotene Wissen zusammengefasst und um Erkenntnisse erweitert, die Aristoteles noch nicht vorlagen.[46] Von diesen sind heute noch etwa 27 überliefert, d.h. Plinius waren offensichtlich mehr zoologische Bücher des Aristoteles als die heute überlieferten bekannt. Seine Kenntnis von Aristoteles' Werken ist ernst zu nehmen. Ich meine nicht, dass sich Plinius' andersgelagerte Organisation und Selektion des zoologischen Wissens nur auf Unkenntnis oder ein grundlegendes Missverständnis seiner Quelle zurückführen lässt.

Plinius ist mit Sicherheit nicht einfach ein „falsch verstandener Aristoteles". Das gilt insbesondere für seine Einteilung des Tierreichs. Diese funktioniert nicht nach den Kategorien des Aristoteles, jedoch keineswegs aus dem Grund, dass er die Kategorien des Aristoteles nicht gekannt oder falsch verstanden hätte. Letztes geht aus einem Abschnitt hervor, der Plinius' Beschreibung des Tierreichs folgte, nämlich der komparatistischen Darlegung der Morphologie, Physiologie und Ernährungsweise der Tiere im 11. Buch (XI, 121–284). In diesem Abschnitt lassen sich die hier angesprochenen Einteilungskategorien

45 Vgl. Jahn, *Geschichte der Biologie* 80.

46 *Naturkunde* VIII, 44: ‚König Alexander d. Gr., von der Begierde entflammt, die Eigenschaften und das Wesen der Tiere (Species) kennenzulernen, beauftragte Aristoteles, den in jedem Fach ausgezeichneten Gelehrten, mit diesen Forschungen, und einige tausend Personen in ganz Asien und Griechenland, die sich durch Jagd, Vogelfang und Fischerei ernährten und Tiergärten, Herden, Bienenstöcke, Fischteiche und Vogelhäuser besorgten, erhielten den Befehl, seinen Wünschen zu entsprechen, damit ihm kein Lebewesen unbekannt bleibe. Nach ihren Berichten verfasste Aristoteles nahezu 50 berühmte Bücher über die Tiere. Diese habe ich mit dem, was ihm unbekannt war, gedrängt zusammengefasst und bitte nun dafür den Leser um wohlwollende Aufnahme [...]'. „Bücher"/ *libri* meint hier die antike Definition von „Buch" = Buchrolle: also jene Textmenge, die auf einer Papyrus-Rolle Platz fand und welche auch vom Autor für eine Buchrolle bestimmt war. Ein bestimmtes Werk setzte sich also in der Regel aus mehreren Buchrollen zusammen, so auch u.a. Aristoteles' *Historia animalium* und *De partibus animalium*. Die *Historia animalium* hat 10 Bücher, *De incessu animalium* 1 Buch, *De motu animalium* ebenfalls 1 Buch, *De partibus animalium* 4 Bücher, *De generatione animalium* 5 Bücher, die *Parva naturalia* jedenfalls 6 Bücher. Daneben sind zahlreiche Bücher nicht mehr überliefert, wie etwa die Sammlungen *Anatomikai* und *Zoika*.

und -klassen sowie auch noch einige weitere erkennen. Plinius betrieb dort eine sehr enge Aristoteles-Nachfolge, während er in der Organisation, die er seiner integralen Beschreibung des Tierreichs (VIII, 1–XI, 120) zugrunde legte, davon abwich. Im komparatistischen Teil des 11. Buches wendet Plinius die Kategorien der blutführenden und blutlosen Tiere an.[47] Nur blutführende Tiere haben die Organe Leber und Milz.[48] Blutlose Tiere haben keinen Magen.[49] Alle Tiere, die mehr als 4 Füße aufweisen, gehören zu den Blutlosen.[50] Dem Blut und den Organen, die damit zu tun haben, ist ein eigener Abschnitt gewidmet.[51] Die Klassen der Fische (‚pisces'), Vögel (‚aves' oder ‚pinnata [animalia]' oder ‚[animalia] pinnas habentia'), Vierfüßer (‚quadripedes') und Schlangen (‚serpentes') werden immer wieder angewendet. Fische, Vögel und Schlangen ‚besitzen keine Nase', sondern Riechlöcher.[52] Fische haben Sägezähne, mit Ausnahme des Papageifisches (‚Scarus'; = *Scarus cretensis* oder *Sparisoma cretense*), der Zahnplatten aufweist und ein Pflanzenfresser ist.[53] Schlangen sollen ‚lange Eingeweide' haben.[54] Vögel weisen—im Unterschied zu Vierfüßern (und Fischen)—einen Schnabel auf,[55] jedoch—ausnahmslos—keine Zähne.[56] Weiter haben Vögel, mit Ausnahme des Uhus und der Waldohreule, keine Ohren.[57] Im gesamten Abschnitt wendet Plinius stets auch die Kategorien der eierlegenden Tieren (‚ova parientia', ‚ova gignentia' oder ‚ova parientia')[58] gegenüber den lebendgebärenden (‚animal generantia', ‚animal parientia') an. Weibliche eierlegende Tiere haben niemals Zitzen und sie geben keine Milch.[59] Die Vierfüßer (‚quadripedes') unterteilt Plinius nach Aristoteles ebenfalls in lebendgebärende und eierlegende Arten. Zu den eierlegenden zählen Krokodile, Eidechsen (‚lacertae') und Schildkröten. Alle lebendgebärenden haben Haare, die eierlegenden jedoch entweder Federn (Vögel) oder Schuppen (vierfüßige Reptilarten) oder einen Panzer (Schildkröte).[60] Darüber

47 U.a. *Naturalis historia* XI, 133; 192; 199; 204; 219–225 (bsd. 223).

48 Ebd. XI, 192 und 204.

49 Ebd. 199.

50 Ebd. 223.

51 Ebd. 219–225.

52 Ebd. 158.

53 Ebd. 162; vgl. Leitner, „Zoologische Terminologie beim älteren Plinius" 217.

54 *Naturalis historia* XI, 197.

55 Ebd. 138.

56 Ebd. 164.

57 Ebd. 137.

58 U.a. ebd. 175; 204; 232.

59 Ebd. 232.

60 Ebd. 228.

hinaus unterscheidet Plinius bestimmte Klassen wie Paarhufer (‚bisulca'), „Einhufer" (‚solipedes') und Mehrzeher (‚digitos habentes'),[61] jedoch außerdem Ruminantia oder Wiederkäuer,[62] ja sogar „Säugetiere" (‚mammas habentes')[63] und Hornträger.[64] So weisen Hornträger niemals Sägezähne[65] auf und sie haben 13 Rippen(paare).[66] Auch den Begriff der Wirbeltiere kennt Plinius.[67] Wenn Plinius den Begriff ‚genus' anwendet, hat er zumeist eine bestimmte Species vor Augen. Manchmal bedeutet ‚genus' bei ihm jedoch eine Familie, wie die der Katzenartigen[68] und die der Affenartigen (‚simiarum genera' und ‚simiarum genus').[69]

Daraus geht hervor, dass Plinius mit Aristoteles' Einteilungskriterien hervorragend vertraut war und dass er ohne weiteres seine Beschreibung des Tierreiches auf sie gründen hätte können. Plinius verzichtete bewusst darauf. Es war für ihn jedoch in den Büchern VIII–XI, 120 nicht reizvoll, mit diesen oder ähnlichen Kategorien zu arbeiten. Mit der vom *mirabile*-Effekt bestimmten Diskursorientierung seiner Zoologie hängt u.a. zusammen, dass Plinius in neue und gewissermaßen „abenteuerliche" Klassifizierungsregionen vorstößt—in Regionen, in denen der Species- oder Artenbegriff als solcher unsicher und flexibel wird. Plinius suggeriert, dass die vorhandenen Arten ein hohes *Kreuzungspotential* besitzen: Im Prinzip können sich ständig und spontan durch Kopulation neue Specimina von Tieren herausbilden, hybride Specimina, denen Plinius offensichtlich zuschreibt, dass sie fruchtbar sind, da diese—vermeintlichen—Hybriden in großer Zahl auftreten, wie im Fall des „Leo-Parden", der Kreuzung zwischen Löwe (*leo*) und Panther (*pardus*). Seine Theorie der spontanen *Hybridität* verbindet Plinius einerseits mit der Tierwelt Afrikas, andererseits versucht er sie durch eine plausible, rationalistische und empirisch nachvollziehbare Erklärung zu untermauern: Wegen des ständigen Wassermangels würden sich in Afrika stets massenhaft ganz verschiedene Tierarten bei denselben Wasserlöchern aufhalten; psychisch angeregt durch die körperliche Nähe und physiologisch stimuliert durch die Hitze, würden sie sich ungestüm paaren, wobei sie regelmäßig die Grenzen der Species überschritten. So würden in

61 Ebd. 254.
62 Ebd. 199.
63 Ebd. 232.
64 Ebd. 164; 207.
65 Ebd. 164.
66 Ebd. 207.
67 Ebd. 177–178.
68 Ebd. 172.
69 Ebd. 246.

Afrika stets ‚neue' (d.h. noch nie dagewesene, unbekannte) Specimina oder Species von Tieren geboren.[70]

Dementsprechend figuriert in der *Naturalis historia* eine Reihe afrikanischer Mischwesen aus Löwe und Panther,[71] Hyäne und Löwe (*corocotta*),[72] Hund und Wolf (*crocota*),[73] Pferden und Vögeln (*pegasus*, VIII, 72), usw. Die Species *eale* hat ‚die Größe eines Flusspferdes [also abermals ein Tier mit einer Länge von über 8 Metern (!), Anm.], den Schwanz eines Elefanten, eine schwarze oder gelbbraune Farbe, die Kinnladen eines Ebers und über eine Elle lange bewegliche Hörner, die es im Kampf abwechselnd nach vorne oder zur Seite hin aufrichtet' (‚Apud eosdem et quae vocatur *eale*, magnitudine equi fluviatilis, cauda elephanti, colore nigra vel fulva, maxillis apri, maiora cubitalibus cornua habens mobilia, quae alterna in pugna sistit variatque infesta aut obliqua').[74] Wenn man von den ‚beweglichen Hörnern' und der Körperlänge von über 8 Metern absieht, könnte die Inspirationsquelle des Ungeheuers *eale* eine afrikanische Nashornart gewesen sein, entweder das Spitzmaulnashorn (*Diceros bicornis*) oder das Breitmaulnashorn (*Ceratotherium simum*).[75] Manche Species/ Specimina werden zwar nicht direkt als Abkömmlinge verschiedener Arten bezeichnet, jedoch wie Mischwesen beschrieben, z.B. das Nilpferd (VIII, 95, *hippopotamius*): Es soll zwei Zehen haben wie das Rind, den Rücken, die Mähne und das Wiehern des Pferdes, jedoch den Schwanz und die Zähne des Wildschweines.[76]

---

70 *Naturalis historia / Naturkunde* VIII, 42.

71 Ebd. VIII, 43: ‚Am Geruche merkt der Löwe an der ungetreuen Löwin die Begattung mit dem Panther (*pardus*)'.

72 Ebd. VIII, 107: ‚Huius generis coitu leaena Aethiopica parit corocottam [...]' (‚Durch Koitus mit dieser Tierart [= Hyäne] gebiert die äthiopische Löwin die *Corocotta*').

73 Ebd. VIII, 72 ‚Crocota' als Kreuzung von Hund und Wolf: ‚crocotas velut ex cane lupoque conceptos, omnia dentibus frangentes protinusque devorata conficientes ventre' (‚Crocotae, die etwa Mischlinge von Hund und Wolf sind, mit den Zähnen alles zermalmen und das Verschlungene sogleich verdauen'). Leitner, „Zoologische Terminologie beim älteren Plinius" 101, identifiziert diese beiden hybriden „Species" miteinander und setzt sie mit der in Äthiopien vorkommenden Fleckenhyäne gleich.

74 *Naturalis historia* VIII, 73. Es ist bezeichnend, dass die *Eale* im komparatistisch-analytischen Teil der Zoologie im 11. Buch nicht vorkommt.

75 Vgl. Macdonald, *Enzyklopädie der Säugetiere* 476–481. Das Spitzmaulnashorn hat eine Körperlänge von zwischen 2.8 und 3 Metern; das Breitmaulnashorn von 3.4 und 4 Metern bei den Männchen, von 3.4 und 3.6 Metern bei den Weibchen. VIII, 71 gibt Plinius an, dass das Nashorn dieselbe Körperlänge wie der (afrikanische) Elefant besitze, jedoch kürzere Beine habe. An der nämlichen Stelle gibt Plinius dem Tier freilich nicht den Namen ‚eale', sondern ‚rhinoceros'.

76 Plinius, *Naturalis historia / Naturkunde* VIII, 95: ‚Maior altitudine in eodem Nilo belua hippopotamius editur; ungulis binis quales bubus, dorso equi et iuba et hinnitu, rostro resimo, cauda et dentibus aprorum aduncis, sed minus noxiis [...]'.

Die in Äthiopien lebende *leucrocota* (VIII, 72) soll die Größe eines Wildesels, die Beine eines Hirschen, Hals, Schwanz und Brust eines Löwen, den Kopf eines Dachses, gespaltene Hufe, ein bis zu den Ohren aufgerissenes Maul und statt aneinandergereihten (Kiefer)Zähnen einen einzigen langen, messerartigen Zahn oder zahnartigen Knochen aufweisen.[77] Erstaunlicherweise vereint dieses Wesen, dessen „Inspirationsquelle" vielleicht eine Hyänenart war, Merkmale von Pflanzenfressern und Paarhufern mit denen von Raubtieren, vermehrt mit dem mehr als kuriosen Merkmal des langen „Schneideknochens" statt eines Gebisses mit Zähnen. Dies mag an die Phantasiegeschöpfe erinnern, die Hieronymus Bosch entworfen hat. Das verrückte Merkmal des „Messers" statt eines Gebisses schreibt Plinius im übrigen noch einem weiteren Mischwesen, der *corocotta*, einer Kreuzung zwischen weiblichem Löwen und männlicher Hyäne, zu.[78] Dort erfahren wir weitere kuriose Details: Das Messer soll sich nur dort befinden, wo andere Tiere das Obergebiss haben; statt eines Untergebisses soll das Tier eine Art Scheide besitzen, in der das Messer des Obergebisses geborgen wird, um es—in der Tat nach dem Prinzip einer Messerscheide—scharf zu erhalten. Diese wundersame Konstruktion hat sich die Natur ausgedacht, um zu verhindern, dass das Messer—etwa durch den Gegendruck eines Untergebisses—stumpf werde.

Plinius' Beschreibungssystem wirkt in die Richtung, dass er die einzelnen Species als möglichst seltsam und wunderbar präsentiert. Es sind nicht selten geradezu *monstra*, die darin zur Schau gestellt werden. Dies betrifft eben jenen Grundzug von Plinius' Zoologie, dass er darin stets das Exotische, Erstaunliche und Wunderbare betont. In dieser Hinsicht ähnelt seine Enzyklopädie der Wunderkammer der Frühen Neuzeit, die Barbara Findlen überzeugend als Instrument zur Bemächtigung und Bemeisterung der Natur beschrieben hat.[79] In Wunderkammern wurden *Exotica*, Missgeburten (etwa Tiere mit zwei

77 Plinius, *Naturalis historia / Naturkunde* VIII, 72: ‚leucrocotam, pernicissimam feram, asini fere magnitudine, clunibus cervinis, collo cauda pectore leonis, capite melium, bisulca ungula, ore ad aures usque rescisso, dentium loco osse perpetuo'.

78 Ebd. VIII, 107: ‚Huius generis coitu leaena Aethiopica parit corocottam [...]. Acies ei perpetua in utraque parte oris nullis gingivis, dente continuo, ne contrario occursu hebetetur, capsarum modo includitur' (‚Durch Kreuzung mit dieser Tierart [= Hyäne] gebiert die äthiopische Löwin die corocotta [...] Sie hat ein durchgehendes Schneideinstrument an beiden Seiten des Maules, wobei sie kein Zahnfleisch, sondern einen durchgehenden Zahn besitzt, der in eine Art Kapsel eingeschlossen ist, damit er nicht durch den Gegendruck abgestumpft werde'. König übersetzte ‚acies ei perpetua' mit ‚Sie hat ständig geöffnete Augen'; m.E. bezieht sich ‚acies ei perpetua' jedoch auf die durchgängige „Zahnschneide" bzw. das „Zahnmesser".

79 Findlen P., *Possessing Nature. Museums, Collecting and Scientific Culture in Early Modern Italy* (Berkely: 1999).

Köpfen) und alle möglichen Raritäten gesammelt. Der Vergleich mit der Wunderkammer mit ihrem Weltbemeisterungseffekt besitzt für Plinius' Zoologie keine nur beiläufige Bewandtnis. Denn die von ihm beschriebenen exotischen Tiere blieben nicht in Afrika: Sie wurden in Rom dem staunenden Volk zur Schau gestellt, vorgeführt und schließlich—zur symbolischen, jedoch unmissverständlichen Darstellung der Bemeisterung—getötet.[80] Etwa das monströse Wesen *camelopardalis* (Giraffe, VIII, 69) war bereits unter dem Diktator Caesar in Rom vorgeführt und getötet worden. Wenn Plinius den sprichwörtlichen Ausdruck ‚aus Afrika kommt stets etwas Neues' (VIII, 42) auf seine Theorie der Hybriden anwendet, steht ihm ein doppelter Effekt des Staunens vor Augen: Erstens einmal soll der Leser die Merkwürdigkeit und Ausgefallenheit der Species bewundern, zweitens die Tatsache, dass es dem Römischen Reich gelingt, all dieser monströsen Geschöpfe, der messergebisstragenden Löwenhyänen, der reißenden gefleckten Pantherlöwen, schwimmenden Riesenpferde usw., Herr zu werden.

Die vom *mirabile*-Effekt bestimmte Diskursorientierung bringt jedoch nicht mit sich, dass Plinius in seiner Zoologie völlig unsystematisch gearbeitet hätte, wie etwa in Jahns *Geschichte der Biologie* behauptet wird.[81] Plinius ordnete die Tiere nach 3 Lebensbereichen an: Landtiere, Wassertiere, Tiere der Luft (= Vögel und Insekten). Diese Einteilung war für ihn sinnvoller, weil sich anhand der verschiedenen Lebensbereiche die Herrschaft des Menschen über die Tiere am besten kategorisch vorführen ließ. Im Buch über die Landtiere lässt sich eine weiterführende Systematik erkennen: Plinius erstellte eine *Hierarchie* des Tierreichs zu Lande, die er auf 3 Ordnungskategorien gründete: 1. von groß nach klein, 2. von exotisch (d.h. Afrika, Asien, Nordeuropa/ hoher Norden) nach einheimisch (d.h. Süd- und Westeuropa), 3. von wild nach zahm. Die größeren, exotischen und wilden Tiere stehen in der Hierarchie je an übergeordneten Stellen. Buch VIII der *Naturalis historia* fängt mit dem größten Landtier, dem Elefanten, an, wird von den Riesenschlangen Indiens und Äthiopiens, sowie den großen Raubkatzen (Löwen, Panthern und Tigern), weiter von Kamel, Giraffe und Nashorn fortgesetzt, und endet mit dem kleinsten und unbedeutendsten Tier, der Maus. Der Hauptteil des Buches der Landtiere,

80 Vgl. zu den römischen Tierkämpfen grundlegend Wiedemann Th., *Emperors and Gladiators* (London – New York: 1992), ch. 2, 55–101; Jennison G., *Animals for Show and Pleasure in Ancient Rome* (Manchester: 1937).

81 *Geschichte der Biologie* 79–80: ‚Dieser Einteilung [des Plinius, Anm. K.E.] in Landtiere, Wassertiere, fliegende Tiere und Insekten liegt kein systematisches Prinzip zugrunde [...]'. Vgl. oben.

etwa zwei Drittel (VIII, 1–141), ist bezeichnenderweise exotischen und wilden Tieren gewidmet, nur etwa ein Drittel den zahmen und einheimischen (VIII, 142–224), obwohl Plinius über diese Species zweifellos mehr bekannt war.

Diese Hierarchie des Tierreichs zeigt eine *imperialistische* Diskursorientierung. Diese eignete sich besonders gut, die Errungenschaften bzw. die Macht des Menschen zu demonstrieren, dem es gelungen war, das Tierreich zu unterwerfen. Es ist ein Statement, wenn Plinius mit dem größten und stärksten Landtier, das zugleich wild lebend und exotisch ist, anfängt: dem Elefanten. Damit führt er exemplarisch die imperialistische Machtausübung des Menschen vor, dem es gelingt, sogar dieses Tier zu fangen, bändigen, zähmen, zu unterwerfen und sich nutzbar zu machen. Unter „dem Menschen" ist immer prägnant zu verstehen: die Römer bzw. die Bürger und Untertanen des Römischen Reiches.

Aufgrund dieser Diskursausrichtung erübrigt es sich, zu erwarten, dass Plinius einer etwaigen Gefährdung wildlebender Tiere viel Aufmerksamkeit entgegenbringen würde. Das liegt nicht daran, dass eine tatsächliche Gefährdung nicht vorlag. Gerade zwischen ca. 100 v. Chr. und 200. n. Chr. wurden bestimmte wildlebende Tierarten in Teilen des Römischen Reiches stark dezimiert bzw. von Ausrottung bedroht: z.B. der Löwe und der Leopard in weiten Teilen Nordafrikas, besonders in Libyen. Der Hauptgrund für diesen Raubbau an der Population bestimmter Wildtiere lag gerade in der Tatsache, dass man seit dem 2. Jh. v. Chr., v.a. aber seit etwa 60 v. Chr., bestimmte Species —am liebsten in großer Zahl— in Rom präsentierte. Dazu zählten v.a. Löwen und Leoparden, die man seit 186 v. Chr. zeigte (M. Fulvius Nobilior).[82] 169 v. Chr. wurden im Circus Maximus 63 afrikanische Raubkatzen und außerdem 40 Bären und einige Elefanten;[83] am selben Ort bei einer Show, die der Diktator Lucius Sulla veranstaltete, ‚100 langmähnige (*iubati*) Löwen' getötet;[84] weiter 58 v. Chr. 150 afrikanische Leoparden.[85] Pompeius präsentierte dem Volk 55 v. Chr. in Rom u.a. sogar 600 Löwen, 410 Leoparden und 20 Elefanten; Caesar 46 v. Chr. 400 Löwen und eine Giraffe. Augustus ließ allein bei Spielen des Jahres 13 v. Chr. 600 afrikanische Wildtiere; während seiner Regierung insgesamt, wie er in seinem Tatenbericht *Res gestae* nicht ohne Stolz festhielt, 3.500 gefährliche afrikanische Wildtiere[86] töten. Bei den langen Spielen zur

82 Vgl. Wiedemann, *Emperors and Gladiators* 59.

83 Ebd. 60.

84 Vgl. Plinius, *Naturalis historia* / *Naturkunde* VIII,53.

85 Für diese und die folgenden Daten vgl. Wiedemann, *Emperors and Gladiators* 60–61.

86 Also keine „harmlosen" Pflanzenfresser wie Antilopen, Zebras, Wildesel, Gnus und Gazellen.

Einweihung des Colosseum wurden insgesamt 9000 Tiere getötet, bei den Spielen zur Ehren Trajans 106 n. Chr. sogar 11.000. Es ist klar, dass sich diese massive, überdimensioniert erscheinende und einseitige imperialistische Auseinandersetzung mit der Natur auswirkte. Schon die obige, kurze Liste, die nur einige Zurschaustellungs- und Tötungsmomente in der Stadt Rom verzeichnet, deutet darauf hin, dass seit etwa 60 v. Chr. in jedem Jahrzehnt mindestens 1.000 afrikanische Raubkatzen gefangen und getötet wurden. Die Gefährdung der Raubkatzen wird noch eklatanter, wenn man sich vor Augen hält, dass an diversen Orten des Römischen Reiches jahrhundertelang solche Zurschaustellungen und Tötungen stattfanden. Dann verwundert es nicht, dass, wie Textstellen u.a. in den *Scriptores Historiae Augustae* zeigen, es im dritten Jh. n. Chr. bereits Schwierigkeiten bereitete, genügend Raubkatzen zu finden, die man präsentieren und töten konnte.[87] In einem Gedicht der *Anthologia Palatina* wird angesprochen, dass der Löwe in Libyen so gut wie ausgerottet sei (VII, 626): Das anonyme Epigramm lobt den römischen Kaiser dafür, dass er Libyen von den reißenden Bestien befreit habe, indem er sie fangen und nach Rom bringen ließ.[88]

Es ist kein Zufall, dass die meisten der hier genannten Daten aus Plinius' Zoologie stammen. Denn Plinius liefert darin nicht nur eine *Natur*geschichte, sondern zugleich eine *Kultur*geschichte—*eine Geschichte der Bemeisterung des Tierreichs durch den Menschen* (d.h. v.a. den Römer). Bei den größeren und ‚gefährlicheren' Wildtier-Species vermeldet er jeweils, wann, von wem und in welcher Zahl sie in Rom zur Schau gestellt und getötet worden waren. Zuweilen gerät die Zurschaustellung der Species in Rom zu einem zoologischen Ordnungsprinzip: Die Nennung der jeweiligen Species wird durch das Fest/ den Anlass bestimmt, bei dem man sie *zuerst in Rom* zu Gesicht bekam. Die Spiele des Pompeius 55 v.Chr. inszeniert Plinius demnach als ein wahres „Fest unbekannter Tierarten" (VIII, 70–71): Da sah man in Rom zum ersten Mal die Species ‚chama' (wahrscheinlich der Luchs, *Lynx lynx*),[89] zum ersten Mal eine auffällige, jedoch nicht genau bestimmbare Affenart aus Äthiopien

87 Vgl. *Scriptores Historiae Augustae, Gordiani* 3.

88 Für dieses Gedicht vgl. Wiedemann, *Emperors and Gladiators* 64: ‚Borders of the Libyan Nasamonians, your plains are no longer impassable because of the races of wild beasts; no longer will you echo to the roaring of lions in the desert, way beyond the sands which belong to the Nomads, since the lad Caesar has captured a group without number and made them all face his fighters. The mountains which were once home to wild beasts now provide cattle pasturage for men'. Vgl. dazu auch Norman D., *Birds and Beasts of the Greek Anthology* (ebook: 2009), Abschnitt „Mammalia".

89 So auch Leitner, „Zoologische Terminologie beim älteren Plinius" 81.

(‚cepus' oder ‚cephus'),[90] die sich anscheinend auf den Hinterbeinen fortbewegte, und zum ersten Mal ein afrikanisches Nashorn (von Plinius *rhinoceros* genannt). Vor allem das Nashorn scheint einen unvergesslichen Eindruck hinterlassen zu haben. Im Hinblick auf den ‚cepus' oder ‚cephus' hebt Plinius seinen Seltenheitswert hervor: Man habe ihn weder vor noch nach diesen Spielen je noch in Rom gesehen. Es kommt hier also das zoologische Ordnungsprinzip der *Erstzurschaustellung in Rom* zum Tragen: Eine Raubtierart (*Lynx lynx*), ein Pflanzenfresser und Unpaarhufer (Nashorn) und eine Primatenart werden zusammen behandelt, nur weil sie bei derselben Gelegenheit zum ersten Mal zur Schau gestellt wurden.

Da man bei Pompeius' Spielen gleich drei Species zuerst sah, stellen sie einen Meilenstein in der Bemeisterung der Natur dar. Es gab mehrere solche Meilensteine. Marcus Scaurus etwa hatte nur drei Jahre zuvor bei seinen Spielen zuerst die Species Nilpferd (*Hippopotamus amphibius*) und Nilkrokodil (*Crocodylus niloticus*) gezeigt, wie Plinius genau verzeichnet (VIII, 96). Dazu ließ Scaurus eigens einen Wassergraben (‚epirus') auswerfen. Die Species Giraffe (*Giraffa camelopardalis*) hatte ihren Erstauftritt in Rom bei den Spielen, die der Diktator Julius Caesar anlässlich seines vierfachen Triumphes (u.a. in bezug auf den afrikanischen Krieg) veranstaltete. Der afrikanische Löwe (*Panthera leo*) hatte seinen Erstauftritt bereits 104 v. Chr. unter dem kurulischen Ädil Quintus Mucius Scaevola (VIII, 53), der Elefant schon viel früher, 275 v. Chr. anlässlich des Triumphes des Consuls Manlius Curius Dentatus über König Pyrrhos von Epirus (VIII, 16)[91] und, auf besonders spektakuläre Weise, im Jahre 250 v. Chr.: Nach seinem Sieg über die Karthager in Sizilien (Hasdrubal) präsentierte der Pontifex Maximus Lucius Caecilius Metellus eine große Anzahl von Exemplaren dieser Species, über 100, vielleicht sogar 142 (VIII, 16). Die seltenen Riesentiere, die plötzlich in solchen Mengen auftraten, erregten ungeheures Staunen. Manche Zuschauer konnten kaum glauben, dass es sich dabei um Kriegsbeute handelte; sie dachten, dass Metellus eine Elefantenjagd veranstaltet hatte.[92] Bei der Beschreibung des Löwen liefert Plinius sogar eine bündige, chronologische Geschichte der „Karriere" dieser Species in Rom: ‚In Rom ließ zuerst Scaevola, Sohn des Publius, während er kurulischer Ädil war [= 104 v. Chr.], mehrere Löwen zugleich kämpfen. Einen Kampf von 100 langmähnigen Löwen (‚iubati') veranstaltete aber zuerst Lucius Sulla, der spätere

90 Vgl. ebd. 76.

91 Zu dem Triumph des Manlius Curius Dentatus vgl. Broughton T.R.S., *The Magistrates of the Roman Republic*, 2 Bde. (1951–1952), Bd. I, 195. Manlius Curius Dentatus bekleidete damals zum dritten Mal das Konsulat.

92 Florus, *Epitome* I, 13,28.

Diktator, während seiner Prätur [= 93 v. Chr.]. Nach ihm präsentierte Pompeius der Große im Zirkus 600 Löwen, worunter 350 langmähnige [= 55 v. Chr.], und der Diktator Caesar 400 [= 46 v. Chr.]'.[93]

Zu dieser Diskursausrichtung, die die Zurschaustellung und Tötung der Tiere in Rom hervorhob, gehört, dass Plinius bei der Beschreibung wildlebender Species mehrfach einen eigenen Abschnitt dem Fang widmet, z.B. beim Elefanten (VIII, 24–26), Löwen (VIII, 54), Tiger (VIII, 66), Krokodil (VIII, 92–93) usw. So kann man Elefanten mit Hilfe gezähmter Lockelefanten, wie dies in Indien geschah, oder in Fallgruben, oder durch Anschießen mit einem Pfeil mit anschließender Hetzjagd, wie es anscheinend in Afrika praktiziert wurde, oder, wie bei dem Stamm der Trygloten in Äthiopien, durch Durchschneiden der Sehne eines Knies fangen. Eine ähnlich idiosynkratisch anmutende Fangmethode beschreibt Plinius für den Tiger: Tiger soll man am besten fangen, indem man einen ganzen Wurf raubt, diesen zu Pferde im Galopp zu einem bereitstehenden Schiff transportiert und der nachsetzenden Tigerin zur Ablenkung ein Junges zuwirft, das sie dann quasi vorprogrammiert zuerst in ihr Nest zurückbringen soll, was ihr die Möglichkeit nimmt, die berittenen Tierfänger noch einzuholen (VIII, 66). Dasselbe gilt für eine von Plinius als neuartig hingestellte Fangmethode des Löwen, die ein gätulischer Hirte während der Regierungszeit des Kaisers Claudius (41–54 v. Chr.) erfunden haben soll. Wenn man einem Löwen einen Mantel oder eine Decke über den Kopf wirft, soll er erlahmen und man kann das wehrlose Tier abtransportieren (VIII, 54). Krokodile kann man laut Plinius unterwerfen, indem man sich ihnen unerschrocken nähert und ihnen nachsetzt (VIII, 92–93), wie dies die Leute von Dendera (Tentyrus, eine Insel am westlichen Nilufer) praktiziert haben sollen. Diese mutigen Ägypter sollen Krokodile fangen, indem sie ihnen nachschwimmen, sich auf ihren Rücken setzen, ihnen einen Holzstock in den Rachen schieben und die Tiere dann, wie mit einem Zügel, an Land steuern, wo sie sie in Käfige sperren (VIII, 93).

Dem modernen Leser mag es als ein perverser Zug erscheinen, dass Plinius in seiner Zoologie zugleich auch eine Geschichte von der Vernichtung der Tiere geliefert hat. Jedenfalls fällt es dem modernen Leser schwer, dieser Tatsache ins Auge zu blicken bzw. sie emotional nachzuvollziehen. Dabei ist natürlich zu berücksichtigen, dass vorindustrielle Gesellschaften generell eine andere Haltung zum Tier aufweisen als moderne, westliche und postindustrielle Gesellschaften. In vorindustriellen Gesellschaften wird das Tier sehr häufig als ein Wesen betrachtet, das der Mensch sehr weitgehend ausnutzen, ja „ausschlachten" darf. Ein viel größerer Prozentsatz der Bevölkerung als etwa

93 *Naturalis historia* / *Naturkunde* VIII, 53.

im 20. oder 21. Jh. war mit dem regelmäßigen Schlachten von Tieren vertraut, z.B. in der Landwirtschaft. Dennoch betreffen diese Tatsachen wahrscheinlich nicht den wichtigsten Aspekt von Plinius' zoologischer Darbietungsweise. Dieser Aspekt hängt mit der *politischen* und *religiösen* Rolle der Tierhetzen (*venationes*) im antiken Rom zusammen, der mit zwei politisch-religiösen Institutionen verbunden ist, zum einen mit dem Triumph, zum anderen aber besonders mit den Gladiatorenspielen (*munera*).[94]

In der Römischen Republik hingen die *venationes* weder ursprünglich noch ursächlich mit den Gladiatorenspielen zusammen, obwohl sie mehrfach im Verein mit denselben aufgeführt wurden. Häufiger wurden *venationes* im Zusammenhang oder im Zusammenspiel mit dem Triumph eines siegreichen Feldherrn gegeben (Manlius Curius Dentatus, Marcus Fulvius Nobilior, Pompeius, Caesar; in der Kaiserzeit u.a. Augustus, Claudius, Traianus, Probus, Gordianus). Die Verbindung mit dem Triumph zeigt an, dass es sich um eine politisch relevante, ritualisierte Machtdemonstration handelte, in der Rom nicht nur einen erfolgreichen Feldherren (*imperator*) ehrte, sondern zugleich einen imperialistischen Anspruch zur Geltung brachte. Das Kernstück der rituellen Ausgestaltung dieser römischen Siegesfeier war eine *politisch-religiöse Prozession*, in der der Triumphator auf der „heiligen Straße", *via sacra*, über das Forum zum Kapitol zog, wo sich das Staats- und Machtheiligtum *par excellence*, der Tempel des Iupiter Optimus Maximus, befand. In der Prozession marschierten die Kriegsgefangenen, meist Angehörige unterworfener Völkerschaften, mit; weiter wurde auch anderes mitgeführt, z.B. Kriegsbeute, Kunstgegenstände oder Objekte, die die eroberten und nunmehr dem Römischen Reich (etwa als Provinzen) einverleibten Gebiete symbolisierten. Dazu zählten auch exotische Tierarten, welche sozusagen mitunterworfen worden waren, z.B. Elefanten oder Raubkatzen. Es gehörte zum Ritual, dass nach dem Ende der Prozession mitgeführte Kriegsgefangene als Symbol für die Endgültigkeit der Unterwerfung erdrosselt wurden. Den mitgeführten Tieren harrte ein vergleichbares Los, jedoch wurden sie vielfach nicht gleich getötet. Man hatte dazu ein gesondertes Schauspiel entworfen, die Tierjagd (*venatio*) bzw. den Tierkampf. An einem dafür geeigneten Platz, z.B. dem Circus Maximus, mussten die Tiere entweder mit dafür speziell ausgebildeten Tierkämpfern (*bestiarii*) oder gegeneinander vor versammeltem Publikum (*populus Romanus*) kämpfen. Dabei konnte es den Tierkämpfern natürlich

94 Vgl. dazu Wiedemann, *Emperors and Gladiators* 54–67. Wiedemanns Studie ist grundlegend. Andere haben oftmals die *venationes* lediglich als Volksbelustigung betrachtet, vgl. u.a. die Monographie von Jennison G., *Animals for Show and Pleasure in Ancient Rome* (Manchester: 1937).

widerfahren, dass sie schwer verletzt wurden oder unter Umständen ihren Verletzungen erlagen, vor allem war aber der Tod der Tiere vorprogrammiert. Der *ritualiserte Tod der gefährlichen „Bestien"* symbolisierte dabei die imperialistische Herrschaft und Macht Roms.

Eine vielleicht noch wichtigere Sinngebung der *venationes* ergab sich durch ihre Zusammenlegung mit den Gladiatorenspielen (*munera*),[95] die im übrigen einen Zusammenhang mit dem Triumph nicht ausschloss, da Gladiatorenspiele auch anlässlich von Triumphen organisiert wurden. Die Zusammenlegung der *venationes* mit *munera* geschah in der Römischen Republik mehrfach, seit Kaiser Augustus wurde sie jedoch institutionell verankert: Seit Augustus fanden die *venationes* in Rom immer im Verein mit Gladiatorenspielen statt, und zwar jährlich in der Regel jedenfalls ein Mal, zu den *Saturnalien* im Dezember. Dabei spielte der ritualisierte Kampf auf Leben und Tod eine zentrale Rolle, sowohl in den Gladiatorenkämpfen als auch in den Tierkämpfen. Die Tiere wurden diesbezüglich grundsätzlich als „Kämpfer" betrachtet und als solche in die Arena geführt. Sie sollten es dort entweder mit anderen Tieren oder mit dem Menschen aufnehmen. Damit hat die große Vorliebe zu tun, die man für Raubtierspecies (‚ferae', ‚ferae dentatae') oder andere „gefährliche" (‚noxia') Tierarten hatte, d.h. Tierarten, die den Menschen bedrohen konnten. Die riesige Anzahl von Raubtieren, die bei den Shows jeweils präsentiert wurden, legt davon Zeugnis ab. Je wilder, aggressiver, gefährlicher und größer eine Species war, desto höher ihr Rang in der Hierarchie des Tierreichs. Wie u.a. ein erhaltener „Schauspielzettel" aus Beneventum zeigt, hantierte man folgende Hierarchie:

1. Raubkatzen (‚ferae'; an anderer Stelle: ‚ferae dentatae').
2. Bären (‚ursi').
3. „gefährliche" Pflanzenfresser (‚noxia' [sc. animalia']: z.B. Elefant, Nashorn, Stier, Wildschwein).
4. ‚die übrigen Pflanzenfresser' (‚cetera herbaria' [sc. animalia]).[96]

Damit wurden die Tiere im Grunde ähnlich bewertet wie Gladiatoren. Wenn sie tapfer und aufopfernd gekämpft, *virtus* gezeigt bzw. obsiegt hatten, durften sie auf Geheiß des präsidierenden Organisators oder auf Wunsch der Zuschauer

95 Auch Beagon, *Roman Nature* 148, erwähnt die Gladiatorenspiele im Zusammenhang mit der Zoologie. Jedoch reduziert sie den Sinn von Tierhetzen auf den militärischen Bereich, insofern ‚hunting was recognized as a useful training for soldiers' und insofern die Tierjagd dazu dient, das ‚military ethos' zu unterstützen.

96 Vgl. Wiedemann, *Emperors and Gladiators* 59.

sogar einen Kampf überleben. Schauspielzettel gaben bei den Namen der Gladiatoren durch die wiederholten Buchstaben **V** (,victor') und **M** (,missus') an, wie oft der betreffende Gladiator als Sieger hervorgegangen bzw. wie oft er lebend ,entlassen' (,missus') worden war (ohne gesiegt zu haben, jedoch nach tapferem Kampf).[97] Wenn es sich um Kämpfe Tier gegen Tier handelte, war es auch für Tiere im Prinzip möglich, die Arena siegreich zu verlassen. Als besonders aufregend wurden Kämpfe betrachtet, in denen es verschiedene Species mit einander aufnahmen: z.B. ein Löwe gegen zwei oder mehr Leoparden oder Geparden; Löwen gegen Elefanten; Stier gegen Bären; Bären gegen Löwen, ein Tiger gegen mehrere Bären usw. Jedoch wartete nach einem Sieg schon bald der nächste Kampf. Wenn ein Tier eine beeindruckende Körpergröße hatte, jedoch nicht kämpfen wollte, war man enttäuscht. Das war der Fall mit der schaurigen Giraffe. Von der Körpergröße her versprach sie alles, und ihr Name (*Camelopardalis* = Kamelleopard) sowie ihr geflecktes, an Leoparden erinnerndes Fell suggerierten die prickelnden und faszinierenden psychischen Eigenschaften eines Raubtiers,[98] jedoch stellte sich das Tier als ausgesprochen ängstlich und feige heraus. Statt zu kämpfen flüchtete es. Deswegen nannte es ein kritischer Zuschauer spottend ,wildes Schaf' (,ovis fera').[99] Tiere, die in ihrem Arenaverhalten wenig Kampfesmut zeigten, wurden—wie schlechte oder feige Gladiatoren—sofort beim ersten „Einsatz" getötet. Bezeichnend dafür ist die Angabe, dass bei einer bestimmten *venatio* ,alle Pflanzenfresser' getötet wurden.[100] Diese Angabe macht nur Sinn, wenn von den Raubtieren manche überlebten. Mutige oder sehr aggressive Tiere konnten es auf mehrere oder sogar viele Kämpfe bringen. Bezeichnend dafür ist, dass die kämpfenden Tiere manchmal Gladiatorennamen bekamen. Z.B. zeigt ein Mosaik aus Smirat in Tunesien u.a. vier Leoparden und vier gladiatorische Tierkämpfer (*bestiarii*).[101] Die Aufschriften verraten ihre Namen. Die Leoparden hießen VICTOR, CRISPINUS, LUXURIUS und ROMANUS. Wie schon der Name VICTOR besagt, muss man diesen Leoparden als Siegertyp betrachtet haben. Überhaupt würden die Namen der Leoparden wenig sinnvoll gewesen sein, wenn man sie gleich beim ersten Kampf abschlachten hätte wollen. Dasselbe galt zweifellos auch für Löwen. Entscheidend war, wie tapfer sie kämpften. Wenn dem so war, wollte man sie mehrfach kämpfen sehen. Zweifellos werden auch manche Löwen den Namen VICTOR bekommen haben.

---

97 Ebd. 58.

98 Plinius, *Naturalis historia / Naturkunde* VIII, 69.

99 Ebd.

100 Wiedemann, *Emperors and Gladiators* 59.

101 Vgl. Wiedemann, *Emperors and Gladiators* 16.

Die Performanz der Löwen im Tierkampf brachte Plinius interessanterweise dazu, die Species Löwe in zwei Unterarten (Plinius nennt dies: ‚genera') zu klassifizieren: ‚leones iubati' (1), d.h. Löwen mit einer langen, mächtigen, gerade herabfallenden Mähne und mit einem großen Körper, und ‚leones compactiles crispioribus iubis' (2), d.h. kurz- und krausmähnige Löwen mit einem kleineren, gedrungenen Körper.[102] Die kurz- und krausmähnige Löwenart soll in der Regel weniger tapfer gekämpft haben, die langmähnigen Löwen sollen jedoch die „wahren Gladiatoren" gewesen sein, d.h. ‚Verächter der Wunden' (‚contemptores vulnerum'): Auch wenn Sie schwer verletzt wurden, kämpften sie weiter. Dann oblag es dem Organisator zu bestimmen, ob der Kampf stillgelegt werden oder die Tiere bis zum Tode weiterkämpfen sollten. Es ist klar, dass sich diese Klassifizierung des Plinius im Lichte neuerer zoologischer Erkenntnisse nicht halten lässt. Es scheint, dass es sich bei den ‚kurzmähnigen' und kleineren Löwen um männliche Jungtiere, vielleicht sogar um Weibchen gehandelt hat. Nur erwachsene Männchen tragen die lange, charakteristische Mähne. Aber auch für erwachsene Löwen war es nicht leicht, sich gegen gut trainierte und gut bewaffnete menschliche Tierkämpfer durchzusetzen. In dieser Beziehung erhöhte man den Reiz des Kampfes, indem man den *bestiarii* eine eher geringe Ausrüstung beigab: nur eine Tunika und einen Speer; in vielen Fällen verbot man Verteidigungswaffen (Schild, Brustpanzer, Helm, Arm- und Beinschienen). Denn es sollte—jedenfalls soweit es sich realistischer Weise bewerkstelligen ließ—ein Kampf auf Leben und Tod sein. Wollte ein Tier einen Kampf überleben, war es nicht unbedingt notwendig, dass es den ihm zugeordneten *bestiarius* tötete; es „genügte", wenn es diesen verletzte, in die Flucht jagte oder sich einfach nicht töten ließ, ja, dass es nach Meinung der Zuschauer „tapfer gekämpft" hatte. In solchen Fällen konnte ein Tier begnadigt werden und wurde für einen weiteren Kampf bewahrt. Es konnte aber auch sein, dass die Gnade ausblieb, z.B. weil der Organisator nicht entsprechend reagierte, was natürlich auch vorkam. Bezeichnend für letztes ist, was Plinius in seinem Kapitel über den Elefanten mitteilt. Dort schildert er die „Karriere" des Elefanten im Rom als gladiatorähnlicher Kämpfer im Zirkus:

> In Rom kämpfte nach dem Bericht Fenestellas erstmals ein Elefant unter dem kurulischen Ädil Claudius Pulcher während des Konsulats des Marcus Antonius und des Aulus Postumius [= 99 v. Chr.], und sodann kämpften 20 Elefanten gegen Stiere während der kurulischen Ädilität der

102 *Historia naturalis* / *Naturkunde* VIII, 46: ‚Leonum duo genera: compactile et breve crispioribus iubis; hos pavidiores esse quam longos simplicique villo: eos contemptores vulnerum'.

> beiden Luculli [= 79 v. Chr.]. Auch im 2. Konsulat des Pompeius [= 55 v. Chr.], als der Tempel der Venus Victrix eingeweiht wurde, kämpften im Zirkus 20, nach anderen 17 Elefanten gegen mit Wurfspießen bewaffnete Gätuler, wobei besonders ein Elefant durch seine Kampfesweise Bewunderung erregte: Nachdem ihm die Füße durchbohrt waren, kroch er auf den Knien zu dem Haufen der Angreifer, riss ihnen die Schilde weg und warf sie in die Höhe, sodass sie beim Herabfallen zum Ergötzen der Zuschauer im Kreise herumwirbelten, gleichsam als hätte der Elefant sie mit einer einstudierten Kampftechnik und nicht mit viehischer Wut in die Luft geschleudert.

> Romae pugnasse Fenestella tradit primum omnium in circo Claudi Pulchri aedilitate curuli M. Antonio A. Postumio coss. anno urbis DCLV, item post annos viginti Lucullorum aedilitate curuli adversus tauros. Pompei quoque altero consulatu, dedicatione templi Veneris Victricis, viginti pugnavere in circo, aut, ut quidam tradunt, XVII, Gaetulis ex adverso iaculantibus, mirabili unius [sc. elephanti] dimicatione, qui pedibus confossis repsit genibus in catervas [sc. Gaetulorum], abrepta scuta iaciens in sublime, quae decidentia voluptati spectantibus erant in orbem circumacta, velut arte non furore beluae iacerentur.[103]

Die Zuschauer gerieten völlig aus dem Häuschen, weil sich dieser Elefant erstens als ein so tapferer Kämpfer erwies, dass er nicht einmal aufgab, nachdem ihm die Füße durchbohrt worden waren, und zweitens über eine solche beachtliche Kampftechnik verfügte. Sein Verhalten wurde im Grunde nicht anders goutiert als das eines Gladiators, der den Tod verachtend bis zum letzten Blutstropfen zu kämpfen bereit war und zudem ein hervorragendes technisches Können zeigte. Auch wenn sie unterlagen, konnten tapfere Tiere—wie Gladiatoren—die Gunst und das Mitleid des Volkes gewinnen, das unter Umständen in Rage geraten konnte, wenn man die Tiere dann nicht „ordentlich" behandelte. Auch dies zeigte sich bei den denkwürdigen Spielen des Pompeius im Jahre 55 v. Chr.; Plinius berichtet: ‚als die pompejanischen Elefanten keine Hoffnung mehr [...] hatten, erflehten sie das Mitleid des Volkes in unbeschreiblicher Haltung gleichsam sich selbst beklagend, wodurch das

---

103 *Naturalis historia* / *Naturkunde* VIII, 19–20 (Übersetzung mit leichten Adaptionen). Beagon, *Roman Nature* 149, interpretiert die Stelle überraschender Weise in dem Sinn, dass der verwundete und auf den Knien kriechende Elefant die Gätuler selbst in die Luft geworfen habe. M.E. geht es um die Schilder, die in Kreisen aus der Luft herabfielen, nachdem sie der Elephant emporgeschleudert hatte.

Volk so schmerzlich bewegt wurde, dass es vergaß, dass es den *imperator* vor sich hatte und dass dieser ihm durch die Ausrichtung der Spiele eine außerordentliche Ehre erwiesen hatte, und es [das Volk] sich insgesamt weinend erhob und Verwünschungen gegen Pompeius ausstieß [...]'.[104]

Die rührende Szene legt freilich zugleich die grausame Wirklichkeit des Spektakels frei. Letztendlich erwartete alle Tiere im Verlaufe der *venationes* irgendwann ihre als Ritual konzipierte Tötung. Im Tierkampf wurde die imperialistische Allmacht des römischen Zivilisationsreiches über die Natur symbolisch zum Ausdruck gebracht. Die Performanz der Natur und ihrer Bemeisterung, besonders der exotischen Landtiere, spielte in Rom eine wichtige Rolle. Der imperialistische Triumphalismus des Römischen Reiches bewirkte, dass man bei den *venationes* oder Gladiatorenspielen immer aufs neue das identitätsstiftende Spektakel inszenierte, das den Zuschauern die Bemeisterung der Natur leibhaftig vor Augen führte. Dieses politisch-religiöse Spektakel hat einen außerordentlich großen Einfluss auf die Anordnung, Klassifizierung und Beschreibung der Species in Plinius' Zoologie, Buch VIII. Man geht nicht fehl, wenn man Buch VIII über weite Strecken als literarische und wissenschaftliche Nachinszenierung dieses Spektakels betrachtet. Das Buch führt die Bemeisterung der Natur durch den triumphierenden römischen Imperialismus vor Augen. Die Klassifizierung der Species von groß nach klein, von exotisch nach einheimisch (d.h. Süd- und Westeuropa), und von wild nach zahm, wobei die größeren, exotischen und wilden Tiere in der Hierarchie übergeordnet sind, dient diesem Zweck. Besonders auffällig ist, dass das Reich der Landtiere in erster Linie von den *afrikanischen* Species beherrscht wird: Diese nehmen die wichtigsten Stellen ein und erhalten die ausführlichste Behandlung. Dies erklärt sich aus der *politischen Relevanz*, welche die Demonstration der Dominanz gerade über diesen Erdteil hatte: Das von ungezähmter Natur überquellende Afrika war ja lange Zeit das Sinnbild des Feindes der Römer, der Karthager, gewesen. ‚Africa', ‚Libya', das waren ideologische Reizbegriffe. Es ist kein Zufall, dass sie im Nationalepos der Römer, Vergils *Aeneis*, ein zentrale Stellung einnahmen. Der römische Held, Aeneas, muss, bevor er das Römische Reich gründen kann, zuerst *Africa* (= Libyen, karthagisches Reich, Dido) überwinden. Es ist ein symbolischer Akt, dass sich dabei Africa = Dido selbst vernichtet. Zur Erstellung der römischen Identität musste Afrika sterben.

104 *Naturalis historia* / *Naturkunde* VIII, 21: ‚Sed Pompeiani missa fugae spe misericordiam vulgi inenarrabili habitu quaerentes supplicavere quadam sese lamentatione conplorantes, tanto populi dolore ut oblitus imperatoris ac munificentiae honori suo exquisitae flens universus consurgeret dirasque Pompeio quas ille mox luit inprecaretur'.

Kommen wir abschließend zu den Anekdoten des Plinius, die mehrfach zu abschätzigen Urteilen geführt haben.[105] Nach Murphy haben die Anekdoten in der Enzyklopädie die Aufgabe, sozusagen über die Grenzen des Wissens hinwegzutäuschen, indem sie Schwieriges und Entlegenes erklären, sozusagen die Leerstellen auf der Landkarte des Wissens füllen sollen.[106] Diese Auffassung hat einiges für sich, aber ich kann sie dennoch nicht ganz teilen. Meiner Ansicht nach sollen die Anekdoten v.a. Belege, Evidenzmaterial für die Grundlinien der plinianischen Zoologie liefern. Das lässt sich besonders gut im Hinblick auf den folgenden bemerkenswerten Grundzug zeigen, dass Plinius den Tieren sehr häufig *menschliche Eigenschaften* zuschreibt.[107] Diese Interpretationslinie wird mit Hilfe besonders vieler Anekdoten untermauert. Es handelt sich dabei nicht um Grenzfälle, sondern um den Mainstream des von Plinius interpretativ aufbereiteten zoologischen Wissens.[108]

Das hat mit der ideologischen Rolle zu tun, die Plinius der Tierwelt, insbesondere den Landtieren, in politisch-religiöser Hinsicht zuschreibt. Die Zoologie der Landtiere dient dazu, das Römische Reich und seinen Imperialismus zu bestätigen und zu rechtfertigen. Die Tiere werden dazu zum Teil ähnlich eingesetzt wie die Vertreter eroberter Völker bzw. Völker, die potentiell noch erobert werden oder doch wenigstens Tribut abstatten sollen. Die Species müssen sich untertänig verhalten und die Dominanz des Römischen Reiches performativ anerkennen. Sie müssen an politisch-religiösen Feiern teilnehmen, bei den dazu organisierten Spielen auftreten, wie Gladiatoren kämpfen und letztlich bereit sein, für den Kaiser und das von ihm regierte Reich ihr Leben hinzugeben. Wie die Performanz dieser Untertänigkeit funktioniert, möge das Beispiel des Elefanten zeigen (VIII, 1–34).

105 Vgl. u.a. Lloyd, *Science, Folklore and Ideology* 56.

106 Murphy, *Pliny the Elder's* Natural History *passim*.

107 Dass zahlreiche Anekdoten des Plinius in diese Richtung gehen, hat auch Beagon, *Roman Nature* 137–144, festgestellt. Jedoch zieht sie daraus andere Schlussfolgerungen. Für den betreffenden Abschnitt in ihrem Buch (137–144) verwendet sie den etwas kuriosen Begriff „Animalitavianism", der von Lovejoy und Boas geprägt worden war. Beagon beschäftigt sich vornehmlich mit der Frage, inwiefern Plinius in seinen „Anekdoten" das Tier in der Tat mit dem Menschen gleichsetzt oder eben nur „vergleicht". Dieser Frage messe ich eine eher geringe Relevanz zu.

108 Im Vergleich zu Plinius ist Aristoteles sehr zurückhaltend, wenn es darum geht, den Tieren menschliche Eigenschaften zuzuschreiben. Z.B. ist er der Ansicht, dass rationales, kausales und logisches Denken ebenso wie ethische Überlegungen dem Menschen vorbehalten bleiben. Vgl. Beagon, *Roman Nature* 134.

Zunächst figuriert der Elefant bei Plinius nicht nur als die stärkste und größte Wildtierspecies und eine der tapfersten im rituellen Tierkampf, sondern zudem als jene Species, die dem Menschen geistig am ähnlichsten sein soll. Der Elefant bildet in der *Naturalis historia* die Schaltstelle zwischen Menschen und Tier (Buch VII = Mensch/ Anthropologie; Buch VIII = Landtiere). ‚Maximum (animal) est elephans proximumque humanis sensibus' lautet der einleitende Satz (VIII, 1). Der Elefant soll die menschliche Sprache verstehen, ja sogar schreiben können; ein ausgezeichnetes Gedächtnis, weiter die geistig-seelische Disposition zu liebevoller Zuneigung zum Menschen sowie Ehrgeiz und Ruhmstreben besitzen. Die Freude am Ruhm wird dabei als hohe Tugend aufgefasst, die nicht nur für den Menschen an sich, sondern besonders für den Römer bezeichnend ist. Damit stimmt überein, dass Plinius dem Elefanten bestimmte menschliche Tugenden zuschreibt: moralische Rechtschaffenheit (*probitas*), Klugheit (*prudentia*), ein Gefühl für Gerechtigkeit (*aequitas*) und Schamgefühl (*pudor*, VIII, 12–13). Elefanten sollen sogar religiöse Gefühle besitzen bzw. zur Performanz religiöser Verehrung, ja von Ritualen imstande sein. Um dies zu belegen, bringt Plinius eine Anekdote, nach der Elefanten den Mond anbeten sollen:

> [Elefanten zeigen] eine religiöse Verehrung der Gestirne, besonders der Sonne und des Mondes. Nach den Berichten von Gewährsmännern kommen Herden von Elefanten in den bewaldeten Gebirgen Mauretaniens beim Schimmer des Neumonds zu einem Fluss herab, dessen Name Amilus ist, reinigen sich dort, indem sie sich ringsum feierlich mit Wasser besprengen, und kehren dann, wenn sie das Gestirn auf diese ihre Weise begrüßt haben, in die Wälder zurück, wobei sie ihre ermüdeten Jungen vor sich hertragen [...].

> [...] (elephantis) religio quoque siderum solisque ac lunae veneratio. Auctores sunt in Mauretaniae saltibus ad quendam amnem, cui nomen est Amilo, nitescente luna nova greges eorum descendere ibique se purificantes sollemniter aqua circumspergi atque ita salutato sidere in silvas reverti, vitulorum fatigatos prae se ferentes.[109]

Dabei hat Plinius (bzw. seine Quelle) offensichtlich das normale Verhalten der Elefanten, dass sie sich gerne ins Wasser begeben und sich mit dem Rüssel bespritzen, als religiöses Ritual gedeutet. Der Befähigung zu religiöser

109 *Naturalis historia / Naturkunde* VIII, 1–2. Für die Befähigung des Elefanten zu Religion und religösem Ritual vgl. Plutarch, *De sollertia animalium* 17, 972 BC; Aelian, *De natura animalium* IV, 10 und VII, 44.

Verehrung entspricht die Befähigung zur Performanz politischer Verehrung: Elefanten sind sowohl zum Kniefall als symbolischer Demonstration der Untertänigkeit als auch zu anderen symbolischen Handlungen imstande. Z.B. sollen sie einem König die Krone dargereicht haben.[110] Auch soll der Elefant zum Zeichen seiner Untertänigkeit einen Zweig, der ihm vom Herren hingehalten wird, feierlich mit dem Rüssel annehmen (VIII, 25). Er ist sogar bereit und imstande, Untertänigkeit schriftlich zu bestätigen. Dies bescheinigt Plinius durch die Anekdote, dass ein Elefant in griechischer Sprache folgende Worte in den Sand gekratzt haben soll: ‚Ich selbst habe dies geschrieben und die keltische Beute geweiht'—von Plinius wie folgt ins Lateinische übersetzt: ‚Ipse ego haec scripsi et spolia Celtica dicavi' (VIII, 6). Dieser Elefant soll auf diese symbolische Handlung so versessen gewesen sein, dass er die entsprechenden Worte sogar mehrfach nebeneinander in den Sand schrieb. Elefanten sind nicht nur bereit, ihrem Herren aufopfernd zu dienen, ja sie rechnen sich dies individuell zur Ehre an. Zum Beleg dessen erzählt Plinius die Anekdote, dass sich der Leitelefant des Antiochos d. Gr., Ajax, beim Überqueren eines tiefen Flusses als weniger tapfer als der Elefant Nummer 2, Patroklos, bezeigt habe. Als Antiochus daraufhin Patroklos mit einem Kopfschmuck beschenkte, mit dem er ihn symbolisch zum Anführer kürte, habe Ajax aus gekränktem Ehrgefühl und frustriertem Ehrgeiz den Hungertod dem Verlust der Ehre vorgezogen (VIII, 12). Massiv unterbaut durch diese schlagenden anekdotischen Belege konstruiert Plinius also ein geistiges Profil des Elefanten, das dem eines gewissensvollen Untertanen erstaunlich ähnlich ist.

Wie sich im weiteren Verlauf des Kapitels und in den dort angebotenen Anekdoten herausstellt, ist dies nicht zuletzt imperialistisch gedacht: Durch die Zuschreibung der genannten Eigenschaften konstruiert Plinius den Elefanten als Untertanen des Römischen Reiches. Seine Sprachkenntnisse und sein Gedächtnis sind insofern nützlich, als sie ihn befähigen, Befehlen zu gehorchen. Dasselbe gilt *mutatis mutandis* für moralische Rechtschaffenheit, Klugheit, das Gefühl für Gerechtigkeit und die Performanz religiöser und politischer (Untertänigkeits)Rituale. Durch diese Eigenschaften bestätigt der Elefant, dass es gut und gerecht ist, sich zu unterwerfen, und er ist dazu bereit und imstande. Sein Pflichtbewusstsein und Gedächtnis belegt Plinius mit mehreren Anekdoten, von denen jede einzelne einprägsam ist, z.B.: ‚Sicher ist, dass ein Elefant, der bei der Dressur etwas ungelehriger war und öfters Schläge erhalten hatte, des Nachts beim Üben seiner Aufgabe angetroffen wurde' (‚Certum est unum tardioris ingenii in accipiendis, quae tradebantur,

110 *Naturalis historia / Naturkunde* VIII, 3: ‚regem adorant, genua submittunt, coronas porrigunt' (‚[so] verehren sie einen König, beugen die Knie und reichen Kronen dar').

saepius castigatum verberibus, eadem illa meditantem noctu repertum').[111] Lässt sich ein besserer Untertan denken? Hier wie sooft dient die Anekdote nicht dazu, die Grenzen des Wissens aufzubrechen, sondern zum Beleg des Hauptgedankens.

Ich hoffe, in diesem einleitenden Beitrag einen Einblick geboten zu haben, auf welche Weise Plinius' Zoologie kulturell konstruiert ist; wie Plinius das Tierreich dem Römischen Reich sinnfällig einordnete und einverleibte. Die Prämissen und Hauptlinien waren sorgfältig gewählt: dergestalt, dass sie gewährleisteten, dass die römischen Leser das aufbereitete Wissen optimal aufnehmen, verarbeiten und in ihrem Gedächtnis speichern konnten. Das Interessante liegt in der kulturellen Verfasstheit des riesigen und einflussreichen Wissensspeichers. Wenn man diese versteht, vermag man besser zu orten, wie sie sich auf die Formation zoologischen Wissens in der Frühen Neuzeit ausgewirkt hat.

111 *Naturalis historia*/ *Naturkunde* VIII, 6.

English Summary

Pliny's *Naturalis historia* was of the greatest importance for the formation of knowledge in the early modern period. However, it was not always as easy for readers in the 19th and 20th centuries to find their way to the text: they criticised Pliny's lack of systematic approach and scientific judgement, his anecdotal presentation, his preference for the literary tradition above empirical observation, and his poor Latin style. With respect to zoology, Pliny has been reproached for not having made any progress since Aristotle. The present paper tries to explain the outlines, principles of composition, selection of topics, and manner of presentation of Pliny's zoology in its Roman context, especially in relation to its political and religious background.

Importantly, Pliny was not inclined to present his animal descriptions in the framework of a clear and plausible taxonomical system that would explain the common morphological and physiological features of certain families or classes. On the contrary: his zoology seems to break apart into single independent species, and Pliny always does his best to stress the species' uniqueness. His topical presentation always departs from the same question: what is special/unique about a certain species? In general, Pliny emphasises the miraculous character of God's creation, and nature's inherent intelligence, wisdom, and creative power, which according to him can be demonstrated by the big animals as well as by the smallest creatures—mice and insects. The category of the *mirabile* (i.e. information that causes admiration and astonishment) is of the greatest importance for Pliny's zoology: in fact, the zoology is constructed as a lemmatical collection of *mirabilia naturae*.

With respect to this principle of presentation, one can understand why Pliny generally refrains from giving exact figures for the species' size, weight, average age, length of pregnancy etc., and detailed descriptions of the animal's morphology and physiology. When he does give such information, it is always limited to cases where he has something extraordinary, remarkable, or strange in store. For example, he does not give the average length and weight of cattle, horse, pig, or dog (although he surely knew them), but he does give those of the crocodile (*Crocodylus niloticus*) and the *Hippopotamus*: Pliny says that the crocodile often grows longer than 18 cubits, i.e. 7.92 metres. In reality, grown-up female crocodiles have an average length of 3.1 metres, males of 4.2 metres. Pliny made the crocodile twice as long. The same is true for the *Hippopotamus*: in Pliny's account it measures more than 8 metres, whereas the average animal is about 3–5 metres. In the *Naturalis historia*, the elephant reaches an average age of 200–300 years, and its sexual maturity is supposed to start with its 60th year. In reality, elephants live ca. 60 years, whereas their maturity starts at the age

5 years (males) or 10 years (females). The reason for Pliny's exaggerated manner of presentation is probably that he wanted to render the species more impressive and miraculous. One must take into account that the reason he does not compose his zoology on the basis of a taxonomical system that focuses on morphology and physiology, is not due to a lack of knowledge. He was very well acquainted with Aristotle's works on zoology, among others the *Historia animalium*, and therefore also with Aristotle's analytical approach and taxonomical categories, as can be demonstrated, among other ways, by a closer look at book XI. Therefore, it is not plausible to explain Pliny's zoology as a kind of „misunderstood Aristotle". Much more so, it was a conscious choice that Pliny composed his zoology in the way he did.

However, Pliny's discursive orientation toward the miraculous does not at all imply that his zoology lacks method and system: he divides the species according to their habitats: land (book VIII), air (book IX), and water (book X). Moreover, the land animals are presented according to three principles: 1) from large to small; 2) from exotic (i.e. Africa, Asia, northern hemisphere) to indigenous; and 3) from wild to domestic. First he treats the big, exotic, and wild species, and then the domestic and smaller ones. Book VIII starts with the biggest wild land animal, the elephant, and continues with the giant snakes of India and Ethiopia, and the big *Felidae*, such as lion, tiger, and panther. It ends with the smallest land animal, the mouse. It is a statement that two thirds of the book is dedicated to the large, exotic, and wild animals, and only one third to the indigenous and domestic species (although Pliny must have known much more about them).

Pliny's order of species is fully based on imperialistic patterns of thought. Via this order of species, he tries to demonstrate man's domination of nature. It is a paradigmatic statement for his whole zoology when Pliny starts with the elephant and shows in detail that man (i.e. the Roman, the citizen of the Roman Empire) is able to catch, domesticate, and use for his purposes (i.e. to civilise) even the biggest and strongest wild animal. Pliny's zoology is very much directed by the ideology of the Roman Empire. It is not only composed as a *natural* history, but as a *cultural* history: it tells the history of man's, or Rome's, domination of nature. The wild species were literally brought to Rome and shown to the Roman people. Pliny meticulously registers in which year and at which event a certain species was shown first to the *populus Romanus*: the elephant in 275 BC in the triumph of consul Manlius Curius Dentatus over King Pyrrhus of Epirus; the African lion in 104 BC; the hippopotamus and the crocodile in 58 BC by Marcus Scaurus; the rhinoceros in 55 BC by Pompey; etc. Another peculiarity caused by this formation of discourse is that in his descriptions of the larger wild species, Pliny explains in which way the animal can best be caught.

It is very odd that Pliny in his zoology gives a history of the species being killed, and sometimes even that of its extinction. This is directly connected to the religious and ideological context in which the Romans presented these animals: either in triumphs or hunting shows (*venationes*; sometimes in the framework of gladiatorial games, *munera*). The Roman triumph was a very important political and religious performance: the official celebration of a major military victory. The triumphal procession crossed the Forum on the Holy Street (*via sacra*) to end up at the holiest sanctuary of Rome, the temple of Iupiter Optimus Maximus on the Capitoline Hill. The animals were shown in the same way as war prisoners (the enemies), and after the procession they were killed, either directly or after an additional animal show. The ritualised death of the dangerous wild beasts symbolises Rome's, i.e. the civilisation's victory over wild nature. The combination of animal shows (*venationes*) and gladiatorial games became institutionalised during the time of the Emperor Augustus, and they usually took place in December, at the *Saturnalia*. In these shows animals (and animal species) were in fact treated like gladiators: they were highly esteemed if they fought bravely, without regard for their own life. If they obtained a victory or if they fought bravely, they were usually preserved for another fight. In the end, however, all animals were killed, once again as a symbol of Rome's imperial power and domination. In particular, the killings of the African species reminded the Roman of Rome's most decisive victory over the Carthagian empire. One must take into account that the very existence of Rome depended on this victory. Thus, the ritual performance of man's domination of (African) nature was central to the formation of Roman identity, and this is one of the reasons why this spectacle was so successful. With respect to the wild species, Pliny's zoology of the land animals is a literary and scholarly variant of this ritual domination of nature. Time and again, Pliny treats animals like gladiators: he stresses their strength, size, dangerousness, bravery, and fighting spirit; or, on the contrary, he may make a remark if they lack fighting spirit, as was the case with the giraffe. The giraffe was characterised as 'wild sheep', not because it resembled the sheep in shape but because it refused to fight. Pliny's contextualisation of zoology in the framework of the religion and ideology of the Roman Empire is crucial for the understanding of his work, as is his principal *strategy* of the *mirabile* demonstration. It was not in Albertus Magnus's scholasticism, but rather much earlier, in Pliny's *Naturalis historia*, that zoology was first conceived as a religious and theological exercise: its main aim is to demonstrate the godly nature of the creation viz. its wisdom, beauty, intelligence, and creative power.

Bibliographie

Arnar A.S., *Encyclopedism from Pliny to Borges* (Chicago: 1990).

Baldwin B., „The Composition of Pliny's Natural History", *Symbolae Osloenses: Norwegian Journal of Greek and Latin Studies* 70 (1995) 72–81.

Bäumer A., *Die Biologie von der Antike bis zur Renaissance*, in Dies., *Geschichte der Biologie*, Bd. 1 (Frankfurt a.M. – Bern – New York – Paris: 1991).

Beagon M., *Roman Nature: the Thought of Pliny the Elder* (Oxford: 1992).

Boas M., *The Scientific Renaissance 1450–1630* (New York: 1962).

Bodson L., „Le témoignage de Pline l'Ancien sur la conception romaine de l'animal", in Cassin B. – Labbarière J.-L. (Hrsg.), *L'animal dans l'Antiquité* (Paris: 1997) 325–354.

Borst A., *Das Buch der Naturgeschichte: Plinius und seine Leser im Zeitalter des Pergaments* (Heidelberg: 1994).

Cassin B. – Labbarière J.-L. (Hrsg.), *L'animal dans l'Antiquité* (Paris: 1997).

Chibnall M., „Pliny's Natural History and the Middle Ages", in Dorey T.A. (Hrsg.), *Empire and Aftermath: Silver Latin II* (London – Boston: 1975) 57–78.

Citroni Marchetti S., *Plinio il Vecchio e la tradizione del moralismo Romano* (Pisa: 1991).

Collison R.L., *Encyclopedias: Their History Troughout the Ages* [...] (2. Aufl. New York: 1966).

Conte G.B., „The Inventory of the World: Form of Nature and Encyclopedic Project in the Work of Pliny the Elder", in ders., *Genres and Readers: Lucretius, Love Elegy, Pliny's Encyclopedia* (Baltimore: 1994).

Dahl F., „Zur Geschichte der Zoologie. Von Aristoteles bis Plinius", *Sitzungsberichte der Gesellschaft Naturforschender Freunde zu Berlin* 50 (1924) 62–104.

Davies M., „Making Sense of Pliny in the Quattrocento", *Renaissance Studies* 9 (1995) 240–257.

Detlefsen D., *Untersuchungen über die Zusammensetzung der Naturgeschichte des Plinius* (Berlin: 1899).

Doody A., *Pliny's Encyclopedia: The Reception of the Natural History* (Cambridge: 2010).

Driesch A. von den, *Geschichte der Tiermedizin: 5000 Jahre Tierheilkunde* (München: 1989).

Enenkel K.A.E., „Zur Konstituierung der Zoologie als Wissenschaft in der Frühen Neuzeit: Diskursanalyse zweier Grossprojekte (Wotton, Gesner)", in: ders. – Smith P.J. (Hrsg.), *Early Modern Zoology. The Construction of Animals in Science, Literature and the Visual Arts*, Brill, Leiden – Boston 2007, 2 vols., vol. I, 15–74.

Enenkel K.A.E. – Smith P.J. (Hrsg.), *Early Modern Zoology. The Construction of Animals in Science, Literature and the Visual Arts*, Brill, Leiden – Boston 2007, 2 vols.

Findlen, P., *Possessing Nature. Museums, Collecting and Scientific Culture in Early Modern Italy* (Berkely: 1999).

Fineman J., „The History of the Anecdote: Fiction and Fiction“, in Aram Veeser H. (Hrsg.), *The New Historicism* (New York: 1989) 49–76.

French R.K., *Ancient Natural History: Histories of Nature* (London: 1994).

French R.K. – Greenaway F. (Hrsg.), *Science in the Roman Empire: Pliny the Elder, his Sources, his Influence* (London: 1986).

Gigon O., „Plinius und der Zerfall der antiken Naturwissenschaft“ , in ders., *Studien zur antiken Philosophie* (Berlin: 1972) 370–390; urspr. in *Arktos* 4 (1966) 23–45.

Gudger E.W., „Pliny's *Historia Naturalis*: The Most Popular Natural History Ever Published“, *Isis* 6 (1924) 269–281.

Guerrini A., *Experimenting with Humans and Animals. From Galen to Animal Rights* (Baltimore – London: 2003).

Healy J.F., *Pliny the Elder on Science and Technology* (Oxford: 1999).

Heck L., *Die Hauptgruppen des Thiersystems bei Aristoteles und seinen Nachfolgern, ein Beitrag zur Geschichte der zoologischen Systematik* (Diss. Leipzig: 1885).

Howe N.P., „In Defense of the Encyclopedic Mode: On Pliny's Preface of the *Natural History*“, *Latomus* 44 (1985) 561–576.

Hünemörder Ch., „Aims and Intentions of Botanical and Zoological Classification in the Middle Ages and in the Renaissance“, *History and Philosophy of the Life Science* 5 (1983) 53–67.

Jahn I., *Grundzüge der Biologiegeschichte* (Jena: 1990).

Jahn I. (Hrsg.), *Geschichte der Biologie* (Hamburg: 2004; nach dritter Auflage Jena: 1998)

Jardine N. – Secord J.A. – Spary E.C. (Hrsg.), *Cultures of Natural History* (Cambridge: 1996).

Jennison G., *Animals for Show and Pleasure in Ancient Rome* (Manchester: 1937).

Kaufmann T.D., *The Mastery of Nature. Aspects of Art, Science and Humanism in the Renaissance* (Princeton: 1993).

Keller O., *Die antike Tierwelt*, 2 Bde., I (Leipzig: 1909), II (Leipzig: 1913), Register (1920) (Nachdruck, Hildesheim: 1963).

Keller O., *Thiere des classischen Alterthums in culturgeschichtlicher Beziehung* (Innsbruck: 1887; Nachdruck Hildesheim: 1983).

Klotz A., „Die Arbeitsweise des Älteren Plinius und die Indices auctorum“, *Hermes* 42 (1907) 323–329.

Köves-Zulauf Th., „Die Vorrede der plinianischen Naturgeschichte“, *Wiener Studien* 86 (1973) 134–184.

Köves-Zulauf Th., „Plinius der Ältere und die Römische Religion“, in *Aufstieg und Niedergang der Römischen Welt* [...] II, 16.1 (Berlin – New York: 1978) 187–288.

Leitner H., *Zoologische Terminologie beim älteren Plinius* (Hildesheim: 1972).

Lloyd G.E.R., „The Development of Aristotle's Theory of the Classification of Animals“, *Phronesis* 6 (1961) 59–81.

Lloyd G.E.R., *Science, Folklore, and Ideology. Studies in the Life Sciences in Ancient Greece* (Cambridge: 1983).

Lloyd J.E. Barclay, *African Animals in Renaissance Literature and Art* (Oxford: 1971; New York: 1972).

Locher A., „The Structure of Pliny the Elder's *Natural History*", in French – Greenaway (Hrsg.), *Science in the Roman Empire* 20–29.

MacDonald D. (Hrsg.), *Enzyklopädie der Säugetiere* (Königswinter: 2003; urspr. englisch, Oxford: 2001).

Mayr E., *Systematics and the Origin of Species* (New York: 1942).

Mayr E., *Animal Species and Evolution* (Cambridge Mass.: 1963).

Murphy T., *Pliny the Elder's* Natural History. *The Empire in Encyclopedia* (Oxford: 2004).

Naas V., „Réflexions sur la méthode de travail de Pline l'Ancien", *Revue de Philologie* 70 (1996) 305–322.

Naas V., „Extraordinaire et merveilleux dans l'Historia Naturalis: Réflexions sur la définition de ‚l'encyclopédie' plinienne", *Littérales* 21 (1997) 139–168.

Naas V., *Le projet encyclopédique de Pline l'Ancien* (Rom: 2002).

Nauert Ch., „Humanists, Scientists, and Pliny: Changing Approaches to a Classical Author", *American Historical Review* 84.1 (1974) 72–85.

Ogilvie B., *The Science of Describing. Natural History in Renaissance Europe* (Chicago – London: 2006).

Pellegrin P., *La classification des animaux chez Aristote* (Paris: 1982).

Plinius, Gaius Plinius Secundus der Ältere, *Naturkunde. Lateinisch—deutsch. Buch VIII. Zoologie: Landtiere.* Herausgegeben und übersetzt von Roderich König in Zusammenarbeit mit Gerhard Winkler (Kempten: 1976).

Plinius, Gaius Plinius Secundus der Ältere, *Naturkunde. Lateinisch—deutsch. Buch IXI. Zoologie: Insekten. Vergleichende Anatomie.* Herausgegeben und übersetzt von Roderich König in Zusammenarbeit mit Joachim Hopp (München-Zürich: 1990).

Seeck G.A., „Plinius und Aristoteles als Naturwissenschaftler", *Gymnasium* 92 (1985) 419–434.

Steier A., *Aristoteles und Plinius. Studien zur Geschichte der Zoologie* (Würzburg: 1913).

Steier A., *Der Tierbestand in der Naturgeschichte des Plinius* (Würzburg: 1913).

Steiner G., „The Skepticism of the Elder Pliny", *Classical Weekly* 48 (1955) 137–144.

Vegetti M., „Zoologia et antropologia in Plinio", in *Plinio il Vecchio sotto il profile storico e letterario: Atti del Convegno di Como 1979* (Como: 1982) 117–131.

Wiedemann Th., *Emperors and Gladiators* (London – New York: 1992).

The Order of Nature: Early Modern Views on Classification and Generation, and Their Theological, Ideological and Empirical Background

CHAPTER 2

The Species and Beyond: Classification and the Place of Hybrids in Early Modern Zoology

Karl A.E. Enenkel

Introduction

In the history of science it is always a bit seductive to follow a teleological approach—to assess the achievements of the past with respect to the present status quo of scientific knowledge.[1] If one does so, lines of linear and progressive development inevitably appear. This has also happened with the history of early modern zoology. In the 16th and 17th centuries zoology brought forth monumental works, including those by Conrad Gessner (1551–1560),[2] Edward Wotton (1552),[3] Ulisse Aldrovandi (1599 ff.),[4] Wolfgang Franzius (Frantze;

1 Cf. my first draft on the topic in Dutch, for a popular audience, "Mengwezens. Hybridisatie in vroegmoderne zoölogische werken", in Rikken M. (ed.), *Veranderlijke dieren van Conrad Gesner*, in *De boekenwereld* 29.1 (2012) 44–51. In the present contribution, however, only small parts go back to this first draft.

2 *Historiae animalium* [...] 4 vols. (Zurich, Christoffel Froschauer: 1551–1558); *Icones animalium* [...] *editio secunda* (ibidem, idem: 1560).

3 *De differentiis animalium* (Paris, Michel de Vascosan: 1552).

4 *Ornithologiae, hoc est de avibus historia libri XII* (Bologna, Franciscus de Franciscis: 1599; Bologna, Antonio Bernia: 1637); *Ornithologiae tomus alter cum indice copiosissimo* (Bologna, Giovanni Battista Bellagamba: 1600; Bologna, Antonio Bernia: 1637); *Ornithologiae tomus tertius, ac postremus* (Bologna, Antonio Bernia: 1603; 1637); *De animalibus insectis libri septem, cum singulorum iconibus ad vivum expressis* (Bologna, Giovanni Battista Bellagamba: 1602; Bologna, Clemente Ferronio: 1638); *Quadrupedum omnium bisulcorum historia* [...] *Ioannes Utervius Belga colligere incaepit* [...] *Thomas Dempsterus Baro a Muresk Scotus* [...] *absolvit. Hieronymus Tamburinus in lucem edidit* [...] (Bologna, Sebastianus Bonhommius: 1621; Frankfurt a.M., Caspar Röteli: 1647); *De quadrupedibus digitatis viviparis libri tres, de quadrupedibus digitatis oviparis libri duo Bartholomaeus Ambrosinus* [...] *collegit* [...] (Bologna, Antonio Bernia – Nicolao Tebaldeo: 1645); *De reliquis animalibus exanguibus libri quatuor post mortem eius editi, nempe de mollibus, crustaceis, testaceis et zoophytis* (Bologna, Giovanni Battista Bellagamba: 1606); *De piscibus libri V et De cetis liber unus* [...] *Ioannes Cornelius Utervius collegit. Hieronymus Tamburinus in lucem edidit* [...] (Bologna, Giovanni Battista Bellagamba: 1613); *Serpentum, et draconum historiae libri II. Bartholomaeus Ambrosinus* [...] *summo labore opus concinnavit* [...] (Bologna, Antonio Bernia, Clemente Ferronio: 1640).

 | DOI 10.1163/9789004279179_004

1612),[5] John Jonston (1650–1652), and Caspar Schott, S.J. (1662).[6] For example, in Änne Bäumers's *Geschichte der Biologie* (1991),[7] the big players of early modern zoology are hailed as champions of scientific progress and as precursors of modern biology: Gessner, Aldrovandi, and others for their empirical observations; Coiter for founding comparative anatomy;[8] and Wotton and Aldrovandi as forerunners of modern taxonomy. Taxonomy, the classification of animals based on evolutionary biology, is indeed of pivotal importance for modern zoology.[9] The fact that the English physician Edward Wotton dedicated a whole work to the classification of animals, *De differentiis animalium* (1552), was perceived as a sign of great scientific progress. For example, Änne Bäumer regarded Wotton as a founding father of modern biology.[10] Among Wotton's achievements are that he purged zoology from superstitious beliefs, monsters and animals of phantasy.[11]

This teleological approach, however, may turn out to be somewhat tricky, if not misleading. Its biggest drawbacks are its lack of historical sense, a certain blindness with respect to phenomena that do not fit the idea of progress, its lack of contextualisation, and an overall simplification that may lead to misinterpretations of phenomena. For example, it seems inadequate to understand 16th- and 17th-century zoology as a linear progression from a bookish to an empirical foundation of knowledge.[12] Empirical knowledge was surely there, but on the whole it played a less important part than some scholars suggest. In early modern zoology, the literary tradition was still the most important source

5 *Historia animalium sacra* (ed. pr. Wittenberg, Zacharias Schürer – Johannes Gormann: 1612).

6 Schott Kaspar S.J., *Physica curiosa, sive Mirabilia naturae et artis Libri XII.* [...] *variis* [...] *disquisitionibus excutiuntur et innumeris exemplis illustrantur* [...] *Cum figuris aeri incisis* (Nuremberg, Moritz Endter: 1662).

7 *Zoologie der Renaissance—Renaissance der Zoologie*, in eadem, *Geschichte der Biologie*, vol. II (Frankfurt a.M. – Bern – New York – Paris: 1991).

8 Ibidem, 222 ff.

9 Cf. Enenkel K.A.E., "Zur Konstituierung der Zoologie als Wissenschaft in der Frühen Neuzeit: Diskursanalyse zweier Grossprojekte (Wotton, Gesner)", in idem – Smith P.J. (eds.), *Early Modern Zoology. The Construction of Animals in Science, Literature and the Visual Arts*, Intersections 7 (Leiden – Boston 2007), 2 vols., vol. I (15–74) esp. 22–32.

10 "Das erste zoologische Kompendium in der Zeit der Renaissance: Edward Wottons Schrift 'Über die Differenzen der Tiere'", *Berichte zur Wissenschaftsgeschichte* 13 (1990) (13–29) esp. 28.

11 Cf. Bäumer, *Geschichte der Biologie*, vol. II, e.g. 401: 'Wotton lehnte fast alle Fabeltiere als "blosse Erfindungen" ab'. Cf. however, below the section on Wotton.

12 Cf. my "Zur Konstituierung der Zoologie als Wissenschaft in der Frühen Neuzeit".

of biological knowledge. For the majority of topics, empirical facts were simply lacking. In many cases, one gets the impression that the zoological authors were not that interested in actively searching for new empirical facts, and even if they had empirical information at hand, the authority of authors from antiquity, such as Aristotle, Aelianus, Pliny, and Oppianus, frequently outshone the evidence of empirical observation.

As one may suppose, this feature of early modern zoology affected animal classification. To be sure, classification in itself does not equal scientific progress. In the 16th century Wotton and Aldrovandi were interested in an integrative taxonomy, but in the 17th century Franzius, Jonston in his *Thaumatographia naturalis* (1633), Kircher (1683), Caspar Schott (1662), and some others were less interested. Above all, it is important to understand that early modern classification was departing from paradigms that largely differ from the ones of the 19th and 20th centuries.[13] Morphology and physiology, of course, played a role; however, the most important subdisciplines of taxonomy were probably *philology* and *theology*. The tasks of philology and theology were to adequately use the most important sources of animal description, viz. the classical authors and the Bible. Other sources, such as empirical observations, were in many cases only used additionally. The importance of philology and theology was not restricted to few zoological works with a 'special' character. Änne Bäumer labelled a group of works as 'Biblical zoology' ('Biblische Zoologie');[14] however, theology, the use of the Bible as a major source, and philology characterise a big amount of other early modern zoological works as well.

These features of early modern zoology are connected with one of its most striking paradigms: a profound belief in *interspecific hybridisation*, i.e. mating of different animal species that produces offspring; *hybrid speciation*, i.e. cross-breeding that brings forth new species; and the real existence of all kinds of monstrous creatures. This belief has deep roots in the zoological writings of antiquity and of the Middle Ages, especially Pliny's *Naturalis historia*, books VIII–XI (1st century AD), and Albertus Magnus's *De animalibus libri XXVI* (13th century).[15] In both zoologies the attention

13 For a modern perspective cf. Mayr E., *Systematics and the Origin of Species from the View Point of a Zoologist* (Cambridge, Mass. – London: 1999); idem, *Animal Species and Evolution* (Cambridge Mass.: 1963).

14 *Geschichte der Biologie*, vol. II, 156 ff.: 'Teil 2: Spezielle zoologische Schriften. 11. Biblische Zoologie, Fortsetzung der Aelian- und Physiologus-Tradition'. She records in this group Hermann Heinrich Frey, Wolfgang Franzius, Heinrich von Hoevel, and Caspar Dornavius.

15 See below, the section "More Philosophical and Theological Explanations, and Early Modern Treatises on Monsters".

to interspecific hybridisation, hybrid speciation, and monstrous creatures depends on religious beliefs: Pliny had a philosophical belief in the unlimited, miraculous power and artistry of nature (*natura artifex*), and in his zoological descriptions he always stresses the wondrous or *mirabile*.[16] Albertus Magnus shows a similar attitude, but as one may expect, in his case everything is focused on God the Creator. Nature proves not only God's wisdom but, maybe even more importantly, his unlimited power. Both Pliny and Albertus Magnus were in their zoological works admirers and followers of Aristotle. Aristotle himself, however, was less optimistic about interspecific hybridisation and hybrid speciation than Pliny or Albertus were, and he formulated a number of serious limitations.[17]

The Creative Power of the New Aristotelian Taxonomy: Hybrids and Miraculous Creatures in Wotton's *De differentiis animalium* (1552)

The English physician Edward Wotton (1492–1555), son of the senior bedel of theology of the University of Oxford, represents an interesting case, since with his *De differentiis animalium* he authored the first zoology that offered an integrative classification of all *species*.[18] In this project, on which he worked for many years and in which he profited much from his excellent knowledge of Greek,[19] his medical expertise,[20] and his studies in Padua, he was clearly

16 For these features, cf. my contribution in the present volume, "Die antike Vorgeschichte der Verankerung der Naturgeschichte in Politik und Religion: Plinius' Zoologie und der römische Imperialismus", *passim*, and Beagon M., *Roman Nature. The Thought of Pliny the Elder* (Oxford: 1992), chapter I "Divina Natura: the Roots of Pliny's Thoughts" (26–54). According to Beagon, the stoic concept of *natura-artifex* is an "overall theme" of Pliny's *Naturalis historia* (Beagon, *Roman Nature* 131 and *passim*). Cf. also Caprotti E., "Animali fantastici, fantasie zoogoliche e loro realtá in Plinio", in *Plinio e la natura. Atti del ciclo di conferenze sugli aspetti naturalistic dell'opera Pliniana, Como 1979* (Como: 1982) 39–61.

17 See *infra*.

18 On Wotton's zoology, cf. my "Zur Konstituierung der Zoologie als Wissenschaft in der Frühen Neuzeit" (2007); Bäumer, "Das erste zoologische Kompendium in der Zeit der Renaissance" (1990); eadem, *Geschichte der Biologie*, vol. II, 32–41; Jahn (Hrsg.), *Geschichte der Biologie* 186; Hoeniger F.D. and J.F.M., *The Development of Natural History in Tudor England* (Charlottesville, VA: 1969). For Wotton's biography, cf. Pollard A.F., art. "Wotton, Edward (1492–1555)", in *Oxford Dictionary of National Biography* (Oxford: 2004; [http://www.oxforddnb.com/view/article/29999, accessed 8 Feb 2014]).

19 He had been the first reader in Greek of Corpus Christi College (after 1520; officially appointed in 1524); cf. Pollard, "Wotton, Edward".

20 He was a fellow of the College of Physicians from 1528 to 1555; cf. ibidem.

inspired by Aristotle, and he carefully used Aristotle's ground-breaking analyses of the morphology and physiology of animals[21] in order to construct his integrative classification. In his reconstruction—in fact, new construction—of an Aristotelian system of classification, Wotton collected a large amount of data on the species; although—as he said—he 'only' wanted to present to the reader a handy 'compendium' or companion, he ended up with a monumental zoology in 10 books that comprised the zoological knowledge of most Greek and Latin authors of antiquity. The most important feature is, of course, the systematic organisation of this new zoology. After an in-depth discussion of the criteria of his taxonomy, viz. of the relevant 'differences' (books 1 and 2), he applies Aristotle's categories of "animals with blood" (books 3–8) and "bloodless animals" (books 9 and 10). The "animals with blood" he divides, after the example of Aristotle, into the classes *homo sapiens* (book 4), viviparous quadrupeds (book 5), oviparous quadrupeds and serpents (book 6), birds (book 7), and fish (book 8). The "bloodless animals" he divides into the classes of insects (book 9), and *Mollia, Crustacea, Testacea*, and *Zoophyta* (book 10).

Let's have a closer look at the important Aristotelian class of the *viviparous quadrupeds*, to which Wotton dedicated a whole book (book 5). The viviparous quadrupeds are, in principle, mammals, and exclude oviparous quadrupeds, such as reptiles, toads, frogs, and turtles. Following Aristotle, Wotton construed this class on a morphological (four legs) and a physiological (viviparous) criterion.[22] Most interestingly, in his systematic taxonomical approach Wotton went much further than Aristotle. First, he divided the viviparous quadrupeds into three subcategories or larger groups: 1) *multifida* (that is, uneven-toed animals with three or more toes);[23] 2) *bisulca* (artiodactyls or even-hoofed animals);[24] and 3) *solipedes* (that is, odd-toed ungulates).[25] Second, he divided all three subcategories into certain families, in which he collected related species or subspecies; for example, the family of the "wolves" (consisting of about

21 On Aristotle's efforts with respect to animal classification, cf. Pellegrin P., *La classification des animaux chez Aristote* (Paris: 1982); Lloyd G.E.R., "The Development of Aristotle's Theory of the Classification of Animals", *Phronesis* 6 (1961) 59–81; Heck L., *Die Hauptgruppen des Thiersystems bei Aristoteles und seinen Nachfolgern, ein Beitrag zur Geschichte der zoologischen Systematik* (Diss. Leipzig: 1885); Hünemörder Ch., "Aims and Intentions of Botanical and Zoological Classification in the Middle Ages and in the Renaissance", *History and Philosophy of the Life Science* 5 (1983) (53–67) 53–56.

22 Cf. my "Zur Konstituierung der Zoologie als Wissenschaft in der Frühen Neuzeit" 30.

23 *De differentiis animalium*, book V, fol. 56v–71v.

24 Book V, fol. 72r–83v.

25 Book V, fol. 84r–88v.

10 different species or subspecies),[26] the family of the "lions" (also consisting of about 7 to 10 species or subspecies),[27] the family of the ferrets, martens, weasels, and small cats (consisting of about 8 species/subspecies),[28] the family of quadrupeds that live in the water (consisting of about 7 species),[29] the family of *Bovidae* (consisting of about 20 species/subspecies),[30] or the family of the deer or *Cervidae* (consisting of about 13 species/subspecies).[31] Wotton's families or related species are rarely identical to the families of modern taxonomy, and they are constructed by the application of different criteria. Sometimes they depart from very narrow species or even subspecies markers, such as in the chapter on the "dogs"; sometimes they are constructed around random morphological similarities, for example in the chapter on the "wolves" (which also contains *Felidae*); sometimes they bring together animals with a (seemingly) similar size; and sometimes they unite animals that live in the same habitat, for example the "viviparous quadrupeds that live in the water".[32]

With respect to the subcategories or larger groups, the most intriguing thing is that Wotton constructed his classification on the basis of *one single, clearly perceptible morphological detail*—one, two, or more toes. This means that, in

26 "De lupis", ibidem, fols. 62r–63r. The terms "species" and "subspecies" are used here in a modern sense. When one looks closer at Wotton's families (to which he normally dedicates exactly one chapter), it is sometimes hard to say whether a certain animal name represents a species or a subspecies; sometimes the species names he lists do not belong to the described family. In modern taxonomy, the relevant family here would be *Canidae* (species that resemble the dog); a species of this family would be the wolf (*canis lupus*): this species is divided into 39 subspecies, among them the domestic dog (*Canis lupus familiaris*), the Australian dingo (*Canis lupus dingo*), the Tundra Wolf (*Canis lupus albus*), the Steppe Wolf (*Canis lupus campestris*), the African Wolf (*Canis lupus lupaster*), the Eurasian Wolf (*Canis lupus lupus*), the Arabian Wolf (*Canis lupus arabs*), the Red Wolf (*Canis rufus*), and so on. When Wotton lists the *Lupus aureus* (Golden Wolf) of Asia Minor it is difficult to say whether this refers to a subspecies of the Wolf, to the species Wolf itself (*Canis lupus*), or to another species. Other "wolves" collected by Wotton certainly do not belong to the species *Canis lupus*, or even to the *Canidae*; for example, the *Lupus cervarius*, the *Thos*, the *Chaus*, and the *Lynx* all refer to the same *Felidae* species, the lynx (*Lynx lynx*). Also, Wotton's *Lycaon* represents a *Felidae* species, even a large feline, the cheetah (*Acinonyx jubatus*).

27 "De leonibus", ibidem, fols. 63v–64r. Cf. Enenkel, "Zur Konstituierung der Zoologie als Wissenschaft in der Frühen Neuzeit" 32–35; 57.

28 "De viverris, mustelarum generibus, et de fele", ibidem, fols. 67v–68v.

29 Ibidem, fol. 70r–v.

30 "De bove et boum generibus", ibidem, fols. 72r–75r.

31 "De cervis", ibidem, fols. 79v–81r.

32 Ibidem, fol. 70r: "De quadrupedibus aquaticis, ut de fibro, lutra, et quibusdam aliis".

principle, everybody could *empirically* observe and check whether the classification of a certain species was correct. As it seems, no misunderstandings were possible. In this taxonomical division, Wotton deliberately excluded other criteria: for example, whether the animals are carnivores, plant eaters, or omnivores (which, of course, would have morphological effects). In comparison, in modern taxonomy the difference between "even-toed" and "uneven-toed" species refers exclusively to ungulates, such as horses, goats, and deer; other viviparous quadrupeds (mammals) are not classified by the (main) criterion of the number of toes. By consequence, Wotton's group of *multifida* (more than two toes) is very big[33] and includes genetically different genera of animals: elephants (*Proboscidea*);[34] primates;[35] carnivores;[36] rodents (*Rodentia*), such as the mouse and the rabbit,[37] but also the mole (*talpa*) and the bat (*vespertilio*);[38] and even water animals, such as the *Vitulus marinus* (seal). The great variety of genera that constitute Wotton's *multifida* appears from a brief look at the modern stemma of the *mammalia*.[39]

Because of Wotton's concentration on a single, easily perceptible morphological detail, one may suppose that he made an effort to stress the importance of empirical observation. This, however, is questionable. In his entire book on viviparous quadrupeds, Wotton hardly ever talks about empirical observations, let alone autopsy.[40] Instead of personal observations, Wotton normally provides evidence by listing his most important sources from antiquity (names of authors and works) at the beginning of each section, by quoting them, often verbatim, in the main section, and finally by adding some details or remarks in footnotes at the end of each chapter.

Illuminating test cases with respect to his method of classification can be found among the species that he certainly did *not* see—e.g. because they do not exist, were described the wrong way, or live in faraway regions. For example, among the *solipedes* (odd-toed ungulates) and in the family of the horses (*equi*) Wotton classifies a horse with wings and horns, a hybrid species

33 Book V, fols. 56r–72r.

34 Book V, fols. 56v–57v.

35 Book V, fols. 57v–58v.

36 Book V, fols. 58v–68v, and 70v.

37 Book V, fols. 66v–67v; 68v–70r.

38 Book V, fol. 71r–v.

39 MacDonald D. (ed.), *Enzyklopädie der Säugetiere* (Königswinter: 2003; orig. Engl., Oxford: 2001) XVI.

40 Cf. Enenkel, "Zur Konstituierung der Zoologie als Wissenschaft in der Frühen Neuzeit", esp. 31–51.

that supposedly lived in Ethiopia (*Pegasus*).[41] As appears from his argument, Wotton considered it more correct to list this species among the viviparous quadrupeds than among the birds, and he obviously disagreed with writers who classified it in the latter category. Wotton's way of arguing implies that he did not doubt the winged horse's real existence.[42] But, of course, it is clear that he had never personally observed the animal. Instead of empirical observations, he gave Pliny's *Naturalis historia* as "the" authoritative evidence.[43] As the classification of the *Pegasus* suggests, Wotton's taxonomical method was largely based on philology, not on empirical knowledge.

An animal that represents a real species and lives in Europe—next door, so to speak—is the elk (*Alces alces*) [Fig. 2.1, elk].[44] The elk belongs to the family of the deer (*Cervidae*), and as such to the even-hoofed ungulates [Fig. 2.2, footprint of an elk].[45] Wotton, however, classifies it among the *solipedes* (odd-toed ungulates).[46] The reason is that Wotton relied solely on his source Pliny, who stated that, in general, the elk resembles the horse, except for his ears and neck.[47] Wotton took this misleading information literally, and thus attributed to the species odd-toed feet, such as horses would have. Another existing animal is the rhinoceros (family *Rhinocerotidae*, consisting of five species living in Africa and Asia).[48] Rhinoceroses have three toes, as Dürer had already depicted [Fig. 2.3], and therefore—in Wotton's system of classification—they should belong to the *multifida*. Wotton, however, struggled with this species.

41 *De differentiis animalium*, book V, fol. 87r.

42 Änne Bäumer claimed that Wotton 'lehnte fast alle Fabeltiere als "blosse Erfindungen" ab' (*Geschichte der Biologie*, vol. II, 401). Her claim, however, does not correspond with the facts; see below.

43 *Naturalis historia* VIII, 72.

44 MacDonald (ed.), *Enzyklopädie der Säugetiere* 513. It is about the 'elk' in British English ('moose' in American English), not about the wapiti.

45 Ibidem 504–519.

46 *De differentiis animalium*, book V, fol. 87r.

47 *Naturalis historia* VIII, 39. Cf. Leitner H., *Zoologische Terminologie beim älteren Plinius* (Hildesheim: 1972) 15; Keller O., *Die antike Tierwelt*, 2 vols. (vol. I, Leipzig: 1909; vol. II, Leipzig: 1913; reprint, Hildesheim: 1963), vol. I, 281 ff. To be sure, in his description of the elk Plinius did not use the name of the species ('horse', *equus*), but the more general term 'iumenta', which primarily refers to donkeys and horses but may refer to oxen (and thus to even-hoofed ungulates) as well.

48 The rhino consists of five species: the Black rhinoceros (*Diceros bicornis*), and the White rhinoceros or Square-lipped rhinoceros (*Ceratotherium simum*), both living in the southern part of Africa; and the Indian (*Rhinoceros unicornis*), Javan (*Rhinoceros sondaicus*), and Sumatran (*Dicerorhinous sumatrensis*) rhinoceroses, all living in Asia; cf. MacDonald (ed.), *Enzyklopädie der Säugetiere* 476–481.

FIGURE 2.1 *The elk* (Alces alces), *grownup male.*

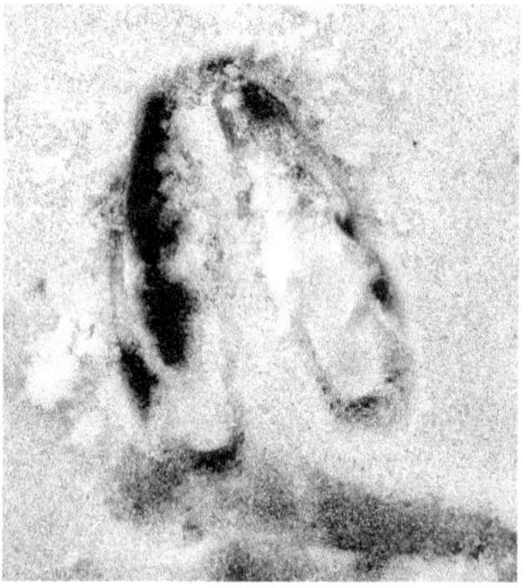

FIGURE 2.2 *Footprint of an elk* (Alces alces).

From his description it appears that he attributed to the animal the features of very different species: horse, deer, elephant, and wild boar.[49] In a first attempt,

49 *De differentiis animalium*, book V, fol. 71v: 'Sed atrocissimum monstrum monoceros est, quem cartazonon vocant: reliquo corpore equo simile, capite cerv[in]o, pedibus

FIGURE 2.3 *Rhinoceros by Dürer. From: Gessner,* Icones animalium *(2nd ed., Zurich, Christoph Froschauer: 1560), p. 60.*

Wotton classified it among the *multifida*; he did not know for sure how many toes the animal had, but he stated firmly that its toes were not split, as he supposed was the case with the elephant (which, as one may know, is not correct). In another passage in his *De differentiis animalium*, however, Wotton classified the rhino among the *even-hoofed animals* (sic),[50] probably because this time he stressed the "bull-element" in the "composite animal" he offers in his description. In the relevant passage he states—maybe to the surprise of the reader—that the rhino is common in Germany, in the 'Hercynian Forest' to the north of the Danube (i.e. Bayerischer Wald and/or Oberpfälzer Wald). From Wotton's description it appears that he construed the German rhino as a *mixtum compositum* of wild boar and bull.[51] Although in this second classification the rhino has become a European animal as well, it is not likely that Wotton

elephanto (digitis scilicet indivisis), cauda apro, mugitu horrido; cornu unum nigrum e media eius fronte protenditur [...]' ('But the unicorn is an outrageous monster which is called cartazonon. With respect to its body it resembles a horse, its head a deer, its feet an elephant (because its toes are not split), and its tail a wild boar, and it produces a horrible sound. On the middle of its face rises a single black horn'). The 1552 printed edition, on fol. 70v, has 'cervino', which should be corrected to 'cervo'.

50 Ibidem, book V, fol. 75r: 'in Hercyna silva gignuntur: rhinocerota vocant et taurum Aethiopicum' ('it exists in the Hercynian Forest. It is called rhinoceros or Ethiopian bull').

51 Ibidem, book V, fol. 75v: '[...] magnitudo tauri est, forma vero apri, praeterquam quod a summis naribus, ut iam diximus, cornu est' ('[this animal has] the size of a bull, but the

ever observed the Danube rhino. Besides, rhinos with one horn exist only in Eastern Asia. One also wonders why Wotton has identified the Bavarian rhino with the 'Ethiopian bull' and why he has ascribed to the animal the capacity to kill elephants by tearing apart their soft belly with its sharp horn. Did Wotton think that there were elephants in Germany?[52]

The "unicorn" filiation in Wotton's attempt to classify the rhino leads to another puzzling problem of his zoology: the abundance of *unicorn species*. Besides the rhino, Wotton lists, among others, a species of unicorn horses in India (*equus Indicus*)[53] and one of Indian unicorn donkeys (*asinus Indicus*).[54] Interestingly, he ascribed the unicorns to different groups: some he classified among the *multifida*, some among the *bisulca* (even-hoofed ungulates), and some among the *solipedes* (odd-toed ungulates). For example, Wotton was well aware that the family of the '*Boves*' (*Bovidae* in modern terms) should be classified among the even-hoofed ungulates (*bisulca*), and that the '*Boves*' are even the "constituting" family of *bisulca*, as is also indicated by the title of the relevant chapter: "De quadrupedibus bisulcis, et primo de bove et boum generibus".[55] However, among the even-hoofed ungulates he classifies one species of unicorn 'Indian bulls' (*boves Indici*) and another one 'Aeonian bulls' (*boves Aeonii*), both of which are odd-toed ungulates.[56] The Aeonian bull bears a single horn, located 'in the middle of its head'.

In the same chapter, Wotton lists another species of unicorn bulls that was supposed to live in the 'Hercynian Forest' of Bavaria: this animal appears to be an even-hoofed ungulate, just like the "German rhino". Its single horn, however, has very strange features: it is much longer and is straighter than other horns, but on its top it has a rare ramification resembling the palm of a human hand.[57] The form of the horn suggests that the ultimate "source of inspiration" behind this composite animal or *mixtum compositum* was the elk [Fig. 2.1]; the

---

shape of its body resembles a wild boar, except the fact that it has on the top of its nose, as we already stated, a horn').

52 However, from a footnote it appears that Wotton identified the 'Ethiopian bull' with the animal Solinus had called 'Indian bull' and Pierre Gilles (Gyllius) 'Wild bull'; cf. *De differentiis animalium*, book V, fol. 75r, footnote 9.

53 *De differentiis animalium*, book V, fol. 86v.

54 Ibidem, fol. 88r–v.

55 Ibidem, fol. 72r.

56 Ibidem: 'Sunt et Indici boves unicornes [...], solidis ungulis nec bifidis. Aeonii quoque boves solipedes sunt, singulaque habent cornua, et haec medio capite' ('There are also unicorn Indian bulls [...]; they are odd-toed, not even-hoofed. The Aeonian bulls are odd-toed as well, and they have a single horn in the middle of the head').

57 Ibidem.

elk (*Alces alces*), however, as was previously mentioned, was wrongly classified by Wotton among the odd-toed ungulates[58] because he interpreted Pliny's misleading information (that the elk almost entirely resembled a horse) in a literal sense. It is even more surprising that Wotton—after having classified at least four species of unicorn odd-toed ungulates—says all of a sudden, with respect to the 'Indian donkey', that he 'is the only unicorn species of odd-toed ungulates'.[59] A footnote reveals the background of this surprising move. It was a philological problem: Wotton was not sure whether 'bos Indicus', 'bos Aeonius', 'asinus Indicus', and 'equus Indicus' all referred to different species. His doubts were raised by Aelianus's *Historia animalium*, where 'equus Indicus' and 'bos Indicus' may mean the same animal.[60]

The method Wotton used in the classification of the Ethiopian winged horse, the elk, the rhino, and the various "unicorn" species indicates that—although he worked strictly along Aristotelian systematical lines—he was inclined to take over and incorporate the miraculous and strange animals described by Pliny and other ancient zoologists, and that he left the door open for hybrid species. If one looks closer at the *De differentiis animalium*, most of the hybrids mentioned by Pliny reappear: e.g. among the even-hoofed ungulates, Wotton classifies the Ethiopian *Leucrocuta*[61] [Fig. 2.4], a kind of hybrid of an even-hoofed ungulate (possibly a deer) and a lion, but resembling a hyena; the *Thos*, an interspecific hybrid of a (male) wolf and a (female) panther;[62] the *Eale*,[63] a strange animal with the features of a number of species (such as elephant, wild boar, horse, and bull) and equipped with the miraculous element of moveable horns, but according to Wotton a true 'equus fluviatilis' (*hippopotamus*), or a species closely related to the hippopotamus;[64]

58 Ibidem, fol. 87r.

59 Ibidem, fol. 88r: 'In solipedum genere unicornum asinus tantum Indicus est'.

60 Ibidem, fol. 88v, footnote no. 4.

61 Ibidem, fol. 81r.

62 Ibidem, fol. 62r: 'Thos lupus est Aethiopicus parvus et deformis, ex lupo et panthera prognatus, qui utrunque refert parentem, facie scilicet genitorem, reliquo corpore matrem' ('The Thos is an Ethiopian wolf, which is small and ugly; it is brought forth by mating of a [male] wolf and a [female] panther: by its face it resembles the father [sc. the wolf], by the rest of its body the mother [the panther]').

63 Ibidem. On the *Eale*, cf. in the present volume my "Die antike Vorgeschichte der Verankerung der Naturgeschichte in Politik und Religion: Plinius' Zoologie und der römische Imperialismus"; Pliny, *Naturalis historia* VIII, 73; Leitner, *Zoologische Terminologie beim älteren Plinius* 114.

64 *De differentiis animalium*, book V, fol. 81v: 'eale ut alias equus fluviatilis [...]' ('the Eale, or as it is called elsewhere, hippopotamus'); 'Hippopotamis comparatur, et ipsa sane aquis

the Ethiopian *Crocuta*,[65] an interspecific hybrid of a (male) hyena and a lioness,[66] but sometimes also a cross-breeding of dog and wolf;[67] the "leo minor", i.e. a smaller sort of a lion with curled short hair instead of a long mane, an interspecific hybrid of a (female) lion with another carnivore (maybe the panther),[68] as well as a similar smaller lion that completely lacks a mane, a cross-breeding of a panther and a lioness;[69] the *Bonasus*, with the features of horse and bull;[70] the *Chaus* or (in the Celtic language) *Raphius*, which somehow unites the features of a wolf and a panther;[71] the *Chaonis*, a cross-breeding of the male *Chaus* with a bitch;[72] the enigmatic Ethiopian *Cepus* or *Cephus*, presented in Rome by Pompey in 55 BC and probably a kind of monkey, but classified by Wotton as a carnivore with the features of panther (body) and lion (face);[73] moreover, another *Cepus*, with the features of a monkey (face), and a bear and a dog (body);[74] and the miraculous being *Mantichora*, a man-eater with a human face, the body and legs of a lion, and the tail of a scorpion (including a stinger) [Fig. 2.5].[75] Similar to the *Crocuta*, the *Mantichora* is able to imitate the human voice.

---

fluminum gaudet' ('it is similar to the hippopotamuses, and indeed it likes the water of rivers'). Pliny, however, did not identify the *Eale* with the hippopotamus.

65 *De differentiis animalium*, book V, fol. 63v; cf. Pliny, *Naturalis historia* VIII, 72; Leitner, *Zoologische Terminologie beim älteren Plinius* 154.

66 *De differentiis animalium*, book V, fol. 63v: 'In Aethiopie parte [sc. hyaena] coit cum leaena, unde nascitur monstrum, cui crocutae nomen est; [...] in ore gingiva nulla, dens unus atque perpetuus [...]' ('In a certain region of Ethiopia the [male] hyena mates with the lioness and gives birth to a monster which is called *Crocuta*').

67 Ibidem: 'Crocutas etiam generat Ethiopia ex cane lupoque conceptos, omnia dentibus frangentes, protinusque devorata conficientes ventre' ('Ethiopia also brings forth *crocutae* that are brought forth by the mating of dogs and wolves; they break everything with their teeth, devour it, and digest it immediately in their stomach').

68 Ibidem, fol. 64v: '(leones) breviores et iubis crispi, plerumque ignavi [...]'.

69 Ibidem: 'at hi quos (leones) creant pardi, in plebe remanent iubarum inopes. Adulterinis enim coitibus degenerantur quandoque partus leaenarum [...]' ('but those lions that are brought forth by the mating with panthers remain vulgar and without a mane. Sometimes the kittens of lionesses who were created by the mating with another species, degenerate [...]').

70 Ibidem, fol. 74v; cf. Pliny, *Naturalis historia* VIII, 40.

71 *De differentiis animalium*, book V, fol. 62v; cf. Pliny, *Naturalis historia* VIII, 70.

72 *De differentiis animalium*, book V, fol. 63r: 'Ex Chao conceptas canes chaonidas nominari [...]' ('The dogs brought forth by the Chaus are called *Chaonides* [...]').

73 Ibidem, fol. 66r.

74 Ibidem, footnote 5.

75 *De differentiis animalium*, book V, fol. 71v: 'Apud Indos (si Ctesiae credendum est) belua gignitur, cui nomen Mantichorae: [...] magnitudo et aurium hircitudo et pedes leonis;

FIGURES 2.4 & 2.5 (*top*) *The* Leucrocuta ('Leucurcuta'), *and the* Martichora ('Martigora') (*bottom*), *by Matthaeus Merian. From: John Jonston,* Historiae naturalis de quadrupedibus libri, cum aeneis figuris [...] (*Frankfurt a. M., Matthäus Merian: 1652*), *Tab. LII.*

---

facies et aures hominis, oculi caesii, color rubricus; cauda scorpionis modo terrestris [...]' ('In India—if one may believe Ctesias—lives an animal called Mantichora: [...] its size, hairiness of its ears, and legs resemble a lion. It has the face and the ears of a man. Its eyes are greyish; its colour is red. Moreover, it has the tail of a scorpion'). In his account, Wotton suggests that his source is the *Indika* of the Greek physician Ctesias of Cnidos. However, he drew exclusively on Pliny's *Naturalis historia* VIII, 75: 'Apud eosdem nasci Ctesias scribit, quam mantichoran appellat [...], facie et auriculis hominis, oculis glaucis, colore sanguineo, corpore leonis, cauda scorpionis modo [...]'. The *Mantichora* origi-

If one considers Wotton to be a rationalist and empirical zoologist, one may be even more surprised by the fact that he has added to the Plinian ones more hybrids: e.g. the Horse-panther (*Hippopardium*),[76] a strange hybrid of horse and leopard; the Horse-deer (*Hippelaphus*, also called *Equicervus*),[77] a *mixtum compositum* of horse and deer [Fig. 2.6, *Hippelaphus/Tragelaphus Gesneri*]; and the *Hippager*,[78] a *mixtum compositum* of a horse and a goat. In his classification, Wotton has systematically united most of the cross-breedings of the horse in one family (chapter 97): "De hippelapho, de hippopardio, hippagro, hippopotamo, et eale" [Fig. 2.7].[79] Since horses are odd-toed ungulates, one is puzzled by the fact that in Wotton's classification all horse hybrids are described as *even-toed*, as if they were cows (*boves*)!

With the *Hippopardium* or Horse-panther Wotton—most remarkably—presents a case where a hybrid of an animal with five toes (the leopard) and an animal with one toe (the horse) brings forth a species with two toes ('bisulca').[80] This seems to be a true miracle of nature. However, it is clear that Wotton was not eager to present a mathematical construction of species. Strangely, in his system of classification Wotton did not give a sign that he had problems with this odd hybrid. In an extremely dry and casual way he tells us that the Horse-panther is not only even-toed, but also that he bears two horns![81] Obviously, he is not bothered either by the question of where the horns would come from: as everybody knows, neither leopards nor horses have horns. Aristotle had explicitly excluded odd-toed ungulates from animals bearing two horns, and—as it appears from the opening chapter of book V of *De differentiis animalium*—Wotton was very well aware of this statement.[82] The Ethiopian *Hippager* seemed to be even less of a problem for Wotton. He is described as an even-toed horn-bearer as well; only his behaviour seems special: when imprisoned, he commits suicide by refusing to eat. The Ethiopian *Leucrocuta* seems to be much more difficult to classify: Wotton describes it as 'even-hoofed, with the legs similar to a deer, extremely fast, but with his breast, neck and tail

nates in the tales of Persia. It was called 'martiaxwar' (i.e. 'man eater') and was brought into Western culture by Ctesias, court physician of King Artaxerxes II, who rendered the Persian name in Greek as 'martichora'.

76 Ibidem, fol. 81v.

77 Ibidem, fol. 81r.

78 Ibidem, fol. 81v.

79 *De differentiis animalium*, book V, fol. 81v.

80 Ibidem.

81 Ibidem.

82 Ibidem, fol. 52r–v.

FIGURE 2.6 Hippelaphus *or* Tragelaphus Gesneri. *From: Gessner,* Icones animalium *(2nd ed., Zurich, Christoph Froschauer: 1560), p. 47.*

similar to a lion, and the head similar to a badger'.[83] Here again, as with the *Hippopardium*, a hybrid of carnivores and even-toed ungulates (thus plant eaters) occurs. In Wotton's description, the animal species has more miraculous features: instead of a set of carnivore teeth according to the carnivore tooth formula, it has two horizontal teeth in the shape of a knife; and moreover, the animal is able to imitate the human voice. Wotton locates this monster species in Scythia, thus northeastern Europe and the adherent parts of Asia.

83 Ibidem, fol. 81r: 'Cruribus cervinis et ungula quoque bisulca apud Aethiopes Leucrocuta est pernicissima fera, [...], collo, cauda, pectore leonis, capite melium [...]'.

FIGURE 2.7 Hippelaphus *and* Tragelaphus. *From: John Jonston,* Historiae naturalis de quadrupedibus libri [...] *(Frankfurt a. M., Matthäus Merian: 1650), Tab. XXIV.*

While Wotton was credited with having purged zoology of fantastic elements,[84] his systematic order of species *abounds* with strange creatures, hybrids, and monstrous animals. It is clear that he did not do anything to reduce their number; on the contrary, he listed more hybrids than Pliny did. Most of them he

84 Cf. Bäumer, *Geschichte der Biologie*, vol. II, 401.

locates in far-off areas, such as Africa, especially Ethiopia, and Asia, especially India; but some of them—maybe to the surprise of the reader—are located in Europe, such as the Bavarian rhino and the Bavarian unicorn. The systematic application of taxonomy constructed after the example of Aristotle has not led to the elimination of miraculous animals. It is sometimes hard to say how Wotton imagined that a certain miraculous or monstrous animal came into being; which animals he considered as true species; which as the results of interspecific hybridisation; and which as singular phenomena or "accidents". In some cases, however, he makes it explicit that the animal is the result of interspecific hybridisation, for example with the *Thos, Crocuta, Chaonis*, and "leo minor"; in other cases, this seems highly probable, such as with the *Chaus, Lupus Cervarius, Cepus, Hippopardium*, and various unicorns. In all of these cases, it is not really clear whether Wotton would also think of hybrid speciation—i.e. whether he would regard the offspring of interspecific hybridisation as fertile and capable to initiate a new species. But since his zoology is devoted to the classification of species, it may go without saying that he regards the hybrids as representatives of species. But animals that are presented by Wotton as a *mixtum compositum* are certainly not necessarily the result of interspecific hybridisation. The majority of them he probably regarded as *independent species*. Sometimes this becomes explicit, for example with the *Eale*, which Wotton identifies as a *hippopotamus*. In general, Wotton must have considered it interesting to offer the reader miraculous creatures and hybrid species. The *mirabile* (wondrous) was dear to him, as it was to Pliny, and it was dear to him for more than one reason. Of course, it made his zoology more fascinating, more remarkable, and thus more precious; and of course the *Wunderkammer* function of the literary collection may have played an important role as well. But maybe above all, Wotton wanted to demonstrate the power of his integrative taxonomical approach. The fact that he was able to classify all those strange creatures proved his sublime *mastery of nature*. Whereas he applied to zoology a systematic Aristotelian method, the whole project was motivated and inspired by a Plinian theological attitude to nature.

The Exclusion of Hybrids from the Godly Created Species: The *Holy Zoology* of Franzius (1612[–1712])

If the reduction of hybrids, "composite animals", and monstrous creatures is identified with scientific progress, then one might ascribe a major accomplishment to Wolfgang Franzius (1564–1628), a Lutheran theologian from Wittenberg. Somewhat paradoxically, this tendency appears in a religious, a 'holy' zoology—*Historia animalium sacra*, a work that was written for religious purposes and was primarily meant as a manual for Lutheran priests and Protestant students of theology,[85] especially for preaching and Bible exegesis. The *Holy History of the Animals* turned out to be an extremely successful work: it saw some 15 editions in the 17th century and was still used in the 18th century.[86]

In Franzius's zoology, most of the hybrids and "composite animals", of which Wotton's seemingly scientific order of species abounded, do not occur: the *Leucrocuta, Crocuta, Eale, Bonasus, Chaus, Chaonis, Lupus cervarius, Cepus, Thos*, winged Ethiopian Horse (*Pegasus*), *Mantichora*, etc. Franzius was not happy with those hybrid species, or with the idea that similar hybrid species could come into existence randomly, at any time and place. Also, he had difficulties accepting that they would belong to God's creation. Since the existing species were created by God, it cannot have been His will that they would be contaminated either by spontaneous mating or deliberate cross-breeding initiated by man, and, as Franzius proves with *auctoritates* from the Bible, God forbade cross-breeding. It is a kind of hubris and arrogance ('petulantia') if man

85 Cf. the title page of the first edition: '*Historia animalium sacra/ In qua plero/ rumque animalium/ praecipue proprietates in/ Gratiam Studiosorum Theologiae/ et Ministrorum Verbi ad usum* [...] *breviter accommodantur,/ In Academia Wittebergensi ante annos aliquot dictata/* a Wolfgango Franzio SS. Theologiae Doctore (Wittenberg, Zacharias Schürer-Johann Gormann: 1612). For this work, cf. Roggen V., "Biology and Theology in Franzius's *Historia animalium sacra* (1612)", in Enenkel K.A.E. – Smith P.J. (eds.), *Early Modern Zoology. The Construction of Animals in Science, Literature and the Visual Arts*, Intersections 7 (Leiden – Boston 2007), 2 vols., vol. I, 121–146; Bäumer, *Geschichte der Biologie*, vol. II, 159–164; eadem, "Biblische Zoologie (Hermann Frey, Wolfgang Franz, Heinrich von Hoevel)", in Kattensted H. (ed.), *"Grenzüberschreitung". Festschrift zum 70. Geburtstag von Manfred Büttner* (Bochum: 1993) 3–17.

86 *Historia animalium sacra* (ed. pr. Wittenberg, Zacharias Schürer – Johannes Gormann: 1612; ibidem, iidem: 1613; 1616; 1621; 1624; 1633; *editio quinta*, Wittenberg, Balthasar Mevius – Johannes Röhner: 1642; *editio sexta*, Wittenberg, Balthasar Mevius – Johannes Bauer: 1659; *editio septima*, Amsterdam, Johannes Jansonius: 1643; *editio octava*, ibidem, idem: 1653; *editio nona* ibidem, idem: 1654; *editio novissima*, Amsterdam, Johannes Ravenstein: 1665; [...] *iam denuo emendatius et correctius edita* [...] *novo etiam nominum propriorum indiculo adaucta* [...], *praetermissa* [...] *Augusti Buchneri praefatione*, Frankfurt a.M., Balthasar Christoph Wust d.J. – Martin Gabriel Hübner: 1671); *Historia animalium* [...] *continuatio cum commentario et supplemento observationum ex recentiori historia naturali similitudinum, emblematum, hieroglyphicorum ad usum Oratoriae tum civilis tum ecclesiasticae, opera Joannis Cypriani* (Dresden, Martin Gabriel Hübner – Melchior Bergens Erben: 1687; Leipzig – Frankfurt, Martin Gabriel Hübner: 1688; Frankfurt – Leipzig, Gottofredus Leschius: 1712). The text is henceforth quoted after the first edition of 1612.

aims at "improving" and changing the godly created species by new 'mixturae'.[87] The biblical figure Ana (*Genesis* 36, 24) was (supposedly) the first man who invented the impious act of deliberate cross-breeding, and his invention, the mule, is morally condemned by Franzius.[88] Franzius insists on the conviction that 'cross-breeding is explicitly forbidden in the Holy Scripture, in *Leviticus* 19,19'.[89] Franzius morally condemns hybrids by comparing them to contemporary 'hypocrites' or 'neutrales': people who refuse to make a religious decision, and waver between Protestantism and Catholicism. In this way they never 'will receive fruit from the vineyard of God', 'will never be illuminated by the Lord'.[90]

Franzius, however, considered the problem of the 'hypocrite' hybrids important enough to devote a special chapter to them.[91] The theological mission of Franzius's zoology also seems to have brought forth other "progressive" zoological statements. For example, he was one of the few zoologists of his time who did not accept the concept of "spontaneous generation", the idea that animals can come into existence without coitus, out of dirt, earth, or 'putrefactio' (decomposition).[92]

87 *Historia animalium sacra* (1612) p. 319.

88 Ibidem, pp. 319, 322, and 323.

89 Ibidem, p. 322: 'Levitici a 19. cap<ite>. expresse prohibetur, ne quis faciat brutum suum coire cum altero bruto'.

90 Ibidem, 323: 'Sicut muli sunt infoecundi, ita Hypocritae non illustrantur divinitus neque ornantur donis spiritualibus, et nullum faciunt fructum in vinea Domini' ('As the mules are infertile, the hypocrites will not be illuminated by God, nor will they receive the present of spirituality, nor will they ever receive fruit from the vineyard of God').

91 "De mixtis animalibus quadrupedibus", in ibidem, pp. 316–323.

92 Cf. Roggen, "Biology and Theology" 135–136. The idea goes back to antiquity, especially to Aristotle and his followers: cf. Balme D.M., "Development of Biology in Aristotle and Theophrastus: Theory of Spontaneous Generation", *Phronesis* 7 (1967) 91–104. It retained its relevance in the Middle Ages and the early modern period: cf. Hasse D.N., "Spontaneous Generation and the Ontology of Forms in Greek, Arabic and Medieval Latin Sources", in Adamson P. (ed.), *Classical Arabic Philosophy: Sources and Reception* (London – Turin: 2007) 150–175; Smith J.E.H., *The Problem of Animal Generation in Early Modern Philosophy* (Cambridge: 2006); Mendelsohn E.E., "Philosophical Biology versus Experimental Biology: Spontaneous Generation in the Seventeenth Century", in Grene M. – Everett E.E. (eds.), *Topics in the Philosophy of Biology* (Dordrecht: 1976) 37–65; Castellani C., "Le problème de la 'generatio spontanea' dans l'oeuvre de Fortunio Liceti", *Revue de synthèse* 89 (1968) 323–340; Nardi B., "Pietro Pomponazzi e la teoria di Avicenna intorno alla generazione spontanea nell'uomo", in idem, *Studi su Pietro Pomponazzi* (Florence: 1965) 305–319; Farley J., *The Spontaneous Generation from Descartes to Oparin* (Baltimore – London: 1974). Microscopic research in the second half of the 17th century led to a certain disbelief

Franzius's zoology, however, does not display a deep interest in taxonomy. Franzius is, of course, aware of the taxonomical achievements of his early modern predecessors, such as Wotton, Gessner, and Aldrovandi.[93] But the mastery of nature he aims to achieve does not depend on classification. So he uses the generally known large categories of animals for the overall structure of his zoology, viz. the quadrupeds (part I, pp. 31–323), birds (part II, pp. 324–583), fish (part III, pp. 584–705), and snakes and insects (part IV, pp. 706–888). With respect to the first part, however, he does not divide the species into viviparous and oviparous quadrupeds, nor does he differentiate (as Wotton and Aldrovandi did) between odd-toed and even-toed ungulates. If one looks closer at the description of the quadrupeds, Franzius uses a principle of ordering that very much resembles Pliny's: according to the species' size and strength, from big to small. Exactly like Pliny, he starts with the elephant (chapter 1), and he ends up with the smallest quadrupeds, such as the mouse and the mole (chapter 28).[94] To a certain extent, he even follows Pliny's ordering principle from exotic animals to domestic ones. As is the case with his predecessors, the notion of species ('genus') has different meanings. Similar to Wotton, he devotes to each species or 'genus' a single chapter; but sometimes 'genus' refers to a species in the modern sense, as is the case with the tiger[95] and the elk,[96] and sometimes it refers to a family of species, as is the case with

in and in the end to the abandonment of the *generatio spontanea*, e.g. by Darwin and his followers: cf. Ruestow E.G., "Leeuwenhoeck and the Campaign against Spontaneous Generation", *Journal of the History of Biology* 17 (1984) 225–248; Strick J.E., *Sparks of Life. Darwinism and the Victorian Debates over Spontaneous Generation* (Cambridge, Mass.: 2000).

93 Except, of course, Aldrovandi's volumes that appeared only posthumously and after the composition of the *Historia animalium sacra*: *Quadrupedum omnium bisulcorum historia* [...] *Ioannes Utervius Belga colligere incaepit* [...] *Thomas Dempsterus Baro a Muresk Scotus* [...] *absolvit. Hieronymus Tamburinus in lucem edidit* [...] (Bologna, Sebastianus Bonhommius: 1621; Frankfurt a.M., Caspar Röteli: 1647); *De quadrupedibus digitatis viviparis libri tres, de quadrupedibus digitatis oviparis libri duo Bartholomaeus Ambrosinus* [...] *collegit* [...] (Bologna, Antonio Bernia – Nicolao Tebaldeo: 1645); *De piscibus libri V et De cetis liber unus* [...] *Ioannes Cornelius Utervius collegit. Hieronymus Tamburinus in lucem edidit* [...] (Bologna, Giovanni Battista Bellagamba: 1613); *Serpentum, et draconum historiae libri II. Bartholomaeus Ambrosinus* [...] *summo labore opus concinnavit* [...] (Bologna, Antonio Bernia, Clemente Ferronio: 1640).

94 *Historia animalium sacra*, p. 310 ff.

95 Ibidem, chapter 9, p. 100 ff.

96 Ibidem, chapter 10, p. 106 ff.

the wolf,[97] goats,[98] and sheep.[99] Franzius's families, however, are in most cases not identical to the families of modern taxonomy: for example, he classifies among the family of the wolves the *Lynx* (a feline), the *Lupus cervarius* (also a feline), and the *Hyena*, which constitutes a family of its own (*Hyaenidae*, consisting of four species).[100]

In his critical chapter on the hybrids,[101] Franzius tries to limit interspecific hybridisation and hybrid species. He argues that due to God's providence, hybrids are often infertile. Franzius reduces the number true *hybrid species* to just three: the leopard, the giraffe, and the mule.[102] However, he admits that interspecific hybridisation by mating occurs more frequently: thus, offspring are also produced by bull and donkey,[103] fox and dog, dog and wolf, male wolf and female panther, goat and aries,[104] or tiger and dog.[105] But in his opinion, the offspring of these combinations are in most cases infertile, and do not lead to *proper hybrid species*. Franzius puts forward certain rules for interspecific hybridisation: 1) the mating species must be of the same size; 2) they must have the same duration of pregnancy; 3) they must have the same seasons of fertility; 4) they must be extraordinarily horny; and 5) They must be of a fertile age.[106] For example, man can never produce an offspring with a dog, nor a horse with an elephant, since the length of pregnancy of human beings is nine months, and that of dogs only three months.[107] Franzius does not give figures on the pregnancy of elephants, but he obviously was aware of the fact that their pregnancy lasts much longer than that of horses.[108]

97 Ibidem, chapter 20, p. 209 ff.

98 "De capra et his cognatis", in ibidem, chapter 23, p. 250 ff.

99 "De ovibus", in ibidem, chapter 22, p. 235 ff.

100 Ibidem, p. 214; cf. MacDonald (ed.), *Enzyklopädie der Säugetiere* 140–145. Cf. in the present contribution infra, paragraph on Gessner.

101 *Historia animalium sacra*, pp. 316–323.

102 Ibidem, p. 321: '[...] mixtorum animalium non possunt constitui certa genera, et vix pauca sunt data: Leopardus, Camelopardalis, Mulus [...]'.

103 Ibidem: 'ex tauro et asina'.

104 All ibidem, p. 317.

105 Ibidem, p. 320.

106 Bäumer, *Geschichte der Biologie*, vol. II, 163, surprisingly, mentions another criterion instead of increased sexual activity: that 'the mating animals must be mean and shrewish' ('Sie müssen trügerisch und verschlagen sein'). This is not part of the five criteria given by Franzius.

107 *Historia animalium sacra*, p. 319. These rules go back to Aristotle, and Della Porta's *Magia naturalis*, see below.

108 The length of pregnancy of elephants is almost 22 months (ca. 656 days), that of horses less than half of it.

The rules given by Franzius seem to make sense, even from a modern point of view.[109] But one wonders about the manner in which he himself has applied them. For example, he acknowledges that a dog and a tiger have produced offspring.[110] Both animals may be extraordinarily horny, but their sizes differ completely: a grown-up tiger is about three times as long as a dog (ca. 2.5–3 m vs. 60–90 cm), and about five to eight times as heavy (ca. 150–250 kg vs. 25–35 kg). Also, the duration of pregnancy differs considerably between the two: tigers give birth after ca. 103 days, dogs (and wolves) after ca. 61–63 days. The differences are even more spectacular in the case of the giraffe (*camelopardalis*), which Franzius acknowledges as a proper, i.e. fertile hybrid species, brought forth by the camel and the leopard. The camel is about six times as big as the leopard, and about eight to twelve times as heavy (ca. 450–650 kg vs. ca. 30–70 kg), and its pregnancy lasts about four times as long (13–14 months vs. 3–3.5 months). Of course it is not totally clear whether Franzius was aware of these differences. But the example he gives of the horse and the elephant suggests that he was able to think along these lines. Here again, it is important to understand that his manner of working was not guided by empirical observation, but primarily by theology and philology. He acknowledges the Tiger-hound as a hybrid of dog and tiger because he had authorities from antiquity which transmitted the story of the brave Tiger-hound owned by Alexander the Great.[111]

We also must take into account that Franzius's most important and most authoritative source was the Bible. Species that are not mentioned in the Bible suffer from a lack of authority and "evidence", so to speak, at least in Franzius's eyes. The *Leucrocuta, Crocuta, Eale, Bonasus, Chaus, Chaonis, Lupus Cervarius, Cepus, Thos, Pegasus*, and *Mantichora* do not occur in the Bible. Moreover, the Bible argued against hybrids. Thus, that alone could have been reason enough for Franzius to exclude these hybrid species from his zoology. On the other hand, all three of the hybrids he acknowledges as *hybrid species* are mentioned in the Bible: the leopard, the giraffe, and the mule. It is true that Franzius had no preference for "composite animals" or strange creatures. But, on the other hand, it is a telling detail that he acknowledges unicorn species (*Monoceros*).[112] The authority of the Bible alone was for Franzius conclusive evidence that the unicorn must exist in reality, especially because the

---

109 Cf. below.

110 *Historia animalium sacra*, p. 320.

111 Ibidem.

112 Chapter 11 "De Monocerote et Rhinocerote", ibidem, pp. 109–118.

unicorn was used as an image of Christ.[113] Empirical evidence may be used additionally; for example, Franzius quotes the eyewitness account of a friend of the polymath Julius Caesar Scaliger.[114] This friend also testifies to the fact that the unicorn and the rhinoceros are different species. Whereas the unicorn bears a single long horn in the middle of its head, the rhino has two horns: one smaller, and one longer and thicker at the top of its nose.[115] Small wonder that, in his otherwise critical chapter on the hybrids, Franzius would even be ready to accept as a fact of natural history that a crocodile had intercourse with a woman. The evidence was a mixture of philology and empirical knowledge: Franzius quotes Plutarch, who transmitted that a certain Philinus was an eyewitness to this monstrous act.[116] Furthermore, Franzius mentioned a hybrid of *homo sapiens* and a donkey.[117] But that relates to another story and has to do with another important scholar of natural history, to whom Franzius reacts in his chapter on the hybrids: Giovanni Battista della Porta.[118]

Hybridisation, or the Blessings of Man's Manipulation of Nature: Giovanni Battista della Porta's *Magia naturalis* (1558; 1589)

The south Italian nobleman Giovanni Battista della Porta (1535–1615) devoted his life to scholarship and science. He erected the *Accademia dei Segreti*, which was closed down by the Inquisition in 1578, and was one of the founding members of the Roman *Accademia dei Lincei*, to which Galileo Galilei also belonged.[119] Della Porta authored an extremely successful work on natural history, the *Magia naturalis*, which was translated into a number of languages

113 Ibidem, p. 109: 'Verum cum in Sacra Scriptura eruditas imagines *passim* ex natura Monocerotis sumptas usurpet et tum ad pios tum ad ipsum Christum accommodet cum singulari doctrina et consolatione, necesse est in rerum natura esse unicornem'.

114 Ibidem, pp. 109–110.

115 Ibidem, p. 110.

116 Ibidem, p. 317.

117 Ibidem.

118 Ibidem, p. 323, last line.

119 For Della Porta, cf. Piccari P., *Giovan Battista Della Porta. Il filosofo, il retore, lo scienziato* (Milano: 2007); Torrini M., *Giovan Battista Della Porta nell'Europa del suo tempo*, Atti del Convegno di Vico Equense (29 settembre–3 ottobre 1986) (Naples: 1990); Valente M., "Della Porta e l'Inquisizione. Nuovi documenti dell'archivio del Santo Uffizio", *Bruniana & Campanelliana* 2 (1999) 415–434; Brentano C., "Della Porta, Giovanni Battista", *Dizionario Biografico degli Italiani* 37 (1989); Baldini U. – Spruit L. (eds.), *Catholic church and modern science. Documents from the archives of the roman congregations of the Holy Office*

and was frequently printed in the 17th century.[120] Its augmented version of 1589 saw some 35 editions, either in Latin or in various translations.[121] Della Porta devoted the second book of the *Magia* especially to hybrid speciation.

For Della Porta, hybrid speciation and "spontaneous generation" were among the most important paradigms of nature, and he considered them to be the basic principles of genetic variation. He firmly believed in a kind of *unlimited fertility* or *generative power of hybrid creatures*, and he argued against 'the opinion of a certain philosopher' who thought that hybrids were generally infertile.[122] This philosopher was, of course, Aristotle. According to Della Porta, hybridisation brought forth animals with a 'natura tertii generis', with features different from both parents, i.e. new species.[123] Thus, he interpreted hybridisation as the "via regia" of genetic variation. The generative power of the new hybrid species ensures that they are 'preserved forever'.[124] Contrary to Franzius, he displayed an extremely positive and optimistic attitude toward hybridisation. Whereas Franzius emphasised that hybridisation was forbidden by God, Della Porta considered it to be completely legitimate, and a big and realistic chance to "improve nature". He even challenges his contemporaries to 'invent new species'.[125]

One of the basic thoughts of his *Magia naturalis* is that man is entitled to manipulate nature to his benefit. And it is this aim that determines the structure of Della Porta's chapter on the hybrids. It is not the strange exotic creatures (which were presented in Wotton's classification) that dominate his argument. In a marked difference, Della Porta's relevant chapter is in fact presented as

---

and the Index, vol. I, Sixteenth-century documents (Rome: 2009) 1507–1564; Muraro L., *Giambattista Della Porta. Mago e scienziato* (Milano: 1978).

120 First version in four books: *Magia naturalis sive de miraculis naturalium libri IV* (ed. pr. Naples: 1558; Antwerp: 1562; Italian translation Venice: 1560). For its influence and reception, cf. Balbiani L., "La ricezione della *Magia naturalis* di Giovanni Battista Della Porta. Cultura e scienza dall'Italia all'Europa", *Bruniana & Campanelliana* 2 (1999) 277–303.

121 Much revised and augmented Latin edition in 20 books *Magiae naturalis libri viginti* (Naples: 1589; Frankfurt: 1607; Hanau, David and Daniel Aubry, Clemens Schleicher: 1619; 1644; Leiden, Pieter Leffen: 1650; German translation Christian Zieger: 1713). The Latin text is quoted after the Leiden edition of 1650.

122 *Magiae naturalis libri viginti* (1650), pp. 60–61.

123 Ibidem.

124 Ibidem: 'quum multa animalia nunc videantur ex diversis prognata generibus, quae gignant ab eisque generata alia generent, ut in perpetuum species illa conservetur' ('since many animals appear now that are born from different species, which are fertile and whose offspring brings forth new offspring so that this species is preserved forever').

125 *Magia naturalis*, book II, 5, p. 60.

a small manual on animal breeding. It first discusses the breeding of dogs (II, 6–7), and continues with mules (II, 9); goats and sheep (II, 10); birds (II, 13 ff.), especially various hybrids of chicken (II, 14); birds of prey (II, 15); and fish (II, 16). All these hybrid species are, of course, conceived as domestic animals. Della Porta's aim of hybridisation is always to emend the drawbacks of the parental species, and to get stronger, tougher, and braver animals.[126] He explains under which circumstances man can breed hybrids, and he also gives practical advice on how to increase the sexual drive of the animals, for example with the help of salt, pepper, myrrh, and perfumes.[127] One of his frequently applied methods is to combine a domestic species with a wild one in order to improve the domestic species with the advantageous properties of the wild one. For example, one can make dogs stronger and braver through hybridisation with tigers,[128] lions, and *Crocuta* (the *Crocuta* itself being a hybrid of dog and wolf);[129] faster and stronger through hybridisation with the *Thos*,[130] wolf,[131] and *Chaus*; faster through cross-breeding with foxes,[132] and so on. Mules will get faster if one 'mixes' donkeys with the *Onager*,[133] and stronger through hybridisation of a donkey with a bull.[134] The result of all of these hybridisations will be *new species*. Dog and tiger bring forth the Tiger-hound, *Mastinus*, or *Canis Indicus*;[135] dog and lion, the *Leontomix*, the *Canis Arcas*, and *Canis Tegaeates*.[136]

Only secondly does Della Porta discuss exotic and wild hybrids:[137] the leopard is brought forth from male panther and female lion; the *Crocuta* from male hyena and female lion; the *Lycopantherus* or *Panther-wolf* from male wolf and female panther; the *Thos* from the same parents; the *Thos* also from male wolf and female hyena; the *Chaus* from the same parents; and the *Bactrian camel* (i.e. with two hummocks) from camel and wild boar. Regarding the exotic hybrids, Della Porta considers them all as fertile and as true *new species*.

---

126 Ibidem II, 6, pp. 63–67.

127 Section "Animalia ut ardentius ad coitum incumbent", in ibidem, II, 5, pp. 62–63.

128 Ibidem II, 6, pp. 63–65, section "Indicus canis fortissimus ex tigride".

129 All of them ibidem, II, 6, p. 66.

130 Ibidem, section "Canis ex Thoe fortis et velox".

131 Ibidem, pp. 66–67.

132 Ibidem, p. 67, section "Laconici canes veloces ex cane et vulpe generari possint".

133 Ibidem, p. 72, section "Mulae veloces ex onagro et asina generantur".

134 Ibidem, section "Ex tauro et asino fortissimi muli".

135 Ibidem, p. 63.

136 Ibidem, p. 66.

137 Ibidem II, 11, pp. 74–76.

Interestingly, Franzius had directly copied his rules of hybridisation from Della Porta's book II. In Franzius's case, these rules looked like limitations; Della Porta, however, has formulated them as positive precepts and practical guidelines for animal breeding. Similar to Franzius, Della Porta did not always keep to his rules: for example, camel and wild boar considerably differ with respect to size, weight, duration of pregnancy, and mating period. However, Della Porta more or less respected these rules. In comparison, Wotton did not limit his hybrids in this manner.

Interspecific Hybridisation and Hybrid Speciation in Modern Zoology

Nowadays it is clear that the number of interspecific hybrids, especially among mammals, is very limited. Moreover, in most cases, hybridity leads to sterility.[138] In general, only species with exactly the same number of chromosomes are able to bring forth fertile offspring.[139] This goes, for example, for some *Canidae*, such as wolf, dingo, coyote, dog, African wild dog (*Lycaon pictus*), and jackal. All of them have 78 chromosomes. But hybrids of *Canidae* with other *Canidae*, such as foxes, are infertile. Among the *Felidae*, only hybrids of some of the smaller species give birth to fertile offspring. The hybrids of the larger *Felidae* (tiger, lion, leopard, jaguar)—for them, names such as Liger, Tigon, and Jaglion were coined—are infertile. Cross-breeding of the two camel species of the Old World, *Camelus dromedarius* and *Camelus bactrianus*, produces fertile offspring, but if they mate with one of the camel species of the New World (guanaco, llama, alpaca), their offspring are infertile. As a general rule, the majority of hybrids are infertile and therefore cannot originate new species. The production of fertile offspring, of course, does not automatically lead to hybrid speciation, viz. the creation of a new species. Among animals, and especially mammals, hybrid speciation is a very rare exception, such as in the case of the American red wolf, which is a hybrid of grey wolf and coyote.[140]

---

138 Cf. Keeton W.T., *Biological science* (New York: 1980) 800.

139 However, there may be exceptions, for example, sometimes with mules and hinnies. Cf. Rong R. – Chandley A.C. – Song J. – McBeath S. – Tan P.P. – Bai Q. – Speed R.M., "A fertile mule and hinny in China", *Cytogenet Cell Genet* 47.3 (1988) 134–139.

140 Cf. http://www.savingwildplaces.com/swp-home/swp-glecrocodile/8287793?preview=&psid=&ph=class%2525253dawc-148772.

Because of a number of reasons, cross-breeding is in general very limited: genetically and morphologically, the animals must be closely related, and the same goes for the length of pregnancy, the seasons of fertility, the size of the animals, and their mating behaviour. Because of these limitations, most of the hybridisations that appear so frequently in early modern zoology are impossible. But interestingly, as we have seen in the works of Della Porta and Franzius, some of these limitations were already known (viz. the same size, period of fertility, and length of pregnancy). The first and the third of these were no early modern inventions, but go back to Aristotle.[141] Nevertheless, these rules were applied in varying and creative ways (as we have seen in the zoology by Franzius and in Della Porta's *De magia naturali*), or even largely neglected (as in Wotton's *De differentiis animalium*). It remains to be seen to what degree hybrids appear in zoological treatises, what parameters caused their appearance, and the ways in which they were constructed.

Reductive Taxonomy Based on Philological Criticism: Conrad Gessner's *Historiae animalium de quadrupedis viviparis* (1551)

Gessner's biological works, among others his groundbreaking *Historiae animalium* (1551–1558),[142] were in various ways inspired by theology and fed by philology.[143] When he worked on the *Historiae animalium* he was teaching natural philosophy at the Zwinglian theological university of Zurich, the so-called

141 *De generatione animalium* II, chapt. 5

142 *Historiae animalium* [...] 4 vols. (Zurich, Christoffel Froschauer: 1551–1558), vol. I *de quadrupedis viviparis* (1551; ed. secunda idem: 1560; Frankfurt a.M., Robertus Cambierius: 1603; Frankfurt a.M., Heinrich Lorenz: 1620); vol. II *de quadrupedis oviparis* (1554; facsimile ed. of vols. I and II Hildesheim: 2012; Frankfurt a.M., Johannes Wechel – Robertus Cambierius: 1586); vol. III *De avium natura* (Zurich, Christoffel Froschauer: 1555); idem, *Icones animalium quadrupedum viviparorum et oviparorum quae in historiae animalium* [...] *libro primo et secundo describuntur* [...] (Zurich, Christoffel Froschauer: 1553; ed. secunda idem: 1560); idem, *Thierbuch. Das ist ein kurtze beschreibung aller vierfüssigen Thieren so auf der erden und in wassern wonend* [...] (Zurich, Christoffel Froschauer: 1563); idem, *Thierbuch. Ausführliche beschreibung und lebendige ja auch eigentliche Contrafactur und abmahlung aller vierfüssigen Thieren so auf der erden und in wassern wohnen* [...] *durch den weitberühmten herrn doctor conrad gessnern* [...] (Heidelberg, Johann Lancellot: 1606).

143 Cf. Leu U.B., "Theologia naturalis", in—, *Conrad Gesner als Theologe. Ein Beitrag zur Zürcher Geistesgeschichte des 16. Jahrhunderts* (Bern-Frankfurt a.M. – New York – Paris:

Prophezei: in the early 1540s as a lecturer, and from 1546 on as a full professor for "Physica".[144] In concreto it meant that almost daily he spent one hour conducting a philological close reading and careful explanation of Aristotle's works on natural history, among others, of course, the *Historia animalium*. According to Gessner, biology and natural philosophy, or *naturalis historia* were an indispensable part of a theologian's education, since the contemplation of nature leads directly to the contemplation of its creator; and the understanding of the creation to the cognition of the creator. Moreover, Gessner considered natural history to be of great value for exegesis of the Bible, especially for the book *Genesis*, which describes the creation of the world, man, and the species of plants and animals. A central piece in Gessner's theological thinking is the *admiration of God and His creation*; this goes for the biggest animals as well as the smallest, seemingly negligible beings, such as ants, lice, bees, flies, worms etc.[145] Man takes a pivotal position in nature: God created man as its administrator and contemplator. Man may use nature to his advantage, but just as important is the contemplation of nature: Gessner conceived it as a religious act, a kind of prayer. The *description* of nature, for example in a zoological work, is equivalent to the *praise* of God. In this sense, Gessner characterises main figures of the Old Testament as zoologists: Adam, the first "classifier of species" (the man who gave names to all the created animals); Noah, the saviour and preserver of the species during the Flood; and Solomon, the first author of a zoology.[146]

Gessner saw himself as a successor to these Old Testament zoologists and, just as importantly, to Pliny the Elder. Like Pliny, Gessner aimed at producing a compendious and complete description of the species; and, similar to Pliny, he emphasised the admirable force of nature, although his method was partly different. For example, Gessner aimed at undertaking a complete collection and evaluation of *all* available literary sources; moreover, he eagerly collected eyewitness reports from contemporary scholars and scientists, and, much more so than Pliny, he took into account his own empirical observations and eyewitness accounts as well. Unlike Pliny, he did not just go for the exotic and strange, but he tried to reduce it through plausible and rational explanations. For Gessner, admiration of nature did not necessarily depend on the exotic and miraculous. As a nomenclator of species, Gessner clearly surpassed Adam and Pliny: he collected *all available names* in *all languages* known to him: not only in Hebrew

---

1990; Zürcher Beiträge zur Reformationsgeschichte 14) 31–100; esp. "Zum Verhältnis von Naturwissenschaft und Theologie" 57 ff., and "Ad coginitionem Dei" 59 ff.

144 Ibidem 50 ff.

145 Cf. ibidem 61–62.

146 Ibidem 62; *Historia animalium*, I, fol. a 4v (preface).

and Latin, but also in Arabic, Syrian, Armenian, German, French, Italian, Spanish etc. In the description of each species, he devoted a full-scale section to the name of the species in the various languages. Animal classification in Wotton's sense, however, was not Gessner's goal in his *Historiae animalium de quadrupedis viviparis* of 1551. He did not describe the species in a system of categories, subcategories/classes, families etc., but in *alphabetical* order. This principle of order was not new or revolutionary: it goes back to Albertus Magnus's zoology and to Isidorus of Sevilla's *Etymologiae*. The alphabetical order, of course, only marginally referred to classification—to the main entry in Latin. As a classificator, however, Gessner experienced a certain development: in his latest publications (on fish, 1558 ff.) he used methods and means that show considerable differences from those in his first zoological book (on the quadrupeds, of 1551).[147] In the second edition of the *Icones quadrupedum* of 1560 he demonstrated much more attention to classification than he did in the publication of 1551.

Gessner's methods (and achievements) of animal classification in his book on the quadrupeds of 1551 are characterised by a number of means: first, by a full collection and philological discussion of all available names of the species (in as many languages as possible); second, by a critical attitude to species-splitting as suggested by ancient, medieval, and modern literature—Gessner aimed as much as possible at constructing single *homogeneous species*; third, by a certain common sense approach with respect to very rare features that occurred either in literature or in eyewitness accounts; fourth, by a flexible use of literary sources and empirical observations; fifth, by a combination of biblical sources with other literary and non-literary sources; and sixth, by a kind of self-confident attitude with respect to the greatest authorities of ancient zoology, such as Aristotle and Pliny.

For example, Gessner refused to split the species *lion/leo* (modern: *Panthera leo*) into different species, such as Aristotle, Pliny, and others had done.[148] In comparison, Wotton had tried to maximise the "splitting" of *Panthera leo* into 7 to 10 different species of lions, which, however, led to strange results.[149] Gessner does not hesitate to contradict Aristotle's authoritative remark that all species

147 Gessner, *Historia animalium*, vol. IV *De piscium et aquatilium animantium natura* (Zurich, Christoffel Froschauer: 1558); ed. secunda (Zurich, Christoffel Froschauer: 1560). For this later development, see the contribution by Sophia Hendrikx in the present volume.

148 Cf. Enenkel, "Zur Konstituierung der Zoologie als Wissenschaft in der Frühen Neuzeit" 56–57.

149 Cf. ibidem 32–35.

of lions had yellowish fur.[150] Of course, Gessner's approach did not automatically bring forth correct taxonomical descriptions. For example, he erroneously stated that melanism was a feature of the species *Panthera leo*.[151] But with respect to other larger *Felidae*, this was an important clue for taxonomy. In a number of cases, Gessner's methods lead to useful clarifications. Among other things, he was able to exclude the existence of a species of smaller (and weaker) lions without a mane. Gessner's "reductive" taxonomical approach of 1551 sometimes also led to a reduction of hybrids or hybrid species. For example, the fact that he did not accept a species of lions without a mane meant that he could leave out the Plinian and Aristotelian explanation of this species as hybrids from lion and panther.

An illuminating example of Gessner's methods is his treatment of the dangerous Indian carnivore *Mantichora*, a man-eater with a human face, three rows of extremely sharp teeth (like a shark), the body, feet, and reddish (or yellowish) fur of a lion, the tail of a scorpion (including a stinger), and the ability to run so fast that a man cannot escape it.[152] Its voice resembles a trumpet (or tuba). It had been described by the authoritative sources of Aristotle's *Historia animalium* II, 3 (501a 26) and Pliny's *Naturalis historia* (VIII, 75). Wotton had included it, albeit with some reluctance, under the section of "monstrous animals of India and Ethiopia".[153] Gessner, however, was not inclined to accept this remarkable *mixtum compositum* as a species *sui generis*. In doing so he was inspired by the author of a Greek traveller's guide from the 2nd century AD, Pausanias, who, however, called the animal *Martiora*.[154] According to Pausanias, the strange features of the *Martiora* were based on nothing but the pure fantasy of the Indian people, caused by their enormous fear of the animal. In Pausanias's eyes, the so-called *Martiora* was none other than the tiger (*Panthera tigris*): a man-eater with a kind of round face (in this respect 'human'). Gessner followed Pausanias's sceptical and rational approach but preferred to identify the *Mantichora* with the so-called *Leucrocuta*, which had been described, for example, by Pliny: 'of the size of a donkey, with the legs similar to a deer, extremely fast, but with his breast, neck and tail similar to a lion, with the head of a badger [...], with a huge mouth, and its jaws reaching

---

150 Cf. ibidem 57.

151 Cf. ibidem.

152 Gessner, *Historiae animalium* [...] lib. I *de quadrupedis viviparis* (1551) 631.

153 Cf. *supra*.

154 *Descriptio Graeciae* IX, 21,4; Ctesias, *Historia Indica*, in FGH (*Fragmenta Graecorum Historicorum*) 688 Frgm. 45d.

up to the ears [...]'.[155] So he came to the identification *Mantichora* = *Martiora* = *Leucrocuta*. With respect to the location of the animal, however, Gessner did not hesitate to contradict the authority of Pliny. Whereas Pliny defined it as an 'Ethiopian' animal,[156] i.e. living in Africa south of Egypt, Gessner preferred 'India'[157] as its home, following the ancient Roman scholar and geographer Solinus.[158] He came to this verdict by a philological comparison and harmonisation of sources, a method that took him another step further. Because in Pliny, *Naturalis historia* VIII, 107, the so-called *Corocotta* is described in a similar way as the *Leucrocuta* in VIII, 75, Gessner came to another reduction of species: *Mantichora* = *Martiora* = *Leucrocuta* = *Corocotta*. And critical observations of the text led him to even more identifications, and thus taxonomical reductions: he also identified Albertus Magnus's and Avicenna's species *Maricorion* or *Maricomorion* with the *Mantichora*; he regarded them as textual corruptions caused by scribal errors.[159] The same goes for Albertus Magnus's species *Leutrochoca*, which according to Gessner was caused by a transcription error from 'Leucrocuta'; and for Albertus's *Cirocrothes* or *Cirotrochea*, which in Gessner's eyes were textual corruptions from 'Crocuta'. *Crocuta* is in Greek written as 'Crocotta'. From this observation Gessner concluded that *Crocotta* and *Corocotta* also must refer to the same species. Thus: *Mantichora* = *Martiora* = *Leucrocuta* = *Leutrochoca* = *Corocotta* = *Maricorion* = *Maricomorion* = *Cirocrothes* = *Cirotrochea* = *Crocuta*. The reduction of six or ten animal species to only one may seem elegant; its underlying method, however, was not empirical observation, but philology. Gessner's taxonomical reductions resemble, in a sense, the *emendatio* and *coniectura* of textual criticism. In his argument, Gessner even uses the technical term of textual criticism ('coniiciam'): he says that he is willing to change his 'conjecture' if another zoologist comes up with a new and better solution.[160]

Gessner's philological and reductive approach, however, had certain limitations. If the identification *Mantichora* = *Martiora* = *Leucrocuta* = *Corocotta* =

---

155 *Naturalis historia* VIII, 72.

156 The paragraph VIII, 72, is devoted to 'Ethiopian' animals. For Pliny, Ethiopia means something like Africa south of Egypt.

157 Gessner, *Historiae animalium* [...] lib. I *de quadrupedis viviparis* (1551) 630–631.

158 Ibidem 631, line 1: 'Solinus capite 55 in descriptione Indiae'; cf. Solinus 52,34.

159 Ibidem 631, line 15–16: 'Albertus et Avicenna maricorion vel maricomorion vocibus corruptis habent [...]' ('Albert and Avicenna list with corrupted names a [species] *"Maricorion"* or *"Maricomorion"*').

160 Ibidem 631, line 49: '[Crocottam] [...] leucrocutam potius quam tigrin esse coniiciam, dum alius meliora adferat' ('I guess that it is better to identify the *Crocotta* with the *Leucrocuta* than with the tiger, until another [scholar] comes up with a better solution').

Maricorion = *Maricomorion* = *Crocuta* was correct, what kind of animal is it? According to most descriptions, the *Mantichora* had the face of a man, and the *Leucrocuta* one of badgers ('capite melium', according to Pliny) or camels ('capite camelino', according to Solinus), thus a *non liquet*. Another *non liquet* refers to the teeth: two horizontal teeth above in the shape of a knife (*Leucrocuta*; *Crocuta*), or three rows of sharp teeth in the whole mouth (*Mantichora*), or a tooth formula similar to the wolf or another species of *Canidae*, the badger, or one of the larger *Felidae* (lion, tiger). And what about its feet? Was the animal even-hoofed (*Leucrocuta*) or did it belong to the *multifidae* (*Mantichora*, lion)? Pliny defined the *Corocotta* as a hybrid, brought forth by a male hyena and a female lion,[161] and in a similar way defined the *Crocuta* as a hybrid of male wolf and female dog.[162] Gessner apparently preferred the first definition: he stated that the *Crocuta* comes forth from the mating of a male hyena and a female lion.[163] He does not further investigate the problems of this hybridisation; following one of Pliny's accounts, Gessner presents the animal with six to ten names as a close relative of the hyena, and the alphabetical locus of the H of 'Hyaena' is also the place where he inserts this dangerous and miraculous hybrid in his zoology.

The *hyena* itself (mod. family: *Hyaenidae*, consisting of four species: the Spotted Hyena/ *Crocuta crocuta*; Brown Hyena/ *Parahyena brunnea*; Striped hyena/ *Hyaena hyaena*; and the Aardwolf/ *Proteles cristatus*)[164] was conceived by the natural history of Antiquity as a monstrous creature with very strange features: it was supposed to be a man-eater, hermaphrodite (each animal having male and female sexual organs) or transsexual (for example, one year male and the other year female), lacking the vertebrae of the neck, able to imitate the human voice, having multicoloured eyes or bearing a precious stone in its eye or on his head, digging up dead bodies in graveyards, possessing magic powers, etc.[165] By taking into account the empirical observations of others, Gessner succeeded in excluding the hyena's hermaphroditic and transsexual nature, which he condemned as a lie.[166] By his sceptical approach he reduced

161 *Naturalis historia* VIII, 107.

162 Ibidem VIII, 75.

163 *Historiae animalium* [...] lib. I *de quadrupedis viviparis* (1551) 630; cf. also p. 626, line 38: 'Hyaenae coitu leaena Aethiopica parit crocutam'.

164 MacDonald (ed.), *Enzyklopädie der Säugetiere* 140–145.

165 Cf. e.g. Pliny, *Naturalis historia* VIII, 105–106; cf. inter alia Aristotle, *Historia animalium* VI, 32, 579b 15–16; *De generatione animalium* III, 6, 757a 2; Aelianus, *De natura animalium* I, 25; Diodorus Siculus XXXII, 12,2; Ovidius, *Metamorphoses* XV, 408 ff.

166 Gessner, *Historiae animalium* [...] lib. I *de quadrupedis viviparis* (1551) 626, line 15–16: 'Quod de hyaena fertur, genitale simul maris et foeminae eandem habere, commentitum est'.

the "imitative" quality of the species' voice to the sound of vomiting (as, for example, dogs do), although not rightly so.[167]

In a passage that illustrates Gessner's comparative taxonomical approach he presents the hyena as a close relative of the wolf (*Canis lupus*): 'The wolf and the hyena have much in common: they have the same size, colour, and appetite for meat, and they hunt other animals; the wolf sheep and goats, the hyena dogs and sometimes men; both have teeth like a saw, and similar genitals; [...] and both hunt in the night' [Fig. 2.8].[168] These are plausible observations or, even better, conclusions Gessner drew from his collected sources. If he had known the hyena from autopsy, he would have probably come to the same conclusions. Of course Gessner could not have known that the hyena does not belong to the *Canidae*, but to the *Feliformia*, among which the four *Hyaenidae* species constitute a family of their own. But it is even generally acknowledged nowadays that the *Hyaenidae* have certain things in common with the *Canidae*: for example, both hyenas and canines are 'cursorial hunters that catch prey with their teeth rather than claws; both eat food quickly and may store it, and their calloused feet with large, blunt, non-retractable nails'.[169] In fact, the family *Hyaenidae* consists of two branches: the 'lightly built dog-like hyenas and the robust bone-crushing hyenas'.[170] Nowadays the dog-like hyenas have almost died out, with the exception of the *Aardwolf* (East and South Africa), who lives on insects. Gessner could not have been aware of this, and he certainly did not know of the *Aardwolf*, which was not yet discovered by Europeans. His taxonomical definition of the hyena as a close relative of the wolf was based on literature. This was also the way he came to attribute to the hyena a rather strange morphological detail: that the animal would lack the vertebrae of the neck, and instead would have a single long neck bone. This, of course, does not correspond with the facts. The hyena has, like all *Carnivora*, seven vertebrae

167 This step is methodically correct, although "in re" wrong: The *Spotted hyena* (*Crocuta crocuta*) has an extremely wide range of vocal expressions; it produces many different sounds consisting of whoops, grunts, groans, lows, giggles, yells, growls, laughs, and whines. Cf. http://en.wikipedia.org/wiki/Hyena. Many of them could have been identified by as human-like.

168 Ibidem 625, line 58 ff.: 'Multa igitur lupo et hyaenae communia sunt: magnitudo et color, ut dixi, item voracitas, et insidiae, quae aliis animalibus moliuntur: ille gregibus ovium et caprarum, haec canibus et homini quandoque; dentes utrisque serrati, genitalia utrisque similia [...]; uterque noctu vagatur cibi causa [...]'.

169 Cf. http://en.wikipedia.org/wiki/Hyena.

170 Ibidem.

FIGURE 2.8 *Image of the hyena, not accepted by Gessner in 1560. From: Gessner,* Icones animalium *(2nd ed., Zurich, Christoph Froschauer: 1560), p. 75.*

of the neck. But Gessner found this odd information in Pliny and Solinus.[171] As a physician, Gessner probably was well aware that the dog—a close relative of the wolf—had seven vertebrae of the neck. But nevertheless, he considered the authority of the authors from antiquity as sufficient proof of the wolf's and the hyena's single long neck bone. And the same authorities testified to the hyena's potential to bring forth hybrids/hybrid species: its mating with lions and wolves would produce fertile offspring. Its mating with the wolf was another argument for the animals' close relationship. As their fertile offspring Gessner defined the *Thos*.[172]

This supposed hybrid, the *Thos*, Gessner described as a species closely related to the wolf, and consequently presented it in an appendix to the main entry "Lupus".[173] Again, via his method of comparative philology, Gessner came to a reduction of species: *Thos* (*minor*) = *Panther minor/pantherion* = *Lycopantheros* = *Lupus carnarius*.[174] On the other hand he clearly differentiated the *Thos* from another (supposed) 'cognate' of the wolf, the *Lupus cervarius* or *Lynx*.[175] In Gessner's antique sources the *Thos* appears as a hybrid, either of male wolf and female hyena (Hesychius and Varinus), or of male wolf with

---

171 Gessner, *Historiae animalium* [...] lib. I *de quadrupedis viviparis* (1551) 626, line 10 ff.

172 Ibidem 626, line 38 ff.

173 Ibidem 766 ff. "De feris illis quae lupo congeneres sunt. Et primum de Thoe, Panthere, Lupo carnario, Lycaone etc.".

174 Ibidem 766, line 51–52: 'Videtur autem thos minor, panther minor, lycopantheros et lupus carnarius idem omnino animal esse' ('It seems to me that the *Thos minor*, the *Panther minor*, and the *Lupus carnarius* refer to the same animal').

175 Cf. ibidem 769 ff.

female panther (Oppianus).[176] Agostino Nifo and Theodorus Gaza used the name *Lupus carnarius* because the animal would resemble a wolf, and Gessner anyway does not seem to disagree.[177] But why 'small panther'? This name of the animal appeared only in the Greek, not in the Latin sources.[178] One would suppose that there were a kind of misunderstanding or misconception. With respect to his reductive taxonomical method, however, Gessner considered the name 'small panther' useful in order to discern the species from the species *Panther maior*. The idea to split the *Thos* into two subspecies, a smaller one and a larger one, goes back to Oppianus's poem *On Hunting*.[179] Gessner's species *Thos (minor) = Panther minor/pantherion = Lycopantheros = Lupus carnarius*; however, it displays a very strange blurring of *Felidae* and *Canidae* features. In the end, it is almost impossible to identify the species. Which animal would have a body longer than the wolf but shorter legs, and would change its fur each season?[180] It cannot be true for the jackal (*Canis aureus, Canis adustus, Canis mesomelas*), which is not as long as the wolf and does not change its fur seasonally, or for any of the smaller *Felidae*, and hardly be true for the lynx, who may sometimes be as long as the wolf, but certainly not longer and usually quite a bit shorter. Moreover, Gessner emphasised that the species *Thos* was totally different from the species *Lynx*. Thus, again a manifold *non liquet*. The enigmatic *Thos* is another example from which it appears that Gessner's reductive taxonomical approach may create as many problems as it tries to solve.

New Efforts of Classification, Inclusion of New Eyewitness Reports: Gessner's *Icones animalium quadrupedum viviparorum* [...] (1560)

In the second edition of the *Icones animalium quadrupedum viviparorum et oviparorum* Gessner introduced many changes.[181] Different from the *Historia animalium* Gessner presented his zoological compendium this time not in alphabetical order, but based on a classification of the viviparous quadrupeds

176 Ibidem 768, line 37 ff.

177 Ibidem 767, line 59: 'itaque canarium dixi lupum (hoc prius fecerat Gaza) a canis similitudine' ('and therefore I call [the animal] "Lupus carnarius" because of its similarity to the dog (Gaza did the same before)').

178 Ibidem.

179 Ibidem 767.

180 Cf. König's and Winkler's remarks on Pliny, *Naturalis historia* VIII, 123.

181 [...], *editio secunda, novis eiconibus non paucis et passim nomenclaturis ac descriptionibus auctior* (Zurich, Christoph Froschauer: 1560).

into six classes.[182] From a taxonomical point of view, however, Gessner's system does not seem to be particularly strong: he divides the mammals into two large categories: I) domestic animals (*mansuetae*) and II) wild animals (*ferae*); the domestic animals (I) are divided into two classes, 1) horn-bearing animals and 2) animals without horns. As one may expect, the second class (of category I) is much bigger and much more diverse (horses, camels, cats, dogs, pigs, etc.) than the first class. Category II is subdivided in a similar way, into the classes 1) horn bearers and 2) animals without horns. The second class is subdivided into three groups: large animals without horns (II.2.1); animals of a medium size without horns (II.2.2); and small animals without horns (II.2.3). Behind this rather simplistic classification one again detects Pliny, who had presented the land animals from big to small, and had divided them into wild (+exotic) and domestic species.

Much more important than Gessner's effort to present his animal description in a now systematic order, was his eagerness to adapt and correct his former taxonomical definitions, and the animal descriptions and illustrations as well. This effort was admirable and led to manifold progress, especially since Gessner included new, recent eyewitness accounts. The question is, of course—with respect to each singular species—how he used eyewitness accounts and what conclusions he drew from them. Let's have a look at what happened to the species described above. An important new source was the traveller's account *Descrittione dell'Africa* by al-Hasan ibn Muhammad al-Wazzan al-Fasi or Leo Africanus (ca. 1494–ca. 1554), a highly educated Andalusian Moor who had worked as a diplomat and ambassador for Muhammad II, the Sultan of Fez. Leo Africanus had been captured by pirates in 1518 and sold to Pope Leo X, who baptised him in 1520.[183] The first edition of his *Descrittione dell'Africa* appeared in 1550,[184] and it was Conrad Gessner's brother Andreas who edited Florianus's Latin translation of the Italian version, which was made after the original Arab text (Zurich: 1559).[185] In this work Gessner found a description of an animal that in Arabic was called 'Dabuh' or 'Dabu', and in a local African

182 Cf. Ibidem 8 ("enumeratio ordinum [...]").

183 For Leo Africanus cf. Davis N.Z., *Trickster Travels: a sixteenth-century Muslim between worlds* (New York: 2007).

184 *Della descrittione dell'Africa et delle cose notabili che ivi sono, per Giovan Lioni Africano* (Venice, Giovanni Battista Ramusio: 1550).

185 Leo Africanus, *De totius Africae descriptione li. IX* [...] *recens in Latinam linguam conversi Ioanni Floriano interprete* [...]. *Arabice primum scripsit Author, deinde Italico sermone reddidit, Ioannes Florianus ex Italico Latinum fecit* (Zurich, Andreas Gessner: 1559).

FIGURE 2.9 *Gessner's Hyena from 1560 (actually the baboon). From: Gessner,* Icones animalium *(2nd ed., Zurich, Christoph Froschauer: 1560), p. 76.*

language 'Jesef'.[186] This animal had the 'size and shape of a wolf, and feet and legs like a man; [...] it digs out dead bodies in graveyards and devours them [...]'.[187] Florianus rendered the *Dabuh* or *Jesef* in Latin with '*Hyaena*'. For Gessner this passage was an eye-opener because it ascribed to the hyena 'feet and legs like a man'. It led him to identify the hyena with the *Babuinus* or baboon (mod. *Papio hamadryas*): 'After I read this passage', Gessner tells us, 'I understood all of a sudden that the animal which is commonly called *Papio* or *Babuinus*, is precisely the *hyena*'.[188] Formerly Gessner had thought that the *Papio* or *Babuin* was the Monkeybear (*Arctopithecus*), an 'animal compositum' of monkey (*simia*) and bear (*ursus*). As pivotal evidence Gessner now adds a woodcut illustration of a baboon, drawn from an exemplar that was shown in 1551 in Augsburg [Fig. 2.9], and a description made by a German eyewitness who called the animal in German '*Pavyon*'. Apparently it did not bother Gessner that this animal appeared to have a strong prefer-

186 Ibidem, p. 503 (chapt. 41).

187 Ibidem: '"Dabuh" Arabica appellatione, Africanis "Iesef" dicitur: animal et magnitudine et forma lupum refert, pedes et crura homini similis. [...] humana corpora sepulchris evellit ac devorat [...]'.

188 Gessner, *Icones animalium* [...], *editio secunda* (1560) p. 76.

ence for fruit and bread, which of course did not qualify it as a carnivore. Gessner sent this illustration for a check to another eyewitness of a baboon, the Antwerp pharmacist Pieter Coudenberg, who approved it.[189] There was, however, another problem: on the same page as the hyena, Leo Africanus had mentioned the *Babuinus* as an African monkey species (*simia*). Gessner solved this problem by splitting the species *Babuinus* and *Papio*. He compared Leo Africanus's remark that the *Babuinus* lacked a tail with the Augsburg illustration where the animal had a short tail. He concluded that thus, the Augsburg animal called *Papio* cannot be a *Babuinus*. So he found another argument to identify the Augsburg *Papio* with the *Hyaena*. Thus Gessner proudly represented the Augsburg *Papio* as the archetypical image of the hyena: 'Because of these reasons I conclude that the *Papio* is the *hyena*, an animal whose form, real being and name were hitherto—for so many centuries—unknown even to the most learned men'.[190]

This new finding made Gessner rethink his taxonomical definition of the hyena's relatives as he gave it in the *Historia animalium* [...] *quadrupedum viviparorum* of 1551. He was no longer fully convinced of his ingenious species equation of *Mantichora* = *Martiora* = *Leucrocuta* = *Leutrochoca* = *Corocotta* = *Maricorion* = *Maricomorion* = *Cirocrothes* = *Cirotrochea* = *Crocuta*. Of course, the reduction of species based on the identification of transmission or scribal errors retained its value. But the finding of the Augsburg *Papio* was for Gessner proof that the ancient descriptions of the hyena were incorrect, or at least questionable: 'It seems that the ancients blurred in their descriptions the *Hyaena, Crocuta, Crocotta* (sic), *Leucrocuta*, and *Mantichora* or *Mantiora* (sic)'.[191] He used as a new clue Porphyrius's remark that 'the Indians call the hyena *Crocuta*'.[192] If the *Crocuta* is a hyena, and the hyena is the Augsburg *Papio*, the *Crocuta* cannot possibly be the same as the *Leucrocuta* and the *Mantichora* (or *Mantiora/Martiora*). And furthermore, there seemed to be certain differences between the Augsburg *Papio* and the ancient descriptions of the *Crocuta*.

Thus, Gessner rearranged his classification. He split the species *Hyaena* into two subspecies: 1) a smaller and less dangerous one he identified with the Augsburg *Papio* and located in Syria, and 2) a stronger and very dangerous one he ascribed to Ethiopia and India, and identified with the *Crocuta* or *Crocotta* (sic). This second subspecies he gave a new name: *Hyaena Indica* or *Hyaena Aethiopica*.[193] In the framework of this new taxonomical arrangement,

189 Ibidem, pp. 76–77.
190 Ibidem, p. 77.
191 Ibidem, p. 78.
192 Ibidem.
193 Ibidem.

Gessner was no longer convinced that the *Hyaena Indica* or *Hyaena Aethiopica* or *Crocuta* was a hybrid. He says that it is '*if it were* a composite animal from wolf and dog', which means: a species of its own. It is bigger, stronger, more cruel, and more voracious than both wolf and dog, with an extraordinary power of its jaws to crush bones.

Gessner still denies the hyena's ability to produce human sounds. But he now corrects his former erroneous morphological description, which stated that the hyena would lack vertebrae of the neck and would have instead a long, single bone. Gessner rightly states: 'I believe that all animals who have a proper neck, also possess vertebra's'.[194] Nevertheless, Gessner did not give up the idea that hyena hybrids or hybrid species existed. He (still) lists the *Thos* as a hybrid from female hyena and male wolf,[195] and adds even another hybrid: the *Onolycus* or *Monolycus*, brought forth from the same combination of species. In the second edition of the *Icones animalium*, Gessner does not add new information to the *Leucrocuta*, and *Mantichora* or *Mantiora*. This does not mean that he entirely abandoned these or similar hybrid species by now; it was probably caused solely by the fact that he had no new information, and no (new) image to offer. For example, in the second edition of the *Icones animalium* Gessner included a hybrid from man, cock, and carnivore that was caught in 1531 in the bisdom of Salisburg (Salzburg in Austria) [Fig. 2.21], or an ape-like monster called *Satyr* [Fig. 2.10].

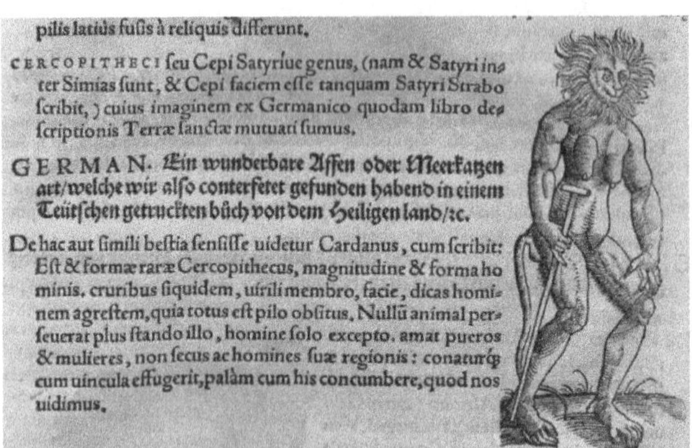
pilis latiùs fusis à reliquis differunt.

CERCOPITHECI seu Cepi Satyriue genus, (nam & Satyri inter Simias sunt, & Cepi faciem esse tanquam Satyri Strabo scribit,) cuius imaginem ex Germanico quodam libro descriptionis Terræ sanctæ mutuari sumus.

GERMAN. Ein wunderbare Affen oder Meerkatzen art/welche wir also conterfetet gefunden habend in einem Teütschen getruckten büch von dem Heiligen land/zc.

De hac aut simili bestia sensisse uidetur Cardanus, cum scribit: Est & formæ raræ Cercopithecus, magnitudine & forma hominis, cruribus siquidem, uirili membro, facie, dicas hominem agrestem, quia totus est pilo obsitus. Nullū animal perseuerat plus stando illo, homine solo excepto. amat pueros & mulieres, non secus ac homines suæ regionis: conaturq́ cum uincula effugerit, palàm cum his concumbere, quod nos uidimus.

FIGURE 2.10 *Gessner's* Satyr=Cercopithecus=Cepus. *From: Gessner,* Icones animalium *(2nd ed., Zurich, Christoph Froschauer: 1560), p. 95.*

194 Ibidem.

195 Ibidem, pp. 77–78.

The New Species of the New World, Their Effect on Taxonomy, and Theological Problems Related to Them: Nieremberg and Kircher

Neither Wotton (1552) nor Franzius (1612) nor Della Porta (1558; 1590) included the newly discovered species of the Americas: Franzius because his main focus was biblical exegesis and preaching with biblical examples; Wotton because he focused solely on Aristotle and the ancient writers; and Della Porta because he offered a manual for cross-breeding for Europeans. However, the species of the New World were inevitably there, and to an increasing degree they found their way into contemporary geography, history, travel accounts, scholarly and other correspondence, and zoology.[196] In zoology, for example, Gessner had described and depicted a couple of them in his *Icones animalium* (1560), such as the llama [Fig. 2.11], armadillo, Sagoin (small Brasilian monkey), sloth [Fig. 2.12], guinea pig [Fig. 2.13], and *Su*.

FIGURE 2.11 *The llama* (Allocamelus Scaligeri). *From: Gessner,* Icones animalium *(2nd ed., Zurich, Christoph Froschauer: 1560), p. 43.*

196 Cf. Asúa M. De – French R., *A New World of Animals. Early Modern Europeans on the Creatures of Iberian America* (Aldershot: 2005); George W., "Sources and Background to Discoveries of New Animals in the Sixteenth and Seventeenth Centuries", *History of Science* 18 (1980), 70–104; Mason P., *Before Disenchantment: Images of Exotic Plants and Animals in the Early Modern World* (London: 2009); idem, *Infelicities. Representations of Exotic* (Baltimore – London: 1998); Morison S.E., *The European Discovery of America.* 2 vols. (New York: 1971–1974); Salas A.M., *Para un bestiario de las Indias* (Buenos Aires: 1985).

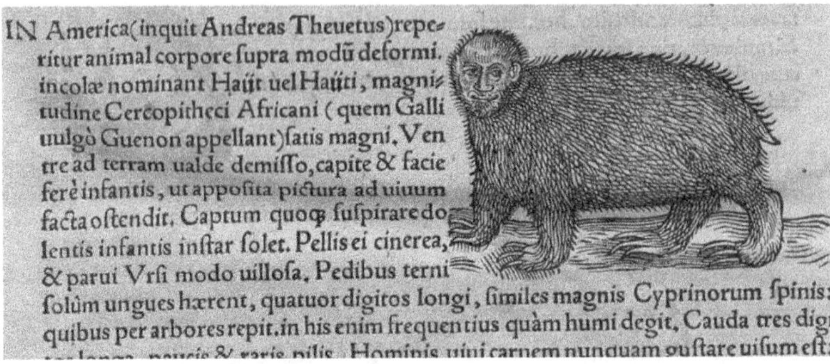

FIGURE 2.12 *The sloth* (Haut). *From: Gessner,* Icones animalium *(2nd ed., Zurich, Christoph Froschauer: 1560), p. 96.*

FIGURE 2.13 *Guinea pig* (Cuniculus Indicus). *From: Gessner,* Icones animalium *(2nd ed., Zurich, Christoph Froschauer: 1560), p. 106.*

For early modern Europeans, these new animals formed more than a little bit of a problem for animal classification. How should one define species that were hitherto unknown, and how should one explain their existence? According to the common Christian dogma, and the *communis opinio*, God was responsible for the species and had created them at the origin of all life, viz. when he had created heaven and earth and man (as it is described in *Genesis*); and he had created them in the Earthly Paradise, which was located in the "Old World", in Asia Minor. From this perspective, it was not easy to understand why the animals/species of the New World differed greatly from the Old World. People were not inclined to assume that God had created a whole set of animals a second time, in another place. It was commonly accepted that the species originated from a single act of creation, and from the Paradise they spread all over the world. But obviously there were a large number of animals in the New World that did not appear in the Old World. How come? The concept of hybridisation was a possible answer, and so was the concept of *generatio spontanea*. Via interspecific hybridisation of the well-known "old" species, new species could come into being. If one argued along these lines, the miraculous thing was, of course, that a whole fauna of hybrid species opened up to eyes of the Europeans. The *generatio spontanea* was another means to explain the existence of the strange animals. *Generatio spontanea* could take place at any place, at any time, and it could bring forth animals that were hitherto unknown. Since the *generatio spontanea* was considered to be created by God, it was theologically acceptable. The problem was only that it normally related only to small and very small species, such as insects, toads, moles, and mice. It was difficult to believe that it brought forth animals such as the armadillo, jaguar, sloth, and llama. Another great problem was the big Flood, one of the major events of *Genesis*. According to the Bible God had decided to kill all life and to preserve only good man (Noah and his family) and pairs of the known animal species on Noah's ark. When the water receded, the ark landed on the mountain of Ararat (Ağrı Dağı) in Turkey, and the preserved animals procreated immediately (it was springtime) and spread themselves again over the whole earth. If God had created the rare American animals in his original creation, how did they get into Noah's ark, and then, after the Flood, back to South America? Almost all quadrupeds were unable to swim long distances, let alone over the ocean. If they were unable to cross the ocean, how could they have survived? It was generally accepted that no quadrupeds survived the Flood except the species taken by Noah into his ark. Possible answers were, again, hybridisation and spontaneous generation, but the problems remained the same as they were with respect to the original creation. There were even more problems with hybridisation: was it foreseen by God when he created the species? Did he not explicitly forbid it, as (seemingly) indicated in the Bible (*Leviticus* 19, 19)? If hybrids do not belong to God's creation, they must be "contra naturam". Were they sinful? Did they come into existence after the fall of Adam and Eve?

Angels, the Guardians of the American Species: Nieremberg's *Historia naturae, maxime peregrinae* (1634/1635)

Juan Eusebio Nieremberg, S.J. (1595–1658), a theologian, prolific author of devotional treatises and various scholarly works, and a professor who for many years taught natural history and humanities at the Colegio Imperial Royal at

Madrid,[197] wrote an impressive natural history focusing on exotic regions, especially the New World: *Historia naturae, maxime peregrinae*.[198] Nieremberg believed in both hybrid speciation and spontaneous generation.[199] In his opinion, neither was "contra naturam"; both were part of God's creation,[200] and hybrids were not sinful. His main argument is fertility. If hybrids are fertile, it is proven that they are part of God's creation.[201] Fertility is a gift of God and can only belong to the godly created nature. Accordingly, he was not inclined to believe that hybridisation was invented by man—for example, by Ana—as the Bible seems to tell us. Nieremberg uses Bible philology to contradict the widespread opinion that Ana was the inventor of cross-breeding. In a

197 For Nieremberg, cf. Calvo M.J.Z., "Muerte, alma y desengaño: las obras latinas del padre Nieremberg", *Revista de Humanidades: Tecnológico de Monterrey* 21 (2006) 15–121; Didier H., *Vida y pensamiento de Juan E. Nieremberg* (Madrid: 1976; original ed. in French 1974); and Pérez Goyena A., "Juan Eusebio Nieremberg y Otin", in *Catholic Encyclopedia* 11 (1913). Nieremberg joined the Jesuit order in 1614. Among the devoted treatises he composed before his natural history are *Sigalion sive de sapientia mythica* (Madrid: 1629); *De adoratione in spiritu et veritate* (Antwerpen: 1631); *De arte voluntatis* (Lyon, Laurentius Anisson: 1631); *De la afición y amor de Jesus*, and *De la afición y amor de María* (both Madrid: 1632); *Vida Divina y Camino Real de Grande Atajo para la Perfección* (Madrid: 1633); *Libro de la vida de Jesús crucificado, impreso en Jerusalén con su sangre* (Barcelona: 1634). Among his works as a naturalist and natural philosopher are: *Prolusión a la doctrina e historia natural* (Madrid: 1629); *Curiosa Filosofía y cuestiones naturales* [...] (Madrid: 1630); *Oculta Filosofía* (Barcelona: 1645); *Curiosa y oculta filosofia: primera y segunda parte de las maravillas de la naturaleza, examinadas en varias questiones naturales* [...] *Tercera impression añadida por el mismo autor* (Alcalá, María Fernández: 1649); *Obras filosóficas del Padre Juan Eusebio Nieremberg* (Madrid: 1664).

198 *Historia naturae, maxime peregrinae* (Antwerp, Balthazar Moretus: 1635); for Nieremberg's zoology, cf. De Asúa – French, *A New World of Animals*, the passage on "Juan Eusebio Nieremberg and his *Historia naturae maxime peregrinae*" 162 ff.

199 *Historia naturae, maxime peregrinae*, esp. book V, chapt. 21, pp. 88–89; chapt. 24 "De adulterinis animalibus", pp. 90–91, and book VI, chapt. 8, p. 95 "De spontanea genesi animalium".

200 Ibidem, pp. 88–89.

201 Esp. book V, chapt. 21 "An ulterinae naturae a Deo conditae in prima productione": 'Mihi videtur lis simili dirimenda regula qua usus sum in sponte genitis animalibus: si quae omnino sterilia sunt, nulla naturae vi producere genus suum valentia, ista opinor tum non erupisse, cum divinae benedictionis, secundantis naturas, sint expertia; quae autem effeta visa, absque dubio effusa pridem a terra fuere [...]' ('In my opinion, this question should be solved according to the same principle I applied to the question of spontaneous generation. If certain animals are completely sterile, and are unable to procreate in a natural way, they were, as I think, no part of the creation, since they lack God's benediction which favours nature. Fertile animals, however, doubtless belong to the creation [...]').

Hebrew version the Bible has that Ana 'brought his donkeys to the horses (jemin)'; Nieremberg argues that the correct Hebrew text should not be 'jemin' ('horses'), but 'jamin' ('waterholes'); and that 'jamin' would make much more sense because Ana lived in the desert,[202] ergo: 'Ana brought his donkeys to the water holes'. Thus, cross-breeding is not an invention by man, but part of nature or of God's creation.

In this vein, Nieremberg welcomes a considerable number of "Old World hybrids" as part of God's creation: the *Leopard*, the *Lynx* (brought forth by male wolf and hind [*cerva*]),[203] the *Arcadian dog* or *Indian dog* (a cross-breeding of female dog and male tiger), the *Bactrian Camel* (*Camelus Bactrianus*, brought forth by camel and wild boar), the *Hippelaphus* (cross-breeding of horse and deer), and the *Leontomix* (brought forth by female dog and lion).[204] Their partly horrifying appearance is no proof that they were conceived in sin, although Nieremberg calls them 'adulterous animals'. Original sin is no prerequisite for hybridisation.

After this strong argument in favour of hybridisation one might expect that Nieremberg would primarily use this concept to describe the species of the New World. Interestingly enough, this is not the case. In general, he acknowledges the unknown and strange animals as *species in their own right* (!), without depicting them as a mixture of different parent species. This is, of course, a great achievement in the field of taxonomy. Nieremberg even installs them with a proper name, mostly more than one. This is because he normally first mentions their Indian name. In this way, a big number of new species with new, strange names occur: the *Ocotochtlo*, a smaller feline of South America; the *Ocelot* (rendered in Latin by Nieremberg as '*Pinuum dasypos*', mod. *Leopardus pardalis*);[205] the *Hoitzaquatzin* (one of the *Ererhizontidae* species);[206] the *Tlaquatzin*, a species of the large *Opossum* family (mod. family name: *Didelphidae*),[207] a large order of marsupials common in North and Central America; the *Aiatochtli* or *Tatou* (the Armadillo);[208] the Haut (Sloth);[209] and so on.

---

202 Ibidem, p. 89.

203 Both ibidem, p. 88.

204 All ibidem, p. 89.

205 Ibidem, IX, 1, p. 153. The ocelot lives in the rainforests of the middle and northern parts of South America, and further those of Central America, as well as some parts of Mexico, in 10 subspecies. Its length varies from 68 to 100 cm, its weight from 8 to 18 kg.

206 Ibidem, IX, 2, p. 154.

207 Ibidem IX, 4, p. 154.

208 Ibidem, IX, 6, p. 158 ff.

209 Ibidem, IX, 13, pp. 163–166.

According to Nieremberg, they were all part of God's creation. But if so, how did they come to South America? Nieremberg's answer is: they were brought by God's angels.[210] He points to the fact that not all animal species are present in all of the countries all over the world. For example, the *Sphinx*, the *Cepus*, the Giraffe, and the even-toed wild horses live only in Africa.[211] Thus, God's angels must have brought them from Paradise to their proper regions, in this case Africa. When the Flood came, the angels took action again and brought a number of exemplars of the various "regional" species to the ark. After the Flood was over, the angels flew them back to South America, Africa, and other faraway regions. In Nieremberg's zoology, angels function as a kind of powerful, large airplanes that unite the continents, especially to regulate the fauna and flora. Nieremberg calls the angels the guardians of the natural species ('angelos *specierum naturalium* custodes').[212] Without the angels, many species would have been extinguished. There is also a second variant, in which, however, the angels play the same role. This is connected to the climate theory, according to which different climatic zones bring forth different faunas and floras. In this variant, the South American species ori.ginated from the Old World species. The angels transported Old World species to America: under the influence of a totally different environment they changed into new variants, sometimes even new species. The angels would eventually offer the new species a return ticket to Noah's ark.

In taxonomy, the acknowledgement of a number of unknown South American animals as new species was an important development. But perhaps equally important is the fact that the unknown creatures sharpened Nieremberg's sense for *common elements* and stimulated him to construct *families* of animals, which united a number of *species*. For example, he brings together nine species of 'dasypodes' in one family:[213] the *Pactli, Elitactochtli, Tuitlatepolli, Tocanthochtli, Quanutochtli, Metochtli, Cacatochtli, Tuitlatepolli alter*, and *Hapaztochtli*. The same goes for, among others, the *Felidae*, of which Nieremberg constructs certain groups, such as "leones Indici", "animalia leoni affinia", or "leonum parvorum feliumque genus".[214] In a sense, the last family resembles the modern classificatory category of the *smaller Felidae*. But, of course, Nieremberg's "families" are not always similar to the families of

210 Ibidem V, 27, p. 91.

211 Ibidem V, 28, pp. 91–92.

212 Ibidem V, 27, p. 91.

213 Ibidem IX, 7.

214 Ibidem, IX, 21–22, p. 169.

modern taxonomy. The modern *Dasypodidae* family lists 20 or 21 *armadillo* species, whereas Nieremberg's 'dasypodes' primarily collect various relatives of the 'Cuniculus Indicus', or the guinea pig.

Interestingly, although Nieremberg frequently applied a classificatory system of "family" and "species", he nevertheless also used the concept of hybridisation for the description of certain New World species. This goes, for example, for the cougar (*Puma concolor*): Nieremberg interpreted this species as a hybrid of lion (mod. *Panthera leo*) and leopard (mod. *Panthera pardus*).[215] Here we get an interesting glimpse of his application of empirical evidence. Nieremberg narrates that a pregnant 'Indian lion' (i.e. cougar) was killed by Indians. When its belly was cut open it turned out that the cubs' fur had a stippled pattern similar to that of a leopard's. Nieremberg regarded this as proof that the father of the whelps was a panther, and thus proof of the animal's hybridity. Whereas Nieremberg classified the frequently misinterpreted giraffe and zebra as species of their own,[216] in his description of American species he included a number of new hybrids: the *Macamitzli*, a hybrid of lion and deer (Lion-deer); the *Cuitlamitzli*, of lion and wolf (Lion-wolf);[217] the *Tlalmitzli*, of a small wildcat and a lion; and the *Su*, a monstrous hybrid creature with the features of lion, man, and squirrel [cf. Fig. 2.14],[218] but possibly inspired by one of the *Opossum* species. The illustration presented in Nieremberg's zoology suggests that the *Su*'s whelps greatly resemble squirrels, whereas the grown-up animal more closely resembles the lion. This illustration of the *Su* had already been published in a zoology more 70 years earlier, in the second edition of Gessner's *Icones animalium* (1560), in the appendix ("additiones").[219] In the German edition of the *Icones animalium* of 1563, in the *Thierbuch*, it was even used as the title page illustration [Fig. 2.14].

215 IX, 21 "De leonibus Indicis", p. 169.

216 IX, 72, p. 191: "De graffa" [sic]. According to Nieremberg, the animal lived (only) on the island of Zanzibar; in chapter IX, 17, Nieremberg argues that the zebra could not be a mule-like hybrid, as many supposed, because the animal was extraordinarily fertile: 'parit quotannis, unde et maxima habetur copia' ('it [the zebra] gives birth each year, and that is why there is an enormous amount of them').

217 IX, 21, "De animalibus leoni affinibus", p. 169.

218 IX, 74 "De su animali", p. 189.

219 P. 127.

Thierbůch

Das ist ein kurtze bschreybung aller vierfüssigen Thieren/ so auff der erdẽ vnd in wassern wonend/sampt jrer waren conterfactur: alles zů nutz vñ gůtem allen liebhabern der künsten/ Artzeten/ Maleren/Bildschnitzern/Weydleüten vnd Köchen/gestelt. Erstlich durch den hochgeleerten herren D. Cůnrat Geßner in Latin beschriben/ yetzunder aber durch D. Cůnrat Forer zů mererem nutz aller mengklichem in das Teütsch gebracht/vnd in ein kurtze komliche ordnung gezogen.

Mit Keyserlicher Maiestat freyheit/in acht jaren nit nachzetrucken/bey peen vnd straaff acht Marck lötigs golds/nach laut deß Originals.

Getruckt zů Zürych bey Christoffel Froschower/im Jar als man zalt M. D. LXIII.

FIGURE 2.14 *The* Su. *From: Gessner, Title page of the* Thierbuch (*Zurich, Christoph Froschauer: 1563*).

The Flood and the Survival of the Species: Athanasius Kircher's *Arca Noe* (1675)

A major achievement of Nieremberg's *Historia naturae* was the inclusion of the American species in animal classification; a less strong point was the explanation of how the American species came to the ark and then went back to their continent (by God's angels). The Jesuit polymath and universal scholar Athanasius Kircher (1602–1680),[220] who had taught mathematics, physics,

220 On Kircher, cf. Findlen P. (ed.), *Athanasius Kircher: The Last Man Who Knew Everything* (New York: 2004); eadem, "Science, History, and Erudition: Athanasius Kircher's Museum at the Collegio Romano", in Stolzenberg D. (ed.), *The Great Art of Knowing: The Baroque*

and Oriental language at the Collegio Romano (from 1638 on), focused his zoological thinking on the problem of the ark. In his *Arca Noe* (1675) he tried to explain which species survived in which ways.[221] Kircher was not inclined to give to the animals of the New World an important status. His zoological thoughts, in fact, are more oriented toward the animals of the Old World. In general, his zoology is very much determined by a Europe-centred, imperialistic ideology. He was not interested in a differentiated taxonomical system like some of his early modern forerunners were, including Wotton, Aldrovandi, and Nieremberg; nor did he aim at a full description of the singular species as offered by Gessner, Aldrovandi, Nieremberg, and Jonston. With the quadrupeds, Kircher's way of ordering the animals very much resembles Pliny's: from larger to smaller species. Kircher starts with elephant and camel,[222] and ends with weasel, rat, hedgehog, and small monkeys (*cercopitheci*);[223] Pliny started with the elephant, and he ended with the mouse. Furthermore, Kircher divides the quadrupeds into a number of subcategories, such as carnivores vs. planteaters; land animals vs. water animals and amphibians; hybrid species;[224] and 'clean' ('mundi') vs. 'unclean' ('immundi') species. These subcategories, however, were not based on morphology or physiology, but according to their keeping on the ark. For example, it seemed wise to separate the plant-eaters from the carnivores, the small and harmless animals from the big and dangerous species, and the clean from the unclean species. For Kircher, taxonomy was

---

Encyclopedia of Athanasius Kircher (Stanford – Fiesole: 2001); Godwin J., *Athanasius Kircher's Theatre of the World* (London: 2009); idem, *Athanasius Kircher. Ein Mann der Renaissance auf der Suche nach verlorenem Wissen* (Berlin: 1994); Fletcher E. (ed.), *A Study of the Life and Works of Athanasius Kircher, "Germanus Incredibilis": With a Selection of His Unpublished Correspondence and an Annotated Translation of His Autobiography* (Leiden: 2011); Merrill B.L., *Athanasius Kircher (1602–1680), Jesuit Scholar* (Brigham Young University Library: 1989); Leinkauf Th., *Mundus combinatus. Zur Struktur der barocken Universalwissenschaft am Beispiel Athansius Kircher, S.J. (1602–1680)* (Berlin: 1993).

221 *Arca Noe in tres libros digesta* [...] *quae omnia nova methodo necnon summa argumentorum varietate explicantur et demonstrantur* (Amsterdam, Johannes Jansonius: 1675, first and only edition); for Kircher's work, cf. De Asúa – French, *A New World of Animals* 170 ff., the passage "Athanasius Kircher: Noah's Ark, a Wonrous Museum and Animal Magnetism"; Cameron A.D., *The Legend of Noah: Renaissance Rationalism in Art, Science, and Letters* (Illinois U.P.: 1963), and for Kircher's palaeontological views, S.J. Gould, "Father Athanasius on the Isthmus of a Middle State: Understanding Kircher's Paleontology", in Findlen (ed.), *The Last Man Who Knew Everything* 207–238.

222 Ibidem, p. 57.

223 Ibidem, pp. 65–67.

224 Ibidem, pp. 49 and 56; hybrid species Kircher defines (p. 49) in the following way: 'quae ex diversarum specierum commixtione originem suam trahunt'.

not a goal in itself. Instead, he developed a number of strategies of zoological explanation that were connected to the ark and its logistics, and aimed at a certain efficiency with respect to the preservation of the species that were part of God's original creation. These strategies were, among others, a maximised *generatio spontanea*, maximised hybridisation and hybrid speciation, and an emphasis on climate theory.[225] If certain species were able to procreate via *generatio spontanea* or hybrid speciation, it was not necessary to take them into the ark.[226] For example, Kircher thought that all insects and most of the reptiles, but also mice, moles, frogs, and other small creatures, came into being via *generatio spontanea*. Thus, he excluded all insects and most reptiles from the ark.[227] Whereas the Bible had mentioned *expressis verbis* the *reptilia* as legitimate passengers, Kircher excludes most of them. He authorises this by 'experiments' of *generatio spontanea* he had presented in his work *De mundo subterraneo*.[228] According to Kircher, all parts of the body of snakes and reptiles contained sperm. If one cuts a snake into small parts, after a few days small worms will creep out of the flesh. The kind of proof he offers (experimental) seems to suggest scientific progress; his conclusions, however, do not point in that direction.

Furthermore, Kircher excluded from the ark species he identified as 'hybrid' (= 'ex diversarum specierum commixtione'). Interestingly, he considered the number of hybrids to be *unlimited*, and thought that India and the New World especially had an abundance of them.[229] He discussed only a relatively small number of them[230]—among others the giraffe,[231] the *Leucrocuta*,[232] the leopard, the *Tragelaphus*, the *Hippelaphus*,[233] and the Horse-panther or Panther-horse

225 On the climate theory, ibidem 49–50.

226 Cf. book I, chapt. 3 "De omnibus et singulis speciebus, Animalium Quadrupedum, quae in Arcam introduci debeant", ibidem, p. 47 ff.

227 On the exclusion of insects, ibidem, pp. 51–52; on the exclusion of reptiles and Kircher's idea that most of them procreate via *generatio spontanea*, ibidem, p. 52 ff.

228 Ibidem, p. 53.

229 Ibidem, p. 67: 'Cum infinitae propemodum animalium species sint ex promiscuo diversarum specierum coitu productae, illas omnes intra Arcam introductas non est verisimile [...]' ('Because there are an endless number of species brought forth from the mating of different species, it is not likely that they all were taken into the Ark').

230 Ibidem, pp. 67–73.

231 *Arca Noe*, p. 68 'Animal ex pardo and camelo natum' ('An animal born from a panther/leopard and a camel').

232 Ibidem, p. 70.

233 All three ibidem, p. 68.

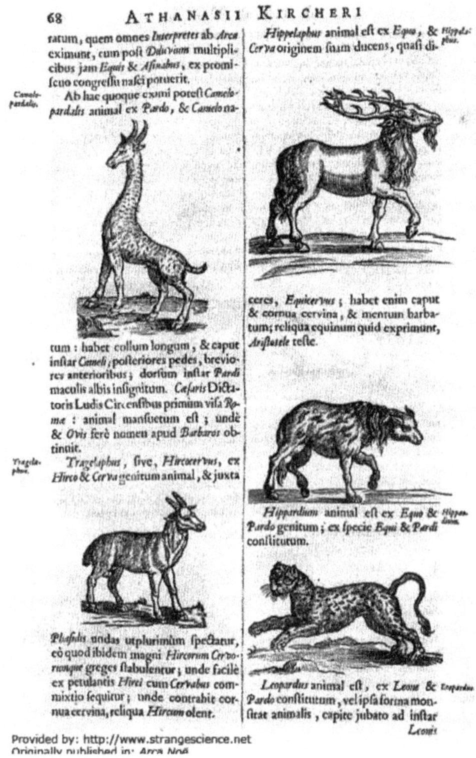

68 ATHANASII KIRCHERI

ratum, quem omnes *Interpretes* ab *Arca* eximunt, cum poſt *Diluvium* multiplicibus jam *Equis* & *Aſinabus*, ex promiſcuo congreſſu naſci potuerit.

Ab hac quoque eximi poteſt *Camelopardalis* animal ex *Pardo*, & *Camelo* natum: habet collum longum, & caput inſtar *Cameli*; poſteriores pedes, breviores anterioribus; dorſum inſtar *Pardi* maculis albis inſignitum. *Cæſaris* Dictatoris Ludis Circenſibus primùm viſa *Romæ*: animal manſuetum eſt; undè & *Ovis* ferè nomen apud *Barbaros* obtinuit.

Tragelaphus, ſive, *Hircocervus*, ex *Hirco* & *Cerva* genitum animal, & juxta *Phaſidis* undas utplurimùm ſpectatur, eò quod ibidem magni *Hircorum Cervorumque* greges ſtabulentur; unde facilè ex petulantis *Hirci* cum *Cervabus* commixtio ſequitur; unde contrahit cornua cervina, reliqua *Hircum* olent.

Hippelaphus animal eſt ex *Equo*, & *Cerva* originem ſuam ducens, quaſi diceres, *Equicervus*; habet enim caput & cornua cervina, & mentum barbatum; reliqua equinum quid exprimunt, *Ariſtotele* teſte.

Hippardium animal eſt ex *Equo* & *Pardo* genitum; ex ſpecie *Equi* & *Pardi* conſtitutum.

Leopardus animal eſt, ex *Leone* & *Pardo* conſtitutum, vel ipſa forma monſtrat animalis, capite jubato ad inſtar *Leonis*

FIGURE 2.15 *Selection of Kircher's hybrids. From: Athansius Kircher,* Arca Noe *(Amsterdam, Johannes Jansonius: 1675), p. 68.*

(*Hippardium*)[234]—but he seems to have used them more or less only as a kind of exemplary evidence [Fig. 2.15]. Apparently he considered the kind and the number of combinations of species to be unlimited: carnivores might mate with plant-eaters, land animals with water animals, and very large species (such as the camel) with much smaller ones (such as the panther). Are hybrids "contra naturam", or are they part of nature? Kircher does not answer this question explicitly. But since he considered their number so big—before and after the Flood—it is clear that he must have accepted hybridisation as an important means of procreation, and thus of God's creation. What God created in this respect was, however, only the power and potential

234 Ibidem: 'Hippardium est animal ex equo et pardo natum' ('The Hippardium is an animal born from a horse and a panther/leopard').

to procreate "inter species", not the various hybrids themselves. According to Kircher, God has created exclusively 'pure' and 'perfect' species; the hybrids had come into existence later on, mostly 'per coitum'.[235] The manner of presentation also suggests that Kircher considered the hybrids as species (not as singular results of hybridisation), and thus to be fertile. It is a telling fact that Kircher considered a large number of the animals of the New World as hybrid species;[236] this was, of course, one of the instruments used to explain their existence, and at the same time a means to devaluate them as 'imperfect' and 'impure' species. For example, Kircher regarded the *Armadillo* as a hybrid of hedgehog and turtle (!), and the *Marmota* (*Alpine Marmot*) one of badger and squirrel.[237] Sometimes Kircher lists a New World animal among the hybrids, although he could not say what its different "species parents" were, for example the *Corcobado* (i.e. the American Bison, mod. *Bison bison*). In such cases, Kircher ascribes the strange, in his eyes degenerated, features of the animal to influences of climate and environment.[238]

Indeed, another way of reducing the new species to a minimum was the *climate theory*. According to this traditional model of biological explanation the climate zones were of pivotal importance to the fauna. Climate was supposed to influence the kind, size, and shapes of the animals. Kircher applied this stratagem in the following way: the "true" and real species belonged only to the moderate climate zone of the Old World; they originated in God's creation, which took place in the ideal climate of Paradise. God ordained to the animals to spread all over the world. In other, more extreme climate zones, however, the species took on different features: under the influence of different environmental and living conditions, the animals changed their features and habits.

---

235 Ibidem, p. 67 ff.

236 Ibidem, pp. 69–70.

237 Ibidem.

238 Ibidem, p. 70: 'Ego sane non facile assererem, id ex diversae speciei congressu natum, sed esse bovinae speciei animal, vel coeli influxu peculiari, aut climatis locique natura, uti fere omnia animalia Americae, in hanc formam transmutatum fuisse' ('[In this case] I could not easily make plausible that the animal was brought forth by the mating of two different species, but rather [I could argue] that the animal [bison] belongs to the species of the *Bovinae*, and that it had changed its form either by the influence of the stars or by the character of the region or the climate, as have almost all the species of America'). Ironically, in the evolutionary history of the *Bison bison*, cross-breedings between various species of *Taurids* and *Bisonids* play an important role. Cf. "Maternal and Paternal Lineages in Cross-Breeding Bovine Species. Has Wisent a Hybrid Origin?", in Mbe.oxfordjournals.org. 2004-01-22; Groves C.P., "Systematic relationships in the Bovini (Artiodactyla, Bovidae)", *Zeitschrift für Zoologische Systematik und Evolutionsforschung* 4 (1981) 264–278.

For example, the cow, transferred to North America, became a bison, and in Scandinavia an elk; the European wolf, transferred to the high North, became a '*gulo*' (i.e. the wolverine, glutton, or skunk bear, mod. *Gulo gulo*);[239] in the New World, the dog (*canis*) lost its hair; the birds changed their colours; etc.[240]

In this way, Kircher frequently explains the new species of the New World as "adaptations" of the old, godly created species to a different environment. Therefore, he considered it unnecessary to take American species on the ark. However, Kircher does not give a positive judgement on the new species: it is true that they represent a kind of adaptation, but more for the worse than for the better. Instead of adaptation, Kircher frequently uses the term 'degeneration'. When the sheep came to the New World it changed its features in such a way that you cannot recognize the animal anymore: it 'degenerated' to such a degree that it got a bump.[241] When they came to the far North, many species became white. It is a telling detail that Kircher considered this clever adaptation of nature as *degeneration*, viz. a *loss* (of colour). For example, for a bear brown would be the appropriate colour; the Polar bear with his white hair Kircher considered to be a degeneration of the European brown bear. Since for Kircher adaptation also meant degeneration,[242] he did not consider it worthwhile to keep the mutated species. When the Flood came, at least in Kircher's zoology, all of the hitherto 'adapted' American species died. When the Flood was over, the old species wandered back into the Americas, and once more they underwent a metamorphosis into newly 'adapted' species.

But how did the animals reach South and North America? Here Kircher preferred a more "natural" explanation than Nieremberg's guardian angels. He combines three explanations: 1) by swimming from the continent to an island, and from one island to the other; 2) through land bridges or isthmuses; and

---

239 Ibidem, pp. 50 and 71. The wolverine or *Gulo gulo* indeed lives in the Northern hemisphere (Canada, Northern Europe, Russia). He belongs, however, to the family of the *Mustelidae* (weasels), not to the wolves (species *Canis lupus*) or to the family of the *Canidae*. Apparently Kircher did not know that the wolf (*Canis lupus*), especially in its subspecies the Eurasian wolf (*Canis lupus lupus*), also populated the most Northern regions of Europe and Russia, and that its population in the Northern regions in the 17th century was already much denser than it was in the mild or hot climate zones.

240 Ibidem, p. 49.

241 Ibidem.

242 For 'degeneration' in the species debate cf. Smith J.E.H., "Degeneration and Hybridism in Early Modern Species Debate: Towards the Philosophical Roots of the Creation-Evolution Controversy", in Wolfe Ch.T. (ed.), *Monsters and Philosophy* (London: 2005) 109–130.

3) brought by human beings on ships.[243] Of course, in 1675 the whole world was not perfectly mapped out, and it was still possible for new islands to be discovered. Obviously, Kircher thought that there might be more islands and land bridges than hitherto known.

Like those of many others, Kircher's zoological thoughts were inspired by and centred around theology. He tried to explain in which ways the Flood determined the species as they occurred in his time. Of course he was still far away from Darwin, and evolution was not yet one of the basic paradigms of zoology. Nevertheless, theological thinking brought him to the concept of 'adaptation' of species as an answer to environmental circumstances, which is fundamental to evolutionary theory. Also, the idea of migration of species via land bridges perfectly fits with modern zoology. For example, the *Felidae* are thought to go back to one single ancestor (*taxon*) who lived some 10–15 mio. years ago and originated in Asia; from there, he spread across continents via land bridges (e.g. the Panama land bridge), with the exception of Australia, of course. Would it be adequate to regard Kircher's zoology—in comparison with that of his forerunners—as progressive? This is not entirely plausible. First, one must take into account the importance he attached to the concepts of hybridisation, hybrid speciation, and *generatio spontanea*. Whereas some of his forerunners—such as Franzius, Aldrovandi, Nieremberg, and Jonston—did their best to restrict hybridisation and hybrid speciation, Kircher acknowledges 'countless' hybrid species. That he even lists the giraffe among the hybrids has more the flavour of an old-fashioned, outdated zoology. Whereas Franzius did not believe in *generatio spontanea*, Kircher maximised it as a relevant way of procreation. Whereas in modern evolutionary theory *adaptation* is seen as "the" means to make a species better, fitter, and stronger, Kircher devalued it as 'degeneration' from the supposedly "true", i.e. godly created, species. In sum, a mixed picture emerges that does not favour a teleological approach, if one looks at the history of early modern zoology. With respect to his main paradigms, Kircher has firm roots in medieval thought, especially with his emphasis on unlimited hybridisation.

243 Ibidem, p. 196. On Kircher's isthmus theory, cf. S.J. Gould, "Father Athanasius on the Isthmus of a Middle State: Understanding Kircher's Paleontology", in Findlen (ed.), *The Last Man who Knew Everything* 207–238.

Extended Range of Hybridisation: More Philosophical and Theological Explanations, and Early Modern Treatises on Monsters

So far, we have dealt with the paradigm of hybridisation via coitus, i.e. the mating of different species, and also marginally via *generatio spontanea*, and we have also observed in which ways Aristotle's limitation "rules" of hybridisation (only between closely related species, species of a similar size, and species with the same length of pregnancy) were applied, creatively adapted, or ignored. But in the early modern period, as well as in the Middle Ages, many naturalists went far beyond these Aristotelian limitations. Not in the least because of Aristotle himself, by his theory of *prenatal epigenesis*: the idea that the distinctive features of a species emerge only during the pregnancy, in the development of the embryo and foetus. Aristotle thought that the egg develops from formless *materia* to a complex, specific organism. In medieval scholasticism *prenatal epigenesis* was a widely respected theory. Most importantly, *prenatal epigenesis* allowed acceptance of a very broad spectrum of variants of organisms or hybrids.

Medieval thinkers and zoologists have elaborated much upon the reasons behind hybrids, monsters, and prodigies.[244] Albertus Magnus (d. 1280), the author of an important zoology,[245] gave a number of reasons:[246] poor disposition of the *materia*, which lacks corrective resistance and simply takes over the formative quality of the sperm; insufficient sperm, which fails to transmit its formative quality; inharmonious division of the humours (*improportionalitas qualitatum*); imaginations of the mother animal; or the influence of the

---

244 Verner L., *The Epistemology of the Monstrous in the Middle Ages* (New York – London: 2005); Williams D., *Deformed Discourse. The Function of the Monster in Medieval Thought and Literature* (Exeter: 1996); Wilson D., *Sign and Portens. Monstrous Births from the Middle Ages to the Enlightment* (London – New York: 1993); Kappler C., *Monstres, démons et merveilles à la fin du Moyen Age* (Paris: 1980); Cohen J.J., *Hybridity, Identity, and Monstrosity in Medieval Britain: On difficult Middles* (New York: 2006); idem, *Of Giants. Sex, Monsters, and the Middle Ages* (London: 1999); Gravestock P., "Did Imaginary Animals Exist?", in Hassig D. (ed.), *The Mark of the Beast. The Medieval Bestiary in Art, Life, and Literature* (New York: 1999) 119–135; South M. (ed.), *Mythical and Fabulous Creatures. A Source Book and Research Guide* (New York – London: 1987).

245 *De animalibus libri XXVI*; cf. Balss H., *Albertus Magnus als Biologe* (Stuttgart: 1947).

246 Cf. Roling B., *Drachen und Sirenen. Die Rationalisierung und Abwicklung der Mythologie an den europäischen Universitäten* (Leiden – Boston: 2010) (Mittellateinische Studien und Texte 42) 137–138.

stars.[247] Other naturalists also believed in the influence of demons. Albertus was convinced that the above-mentioned factors frequently led to hybridisation but denied that they automatically caused hybrid speciation.[248] However, Albertus Magnus's zoology, *De animalibus libri XXVI*, was full of hybrid species. Some of them went back to Pliny's *Naturalis historia*, but Albertus added a considerable number of new creatures as well: among others the *Martimorion* (cross-breeding of various species);[249] *Linciscus* or Dog-wolf (of male dog and female wolf);[250] *Hybris*, an hybrid of wild boar and pig; *Musinuo* ('ex capra et ariete'); *Cirinus* (of a woman and a billy goat);[251] *Onocentaurus* (of man and donkey);[252] *Orasius* (of deer and horse); *Pilosus* (of man and goat); and the *Papio* (of wolf and fox). In a marked difference to Pliny, Albertus also believed in hybrids of man and various animals, for example the Donkey-man (*Onocentaurus*), and the Goat-man (*Cirinus* and *Pilosus*). Albertus's *De animalibus libri* was widely used not only in the 13th and 14th centuries, but also in the age of Humanism, in the form of manuscripts and printed editions as well, and until the first half of the 16th century, it retained its status as an authoritative text. In the 15th and 16th centuries, it was available not only Latin editions, but also in vernacular translations, enriched with illustrations. It is a telling detail that on the title page of the 1545 edition two hybrids appear next to the "kings" of the animals (the eagle and the lion): the basilisk and a kind of Sea-wolf [Fig. 2.16]. Other medieval zoologists, such as the influential Avicenna[253] or Roger Bacon,[254] also believed in hybrids resulting from human beings and all kinds of animals. As a result, in scholastic physics hardly any limits in construing and describing hybrids remained. From the High Middle Ages on, almost every hybridisation was possible.

Interestingly, early modern zoology did not abandon the thoughts of the above mentioned Mediaeval thinkers, but continued to think along these lines. In the very period when early modern zoology flourished (1550–1675), there came into being a hype in the interest in monsters: a veritable science—

247 Albertus Magnus, *Physica*, ed. P. Hossfeld, 2 vols. (Münster: 1987), book II, tract. 3, cap. 3 (pp. 136–138).

248 Roling, *Drachen und Sirenen* 138–139: 'Keine neue Spezies entstand jedoch auf diese Weise, sondern [...] eine mangelhafte Kreatur [...]'.

249 Cf. *De animalibus*, Mantua 1479, fol. CCLXVIIr.

250 Ibidem.

251 All three ibidem, fol. CCLXVIr.

252 Ibidem, fol. CCLXVIIv.

253 *De natura animalium* (e.g. idem, *Opera philosophica* [Venedig: 1508]).

254 Roling, *Drachen und Sirenen* 139–140.

Thierbuch.
Alberti Magni / Von Art Natur
vnd Eygenschafft der Thierer / Als nemlich Von Vier
füssigen / Vögeln / Fyschen / Schlangen oder kriechenden
Thieren / Vnd von den kleinen gewürmen die man
Insecta nennet / Durch Waltherum
Ryff verteutscht.

Mit jhren Contrafactur
Figuren.

Hierinn findestu auch viel Artznei krancker Roß vnd anders haußuiehes
Auch wider die schedliche gifft der Schlangen vnd
anderer gewürme.

Begnadet mit Keyserlicher freiheit in Syben
Jar nit nach zutrucken.

FIGURE 2.16 *The basilisk and the sea-wolf. Title page of Albertus Magnus,* Thierbuch *(1545).*

teratology—developed,[255] and extensive treatises on *monstra* were composed, e.g. Ambroise Paré's (1510–1590) *Le livre des monsters et prodiges* (1573),[256] Martin Weinrich's (1548–1609) *De ortu monstrorum commentarius* (1595),[257] Fortunio Liceti's (1577–1657) *De monstrorum natura, differentiis et caussis* (ed. pr. Padua: 1616),[258] or Ulisse Aldrovandi's (1522–1605) *Monstrorum historia* (1642–[1658])

255 Cf. Céard J., "Tératologie et Tératomancie au XVIe siècle", in Jones-Davies M.T. (ed.), *Monstres et prodiges au temps de la Renaissance* (Paris: 1980) 5–15; Calixte E., *Les monstres d'Ambroise Paré et la tératologie moderne* (Paris: 1946); Jones-Davies (ed.), *Monstres et prodiges au temps de la Renaissance*; Scarani P., "Aldrovandi e la teratologia", in Antonino B. (ed.), *Animali e creature monstruose di Ulisse Aldrovandi* (Milan: 2004) 44–55; Bates A.W., *Emblematic Monsters. Unnatural Conceptions and deformed Births in Early Modern Europe* (Amsterdam – New York: 2005); idem, "The Sooterkin Dissected: The Theoretical Basis of Animal Birth to Human Mothers in Early Modern Europe", *Vesalius* 9 (2003) 6–14; Caprotti E., *Mostri, Draghi e serpenti nelle silografie dell'opera di Ulisse Aldrovandi e dei suoi contemporanei* (Milano: 1980); Bithol-Hespériès A., "Monsters, Nature, and Generation from the Renaissance to Early Modern Period: The Emergence of Medical Thought", in Smith J.E.H. (ed.), *The Problem of Animal Generation in Early Modern Philosophy* (Cambridge: 2006) 47–64; Park K. – Daston L.J., "Unnatural Conceptions: The Study of Monsters in Sixteenth- and Seventeenth-Century France and England", *Past and Present* 91 (1991) 20–54; Park K., "Una historia de la admiración y del prodigio", in Lafuente A. – Moscoso J. (eds.), *Monstruos y seres imaginarios en la Biblioteca Nacional* (Madrid: 2000) 77–89; Ewinkel I., *De monstris. Deutung und Funktion von Wundergeburten auf Flugblättern im Deutschland des 16. Jahrhunderts* (Tübingen: 1995); Schumacher G.-H., *Monster und Dämonen. Unfälle der Natur. Eine Kulturgeschichte* (Berlin: 1996); Moscoso J., "Monsters as Evidence: The Uses of the Abnormal Body during the Early Eighteenth Century", *Journal of History of Biology* 31 (1998) 355–382.

256 *Les Oeuvres d'Ambroise Paré* (12th ed., Lyon, Jean Gregoire: 1664) pp. 644–701; modern edition by J. Céard (Geneva: 1971); cf. Calixte, *Les monstres d'Ambroise Paré et la tératologie moderne*.

257 Bratislava, Heinrich Osthaus: 1595.

258 Other editions appeared in Padua, 1634 (with illustrations), Amsterdam, 1665, and Padua, 1668; 1694; French trans. Leiden, 1708. For the work, cf. Bates A.W., "The *De monstrorum* of Fortunio Liceti: A Landmark of Descriptive Teratology", *Journal of Medical Biography* 9 (2001) 49–54; Zanca A., "Fortunio Liceti e la scienza dei mostri in Europa", in *Atti del XXXII Congresso nazionale della Società italiana di storia della medicina* (Padova-Trieste: 1985) 35–45; Roling, *Drachen und Sirenen* 140–146; for Liceti, cf. Agosto A. et al. (eds.), *IV centenario della nascita di Fortunio Liceti* (Rapallo: 1978); Ongaro G., art. "Liceti, Fortunio", in *Dizionario biografico degli Italiani* 65 (2005); Castellani C., "Le problème de la 'generatio spontanea' dans l'oeuvre de Fortunio Liceti", *Revue de synthèse* 89 (1968) 323–340; De Angelis S., "Zwischen Generatio und Creatio. Zum Problem der Genese der Seele um 1600—Rudolph Goclenius, Julius Caesar Scaliger, Fortunio Liceti", in Danneberg L. et al. (eds.), *Säkularisierung in den Wissenschaften seit der frühen Neuzeit* (Berlin – New York: 2002).

FIGURE 2.17 *Title page of Ulisse Aldrovandi,* Monstrorum historia [...] *(Bologna, Marcantonio Bernia – Nicolao Tebaldini: 1642).*

[Fig. 2.17].[259] These works were partly written by the same authors as the zoologies, by physicians and natural historians, and they belonged to the same discourse. Ulisse Aldrovandi held a professorship in Bologna, Fortunio Liceti held professorships for medicine and philosophy in Padua and Bologna, and

259 *Monstrorum historia cum paralipomenis historiae omnium animalium. Bartholomaeus Ambrosinus* [...] *composuit* (Bologna, Marcantonio Bernia – Nicolao Tebaldini: 1642 [–1658]); cf. Scarani, "Aldrovandi e la teratologia"; Antonino (ed.), *Animali e creature monstruose di Ulisse Aldrovandi*; Olmi G., *Ulisse Aldrovandi. Scienze e natura cinquecento* (Trento: 1976).

Ambroise Paré was royal surgeon to French kings.[260] The work on monsters demonstrated that the number of hybrid combinations was almost unlimited: they include hybrids of man and pig—or, more precisely, monk and pig [Fig. 2.18]; hybrid unicorns [Fig. 2.19]; the Calf-monk [Fig. 2.20]; and sometimes true *polyhybrids*, e.g. mixtures of man, bird, and carnivore [Fig. 2.21], or of donkey and bird [Fig. 2.22]. These works on *monstra*, of course, mostly deal with singular hybrids or teratogenic misfits and only rarely deal with new species, although some of them, such as the Calf-monk or Calf-man[261] and the Pig-man [Fig. 2.23], appeared more frequently. In the Middle Ages and the early modern period, most scholars ascribed to hybrids a greater fertility than is actually possible. Since Aldrovandi includes a number of monk hybrids, it may seem that he wanted to accuse them of immoral behaviour, and one may connect this with his Anabaptist sympathies. However, Aldrovandi was forced to renounce Anabaptist belief as early as 1549, and in his *Historia monstrorum* he actually interpreted the Calf-monk as a presage of Lutheran heresy.[262]

Newly discovered *monstra* appeared in works on zoology, and like those in the histories on *monstra*, they included well-known hybrid species or monstrous species of zoology, for example in Ambroise Paré's *Le livre des monstres et prodiges* from 1573, inter alia, the giraffe[263] [Fig. 2.24], the Sea-lion (described previously by Philippe Forestus),[264] the Sea-horse, Sea-elephant, and Sea-calf,[265] the crocodile,[266] whale,[267] ostrich,[268] and Chameleon,[269] but also newly discovered species, such as the *Huspalis* (also: *Upalis*) from Ethiopia, which—as it had been described by Thevet—somehow resembled the mysterious *Mantichora* or *Martiora*, and combined elements of monkey, man, and a carnivore;[270] the *Toucan*, the Bird of Paradise;[271] or the *Tanacth* from Haiti, a strange hybrid seemingly brought forth from man and a carnivore.[272]

260 Henry II, Francis II, Charles IX, and Henry III.

261 *Monstrorum historia*, p. 371; for the *calf-man*, cf. also Liceti, *De monstrorum natura* (Padua: 1634) p. 190.

262 *Historia monstrorum*, p. 372.

263 In *Les Oeuvres d'Ambroise Paré* (12th ed., Lyon, Jean Gregoire: 1664) p. 695.

264 Ibidem, p. 678.

265 For all three hybrid species, cf. ibidem, p. 679.

266 Ibidem, pp. 680–681.

267 Ibidem, p. 690.

268 Ibidem, pp. 691–692.

269 Ibidem, p. 698.

270 *Les Oeuvres d'Ambroise Paré*, p. 694.

271 For both species, cf. ibidem, p. 693.

272 *Les Oeuvres d'Ambroise Paré*, p. 696.

FIGURE 2.18 *Pig-monk. From: Ulisse Aldrovandi,* Monstrorum historia cum paralipomenis historiae omnium animalium. Bartholomaeus Ambrosinus [...] composuit *(Bologna, Marcantonio Bernia – Nicolao Tebaldini: 1642), p. 438.*

FIGURE 2.19 Unicornu Aldrovandi. *From: Ulisse Aldrovandi,* Monstrorum historia cum paralipomenis historiae omnium animalium. Bartholomaeus Ambrosinus [...] composuit *(Bologna, Marcantonio Bernia – Nicolao Tebaldini: 1642).*

FIGURE 2.20 *Calf-monk* (Vitulomonachus Sorbini). *From: Ulisse Aldrovandi,* Monstrorum historia [...] *(Bologna, Marcantonio Bernia – Nicolao Tebaldini: 1642), p. 371.*

FIGURE 2.21 *Polyhybrid. From: Ulisse Aldrovandi,* Monstrorum historia [...] *(Bologna, Marcantonio Bernia – Nicolao Tebaldini: 1642), p. 577; the same illustration as in Gessner,* Icones animalium *(2nd ed., Zurich, Christoph Froschauer: 1560), p. 95.*

FIGURE 2.22 *Another polyhybrid from: Ulisse Aldrovandi,* Monstrorum historia *[...] (Bologna, Marcantonio Bernia – Nicolao Tebaldini: 1642), p. 377.*

FIGURE 2.23 *Pig-man by Fortunio Liceti. From: Fortunio Liceti,* De monstrorum natura, differentiis et caussis *(Padua: 1616).*

FIGURE 2.24 *The giraffe. From Ambroise Paré,* Le livre des monstres et prodiges *(1573).*

The *Upalis* reappeared with the same features, inter alia, in Aldrovandi's *De quadrupedibus digitatis viviparis* of 1645 [Fig. 2.25].[273] In the same zoology are found the monstrous *Cercopithecus* or *Barbilias*, which was thought to live in China [Fig. 2.26];[274] the *Lupus Marinus or Sea-wolf* [Fig. 2.27];[275] and the small Panther (*Panther minor*) or *Pardalion*, which was conceived as a hybrid.[276] The last three monstrous creatures were already present in Gessner's *Icones animalium* of 1560, which also contained more animals that the naturalist from Zurich characterised as 'monstra', such as the *Su*, a 'horrible' 'monstrous' hybrid with the features of man, lion, and squirrel [Fig. 2.28].

An important element in Paré's *Le livre des monstres et prodiges*, as in the zoologies from Gessner on, is the massive presentation of visual evidence. In fact, almost all of the *monstra* discussed by Paré were proven by intriguing woodcut illustrations to be animals that really existed. This massive collection of visual evidence is even more important for Aldrovandi's *Monstrorum*

273 P. 261.
274 P. 249.
275 P. 176.
276 P. 68.

FIGURE 2.25 *The* Upalis. *From Ulisse Aldrovandi,* De quadrupedibus digitatis viviparis [...] *(Bologna: 1645), p. 261 (private collection).*

historia (1642): in this work the existence of hundreds of monstrous hybrids is backed by carefully carved woodcuts.

The Visual Evidence and Hybrids—A Shift from Verbal Description to Animal Illustrations: John Jonston and Matthaeus Merian (1650 ff.)

The *Historia naturalis animalium* in five volumes by the Scottish physician John Jonston (1603–1675) [Fig. 2.29], published in Frankfurt by the Heirs of Matthaeus Merian the Elder, 1650–1653,[277] was used as a zoological manual

277 *Historiae naturalis de quadrupedibus libri, cum aeneis figuris* [...] (Frankfurt a.M., Matthaeus Merian the Elder: 1650; Amsterdam, J.J. Schipper: 1657); *Historiae naturalis de avibus libri VI cum aeneis figuris* [...] (Frankfurt a.M., Matthaeus Merian the Elder: 1650; Amsterdam, J.J. Schipper: 1657); *Historiae naturalis de insectis libri II, de serpentibus et draconibus libri II, cum aeneis figuris* (Frankfurt a.M., Matthaeus Merian the Elder: 1653; Amsterdam, J.J. Schipper: 1657; idem: 1665); *Historiae naturalis de exanguibus aquaticis libri IV, cum figuris aeneis* [...] (Amsterdam, J.J. Schipper: 1657); *Historiae naturalis de piscibus et cetis libri V, cum aeneis figuris* [...] (Amsterdam, J.J. Schipper: 1657); niederländ. Übers. dieser Werke (Amsterdam, J.J. Schipper: 1660); *Theatrum universale omnium animalium, piscium, avium, quadrupedum, exanguium, aquaticorum, insectorum et anguium* (Amsterdam, R. and G. Wetstein: 1718); *A Description of the Nature of Four-footed Beasts*

FIGURE 2.26 *The* Cercopithecus *or* Barbilias. *From Ulisse Aldrovandi,* De quadrupedibus digitatis viviparis [...] *(Bologna: 1645), p. 249 (private collection).*

FIGURE 2.27 *The* Lupus Marinus. *From Ulisse Aldrovandi,* De quadrupedibus digitatis viviparis *(Bologna: 1645), p. 176 (private collection).*

NOVI Orbis regionem quandam Gigantes dicti(lingua ipsorum Patagones)incolunt. & quoniam cœlo non admodum calido fruuntur, ueſtiunt ſe pellibus animantis illius, quam Su appellant, id eſt, aquam : ab eo nimirum, quòd magna ex parte circa fluuios degat. Eſt autē omnino rapax hæc fera, & formæ monſtroſæ, qualem hîc exhibeo. Cum à uenatoribus urgetur, ſuæ pellis gratia, catulos ſuos in dorſum admiſſos cauda ampla longaq; tegit, & fuga elabitur. Itaq; dolo ſcrobe effoſſa, & frondibus obtecta, unà cum catulis capitur. Cum autem ita incluſam ſe uidet, rabie quadam ſuos catulos obtruncat & occidit; & clamore horribili ipſos etiam uenatores terret : à quibus tandem ſagittis confoſſa excoriatur, Andreas Theuetus cap. 56. Deſcriptionis Americæ.

FIGURE 2.28 *The* Su. *From: Gessner,* Icones animalium *(2nd ed., Zurich, Christoph Froschauer: 1560), p. 127.*

FIGURE 2.29 *Andreas Gryphius, Portrait of John Jonston. Engraving.*

until the end of the 18th century. With respect to taxonomy, Jonston's work was systematical but not very original: In his *De Quadrupe[t]dibus Libri IV* he used a system of classification very similar to Wotton's *De differentiis animalium* (1552), and in fact the same as Aldrovandi in his posthumously edited *Quadrupedum omnium bisulcorum historia* (Bologna: 1621; Frankfurt a.M.: 1547)

with their Figures Engraven in Brass (London, Moses Pitt: 1678); for John Jonston cf. Miller G., "Beasts of the New Jerusalem: John Jonston's Natural History and the Launching of Millenarian Pedagogy in the Seventeenth Century", *History of Science* 46 (2008) 203–243.

and *De quadrupedibus digitatis viviparis* (Bologna: 1645). Jonston's three books on the viviparous quadrupeds were divided into odd-toed ungulates (= book 1), even-hoofed ungulates (= book 2), and 'quadrupedes digitati' (= book 3), which is the same as Wotton's category 'multifida', which brought together all quadruped species with more than two toes. Jonston's fourth book is dedicated to the oviparous quadrupeds.

In Jonston's zoology many of the above-mentioned hybrids reappear, albeit partly in another way. In his verbal descriptions, Jonston is sometimes not very explicit about the traditional hybrids. Sometimes he seems to be a bit reluctant to take responsibility for these species. In such cases he says: 'According to author X or Y, species Z looks like [...]/ has [...]/ lives in [...]'. However, he does not take a theoretical stand against hybridisation or hybrid speciation, and in a number of cases he explicitly acknowledges hybridisation. For example, he clearly incorporates in his classification the hybrids of the lion with various other species: *Leopardus* (female lion with Panther/*Pardus*); the smaller, weaker and maneless lion (female lion with Panther); the *Crocuta* (female lion with male hyena); and the Lion-dog.[278] Furthermore, there are hybrids stemming from female Panther (*pardus*) and male lion, wolf, or dog,[279] much in the same way the hybrid 'smaller Panther'.[280] There are other "composite animals" or hybrids that occur in Wotton that Jonston does not characterise as hybrids, such as *Equus Indicus*,[281] *Asinus cornutus*,[282] *Mantichora* (presented as a tiger),[283] and elk;[284] there are some he does not mention at all, such as the *Eale* and *Hippelaphus*.

In a striking contrast, on the high qualilty copperplate engravings made by the famous engraver (and publisher) Matthäus Merian the Elder

278 *Historiae naturalis de quadrupedibus libri IV* [...] (Frankfurt a.M., Heirs of Matthias Merian: 1652) 115: 'Elapso primo anni tempore maribus ob nimium calorem coire non valentibus leaneae cum pardis—unde leopardi; Pantheris—unde foetus non iubatus; Hyenis [sic]—unde Crocuta, miscentur. Et ipsis cum cane commercium'.

279 Ibidem, p. 119: 'Congreditur [sc. pardus] saepe cum leone, [...], aliquando cum cane, nonumquam cum lupo [...]' ('[The panther] mates often with the lion, [...] sometimes with the dog, and sometimes also with the wolf').

280 Ibidem, p. 156.

281 Ibidem, p. 20.

282 Ibidem, p. 37.

283 Ibidem 124: 'Equidem refert ex Ctesia Pausanias esse apud Indos Mantichora dictum animal, cui triplex dentium sit in utraque maxilla ordo, missiles in cauda aculei, idque Tigrim quidam opinatur'.

284 Ibidem 97: 'ungula bifida, ut bobus'.

(1593–1650),[285] who was born in Basel but worked in Frankfurt a.M., almost all of the traditional hybrids reappear, and they provide overwhelming evidence of their existence: among others are the *Leucrocuta* and the man-eating but human-like *Mantichora* (Tab. LII) [Fig. 2.5]; the Horse-deer (*Hippelaphus*) and Goat-deer (*Tragelaphus*) (Tab. XXXIV) [Fig. 2.6]; the *Eale* (Tab. XXIV) [Fig. 2.30]; the *Bonasus* (Tab. XVIII–XIX); and so on. In a marked difference from Wotton, the *Leucrocuta* is depicted as an even-hoofed species, just as it was described by Pliny [Fig. 2.31]; obviously, Jonston and his illustrator "corrected" Wotton's misconception. The elk is presented in the illustration as a strange creature with an idiosyncratic horn, just as Wotton had described him (Tab. XXXVI) [Fig. 2.32], albeit not odd-toed;[286] but, surprisingly, on another illustration the elk resembles a kind of goat (Tab. XXX) [Fig. 2.33]. In Merian's illustrations, however, the number of unicorn species is even higher than in Wotton's *De differentiis animalium*: in total, eight species of unicorns are depicted (Tab. X–XII) [Figs. 2.34–35, Tab X and XII]. As in the classification by Wotton, the majority of the unicorns are even-toed species (the six on Tab. X

FIGURE 2.30 *The* Eale, *by Matthaeus Merian the Elder. From: John Jonston,* Historiae naturalis de quadrupedibus libri IV *(Frankfurt a. M., Heirs of Matthaeus Merian: 1652), Tab. XXIV.*

285 Wüthrich L.H., *Matthaeus Merian d. Ä. Eine Biographie* (Hamburg: 2007); idem, *Das druckgraphische Werk von Matthäus Merian d.Ä.*, vols. 1–2 (Basel: 1966), vol. 3 (Hamburg: 1993); vol. 4 (Hamburg: 1996).

286 Ibidem 97: 'ungula bifida, ut bobus'.

FIGURE 2.31 *The* Leucrocuta (here 'Leucurcuta'), *by Matthaeus Merian the Elder. From: John Jonston,* Historiae naturalis de quadrupedibus libri IV *(Frankfurt a. M., Heirs of Matthaeus Merian: 1652), Tab. LII.*

FIGURE 2.32 *The elk, by Matthaeus Merian the Elder. From: John Jonston,* Historiae naturalis de quadrupedibus libri IV *(Frankfurt a. M., Heirs of Matthaeus Merian: 1652), Tab. XXXVI.*

FIGURE 2.33 *Another elk, by Matthaeus Merian the Elder. From: John Jonston,* Historiae naturalis de quadrupedibus libri IV *(Frankfurt a. M., Heirs of Matthaeus Merian: 1652), Tab. XXX.*

FIGURE 2.34 *Unicorn species, by Matthaeus Merian the Elder. From: John Jonston,* Historiae naturalis de quadrupedibus libri IV *(Frankfurt a. M., Heirs of Matthaeus Merian: 1652), Tab. X.*

FIGURE 2.35 *Unicorn species, by Matthaeus Merian the Elder. From: John Jonston,* Historiae naturalis de quadrupedibus libri IV *(Frankfurt a. M., Heirs of Matthaeus Merian: 1652), Tab. XII.*

and XI, and one on Tab. XII), but one is odd-toed: the *Onager* on Tab. XII. And there are a number of hybrids in the illustrations that are not at all described in the text, such as the Lion-goat ('*Lea Capra*') (Tab. XLV) [Fig. 2.36], the Camel-deer (Tab. XXIX), or the *Gryphus* (Tab. XLIX).

How should one explain the contrast between text and images? In Jonston's zoology, with respect to the hybrids, the illustrations took over the presentation of evidence. The text does not contain the same argument as the images. Whereas in the text certain hybrids are not even mentioned, they are demonstrated in the images. It could very well be that for the publisher it was interesting to present these images to the reader as a way to sell the book—as eye-catching curiosities or *miracula*—and the author could have had some interest in coming up with a spectacular collection, a kind of *Wunderkammer*: the reader may admire the curious creatures, whereas for the author it was not necessary to testify to their real existence in the text.

FIGURE 2.36 Lea Capra, *by Matthaeus Merian the Elder. From: John Jonston,* Historiae naturalis de quadrupedibus libri IV *(Frankfurt a. M., Heirs of Matthaeus Merian: 1652), Tab. XLV.*

Zoology Between New Thaumatography and Neo-Scholastic Theology: Caspar Schott's *Mirabilia naturae* (1662)

For the German Jesuit Caspar Schott, S.J. (1608–1666)—as for the majority of the early modern naturalists—zoology was a religious and theological exercise. Zoology was directed toward the essential task of man, i.e. to contemplate and to admire God's creation, as Schott argues in the preface of the zoological books (VII–X) of the *Physica curiosa sive Mirabilia naturae et artis Libri XII* (1662).[287] In this respect, not a single godly created species exists that would be essentially useless. Therefore, zoology is a discipline which is not only praiseworthy from a moral point of view, but approved by God, without any restrictions.[288] God only, and nothing else, inspired the zoologists to engage in such a seemingly endless workload that demanded such scrutiny and such an amount of scholarly knowledge. Among his godly inspired forerunners, Schott lists the ancient writers Aristotle, Pliny the Elder, Aelianus, Oppianus, and Plutarch; the church fathers Basilius Magnus and Ambrosius; the scholastic theologian Albertus Magnus; and the early modern zoologists

287 Schott Caspar S.J., *Physica curiosa, sive Mirabilia naturae et artis Libri XII.* [...] *variis* [...] *disquisitionibus excutiuntur et innumeris exemplis illustrantur* [...] *Cum figuris aeri incisis* (Nuremberg, Moritz Endter: 1662), VII, p. 772; on the New World animals in this work cf. De Asúa – French, *A New World of Animals* 178 ff.

288 Ibidem, book VII, pp. 772–773.

Gessner, Aldrovandi, Nieremberg, Jonston, Guillaume Rondelet (1507–1566), Pierre Belon (1517–1564), Wotton, the Wittenberg professor Johann Sperling (1603–1658),[289] the author of the *Zoologia physica*; furthermore, there are the Italian physician, mathematician, and universal scholar Gerolamo Cardano (1501–1576); the Dutch physician and naturalist Willem Piso (Pies, 1611–1678),[290] author of the famous *Historia Naturalis Brasiliae*; and the Leiden physician Jacob de Bondt (Bontius, 1592–1631), who worked for the Dutch East India Company as ambassador in Jakarta and authored the *Historiae naturalis et medicae Indiae orientalis libri VI*.[291]

Schott does not want to repeat the huge amount of work done by his esteemed forerunners, but limits himself to the miraculous and rare aspects ('rariora et curiosiora') only. For the present purpose, Schott decided to leave out the categories of reptiles and insects, probably because these would cause a lot of work, while the results would be less spectacular. He kindly asks the reader not to insist on autopsy as the only valid source of knowledge. There is no contradiction between admiration and truth, and no reason to distrust the ancient and modern zoologists.[292] In this vein, Schott praises the eagerness and assiduity of Aristotle, Pliny, and Aelianus. The first zoologists, however, were God, who invented the species, and Adam, who gave names to all species according to their nature.[293] God was the first magister in zoology, Adam his

289 Johannes Sperling, *Zoologia physica* (posthumously edited by his former pupil, the polymath Georg Kaspar Kirchmaier, Leipzig: 1661; Wittenberg, Heirs of Johann Berger: 1669). Sperling was a naturalist, physician, and zoologist. In Wittenberg he had studied philosophy, theology, medicine, and physics, and had been a professor of physics in Wittenberg.

290 Willem Piso partook in the Brazilian expedition of Prince Maurits of Nassau (1637–1644), and authored (together with Georg Marcgrave) the *Historia Naturalis Brasiliae* (Leiden, Joannes de Laet: 1648).

291 Posthumously edited by Willem Piso (Amsterdam: 1658).

292 Ibidem, VII, p. 774: 'Nec est quod testium, quos proferam, fidem habeas suspectam, Plinii, Aeliani, aliorumque tum antiquorum tum recentium, eorum praesertim, qui novissime de Indiarum animalibus scripsere. Causa nulla subest, cur eos aut fallere voluisse aut falsos esse credamus' ('There is no reason why you should doubt the trustworthiness of Pliny, Aelianus, and the other sources of antiquity and of more recent age, and especially not that of those who very recently wrote about the animals of the Indies. There is no reason why we should think that they would have intended to cheat on us or give us wrong information').

293 Ibidem, VII, p. 776: 'En Adami sapientiam et scientiam, qui singulorum animalium naturas novit eisque congrua indidit nomina' ('Look at the wisdom and scientific knowledge of Adam, who knew about the nature of the singular animal species and gave them names according to their nature').

pupil. The inventor of animal classification was Moses, who divided the species according to their habitats in 'Aquatilia', 'Volatilia', and 'Terrestria'.[294]

Schott had been educated at the Jesuit college in Würzburg and had studied philosophy (including mathematics) and theology at the University of Würzburg (1629–1631), and theology at Caltagirone and Palermo in Sicily (1633–1636).[295] He had joined the Jesuit order in 1627, and had been ordained a priest in 1637. From 1652 on he worked as the assistant of professor Athanasius Kircher at the Collegio Romano. In 1655 he returned to Würzburg, where he composed and edited monumental works on physics (including optics, acoustics, hydraulics), mathematics, and occult philosophy, among others his four volumes on *Magia universalis naturae et artis* (1657–1659),[296] which contained a great number of physical experiments; his *Mechanica hydraulico-pneumatica* (Würzburg: 1657); *Technica curiosa* (Würzburg: 1664); and his *Physica curiosa sive Mirabilia naturae et artis Libri XII* (1662).

Schott turned out to be one of the fiercest defenders of hybrid speciation.[297] In his chapter on the 'adulterous species' ('animalia adulterina ac bigenera'), he gives one of the fullest lists of hybrid species offered by early modern zoologists:[298] the mule ('mulus', 'mula', 'hinnus'); the *Tityrus*, a crossbreeding of billy goat and female sheep; the *Musmus*, of ram and female goat; the *Hybris*, of (male) wild boar and female pig; the *Lynx*, of male wolf and hind ('cerva'); the *Thos*, of male wolf and female dog; the *Crocuta*, of male wolf and female panther; the Arcadian or Indian dog, of dog and tiger (*Panthera tigris*); the *Leontomix*, of dog and lion; the Spartan dog (*Canis Laconicus*); the *Alopex*, of dog and fox; the *Lupus carnarius*, of female wolf and male dog; the smaller and weaker lion, of female lion and male panther ('pardus'); the leopard; the *Camelus Bactrianus*, of wild boar and camel; the Bull-horse, of bull and mare; the Horse-deer, of male deer and mare; etc. Although his list is quite extensive, Schott presents it only *exempli gratia*. He thinks that the number of

294 Ibidem, VII, chapt. 2 "De differentiis animalium", p. 777; cf. *Genesis* 1, 20 und 24.

295 For Caspar Schott, cf. Duhr G., *Geschichte der Jesuiten in den Ländern deutscher Zunge*, vol. III (Munich – Regensburg: 1921) 589–592; Backer A. de – Sommervogel K., *Bibliothèque des écrivains de la Compagnie de Jésus*, 7th ed. (Paris: 1896), 904–912; Léger St., *Notice des ouvrages de G. Schott* (Paris: 1765).

296 *Magia universalis naturae et artis sive recondita naturalium et artificalium rerum scientia, cuius ope per variam applicationem activorum cum passivis, admirandorum effectuum spectacula, abditarumque inventionum miracula, ad varios humanae vitae usus, eruuntur* (Würzburg, Johann Gottfried Schönwetter: 1657–1659); vol. I *Optica*; vol. II *Acustica*; vol. III *Mathematica*; vol. IV *Physica*.

297 Schott, *Physica curiosa, sive Mirabilia naturae*, book VII, chapt. 12 "An animalia adulterina fuerint a Deo in Mundi exordio producta", pp. 821–825.

298 Ibidem, p. 821.

hybrid species is in fact unlimited. All kinds of animal species may mate with each other, and almost anything is possible. He is not bothered by the well-known limitations introduced by Aristotle (same size, same length of pregnancy, closely related species) or extended by early modern zoologists, such as Franzius (same seasons of fertility; "parent species" must be extraordinarily horny). He does not care that the deer is much bigger than the wolf, has a pregnancy almost four times as long as the wolf (ca. 230–240 vs. 61–63 days), and is a plant-eater while the wolf is a carnivore; he is not bothered by the fact that the camel is about six times as big as the wild boar and 8 to 10 times as heavy. With respect to his optimism concerning the range of possible hybridisations Schott resembles Della Porta, although his focus is not on animal breeding, but on wild species.

A most important theological decision is that he ascribes the hybrid species to God's creation and that he regards them in fact as 'perfect animals'. Since he considers them all to be fertile, Nieremberg's limitation is not relevant for him. Schott presents his argument in the form of a scholastic "quaestio": he puts to the fore three arguments contra and three pro, and then takes a clear decision, as a scholastic magister would do. As Schott points out, hybridisation cannot be an invention of man, since wild hybrids occur both all over the world and in any period of history. Of course, Bible philology plays an important role in this respect. Similar to Nieremberg, Schott argues that it is not at all certain whether *Genesis* 36, 24 indeed proves that Ana is the inventor of the mule;[299] that the Vulgate reads that 'Ana invented hot water in the desert'; and that it is not at all plausible if the 'Hebrew Rabbis' want to read 'jemin' instead of 'jamin'.[300] But even if the rabbis were right and Ana was the inventor of the mule, what relevance would it have for the whole of hybridisation? Who would be the inventor of all the wild hybrid species? And there are more arguments belonging to the field of Bible philology or Bible exegesis. Benedict Pereira, a Jesuit theologian, Bible exegete, and natural philosopher (1536–1610) born near Valencia but later teaching in Rome,[301] and Ascanio Martinengo, the author of the *Glossae magnae* to *Genesis* (+ ca. 1600), emphasised that according to *Genesis*, chapter 1, God created all the animals, and thus the hybrids as well.[302]

299 Ibidem, pp. 823–824.

300 Ibidem, p. 824.

301 On Pereira, cf. Blum P.R., "Benedictus Pererius: Renaissance Culture at the Origins of Jesuit Science", *Science & Education* 15 (2006) 279–304; and idem, *Studies on Early Modern Aristotelianism*, chapter 9: "Benedictus Pereirus: Renaissance Culture at the Origins of Jesuit Science" (Leiden: 2012) 139–182.

302 Schott, *Physica curiosa, sive Mirabilia naturae*, p. 823; cf. Peirera Benedict, *Commentariorum et disputationum in Genesim tomi quattuor* (Rome: 1591–1599); Martinengo Ascanio,

Leviticus 19, 19, on the other hand, is not relevant for animal breeding, but directed against the idolatry of Egyptian religion, which venerates hybrid animals as gods.[303]

Most interestingly, Schott takes a totally different stand on hybridisation than his former master, Athanasius Kircher. Kircher excluded the hybrids from God's creation and he severely devalued them by regarding them as 'degenerations' of the 'perfect', i.e. the godly invented species. Schott emphasises that the hybrid species are certainly not worse than the pure species. They are in no way monsters. For Schott hybridisation is more proof of the outstanding quality of God's creation, and of its admirable, in fact unlimited variety: the more variety, the more admiration, the bigger the praise God deserves. In this sense, Schott's Neo-Scholastic Bible exegesis is connected with his new "thaumatography". His zoology is meant to bring together the wonders of God's creation.

This attitude is in fact relevant for the whole presentation of his zoology. Although he is well aware of the taxonomical categories of Aristotle (and of most of his early modern forerunners), he is not inclined to base his zoology on an integrative system of animal classification—such as Wotton, Aldrovandi, or Jonston did—in his *Historia naturalis animalium*. In doing so, he wanted to maximise the admirable variety of nature, and in this he followed, for example, Pliny, Aelianus, Gessner, Franzius, and Jonston in his *Thaumatographia naturalis* of 1633. This also refers to the description of the singular species. Primarily, that is also the reason why Schott presented the species in alphabetical order, as Gessner, Franzius, and Jonston (in his *Thaumatographia*) did. Sometimes it adds to the maximisation of the "admirable" to present the species in question as a scholastic "disputatio": putting together various, contradicting *auctoritates* of both antiquity and recent times.[304] Sometimes, if the animal is in itself strange and exotic, Schott considered it enough to quote one or two descriptions of the species.[305] A likewise differentiated approach goes for animal illustrations: Schott in fact only adds illustrations if the animal has remarkable or amazing features, or if its appearance is doubtful. He would not illustrate, for example, horse, cat, donkey, pig, lion, wolf, etc.; but he illustrates the sloth ('*Ignavus*

Glossae magnae in sacram Genesim [...]: *tomus primus* [...] *authore D. Ascanio Martinengo brixiano* (Padua, Laurentius Pasquatus: 1597).

303 Schott, *Physica curiosa, sive Mirabilia naturae*, p. 824.

304 This is the case, for example, with the elk (VIII, 1, pp. 899–906); the sloth or *Ignavus* (VIII, 2, pp. 907–910); the *Armadillo* (VIII, 3, pp. 911–916).

305 As in the case of the *Bison* (VIII, 11, p. 937), *Dabuh* (VIII, 29, p. 1009), Ocelot (VIII, 30, p. 1010).

FIGURE 2.37 *The bison and the* Bonasus. *From: Caspar Schott, S.J.*, Physica curiosa, sive Mirabilia naturae et artis Libri XII. [...] variis [...] disquisitionibus excutiuntur et innumeris exemplis illustrantur [...] Cum figuris aeri incisis *(Nuremberg, Moritz Endter: 1662), Iconismus XXXVIII.*

maior' and *'Ignavus minor'*),[306] elk,[307] *Bison* [Fig. 2.37],[308] anteater (*Achaus, Tamandoa*),[309] Camel-deer (*Camelo-cervus*, in fact the llama) [Fig. 2.38],[310]

306 Schott, *Physica curiosa, sive Mirabilia naturae*, Iconismus XXV, ad. p. 907.

307 Iconismus XXIV, ad. p. 902.

308 Iconismus XXVIII, ad p. 937.

309 Iconismus XXVII, ad. p. 917.

310 Iconismus XXXII, ad p. 953.

FIGURE 2.38 *The* Camelocervus (*llama*) *and the* Carigueya (*opossum*). *From: Caspar Schott, S.J.*, Physica curiosa, sive Mirabilia naturae et artis Libri XII. [...] variis [...] disquisitionibus excutiuntur et innumeris exemplis illustrantur [...] Cum figuris aeri incisis (*Nuremberg, Moritz Endter: 1662*), *Iconismus XXXII.*

Ichneumon,[311] *Carigueya* (*Opossum*) [Fig. 2.38, same as llama],[312] Jaguar,[313] *Su*,[314] *Armadillo*,[315] etc.

311 Iconismus XLI, ad p. 1045.
312 Iconismus XXXII, ad p. 953.
313 Iconismus XLI, ad p. 1045.
314 Iconismus XLVIII, ad p. 1088.
315 Iconismus XXVI, ad p. 911.

As in Nieremberg's, in Schott's zoology the American species play an important part. In the book on the land animals (all quadrupeds), Schott describes in total ca. 80 species or families; some 10 of them are South American species. Most of them he considers as species in their own right and, similar to Nieremberg, he also gives their name in the local language(s). In a marked difference from Kircher, he does not classify the majority of the American species as hybrids or other forms of degenerations. And because he considers them to be species in their own right, he is convinced that they took a place on Noah's ark. Concerning the question of how they got to the ark and back from Mount Ararat to South America,[316] he seems almost to polemicise against Kircher, whose opinion he might have heard in Rome, 1652–1655. He considers it impossible for them to have made it by swimming or via land-bridges, or where just brought by man, thus refuting Kircher's three explanations. Schott is convinced that the New World was inhabited by animals long before the Europeans discovered it and brought European animals with them by ship, and that these animals were created by God. And Schott considered it totally improbable that the animal species, one after the other, would have found the few land-bridges that man did not discover for so many thousands of years. Thus, with a meticulous Neo-Scholastic argument, he proves in the end that Nieremberg's guardian angels of species were the only possible means for the South American species to have survived.[317]

Conclusion

In the 16th and 17th centuries zoology was a flourishing field of science, with an amazing number of substantial works, and even more in-depth and intriguing discussions. The methods and discourses of early modern zoology, however, were very different from the established science of the 19th–21st centuries. The most important disciplines and discursive formations of early modern zoology were philology and theology. Authors that belonged to different parts of the religious spectrum engaged in zoology, Protestants and Catholics alike. There is no clearly discernible, progressive development, either from bookish scholarship, based on the authority of texts, to empirical science, or in another sense. Taxonomy and species description of the 17th century is no more centred on empirical observation than that of the 16th century. Taxonomy and species description develop in the 16th and 17th centuries as a subtle art that takes into account various and heterogeneous facts. New findings were sometimes, but not

316 Cf. chapt. VII, 15 "Quomodo homines et caetera animalia, in novum orbem insulasque a continente remotas delati fuerint", pp. 834–841, esp. 839–841.

317 Ibidem, pp. 840–841.

always, based on empirical observations; many new solutions that may represent "progress" from a modern perspective were brought forth either by philological improvements—for example, by advanced textual criticism, increased knowledge of Greek, and successful close reading—or by theological argumentation, which was sharpened by the religious conflicts of 16th and 17th centuries, especially by Confessionalisation. The seemingly strange phenomenon of hybrid speciation functions as an important paradigm for the production and procession of zoological information, in both the 16th and the 17th centuries.

Bibliography

Ackermann J.M., "Early Renaissance 'naturalism' and scientific illustration", in Ellenius A. (ed..), *The Natural Sciences and the Arts* (Uppsala: 1985) 1–17.

Agosto A. et al. (eds.), *IV centenario della nascita di Fortunio Liceti* (Rapallo: 1978).

Albertus Magnus, *Physica*, ed. P. Hossfeld, 2 vols. (Münster: 1987).

Aldrovandi Ulisse, *Quadrupedum omnium bisulcorum historia* [...] *Ioannes Utervius Belga colligere incaepit* (sic) [...] *Thomas Dempsterus Baro a Muresk Scotus* [...] *absolvit. Hieronymus Tamburinus in lucem edidit* [...] (Bologna, Sebastianus Bonhommius: 1621; Frankfurt a.M., Caspar Röteli: 1547).

Aldrovandi Ulisse, *De quadrupedibus digitatis viviparis libri tres, de quadrupedibus digitatis oviparis libri duo* (Bologna, Antonio Bernia: 1645).

Aldrovandi Ulisse, *Serpentum, et draconum historiae libri II. Bartholomaeus Ambrosinus* [...] *summo labore opus concinnavit* [...] (Bologna, Antonio Bernia, Clemente Ferronio: 1640).

Aldrovandi Ulisse, *Monstrorum historia cum paralipomenis historiae omnium animalium. Bartholomaeus Ambrosinus* [...] *composuit* (Bologna, Marcantonio Bernia – Nicolao Tebaldini: 1642).

Allan D.E., *The Naturalists in Britain. A Social History* (London: 1976).

Arnar A.S., *Encyclopedism from Pliny to Borges* (Chicago: 1990).

Asúa M. De – French R., *A New World of Animals. Early Modern Europeans on the Creatures of Iberian America* (Aldershot: 2005).

Balme D.M., "Development of Biology in Aristotle and Theophrastus: Theory of Spontaneous Generation", *Phronesis* 7 (1967) 91–104.

Balss H., *Albertus Magnus als Biologe* (Stuttgart: 1947).

Bates A.W., *Emblematic Monsters. Unnatural Conceptions and deformed Births in Early Modern Europe* (Amsterdam – New York: 2005).

Bates A.W., "The De monstrorum of Fortunio Liceti: A Landmark of Descriptive Teratology", *Journal of Medical Biography* 9 (2001) 49–54.

Bates A.W., "The Sooterkin Dissected: The Theoretical Basis of Animal Birth to Human Mothers in Early Modern Europe", *Vesalius* 9 (2003) 6–14.

Bäumer A., *Die Biologie von der Antike bis zur Renaissance*, in eadem, *Geschichte der Biologie*, vol. 1 (Frankfurt a.M. – Bern – New York – Paris: 1991).

Bäumer A., *Zoologie der Renaissance—Renaissance der Zoologie*, in eadem, *Geschichte der Biologie*, vol. 2 (Frankfurt a.M. – Bern – New York – Paris: 1991).

Bäumer A., "Das erste zoologische Kompendium in der Zeit der Renaissance: Edward Wottons Schrift 'Über die Differenzen der Tiere' ", *Berichte zur Wissenschaftsgeschichte* 13 (1990) 13–29.

Bäumer-Schleinkofer A., "Zoologie der Renaissance—Renaissance der Zoologie", in Döring K. – Wöhrle G. (ed.), *Antike Naturwissenschaft und ihre Rezeption*, vol. I und II (Bamberg: 1992) 275–295.

Bäumer-Schleinkofer A., "Biblische Zoologie (Hermann Frey, Wolfgang Franz, Heinrich von Hoevel)", in Kattenstedt H. (ed.), *'Grenzüberschreitung'. Festschrift zum 70. Geburtstag von Manfred Büttner* (Bochum: 1993) 3–17.

Bay J.Ch., "Conrad Gesner (1516–1565), the Father of Bibliography. An Appreciation", *Papers of the Bibliographical Society of America* 10 (1958) 53–88.

Beagon M., *Roman Nature: the Thought of Pliny the Elder* (Oxford: 1992).

Belon Pierre, *Travels in the Levant: The Observations of Pierre Belon of Le Mans on Many Singularities and Memorable Things Found in Greece, Turkey, Judaea, Egypt, Arabia and Other Foreign Countries (1553)*, transl. into English by J. Hogarth, with an Introd. by A. Merle (Hardinge: 2012).

Bithol-Hespériès A., "Monsters, Nature, and Generation from the Renaissance to Early Modern Period: The Emergence of Medical Thought", in Smith J.E.H., *The Problem of Animal Generation in Early Modern Philosophy* (Cambrigde: 2006) 47–64.

Boas M., *The Scientific Renaissance 1450–1630* (New York: 1962).

Bono J., *The Word of God and the Language of Men: Interpreting Nature in Early Modern Science and Medicine*, vol. 1 (Madison: 1995).

Borst A., *Das Buch der Naturgeschichte: Plinius und seine Leser im Zeitalter des Pergaments* (Heidelberg: 1994).

Braun L., *Conrad Gesner* (Geneva: 1990).

Calixte E., *Les monstres d'Ambroise Paré et la tératologie moderne* (Paris. 1946).

Cameron A.D., *The Legend of Noah: Renaissance Rationalism in Art, Science, and Letters* (Illinois U.P.: 1963).

Caprotti E., *Mostri, Draghi e serpenti nelle silografie dell'opera di Ulisse Aldrovandi e dei suoi contemporanei* (Milano: 1980).

Caprotti E., "Animali fantastici, fantasie zoogoliche e loro realtá in Plinio", in *Plinio e la natura. Atti del ciclo di conferenze sugli aspetti naturalistic dell'opera Pliniana, Como 1979* (Como: 1982) 39–61.

Castellani C., "Le problème de la 'generatio spontanea' dans l'oeuvre de Fortunio Liceti", *Revue de synthèse* 89 (1968) 323–340.

Céard J., "Analogie et zoologie chez les naturalistes de la Renaissance", in Christian L.C. (ed.), *Theatrum mundi. The History of an Idea* (New York: 1987).

Céard J., "Tératologie et Tératomancie au XVIe siècle", in Jones-Davies M.T. (ed.), *Monstres et prodiges au temps de la Renaissance* (Paris: 1980) 5–15.

Cherry J. (ed.), *Fabeltiere. Von Drachen, Einhörnern und anderen mythischen Wesen* (Stuttgart: 1997).

Chibnall M., "Pliny's Natural History and the Middle Ages", in Dorey T.A. (ed.), *Empire and Aftermath: Silver Latin II* (London – Boston: 1975) 57–78.

Cohen J.J., *Of Giants. Sex, Monsters, and the Middle Ages* (London: 1999).

Cohen J.J., *Hybridity, Identity, and Monstrosity in Medieval Britain: On difficult Middles* (New York: 2006).

Cook H.J., "Physick and natural history in the 17th century", in Ariew R. – Barker P. (ed.), *Revolution and Continuity: Essays in the History of Philosophy of the Early Modern Science* (Washington: 1991) 63–80.

Cook H.J., "Physicians and natural history", in Jardine N. – Secord J.A. – Spary E.C. (ed.), *Cultures of Natural History* (Cambridge: 1996) 91–105.

Cunningham A., *The Anatomical Renaissance* (Aldershot: 1997).

Dahl F., "Zur Geschichte der Zoologie. Von Aristoteles bis Plinius", *Sitzungsberichte der Gesellschaft Naturforschender Freunde zu Berlin* 5 (1924) 62–104.

Davies M., "Making Sense of Pliny in the Quattrocento", *Renaissance Studies* 9 (1995) 240–257.

Delauny P., *La Zoologie au XVIe siècle*, Histoire de la pensée 7 (Paris: 1962; reprint Paris: 1997).

Doody A., *Pliny's Encyclopedia: The Reception of the Natural History* (Cambridge: 2010).

Enenkel K.A.E., "Zur Konstituierung der Zoologie als Wissenschaft in der Frühen Neuzeit: Diskursanalyse zweier Großprojekte (Wotton, Gesner)", in idem – Smith P.J. (eds.), *Early Modern Zoology. The Construction of Animals in Science, Literature and the Visual Arts*, Intersections 7 (Leiden – Boston: 2007), 2 vols., vol. I, 15–74.

Enenkel K.A.E., "Mengwezens. Hybridisatie in vroegmoderne zoölogische werken", in M. Rikken (ed.), *Veranderlijke dieren van Conrad Gesner*, in *De boekenwereld* 29.1 (2012) 44–51.

Enenkel K.A.E. – Smith P.J. (eds.), *Early Modern Zoology. The Construction of Animals in Science, Literature and the Visual Arts*, Intersections 7 (Leiden – Boston: 2007), 2 vols.

Ewinkel I., *De monstris. Deutung und Funktion von Wundergeburten auf Flugblättern im Deutschland des 16. Jahrhunderts* (Tübingen: 1995).

Faber Kolb A., *Jan Brueghel the Elder: The Entry of the Animals into Noah's Ark* (Los Angeles: 2005).

Farber P.L., "Buffon and the Concept of Species", *Journal of History of Biology* 5 (1972) 259–284.

Farley J., *The Spontaneous Generation from Descartes to Oparin* (Baltimore – London: 1974).

Felfe R., *Naturgeschichte als kunstvolle Synthese. Physikotheologie und Bildpraxis bei Johann Jakob Scheuchzer* (Berlin: 2003).

Field Museum of Natural History, *The Giraffe in History and Art* (Chicago: 1928).

Findlen P., *Possessing Nature. Museums, Collecting and Scientific Culture in Early Modern Italy* (Berkely: 1999).

Findlen P., "Science, History, and Erudition: Athanasius Kircher's Museum at the Collegio Romano", in Stolzenberg D. (ed.), *The Great Art of Knowing: The Baroque Encyclopedia of Athanasius Kircher* (Stanford – Fiesole: 2001).

Findlen P. (ed.), *Athanasius Kircher: The Last Man Who Knew Everything* (New York: 2004).

Fischer H., "Conrad Gesner"; "Conrad Gesner als Bibliograph" and "Conrad Gesner als Arzt", in eadem et alii, *Conrad Gesner 1516–565. Universalgelehrter, Naturforscher, Arzt* (Zurich: 1967) 11–22; 31–37; 39–47.

Fischer H., "Conrad Gesner als Arzt", *Gesnerus* 22 (1965) 205–215.

Fischer H., *Conrad Gesner. Leben und Werk* (Zurich: 1966).

Franzius Wolfgang, *Historia animalium sacra* (ed. pr. Wittenberg, Zacharias Schürer – Johannes Gormann: 1612).

Franzius Wolfgang, *Historia animalium sacra* [...] *iam denuo emendatius et correctius edita* [...] *novo etiam nominum propriorum indiculo adaucta* [...], *praetermissa* [...] *Augusti Buchneri praefatione* (Frankfurt a.M., Balthasar Christoph Wust d.J. – Martin Gabriel Hübner: 1671).

Franzius Wolfgang, *Historia animalium* [...] *continuatio cum commentario et supplemento observationum ex recentiori historia naturali similitudinum, emblematum, hieroglyphicorum ad usum Oratoriae tum civilis tum ecclesiasticae, opera Joannis Cypriani* (Dresden, Martin Gabriel Hübner – Melchior Bergens Erben: 1687; Leipzig – Frankfurt, Martin Gabriel Hübner: 1688; Frankfurt – Leipzig, Gottofredus Leschius: 1712).

French R.K., *Ancient Natural History: Histories of Nature* (London: 1994).

French R.K., *Dissection and Vivisection in the European Renaissance* (Aldershot: 1999).

French R.K. – Greenaway F. (eds.), *Science in the Roman Empire: Pliny the Elder, his Sources, his Influence* (London: 1986).

French R.K. – Wear A. (eds.), *Medical Revolution in the 17th century* (Cambridge: 1989).

Gantenbein U.L., "Am Scheideweg zwischen Fiktion und Realität. Das Phantastische in Conrad Gesners Naturbetrachtung", in Schmutz H.K. (ed.), *Phantastische Lebensräume, Phantome und Phantasmen. Aufsätze des Zürcher Symposions der*

Schweizerischen Gesellschaft für Geschichte der Medizin und der Naturwissenschaften vom 10. und 11. Oktober 1996 (Marburg: 1997) 173–196.

George W., "Alive or Dead: Zoological Collections in the Seventeenth Century", in Impey O. – MacGregor A. (eds.), *The Origins of Museums. The Cabinet of Curiosities in Sixteenth- and Seventeenth-Century Europe* (Oxford: 1986) 179–187.

George W., "Sources and Background to Discoveries of New Animals in the Sixteenth and Seventeenth Centuries", *History of Science* 18 (1980) 70–104.

Gessner Conrad, *Historiae animalium* [...] 4 vols. (Zurich, Christoffel Froschauer: 1551–1558). Vol. I *de quadrupedis viviparis* (1551; editio secunda idem: 1560; Frankfurt a.M., Robertus Cambierius: 1603; Frankfurt a.M., Heinrich Lorenz: 1620); vol. II *de quadrupedis oviparis* (Zurich, Christoffel Froschauer: 1554; Frankfurt a.M., Johannes Wechel – Robertus Cambierius: 1586); vol. III *De avium natura* (Zurich, Christoffel Froschauer: 1555); vol. IV *De piscium et aquatilium animantium natura* (idem: 1558); editio secunda (idem: 1560).

Gessner Conrad, *Historiae animalium* [...] vols. I and II (Hildesheim: 2012; facsimile edition of Zurich, Christoffel Froschauer: 1551 and 1554).

Gessner Conrad, *Icones animalium quadrupedum viviparorum et oviparorum quae in historiae animalium* [...] *libro primo et secundo describuntur* [...] (Zurich, Christoffel Froschauer: 1553; editio secunda idem: 1560).

Gessner Conrad, *Thierbuch. Das ist ein kurtze beschreibung aller vierfüssigen Thieren so auf der erden und in wassern wonend* [...] (Zurich, Christoffel Froschauer: 1563).

Gessner Conrad, *Thierbuch. Ausführliche beschreibung und lebendige ja auch eigentliche Contrafactur und abmahlung aller vierfüssigen Thieren so auf der erden und in wassern wohnen* [...] *durch den weitberühmten herrn doctor conrad Gesnern* [...] (Heidelberg, Johann Lancellot: 1606).

Gilles Pierre, *Ex Aeliani historia per Petrum Gyllium latini facti itemque ex Porphyrio, Heliodoro, Oppiano, tum eodem Gyllio luculentis accessionibus aucti libri XVI de vi et natura animalium. Liber de Gallicis et Latinis nominibus piscium* (Lyon, Sebastian Gryphius: 1533 [4°]; idem: 1535 [4°]; idem: 1555 [4°]).

Gmelig-Nijboer C.A., *Conrad Gesner's "Historia Animalium": An Inventory of Renaissance Zoology* (Diss. Utrecht: 1977).

Gmelig-Nijboer C.A., "Conrad Gesner (1516–1565) Considered as a Modern Naturalist", *Janus* 60 (1973) 41–51.

Godwin J., *Athanasius Kircher. Ein Mann der Renaissance auf der Suche nach verlorenem Wissen* (Berlin: 1994).

Gould S.J., *Das Lächeln des Flamingos. Betrachtungen zur Naturgeschichte* (Frankfurt: 1995).

Gravestock P., "Did Imaginary Animals Exist?", in Hassig D. (ed.), *The Mark of the Beast. The Medieval Bestiary in Art, Life, and Literature* (New York: 1999) 119–135.

Guerrini A., *Experimenting with Humans and Animals. From Galen to Animal Rights* (Baltimore – London: 2003).

Harrison P., *The Bible, Protestantism, and the Rise of Natural Science* (Cambridge: 1998).

Hasse D.N., "Spontaneous Generation and the Ontology of Forms in Greek, Arabic and Medieval Latin Sources", in Adamson P. (ed.), *Classical Arabic Philosophy: Sources and Reception* (London – Turin: 2007) 150–175.

Heck L., *Die Hauptgruppen des Thiersystems bei Aristoteles und seinen Nachfolgern, ein Beitrag zur Geschichte der zoologischen Systematik* (Diss. Leipzig: 1885).

Heller J.L., "The Early History of Binomial Nomenclature", in idem, *Studies in Linnaen Method and Nomenclature* (Berlin – Bern: 1983) 41–75.

Heller J.L., "Classical Mythology in the Systema Naturae", in idem, *Studies in Linnaen Method and Nomenclature* (Berlin – Bern: 1983) 1–30.

Helmcke J.G., "Der Humanist Conrad Gesner auf der Wende von mittelalterlicher Tierkunde zur neuzeitlichen Zoologie", *Physis* 12 (1970) 339–346.

Heuvelmans B., *Les derniers dragons d'Afrique* (Paris: 1978).

Heuvelmans B., *Les bêtes humaines d'Afrique* (Paris: 1980).

Heuvelmans B., "What is Cryptozoology", *Cryptozoology* 1 (1982) 1–12.

Heuvelmans B., "How many animal Species remain to be discovered", *Cryptozoology* 2 (1983) 1–24.

Heuvelmans B., "The Sources and Method of cryptozoological Research", *Cryptozoology* 7 (1988) 1–21.

Heuvelmans B., "The Metamorphosis of unkown Animals into fabulous Beasts and of fabulous Beasts into known Animals", *Cryptozoology* 9 (1990) 1–12.

Heuvelmans B., *On the Track of Unkown Animals* (London – New York: 1995).

Heuvelmans B., *The Natural History of Hidden Animals*, ed. by Hopkins P.G. (London: 2007).

Hoeniger F.D. and J.F.M., *The Development of Natural History in Tudor England* (Charlottesville, VA: 1969).

Hünemörder Ch., "Aims and Intentions of Botanical and Zoological Classification in the Middle Ages and in the Renaissance", *History and Philosophy of the Life Science* 5 (1983) 53–67.

Jahn I., *Grundzüge der Biologiegeschichte* (Jena: 1990).

Jahn I. (ed.), *Geschichte der Biologie* (Hamburg: 2004; 3rd edition Jena: 1998).

Jardine N. – Secord J.A. – Spary E.C. (eds.), *Cultures of Natural History* (Cambridge: 1996).

Jonston John, *Historiae naturalis de quadrupedibus libri, cum aeneis figuris* [...] (Frankfurt a. M., Matthaeus Merian: 1650; Amsterdam, J.J. Schipper: 1657); *Historiae naturalis de avibus libri VI cum aeneis figuris* [...] (Frankfurt a.M., Matthaeus Merian: 1650; Amsterdam, J.J. Schipper: 1657); *Historiae naturalis de insectis libri II, de ser-*

pentibus et draconibus libri II, cum aeneis figuris (Frankfurt a.M.: 1653; Amsterdam, J.J. Schipper: 1657; idem: 1665); *Historiae naturalis de exanguibus aquaticis libri IV, cum figuris aeneis* [...] (Amsterdam, J.J. Schipper: 1657); *Historiae naturalis de piscibus et cetis libri V, cum aeneis figuris* [...] (Amsterdam, J.J. Schipper: 1657); niederländ. Übers. dieser Werke (Amsterdam, J.J. Schipper: 1660).

Jonston John, *A Description of the Nature of Four-footed Beasts with their Figures Engraven In Brass* (London, Moses Pitt: 1678).

Jonston John, *Theatrum universale omnium animalium, piscium, avium, quadrupedum, exanguium, aquaticorum, insectorum et anguium* (Amsterdam, R. and G. Wetstein: 1718).

Kappler C., *Monstres, démons et merveilles à la fin du Moyen Age* (Paris: 1980).

Kaufmann T.D., *The Mastery of Nature. Aspects of Art, Science and Humanism in the Renaissance* (Princeton: 1993).

Keeton W.T., *Biological science* (New York: 1980).

Keller O., *Die antike Tierwelt*, 2 vols. (vol. I, Leipzig: 1909; vol. II, Leipzig: 1913; reprint, Hildesheim: 1963).

Kempe M., *Wissenschaft, Theologie, Aufklärung. Johann Jakob Scheuchzer (1672–1733) und die Sintfluttheorie* (Epfendorf: 2003).

Kircher Athanasius S.J., *Arca Noe in tres libros digsta* [...] *quae omnia nova methodo necnon summa argumentorum varietate explicantur et demonstrantur* (Amsterdam, Johannes Jansonius: 1675), First and only edition.

Knight D., *Zoological Illustration. An Essay towards a History of Printed Zoological Pictures* (Folkstone: 1977).

Koreny F., *Albrecht Dürer and the Animal and Plant Studies* (Boston: 1988).

Leitner H., *Zoologische Terminologie beim älteren Plinius* (Hildesheim: 1972).

Leu U.B., "Theologia naturalis", in idem, *Conrad Gesner als Theologe. Ein Beitrag zur Zürcher Geistesgeschichte des 16. Jahrhunderts* (Bern – Frankfurt a.M. – New York – Paris: 1990; Zürcher Beiträge zur Reformationsgeschichte 14) 31–100.

Ley W., *Konrad Gesner. Leben und Werk* (Munich: 1929).

Lloyd G.E.R., "The Development of Aristotle's Theory of the Classification of Animals", *Phronesis* 6 (1961) 59–81.

Lloyd J.E. Barclay, *African Animals in Renaissance Literature and Art* (Oxford: 1971; New York: 1972).

Lloyd G.E.R., *Science, Folklore, and Ideology. Studies in the Life Sciences in Ancient Greece* (Cambridge: 1983).

Lugli A., *Naturalia et Mirabilia. Il collezionismo nelle Wunderkammern d'Europa* (Mailand: 1983).

MacDonald D. (ed.), *Enzyklopädie der Säugetiere* (Königswinter: 2003; or. Engl., Oxford: 2001).

Magana E., "Paisajes, hombres y animales imaginarios de algunas tribus Gê del Brasil Central", in idem – Mason P. (eds.), *Myth and the Imaginary in the New World* (Dordrecht: 1986) 95–164.

Magoun H.W., "Early Concept of the Anthropomorpha", *Physis* 4 (1962) 39–64.
Mason P., *Infelicities. Representations of Exotic* (Baltimore – London: 1998).
Mason P., *Before Disenchantment: Images of Exotic Plants and Animals in the Early Modern World* (London: 2009).
Mayerhöfer J., "Conrad Gesner als Bibliograph und Enzyklopädist", *Gesnerus* 22 (1965) 176–195.
Mayr E., *Systematics and the Origin of Species* (New York: 1942).
Mayr E., *Animal Species and Evolution* (Cambridge Mass.: 1963).
Mayr E., *Systematics and the Origin of Species from the View Point of a Zoologist* (Cambridge, Mass. – London: 1999).
Mendelsohn E.E., "Philosophical Biology versus Experimental Biology: Spontaneous Generation in the Seventeenth Century", in Grene M. – Everett E.E. (eds.), *Topics in the Philosophy of Biology* (Dordrecht: 1976) 37–65.
Mesnard P., "L'horizon zoologique de la Renaissance. Les animaux anciens et les animaux modernes", in *Sciences de la Renaissance. VIIe congrès international de Tours* (Paris: 1973).
Miller G., "Beasts of the New Jerusalem: John Jonston's Natural History and the Launching of Millenarian Pedagogy in the Seventeenth Century", *History of Science* 46 (2008) 203–243.
Mode H., *Fabeltiere und Dämonen in der Kunst—Die fantastische Welt der Mischwesen* (Stuttgart – Berlin: 1975).
Montalenti G., "Aldrovandi Ulisse", in *Dizionario Biografico degli Italiani* 2 (1960).
Morison S.E., *The European Discovery of America*. 2 vols. (New York: 1971–1974).
Moscoso J., "Monsters as Evidence: The Uses of the Abnormal Body during the Early Eighteenth Century", *Journal of History of Biology* 31 (1998) 355–382.
Müller U. – Wunderlich W. (eds.), *Dämonen. Monster. Fabelwesen* (St. Gallen: 1999).
Münch P., "Verwandtschaft oder Differenz? Zur Theorie des Mensch/Tier-Verhältnisses im 17. Jahrhundert", in Lehmann H. – Trepp A.-Ch. (eds.), *Im Zeichen der Krise. Religiösität im Europa des 17. Jahrhunderts* (Göttingen: 1999) 517–535.
Muraro L., *Giambattista Della Porta. Mago e scienziato* (Milan: 1978).
Murgatroyd P., *Mythical Monsters in Classical Literature* (London: 2007).
Naas V., *Le projet encyclopédique de Pline l'Ancien* (Rome: 2002).
Nardi B., "Pietro Pomponazzi e la teoria di Avicenna intorno alla generazione spontanea nell'uomo", in idem, *Studi su Pietro Pomponazzi* (Florence: 1965) 305–319.
Nauert Ch., "Humanists, Scientists, and Pliny: Changing Approaches to a Classical Author", *American Historical Review* 84.1 (1974) 72–85.
Newton M., *Encyclopedia of Cryptozoology. A global Guide* (Jefferson – London: 2005).
Nieremberg Eusebio, *Historia naturae, maxime peregrinae* (Antwerp, Balthazar Moretus: 1635).
Nissen C., *Die naturwissenschaftliche Buchillustration*, 2 vols. (Bad Münster am Stein: 1950).

Nissen C., *Die zoologische Buchillustration, ihre Bibliographie und Geschichte* (Stuttgart: 1971).

Nutton V., "Conrad Gesner and the English naturalists", *Medical History* 29 (1985) 93–97.

Ogilvie B., "Travel and Natural History in the Sixteenth Century", in idem (ed.), *Sammeln in der Frühen Neuzeit* (Berlin: 1996) 2–28.

Ogilvie B., "Image and Text in Natural History, 1500–1700", in Lefèvre W. – Renn J. – Schoepflin U. (eds.), *The Power of Images in Early Modern Science* (Basel – Boston – Berlin: 2003) 141–166.

Ogilvie B., *The Science of Describing. Natural History in Renaissance Europe* (Chicago-London: 2006).

O'Malley C.D. – Morgan H.W., "Early Concepts of the Anthropomorpha", *Physis* 4 (1962) 39–64.

Olmi G., *Ulisse Aldrovandi* (Trento: 1976).

Oriol-Boyer C., "Les monstres de la mythologie grecque – Réflexion sur la dynamique de l'ambigu", in Burgos J. (ed.), *Le monstre. Présence du monstre – mythe et réalité* (Paris: 1975) 25–50.

Paré Ambroise, *Le livre des monstres et prodiges* (1573), in idem, *Les Oeuvres d'Ambroise Paré* (12th ed., Lyon, Jean Gregoire: 1664) 644–701; modern edition by J. Céard (Geneva: 1971); trans. By J.L. Pallister, *Ambroise Paré, On Monsters and Marvels* (Chicago: 1982).

Park K. – Daston L.J., "Unnatural Conceptions: The Study of Monsters in Sixteenth- and Seventeenth-Century France and England", *Past and Present* 91 (1991) 20–54.

Park K., "Una historia de la admiración y del prodigio", in Lafuente A. – Moscoso J. (eds.), *Monstruos y seres imaginarios en la Biblioteca Nacional* (Madrid: 2000) 77–89.

Peirera Benedict, *Commentariorum et disputationum in Genesim tomi quattuor* (Rome, 1591–1599).

Pellegrin P., *La classification des animaux chez Aristote* (Paris: 1982).

Petit G. – Theodoride S.J., *Histoire de la zoologie des origines à Linné* (Paris: 1962).

Petit G., "Conrad Gesner zoologiste", *Gesnerus* 22 (1965) 195–205.

Petit G., "Conrad Gesner, Zoologiste", in Fischer H. et alii (eds.), *Conrad Gesner 1516–1565. Universalgelehrter, Naturforscher, Arzt* (Zurich: 1967) 49–56.

Plinius, Gaius Plinius Secundus Der Ältere, *Naturkunde. Lateinisch—deutsch. Buch VIII. Zoologie: Landtiere.* Herausgegeben und übersetzt von Roderich König in Zusammenarbeit mit Gerhard Winkler (Kempten: 1976).

Raven Ch. E., *English Naturalists from Neckham to Ray: A Study of the Making of the Modern World* (Cambridge: 1947).

Riedl-Dorn Ch., *Wissenschaft und Fabelwesen. Ein kritischer Versuch über Conrad Gessner und Ulisse Aldrovandi* (Vienna-Cologne: 1989) (Perspektiven der Wissenschaftsgeschichte, Bd. 5).

Roggen V., "Biology and Theology in Franzius's *Historia animalium sacra* (1612)", in Enenkel K.A.E. – Smith P.J. (eds.), *Early Modern Zoology. The Construction of Animals in Science, Literature and the Visual Arts*, Intersections 7 (Leiden-Boston 2007), 2 vols., vol. I, 121–146.

Roling B., *Drachen und Sirenen. Die Rationalisierung und Abwicklung der Mythologie an den europäischen Universitäten* (Leiden – Boston: 2010) (Mittellateinische Studien und Texte 42).

Rowland B., *Animals with Human Faces. A Guide to Animal Symbolism* (Knoxville: 1975).

Ruestow E.G., "Leeuwenhoeck and the Campaign against Spontaneous Generation", *Journal of the History of Biology* 17 (1984) 225–248.

Salas A.M., *Para un bestiario de las indias* (Buenos Aires: 1985).

Sarton G., *The Appreciation of Ancient and Medieval Science during the Renaissance (1450–1600)* (Philadelphia Penn.: 1953).

Scarani P., "Aldrovandi e la teratologia", in Antonino B. (ed.), *Animali e creature mostruose di Ulisse Aldrovandi* (Milan: 2004) 44–55.

Schneider D., "Grenzbetrachtungen und Synthesewesen: Conrad Gesners Meermenschen", in Eickmeyer J. – Soppa S. (eds.), *Umarmung und Wellenspiel. Variationen über die Wasserfrau* (Overath: 2006) 95–107.

Schott Kaspar S.J., *Physica curiosa, sive Mirabilia naturae et artis Libri XII.* [...] *variis* [...] *disquisitionibus excutiuntur et innumeris exemplis illustrantur* [...] *Cum figuris aeri incisis* (Nuremberg, Moritz Endter: 1662).

Schumacher G.-H., *Monster und Dämonen. Unfälle der Natur. Eine Kulturgeschichte* (Berlin: 1996).

Schumacher Y., *Tiermythen und Fabeltiere. Mythologische Streifzüge durch die Tierwelt der Schweiz* (Bern: 2001).

Seeck G.A., "Plinius und Aristoteles als Naturwissenschaftler", *Gymnasium* 92 (1985) 419–434.

Simpson G.G., "Mammals and Cryptozoology", *Proceedings of the American Philosophical Society* 128 (1984) 1–19.

Slaughter M.M., *Universal Languages and Scientific Taxonomy in the Seventeenth Century* (Cambridge: 1982).

Smith J.E.H., "Degeneration and Hybridism in Early Modern Species Debate: Towards the Philosophical Roots of the Creation-Evolution Controversy", in Wolfe Ch.T. (ed.), *Monsters and Philosophy* (London: 2005) 109–130.

Smith J.E.H., *The Problem of Animal Generation in Early Modern Philosophy* (Cambridge: 2006).

South M. (ed.), *Mythical and Fabulous Creatures. A Source Book and Research Guide* (New York – London: 1987).

Stagl J., *A History of Curiosity. The Theory of Travel 1550–1800* (Chur: 1995).

Steier A., *Aristoteles und Plinius. Studien zur Geschichte der Zoologie* (Würzburg: 1913).

Steier A., *Der Tierbestand in der Naturgeschichte des Plinius* (Würzburg: 1913).

Strick J.E., *Sparks of Life. Darwinism and the Victorian Debates over Spontaneous Generation* (Cambridge, Mass.: 2000).

Suutala M., *Tier und Mensch im Denken der Deutschen Renaissance* (Helsinki: 1990).

Thevet André, *Histoire d'André Thevet Angoumoisin, Cosmographe du Roi, des deux voyages par luy faits aux Indes Australes, et Occidentales*, ed. crit. J.-C. Laborie – F. Lestringant (Geneva: 2006).

Thevet André, *Cosmographie de Levant* (Lyon, Jean de Tournes – Guillaume Gazeau: 1554; iidem: 1556).

Thevet André, *La Cosmographie universelle* [...], 2 vols. (Paris, Pierre L'Huillier – Guillaume Chaudière: 1575).

Thomas K., *Man and the Natural World: Changing Attitudes in England 1500–1800* (London: 1983).

Velde A.J.J. van De, "Rond Gesner's *Historiae animalium liber I* van 1551", *Mededelingen van de Koninklijke Vlaamse Academie voor wetenschappen letteren en schone kunsten van België. Klasse der Wetenschappen* XIII,17 (Brussel: 1951).

Verner L., *The Epistemology of the Monstrous in the Middle Ages* (New York – London: 2005).

Whitehead P.J.P., "Georg Markgraf and Brazilian Zoology", in van Den Boogart E. – Houtink H.R. – Whitehead P.J.P. (eds.), *Johan Maurits van Nassau-Siegen 1604–1679. Essays on the Occasion of the Tercentenary of his Death* (The Hague: 1979) 424–471.

Williams D., *Deformed Discourse. The Function of the Monster in Medieval Thought and Literature* (Exeter: 1996).

Wilson D., *Sign and Portens. Monstrous Births from the Middle Ages to the Enlightment* (London – New York: 1993).

Wolfe Ch.T., "The Materialist Denial of Monsters", in idem (ed.), *Monsters and Philosophy* (London: 2005) 187–204.

Wolfe Ch.T. (ed.), *Monsters and Philosophy* (London: 2005).

Wotton Edward, *De differentiis animalium libri decem* (Paris, Michel de Vascosan: 1552).

Wüthrich L.H., *Das druckgraphische Werk von Matthäus Merian d.Ä.*, vols. 1–2 (Basel: 1966), vol. 3 (Hamburg: 1993); vol. 4 (Hamburg: 1996).

Zanca A., "Fortunio Liceti e la scienza dei mostri in Europa", in *Atti del XXXII Congresso nazionale della Società italiana di storia della medicina* (Padua-Triest: 1985) 35–45.

CHAPTER 3

Identification of Herring Species (*Clupeidae*) in Conrad Gessner's Ichthyological Works: A Case Study on Taxonomy, Nomenclature, and Animal Depiction in the Sixteenth Century

Sophia Hendrikx

Between 1551 and 1558 the Swiss physician and naturalist Conrad Gessner (1516–1565) published the very first zoological encyclopaedia, the *Historiae Animalium*. Out of four volumes, Liber IV, the *De Piscium et aquatilium animalium natura* (1558), on fish and aquatic animals, is of special interest for research on taxonomy in the sixteenth century. Since the tenth edition of Linnaeus' *Systema naturae* (1758), taxonomy is understood to be the definition and separation of groups of biological organisms on the basis of morphological characteristics, and the creation of a hierarchical classification. In present-day ichthyology this is done by measuring and counting external physical characteristics, such as the length/height ratio of the body, or the number and position of the fins in relation to the animal's head and tail. Linnaeus' taxonomy makes it possible to distinguish species, establish relationships between species, and create a logical nomenclature. Although it does not meet Linnaean standards in classifying species by measuring and counting external morphological characteristics, Conrad Gessner's ichthyological work sets the groundwork for an organisation of species as falling into groups or genera, and as varying to a greater or lesser extent according to morphological characteristics. Remarkably, very little has been published on *De piscium et aquatilium animalium natura* (1558) or Gessner's subsequent ichthyological work. In contrast, Liber I of Gessner's *Historiae Animalium*, the *De Quadrupedibus viviparis* (1551), received much more attention. Based on Liber I of the *Historiae Animalium*, it is sometimes suggested Gessner failed to take into account whether the material he published had sufficient scientific value, and that he copied text without applying proper criteria for selection.[1] This negative press has largely obscured the scientific value of Gessner's ichthyological work.

1 Kusukawa S., "The Sources of Gessner's Pictures for the Historiae Animalium", *Annals of Science* 67, 3 (2010) 303–328.

 | DOI 10.1163/9789004279179_005

The present article outlines the identification of herring species in Conrad Gessner's work as one of many groups of species Gessner outlined and named correspondingly. Gessner produced three subsequent ichthyological publications: Liber IV of the *Historiae Animalium* (1558), the *Nomenclator Aquatilium Animantium* (1560), and the *Fischbuch* (1563). In addition, he left a collection of hundreds of original drawings (ca. 1543–1565), upon which he modelled many of his woodcuts. These were assembled into albums and annotated by Gessner's associate Felix Platter (1536–1614).[2] Gessner's identification of species progressed from the *Historiae Animalium* to his later ichthyological publications. In the *Historiae Animalium* Gessner already identifies a 'core group' of very similar species—each of which is also described separately—which he outlines in his description of the herring. In the *Nomenclator* and the *Fischbuch* he connects further, separately described species with this group by introducing a consistent nomenclature that is applied to this group only, calling these species *Hering, Häring*, or "*Hering Art*". Combined with the 'core group' of similar species, this identifies and connects a group of eight 'herring-like' species. Physical characteristics listed in the description of the herring are repeated in the descriptions of the species Gessner identifies as similar. The nomenclature in the *Nomenclator* and the *Fischbuch* consequently reflects the group in which a species should be placed based on its morphology, while in the *Historiae Animalium* this is not yet done. These references to a wider family of species take place in the primary German nomenclature. By primary nomenclature is meant names used to describe a species that are not merely cited from other authors, or names that are part of an overview of local names. This can generally be found in the title above the descriptions, in the first lines of the descriptions, and in the *Nomenclator* at the bottom of the descriptions, where Gessner presents primary nomenclature in various languages. Gessner was the first to group related or similar species together on this scale using nomenclature. What does Gessner's identification of species tell us of his conception of classification into groups? Based on which characteristics does Gessner group species together, and does he successfully distinguish between similar species?

In his attention to species' physical characteristics and, in connection with this, the relation between species, Gessner's work is a significant improvement upon the work of other sixteenth-century ichthyologists, such as Pierre Belon

2 Now held at the Special Collections of the University of Amsterdam and identified in 2012, see: Egmond F., "A Collection within a Collection. Rediscovered Animal Drawings from the Collections of Conrad Gessner and Felix Platter", *Journal of the History of Collections* 25, 2 (2012) 149–170.

(1517–1564),[3] Hippolito Salviani (1514–1572),[4] and, most significantly, Guillaume Rondelet (1507–1566).[5] While Rondelet's is arguably the only ichthyological work of the time that can be compared to Gessner's in terms of quality of research and level of detail of information, and although he makes a careful start connecting species based on their physical appearance, in this last respect his work is still very limited. Rondelet's nomenclature often refers to one species only, and there is little consistency in the morphological characteristics he describes. In contrast, in Gessner's later work his vernacular nomenclature almost without exception places species within larger groups of similar fishes, and his descriptions of species include morphological characteristics connecting them to, and distinguishing them from, this larger group. In addition, Gessner often provided illustrations that pay more attention to morphological details than those produced by his predecessors, particularly when he had a chance to study species personally. Gessner's woodcut illustrations tend to be of a high quality and clearly depict those physical characteristics needed to identify a species.

Herring and Related Species in Gessner's Work

Herring species discussed in Gessner's work include the North Sea herring (*Clupea harengus*, Linnaeus, 1758), the Baltic herring (*Clupea harengus membras*, Linnaeus, 1761), the sardine (*Sardina pilchardus*, Walbaum, 1792), the round sardinella (*Sardinella aurita*, Valenciennes, 1847), the Allis shad (*Alosa alosa*, Linnaeus, 1758), the Twaite shad (*Alosa fallax*, Lacépède, 1803), the Macedonian shad (*Alosa macedonica*, Vinciguerra, 1921), and the sand-smelt (*Atherina presbyter*, Cuvier, 1829). With one exception, these species fall within the group identified by modern taxonomy as the Clupeidae family. Drawing from the nomenclature, descriptions, and depictions of these eight species in Gessner's work, modern fish guides were searched for these species. In most cases these were easily identified. Since Gessner centres his group of clupeids around the herring, which provides the nomenclature he uses to connect the species, the herring is discussed first. Next, the 'core group' of species Gessner discusses in his description of the herring in his early work, consisting of the

3 Belon Pierre, *De aquatilibus* [...] (Paris, Carolus Stephanus: 1553); idem, *La nature et diversité des poissons* [...] (Paris, Charles Estienne: 1555).

4 Salviani Hippolito, *De historia aquatilium animalium* (Rome, Hippolito Salviani: 1558).

5 Rondelet Guillaume, *Libri de piscibus marinis* (Lyon, Matthias Bonhomme: 1554); idem, *Universae aquatilium historiae pars altera* (Lyon, Matthias Bonhomme: 1555); idem, *L'histoire entière des poissons* (Lyon, Bonhomme: 1558).

Allis and Twaite shads and the sardine, is discussed. Lastly, the species Gessner adds to this 'core group' in the *Nomenclator* and the *Fischbuch* will be discussed.

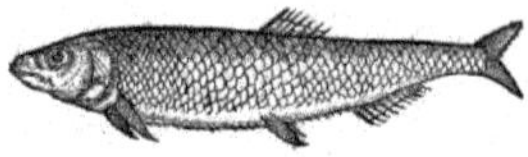

FIGURE 3.1A *Conrad Gessner,* Harengus. *Woodcut illustration to* Historiae animalium liber IV [...] *(Zurich, Christopher Froschauer: 1558) 485.*

FIGURE 3.1B *Conrad Gessner,* Harengus inveteratus. *Woodcut illustration to* Historiae animalium liber IV [...] *(Zurich, Christopher Froschauer: 1558) 486.*

In the *Historiae Animalium,*[6] as well as the *Nomenclator*[7] and the *Fischbuch,*[8] Gessner calls the herring *harengus* or *thrissis.* In the *Historiae Animalium* he describes it as a marine species. While the *Historiae Animalium* is organised alphabetically, the *Nomenclator* and the *Fischbuch* are organised according to habitat, into marine, anadromous,[9] and freshwater fish. In both the *Nomenclator* and the *Fischbuch* Gessner places the herring in the first order of marine species, that of small marine fish. Following Rondelet,[10] in his description of the herring Gessner identifies a group of species which due to their similarity are difficult to tell apart. In addition to the herring, this group includes the Allis shad, the Twaite shad, and the sardine. Gessner describes these species as having silvery scales and black and blue colouring on their backs, lighter keels with a row of sharp scales, and scales that can be quite easily removed. In addition, he states they have similar shapes and similar flavours, as well as similar intestines. These characteristics, particularly the row of sharp scales on

6 Gessner Conrad, *Historiae animalium liber IV* [...] (Zurich, Christopher Froschauer: 1558) 485–486.

7 Gessner Conrad, *Nomenclator Aquatiluim Animantium* (Zurich, Christopher Froschauer: 1560) fols. 4r–5r.

8 Gessner Conrad, *Fischbuch* (Zurich, Christopher Froschauer: 1563) fols. 5r–6r.

9 According to the modern definition, anadromous species live in marine environments and migrate up rivers to spawn, as opposed to catadromous species, which live in freshwater environments and migrate to the sea to spawn. However, in Gessner's work all species that spend part of their lives in marine and part of their lives in freshwater environments are described as anadromous.

10 Rondelet, *Libri de piscibus marinis* 222–223.

the keel, the silvery colouring, and the easily removed scales, are repeated in the descriptions of the other species Gessner identifies as clupeids.

Rondelet also identified a group of similar species in his description of the herring, which he calls *hareng, thrissis, harengo*, or *halec*. In Rondelet's case, this group includes only the Allis shad, the herring, and the sardine. He applies the term *thrissis* to each of these species,[11] but not to other herring-like fish. Consequently, based on the nomenclature, Rondelet's description of the herring can be linked to only two other species. In contrast, Gessner's nomenclature connects the herring with the chalcis Rondeletii, the atherina, the membras, the sardine, and the liparis.[12] These species are thus added to the 'core group' discussed in Gessner's description of the herring. Consequently, although both authors outline a group of herring-like species, Rondelet's group consists of three species, while Gessner's group contains eight. Both Rondelet and Gessner provide ways of telling the herring, the sardine, and the shads apart, listing distinguishing features such as the spots on the back of the sardine and Twaite shad, relative size, and the fact that the row of sharp scales on the keel is softer in the case of the herring.

Although in many cases Gessner did additional research after the *Historiae Animalium* was published, it appears he did his research on the herring early on, as his descriptions in the *Nomenclator* and the *Fischbuch* do not contain information that cannot be found in the *Historiae Animalium*. It is clear he did a fair bit of reading on the herring. Based in landlocked Switzerland he would not often have encountered the species unless preserved; however, Gessner extensively describes the behaviour of the species, pointing out, for example, that herrings swim in schools and spawn near the coast in the autumn or spring, at which time they are caught in great multitudes. He explains the species is caught predominantly in the North Sea, from where it is sent to Switzerland in abundance, presumably pickled, salted, or smoked. That he also relied on eyewitness reports from friends and acquaintances is clear from the fact he reports having heard about a freshwater variety that exists in a lake near Schwerin in Germany.

In addition, Gessner appears to have done research of his own. An illustration of a preserved herring [Fig. 3.1b], which would have been readily available to him, is included alongside an illustration showing a fresh herring [Fig. 3.1a]. The former [Fig. 3.1b] is an original illustration,[13] the drawing upon which this was based was part of Gessner's collection of animal drawings. Gessner states that the latter [Fig. 3.1a] is a composite image based on illustrations from works by

11 Rondelet, *Libri de piscibus marinis* 217–223.

12 Gessner, *Historiae animalium liber IV* [...], pp. 18, 83, 85, 990, and 572.

13 *Platter album*, Special Collections of the University of Amsterdam, C22, fol. 58.

Pierre Belon and Guillaume Rondelet. Close inspection shows that in addition to elements taken from these images, such as Rondelet's excellent depiction of the fins, Gessner has added innovations of his own, quite possibly based on the depiction of the preserved herring. Particularly the head and tail are original innovations that render the image much more useful in identifying the species. While the head and tail of Rondelet's herring are badly depicted, Gessner managed to provide an acceptable taxonomical depiction. Taxonomical depictions are used in biology today, and they display the physical characteristics that can help identify the species and distinguish it from others. Since Gessner explains that herring was imported to Switzerland in great quantities, it is likely he examined herrings, and it is probable he oversaw the creation of both images.

The description, nomenclature, and illustrations [Fig. 3.1a and 3.1b] identify this species as a *Clupea harengus* or herring. The described silvery colouring, with a blue to greenish blue back but no distinctive dark spots on the body or fins, and row of sharp scales on the keel are distinct characteristics of the species, as are the single dorsal fin, absence of a lateral line, and protruding lower jaw, which are visible in the depictions.

FIGURE 3.2 *'Conrad Gessner,* Alausa. *Woodcut illustration to* Historiae animalium liber IV [...] *(Zurich, Christopher Froschauer: 1558) 21.*

The alausa is part of the 'core group' of 'herring-like' species Gessner presents in his description of the herring. He also describes the species separately, here also remarking on its resemblance to the herring. In the *Historiae Animalium,*[14] as well the *Nomenclator*[15] and the *Fischbuch,*[16] the species is called *clupea*; in addition, the terms *thrissa, thrichis,* and *chalcide* are used to describe the species, as are the German words *Elft* and *Alfe*. Rondelet[17] calls the species *trissa* or *alose*. In the Historiae Animalium Gessner described the species as a common fish which is similar to a herring. Like the herring it has a row of sharp scales on its keel, but it is bigger and broader than the herring, and has a cleft mouth. In the number of scales and the position of the fins it resembles the sardine. It is a fish with many bones, in this it resembles the "apua phalerica" (most likely

14 Gessner, *Historiae animalium liber IV* [...], p. 21.

15 Gessner, *Nomenclator* fol. 322r.

16 Gessner, *Fischbuch* fol. 179v.

17 Rondelet, *Libri de piscibus marinis* 220–222.

the sprat), the "membras" (most likely the Baltic herring), and the sardine. The head is more similar to that of a sardine than a herring. This information is repeated in the *Nomenclator*, and in the *Fischbuch* Gessner adds to this that the species is about as large as a barbell, or about a cubit.[18]

Over the course of publishing his works, Gessner did additional research on the alausa. While in the *Historiae Animalium* the species is described and depicted on its own, in the *Nomenclator* and subsequently in the *Fischbuch* it is described and depicted alongside another, similar species. It appears these two species are the Allis and Twaite shad; due to their similarity, these shads have throughout history often been mistaken for one single species.[19] Gessner was the first to describe both while distinguishing between them and at the same time acknowledging their close relation. Rondelet had also described a shad, referred to as *alose* or *thrissa*, which he described as similar to a sardine but larger. This description, however, does not distinguish between the two types of shad. Although the depiction Rondelet provides[20] most likely depicts an Allis shad, due to the relatively wide body, small scales, and lack of dark spots on the back, it is not of sufficient quality to conclusively identify the species. The illustrations [Figs. 3.2 and 3.3] included by Gessner very clearly accentuate the physical differences between the species, such as the spots on the Twaite shad's back and the wider body of the Allis shad. Both woodcuts are made after original watercolours, which have been preserved in the Felix Platter collection [Figs. 3.10 and 3.11]. The woodcuts are more schematic than the originals, and they more clearly display the features that differentiate the two species from each other.

Gessner's description of the Allis shad is further developed over his three ichthyological publications. In the *Historiae Animalium*, although Gessner reports of a popular belief that the Allis shad travels up rivers and the fish return to sea when they hear thunder, the species is described as marine following the description by Rondelet. The texts in the *Platter album*, in the *Historiae Animalium*, and in the *Nomenclator* state the illustration of the Allis shad was made in Venice, most likely when Gessner was there in 1544.[21] The specimen that served as a model for Gessner's depiction of the species

18 Fischbuch fol. 179v *"Ein Elen lang"*. An 'Elen' translates to a cubit. A cubit is a unit of length based on the length of the forearm from the elbow to the tip of the middle finger, and is usually about 46 centimeters. Definition from the *Encyclopaedia Britannica*.

19 Linnaeus did not list the Allis and Twaite shads as separate species. The two species were not officially recognised until 1803 (Lacépède).

20 Rondelet, *Libri de piscibus marinis* 220.

21 In 1544 Gessner spent the summer in Venice. To read more about this, see: Fisher H., *Conrad Gessner 1516–1565. Universalgelehrter Naturforscher Arzt* (Zurich: 1967).

may have been caught in the Mediterranean Sea, and Gessner's initial conclusion that Rondelet was right to categorise this as a marine species may be due to this. However, by the time he compiled the *Nomenclator* Gessner described the species as anadromous. In the *Fischbuch* it becomes clear where Gessner may have done his research, as he points out the species travels up the Rhine to Basel, Gessner's native town.

The description and particularly the illustration indicate without a doubt that this is an Allis shad, as does the nomenclature; today, the species is known by the Latin name *Alosa alosa*. The original image [Fig. 3.11] is a highly realistic depiction, or rather a portrait of an Allis shad. The printed image [Fig. 3.2] highlights crucial characteristics, particularly the relatively small scales of the Allis shad, which make it stand out from other clupeidae and thus can serve to identify the species.

FIGURE 3.3 *Conrad Gessner, 'Additio ad* Alausam*'. Woodcut illustration to* Historiae animalium liber IV [...] *(Zurich, Christopher Froschauer: 1558) 1259.*

Gessner describes a second alausa in the paralipomena of the *Historiae Animalium*,[22] the addendum to the work. This species is not described by Rondelet. In addition to the name alausa, Gessner provides many local names for the species, including *Meienfische, Lußfich, Laußfisch, Alfe, Ziege*, and *Goldfisch*. In the *Nomenclator* Gessner adds the terms *thrissa, trichis, chalcide*, and *clupea* to this list. Gessner primarily refers to the species as *alausa*, or *Alfe*. Across all Gessner's ichthyological publications the species is described as anadromous, and both in the *Nomenclator*[23] and in the *Fischbuch*[24] the species is placed in the order of anadromous fishes. This second Alausa appears to be a Twaite shad, also included in the group of 'herring-like' species Gessner presents in his description of the herring. The species is described as being similar to the herring but much bigger, with a body as wide as a wide carp, long and not very fleshy, with silvery colouring, a high back, a length five times its width, a head not unlike that of a herring, and four black spots on either side of its back. In the *Fischbuch*, Gessner states the species is much like the herring, except that it is different in size and has black spots on its body.

22 Gessner, *Historiae animalium liber IV* [...], p. 1259.

23 Gessner, *Nomenclator* fol. 322r.

24 Gessner, *Fischbuch* fol. 179v.

The descriptions in all three publications state the species migrates from the North Sea up the river Elbe in May. It is likely the entire description is taken from the German physician Johan Kentmann, who published about the species in 1549,[25] brought it to Gessner's attention,[26] and possibly also supplied him with an illustration.[27] While the Allis shad is discussed in the main body of the *Historiae Animalium*, the description of the Twaite shad was placed in the paralipomena. In addition, the original watercolour of the Allis shad [Fig. 3.10] was pasted on paper between depictions of a smelt and a salmon rather than next to the depiction of the Twaite shad [Fig. 3.11], which was also preserved. This indicates the illustration of the Allis shad was pasted into the album before Gessner obtained the depiction of the Twaite shad. Therefore, it appears that Gessner did not acquire the depiction of the Twaite shad until just before the *Historiae Animalium* was published.

As Gessner's descriptions of both species evolve over his ichthyological publications, it appears research was done after publication of the *Historiae Animalium*, possibly inspired by the acquisition of the depiction of the Twaite shad. In the *Historiae Animalium* (1558) Gessner states this species must either be the same as the Allis shad, or be so similar that they are called by the same name. In the *Nomenclator*, published two years later, the Twaite shad is referred to as either the same species as or a similar species to the Allis shad, while in the *Fischbuch* Gessner describes the Twaite shad as another form of the Allis shad, belonging to the same family. The Twaite shad was not officially recognised as a separate species until 1803.[28] The fact that Gessner was able to recognise the close relation between these species while also distinguishing between them seems particularly remarkable, since he was not backed up by predecessors; Kentmann only described the Twaite shad, without mentioning the Allis shad, and Rondelet produced a description that is most likely of the Allis shad, although he does not provide enough information for successful identification. It is likely Gessner did some research on the species even before publication of the *Historiae Animalium*, as he mentions the species also migrates up the Rhine to Basel, and Kentmann's publication only discusses fish in the River Elbe. The woodcut is based on an original watercolour that is present in the *Platter album*; an annotation with the image states it was obtained from Kentmann [Fig. 3.11].

25 Kentmann Johan, *Animalium aquatilium* [...] (unknown publisher: 1549).

26 Gessner, *Historiae animalium liber IV* [...], p. 1259: 'id quod cum ex icone eius, hic addita, tum historia a Io<anne> Kentmano descripta'.

27 'Clupea ex Albi Zige a Kent', *Platter album* fol. 31.

28 Lacépède 1803.

The description, the nomenclature, and particularly the highly accurate illustration indicate this is the Twaite shad, or *Alosa fallax* (Lacépède, 1803), a species very closely related to the Allis shad. The original aquarelle [Fig. 3.11] image clearly depicts the spots on the flanks that make it possible to distinguish the species from the Allis shad, and in addition is a highly realistic and empirical depiction. The woodcut [Fig. 3.3] modelled on it is more schematic and pays even more attention to the species' crucial physical characteristics, including the spots on the flanks but also the relatively large scales, which identify it as a Twaite shad within the genus *Alosa.*

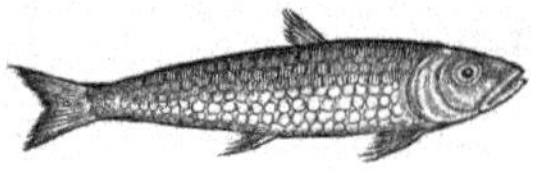

FIGURE 3.4 *Conrad Gessner,* Sardina. *Woodcut illustration to* Historiae animalium liber IV [...] (*Zurich, Christopher Froschauer: 1558) 990.*

Like the Allis and Twaite shads, the sardine is discussed in Gessner's description of the herring as part of a 'core group' of 'herring-like' species. In the *Historiae Animalium,*[29] as well as the *Nomenclator*[30] and the *Fischbuch,*[31] Gessner describes the sardine as *Hering Art.* The terms *trichis, chalcide,* and *thrissis* are also used to describe the species, as is *sardina,* and in the *Fischbuch* the term *Sardein* is used. In the *Historiae Animalium* Gessner describes the sardine as a marine species, and in the *Nomenclator* and the *Fischbuch* he places the species in the order of small marine fishes. The sardine is described as smaller than a herring but otherwise quite similar, with light scales that are easily removed, blue and green colouring on the back, spots on its back, and a row of hard scales on its keel, sharper than those of the herring. Gessner explains that based on the species' physical appearance, and particularly the row of sharp scales on its keel, it could be considered a herring, except that the sardine is smaller. It appears Gessner researched this species early on as the *Nomenclator* and *Fischbuch* contain only information that was already available in the *Historiae Animalium.*

Although he describes the sardine as a marine species, Gessner points out that freshwater varieties exist in lakes, referring to Rondelet, who describes such a variety. Rondelet produced an unrealistic depiction of a sardine and used the

29 Gessner, *Historiae animalium liber IV* [...], p. 990.

30 Gessner, *Nomenclator* fol. 3r.

31 Gessner, *Fischbuch* fol. 3r.

same illustration with both the freshwater[32] and marine variety[33] descriptions. Rondelet refers to the marine variety as *sardina, trichias*, or *trichis*, while he refers to the freshwater variety as *chalcide* or *celerin*. The terms *trichias* and *trichis*, from the Greek *trichias*,[34] connect the marine species with the shad and the herring, which Rondelet calls *thrissis*, while Gessner's description of the species as *Hering Art* connects the sardine with a much wider group of clupeidae. It seems unlikely that Gessner ever saw the freshwater variety, and he uncritically copied both Rondelet's description and illustration of this species.[35] However, he composed his description of the marine species himself, although it contains much information taken from other publications. In addition, Gessner had an original image created for his description of the marine species; this is present in the *Platter album* and was created in Venice[36] [Figs. 3.4 and 3.10], most likely when Gessner stayed there in 1544. Sardines being common in the Mediterranean, he must have seen plenty of fresh specimens of the species there.

The description, nomenclature, and depiction of the species leave no doubt that this is a sardine or *Sardina pilchardus* (Walbaum, 1792). Gessner complains the artist made a mistake when drawing the species, omitting the row of sharp scales in its keel and thereby rendering the image less than realistic. Nonetheless, the image [Fig. 3.4] is useable to identify the species because the relatively large scales of the sardine, which distinguish it from other clupeidae, have been clearly depicted.

FIGURE 3.5 *Conrad Gessner,* Chalcis altera Rondeletii. *Woodcut illustration to* Historiae animalium liber IV [...] *(Zurich, Christopher Froschauer: 1558) 18.*

The 'Chalcis altera Rondeletii' is described in the *Historiae Animalium*[37] as *thrissis* or *chalcide*, terms dating back to Athenaeus and Aristotle[38] that describe the sardine. In the *Nomenclator*[39] and the *Fischbuch*,[40] the term *Hering Art*

32 Rondelet, *L'histoire entière des poissons*, part 1, 181.

33 Rondelet, *L'histoire entière des poissons*, part 2, 105.

34 Τριχιας, see: Cuvier G. – Valenciennes A., *Histoire naturelle des poissons* (Paris: 1847) 20.

35 Following Rondelet, Gessner does not identify this freshwater sardine as a "herring-like" species.

36 *Platter album*, C22, fol. 58.

37 Gessner, *Historiae animalium liber IV* [...], p. 18.

38 Cuvier – Valenciennes, *Histoire naturelle des poissons* 20.

39 Gessner, *Nomenclator* fol. 346r.

40 Gessner, *Fischbuch* fol. 191r.

is added to the nomenclature, conceptually connecting the species with the herring and the 'core group' of 'herring-like' species consisting of the herring, both types of shad, and the sardine. The 'chalcis' is thus placed within a wider group of clupeids. Rondelet[41] also discusses the species, which he calls *chalcide* or *celerin*, providing only a very limited description. In Rondelet's work the term *celerin* is also applied to the Baltic herring, and the term *chalcide* to the sardine. Similarly, many other names applied to clupeid species in Rondelet's work refer to one or two other species, but since there is no consistent nomenclature that links all clupeid species in Rondelet's work, nor is there a consistent comparison with one specific species, such as the herring in Gessner's work, here the chalcis is not placed within the larger family of species.

Gessner's description in the *Historiae Animalium* largely repeats a description by Pierre Belon,[42] upon which Rondelet also based his description. The discussion of the species in the *Historiae Animalium* is much more extensive than the descriptions in the *Nomenclator* and *Fischbuch*, which are based on the description in the *Historiae Animalium*. The illustration [Fig. 3.5] is copied from Rondelet and is largely schematic; it clearly shows the species' features but is not a realistic portrait of an individual specimen. While it is not an empirical depiction, it is a usable taxonomical sketch. The description provides no indication Gessner ever saw or studied this species himself. Following both Rondelet and Belon, Gessner describes this species as a freshwater fish resembling the herring or a small sardine. In the extensive description in the *Historiae Animalium* Gessner quotes Belon, stating these fish are found in several lakes; larger ones can be found in Lake Garda, medium-sized ones in Lake Geneva, and small ones in Lake Como. The *Platter album* contains a series of depictions of species found in Lake Como;[43] therefore, it appears Gessner had contacts who did some research there. In addition, Gessner himself would have been able to do research at Lake Geneva. However, it is clear he did not study the species himself; the fish depicted here is a saltwater fish, and it is not possible that it was found in lakes. Did Gessner do research at Lake Geneva and find himself puzzled when he did not find the species there? Could this be the reason Gessner does not repeat his account about Lake Garda, Lake Como, and Lake Geneva in the *Nomenclator* and the *Fischbuch*?

In the *Nomenclator* and subsequently in the *Fischbuch* Gessner identifies this species as a type of herring, calling it '*Ein Heringart*'. Since Gessner also uses the word *sardanella*, it is possible this is a type of sardinella, a genus

41 Rondelet, *Universae aquatilium historiae* 149.

42 Belon, *La nature et diversité des poisons* 302.

43 *Platter album*, C22, fols. 49–52.

within the clupeidae family that consists of a range of species that look very similar to the image that illustrates the description. In his *La nature et diversité des poissons*, Belon also describes the species as a sardinella. The depiction of the species [Fig. 3.5] was compared to modern depictions of Mediterranean clupeidae.[44] These fall into two morphological categories: fish with high backs, including the Allis shad and Twaite shad, and fish with slender, elongated bodies, including the herring, the sardine, the sprat, and the sardinella. The slender fish depicted by Gessner after Rondelet [Fig. 3.5] should clearly be placed within the second category. Based on other depicted morphological characteristics, it is clear this is not a herring or a sardine. This means this could be a sprat or a sardinella, but since the pelvic fins are positioned further back on the body than the start of the dorsal fin, this can only be a sardinella. Sardinella are extremely common in the Mediterranean and are strictly saltwater fish. The round sardinella or *Sardinella aurita* (Valenciennes, 1847) is the only marine sardinella species that has been caught intensively along the Mediterranean coast.[45] It is very likely this is the species depicted here.

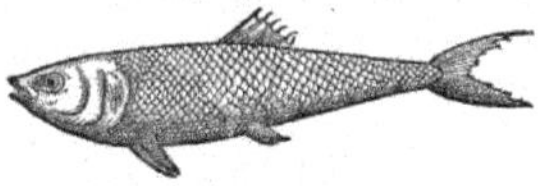

FIGURE 3.6A *Conrad Gessner,* Membras. *Woodcut illustration to* Historiae animalium liber IV [...] *(Zurich, Christopher Froschauer: 1558) 85.*

FIGURE 3.6B *Conrad Gessner,* Apua phalerica. *Woodcut illustration to* Historiae animalium liber IV [...] *(Zurich, Christopher Froschauer: 1558) 80.*

In the *Historiae Animalium*,[46] Gessner uses, amongst others, the terms *alosa, chalcis*, and *celerin* to describe the membras. In the *Nomenclator*[47] he adds the terms *harengus* and *Hering Art* to the nomenclature; only the term *Hering Art* is repeated in the *Fischbuch*.[48] Primarily Gessner refers to the species as membras [Fig. 3.6a], and in the *Historiae Animalium* and the *Nomenclator* he mentions it is sometimes also called *harengade*. That last term is no longer mentioned

44 Bauchot M.L. – Pras A. *Guide des poissons marins de l'Europe* (Paris: 1980).

45 Tsikliras A.C. – Koutrakis E.T. – Stergiou K.I., "Age and Growth of Round Sardinella (*Sardinella aurita*) in the Northeastern Mediterranean", *Scientia Marina* 69 (2005) 231–240.

46 Gessner, *Historiae animalium liber IV* [...], p. 85.

47 Gessner, *Nomenclator* fol. 3r.

48 Gessner, *Fischbuch* fol. 2v.

in the *Fischbuch*, and the description is shortened there. but. The text appears to be based on Rondelet's description of the species.[49] Rondelet calls this species *celerin* or *membradas.* In his work the term *celerin* is also applied to the sardine, while the term *membradas*, which is taken from Aristotle, is not found elsewhere. Consequently, while Gessner's nomenclature places the species in a wider group of clupeids, Rondelet only links the species to the sardine.
Following Rondelet, Gessner states that Aristotle teaches this is the adult state of a species he previously described as *apua phalerica* [Fig. 3.6b]. Aristotle claimed various types of small fish (apua) spontaneously generate out of other types of small fish, the membras coming from the apua phalerica.[50] The reference to the apua phalerica leads to a second description, that of a fish that based on the illustration, although this is notably different from that of the membras, appears beyond doubt to also be a herring-like species.[51] Gessner describes this second fish as small, with sharp scales on the keel, and so high in fat that fishermen can burn their lamps with the oil. This strongly indicates the apua phalerica is in fact the sprat, or *Sprattus sprattus*, which has hard scales on its keel and is one of the oiliest fishes in existence. The illustration [Fig. 3.6b]—in particular the position of the pelvic fin, which is implanted further forward on the body than the dorsal fin, and the small size of the fish—appears to confirm this. Following Rondelet,[52] this species is not identified as a herring-like species but as a type of apua, the category of spontaneously generating small fish described by Aristotle.

Aristotle's presentation of the apua phalerica as the immature form of the membras aside, the illustrations of the two fish [Fig. 3.6a and 3.6b], particularly the shape of the snout and the position of the pelvic fin in relation to the dorsal fin, strongly suggest that even though they are similar, they cannot be the same species. In his description of the herring Gessner mentions two similar small 'herring-like' species that are common in the Baltic Sea;[53] is it possible these are the two species discussed here? The sprat is extremely common in the Baltic Sea, as is a bigger 'herring-like' species, the Baltic herring. These are similar-looking fish, and since the sprat is much smaller, someone who did not pay attention to the position of the fins might mis-

49 Rondelet, *L'histoire entière des poissons* part 2, 182.

50 Aristotle, *History of Animals*, vol. VI, part 15, trans. D'Arcy Wentworth Thompson (Oxford: Clarendon Press: 1910).

51 Gessner, *Historiae animalium liber IV* [...], p. 80.

52 Rondelet, *Libri de piscibus marinis* 212–213.

53 Gessner, *Historiae animalium liber IV* [...], pp. 485–486.

take it for a young Baltic herring. Consequently, it would not be surprising if Gessner connected reports on these two species with Aristotle's descriptions of the membras and apua phalerica, and concluded his source appeared to confirm these, integrating the information he obtained from ancient and contemporary sources. The text does not indicate Gessner studied either species himself, and this may be why, while he replicates and connects information from different sources, he does not add conclusions of his own.

The description of the physical characteristics of the membras, as well as their depiction [Fig. 3.6a] and nomenclature, seem to confirm this identification. While the apua phalerica is very likely the sprat, Gessner's membras is almost certainly a *Clupea harengus* B. membras or Baltic herring. This species is very similar to the herring but smaller. It is often considered a distinct subspecies of the herring, despite the lack of a distinctive genome.[54] Another possibility is that this is a type of sardine. Gessner points out the species is called Harengade in Marseille, and a search in Cuvier's *Dictionnaire des sciences naturelles* (1820) leads to an entry describing a type of sardine.[55] The depiction [Fig. 3.6a], however, does not support this theory, as it does not display crucial physical characteristics of the sardine, such as its relatively large scales, but does display features typical of the Baltic herring, such as the long, angular snout, and pelvic fins that are positioned further back on the body than the origin of the dorsal fin.

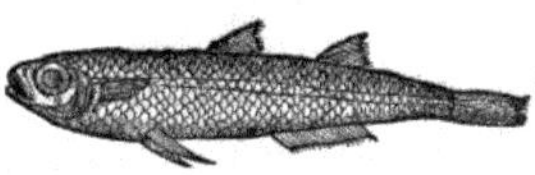

FIGURE 3.7 *Conrad Gessner,* Atherina. *Woodcut illustration to* Historiae animalium liber IV [...] *(Zurich, Christopher Froschauer: 1558) 83.*

The atherina is the only species Gessner describes as 'herring-like' that is not included in the Clupeidae family by modern taxonomists.[56] In his description of the atherina in the *Historiae Animalium*,[57] Gessner uses no nomenclature commonly applied to clupeidae; possibly this is due to the fact the atherina has no herring-like characteristics but does have a number of characteristics

54 Whitehead P.J.P., "Clupeoid Fishes of the World (Suborder Clupeioidei). An Annotated and Illustrated Catalogue of the Herrings, Sardines, Pilchards, Sprats, Shads, Anchovies and Wolf-herrings", *FAO Fisheries Synopsis* 125, vol. 7, part 1 (1985) 115–122.

55 Cuvier G. *Dictionnaire des sciences naturelles* (Strasbourg: 1820) 276.

56 Nijssen, H. – De Groot S.J., *De Vissen van Nederland* 13 (Utrecht: 1987).

57 Gessner, *Historiae animalium liber IV* [...], p. 83.

that render it particularly unlike a herring. However, in the *Nomenclator*[58] and the *Fischbuch*,[59] Gessner describes the species as *Hering Art*. This may be because he copied much of his description from Rondelet, who describes the species as similar to a sardine, which is a clupeid, or an anchovy, which belongs to the order of clupeiformes and is consequently very closely related to the clupeidae.[60] The illustration [Fig. 3.7], a schematic depiction displaying features crucial for identifying the species, is also copied from Rondelet.

Following Rondelet, Gessner describes the Atherina as about the size of a finger, with big eyes, a transparent body, a line from head to tail, and two dorsal fins. As clupeidae never have two dorsal fins, it is remarkable Gessner identified this fish as a type of herring. It may be that Gessner based his identification solely on Rondelet's description of the species as similar to the sardine and the anchovy, because he was not able to research the species himself. However, it appears he was not entirely convinced of this identification; while in the *Historiae Animalium* Gessner is vague about what type of fish this is, in the *Nomenclator* he calls the species a *Spirinchen*, or a small herring.[61] The name *Spirinchen*, which we would now expect to refer to the *Osmerus eperlanus*, or European smelt, was traditionally a vernacular name for the *Atherina presbyter*,[62] or sand-smelt. This identification is correct—this is a sand-smelt; however, in the *Fischbuch* this categorisation has been dropped from the nomenclature. The *Fischbuch* was a popular publication based on the *Historiae Animalium* and the *Nomenclator*, and was not intended for a scholarly audience. Its production was overseen by Conrad Forer, and it is unclear to what extent Gessner was involved in this. Consequently, it is possible the identification of the species as a sand-smelt was dropped by mistake, rather than intentionally.

Based on the characteristics—a small fish with big eyes, a silver-coloured line on the flanks, and two dorsal fins—the atherina appears to be a sand-smelt, or *Atherina presbyter*, a species of marine fish of the Atherinidae family.

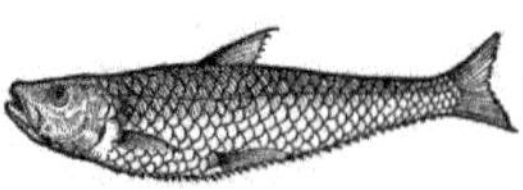

FIGURE 3.8 *Conrad Gessner,* Liparis Lacustris. *Woodcut illustration to* Historiae animalium liber IV [...] *(Zurich, Christopher Froschauer: 1558) 572.*

58 Gessner, *Nomenclator* fol. 3r.

59 Gessner, *Fischbuch* fol. 2v.

60 Rondelet, *Libri de piscibus marinis* 216–217.

61 Gessner, *Nomenclator* fol. 3r.

62 Nijssen – De Groot, *De Vissen van Nederland* 124.

In the *Historiae Animalium*[63] Gessner uses mainly local names to describe this species. However, in the *Nomenclator*[64] he uses the terms *chalcis* and *thrissis*, and in both the *Nomenclator* and the *Fischbuch*[65] he uses the names *agonus, Alfe*, and *Häring*. Rondelet also describes the species[66] and refers to it as *liparis*; while this nomenclature does not link the species to a wider group of clupeids, Gessner's addition of the term *Häring* does. Gessner's description of the Liparis Lacustris seems to be entirely based on a description by Belon,[67] who most likely studied the species himself on a scientific journey he undertook under the patronage of Cardinal François de Tournon between 1546 and 1549.[68] Following Belon, Gessner correctly describes the liparis as an oily fish caught in Greek lakes, similar to the sardine. He states that the head resembles that of a herring, the fish has a row of sharp scales on its keel, and it has silvery scales that can be removed easily. Rondelet also provides a description of a liparis, with which he supplies no other nomenclature, a substandard illustration, and a description that in no way presents this fish as similar to the herring. Gessner characterises the species as a freshwater fish, and in the *Nomenclator* and *Fischbuch* he places it in his order of various freshwater fishes. As he never travelled to Greece, it is unlikely Gessner ever had a chance to study the species himself. The illustration [Fig. 3.8] was copied from Belon, and is consequently not present in the *Platter album*.

The description, nomenclature, and depiction of the liparis strongly suggest the species is a Macedonian shad, or *Alosa macedonica* (Vinciguerra, 1921), otherwise known as liparia. This species is endemic to Lake Volvi in Greece and is currently listed on the IUCN Red List of Threatened Species as being under threat of extinction.[69] *Alosa macedonica* is only found in fresh water and is a member of the *Alosa* genus. It has been suggested that its ancestors inhabited

63 Gessner, *Historiae animalium liber IV* [. . .], p. 572.

64 Gessner, *Nomenclator* fol. 346r.

65 Gessner, *Fischbuch* fol. 191v.

66 Rondelet, *L'histoire entière des poissons* part 2, 215–216.

67 Belon, *La nature et diversité des poissons* 305.

68 See also: Belon Pierre, *Les observations de plvsievrs singvlaritez & choses mémorables, trouuées en Grece, Asie, Iudée, Egypte, Arabie, & autres pays estranges* (Paris, Gilles Corrozet: 1554).

69 Crivelli A.J., "Alosa macedonica", in *IUCN Red List of Threatened Species. Version 2013.2* (www.iucnredlist.org: 2013).

marine regions of the Aegean Sea.[70] It is likely the species was anadromous and at one point became landlocked.

Discussion

What does Gessner's identification of herring species tell us of his concept of classification, and in what way was this innovative? In his descriptions of 'herring-like' species, and particularly that of the herring, Gessner has provided a defined list of their common characteristics as well as a list of characteristics that set this group apart from other species. In doing so he has started to outline how these species can be identified and distinguished on the basis of their morphological characteristics, and on which basis they are grouped together. The key herring-like features he lists are an elongated body, silver-coloured scales that are easily removed, black and blue colouring on the back, and lighter keels with a row of sharp scales. Features he mentions as a means of differentiating between species include, among others, the spots on the back and flanks of the sardine and Twaite shad, size of scales, and habitat. In his description of the herring in the *Historiae Animalium* Gessner already identifies the Allis shad, the Twaite shad, the sardine, and the herring as similar species; in his later publications he adds the round sardinella, the Baltic herring, and the Macedonian shad, as well as the sand-smelt, to this group by adding the word *Häring Art, Haring*, or *Hering* to their nomenclature and by repeating the characteristics he lists in his description of the herring. Consequently, Gessner establishes an organisation of species expressed through a logical nomenclature and based on physical similarities and differences. This delineation of groups of species takes place across a habitat-based type of organisation, which is also used by other ichthyologists of the time, such as Rondelet. In the *Nomenclator* and *Fischbuch* species are organised primarily into marine, anadromous, or freshwater species. Using nomenclature, Gessner links species across these categories. Apart from the sand-smelt, which in the *Nomenclator* is still correctly identified, all herring species Gessner identifies should according to modern taxonomy be placed within the order Clupeidae.

70 Economidis P.S. – Sinis A.I., "Alosa macedonica (Vinciguerra 1921)", in Hoestland H. (ed.), *The Freshwater Fishes of Europe*, vol. 2 (Wiesbaden: 1991) 297–308. Faria R. – Weiss S. – Alexandrino P., "A Molecular Phylogenetic Perspective on the Evolutionary History of *Alosa* spp. (Clupeidae)", *Molecular Phylogenetics and Evolution* 40 (2006) 298–304.

FIGURE 3.9 Sardina. *Aquarelle drawing from the* Platter album. *Special Collections of the University of Amsterdam, C22, fol. 58. Reproduced with the kind permission of the Special Collections of the University of Amsterdam.*

Gessner's focus on morphology is evident from the depictions he used to illustrate his ichthyological works. His woodcuts clearly display physical characteristics that identify the species. As such, the illustrations can be considered good taxonomical depictions, comparable to those used in biology today. The characteristics are displayed schematically and depict the species in general, rather than as individual specimens. In some cases it is clear the image was never drawn after a real specimen; the illustration of the fresh herring, for example [Fig. 3.1a], is a composite image containing elements of other illustrations. As a taxonomical depiction Gessner's herring is a leap forward from that by Rondelet, and particularly Belon, with the species' fins, head, and tail being much more accurately depicted.[71] In the case of the sardine, both the original watercolour [Fig. 3.9] depiction and the woodcut based on it [Fig. 3.4] are taxonomical depictions. Although this depiction is missing the row of sharp scales on the keel, which Gessner complains about, the species is nonetheless clearly depicted, and since its relatively large scales are prominently shown, it can be identified as a sardine within the group of clupeidae.

Even when the original watercolours were drawn after nature and can be considered good empirical depictions that depict an individual specimen, the printed images modelled upon them were also made into abstract representations of the species, usable as taxonomical depictions. This is the case with the depiction of the Allis shad. The original watercolour [Fig. 3.10] is beautifully executed and at first sight a highly realistic portrait, while the woodcut [Fig. 3.2] is much more schematic and can be used as a taxonomical depiction.

71 Gessner, *Historiae animalium liber IV* [...], pp. 485–486; Rondelet, *Libri de piscibus marinis* 222–223; Belon, *La nature et diversité des poissons* 168–169.

FIGURE 3.10 Clupea. *Aquarelle drawing from the* Platter album. *Special Collections of the University of Amsterdam, C22, fol. 30. Reproduced with the kind permission of the Special Collections department of the University of Amsterdam.*

FIGURE 3.11 Clupea ex Albi. *Aquarelle drawing from the* Platter album. *Special Collections of the University of Amsterdam, C22, fol. 31. Reproduced with the kind permission of the Special Collections department of the University of Amsterdam.*

Features that help identify the species are clearly depicted, such as the relatively small scales in relation to the length and height of the fish, which is a crucial characteristic within the Clupeidae family to identify the fish as an Allis shad. The depiction of the Twaite shad [Fig. 3.11] is another example. The colouring has been applied empirically and is so detailed it is clear a real specimen is depicted here. The woodcut in Gessner's publications [Fig. 3.3] is more schematic and pays much attention to features crucial for identifying the species, such as the relatively large scales and the spots on the flanks of the fish, which identify it within the group of the clupeidae as a Twaite shad. This image no longer depicts an individual specimen but the species as a whole.

Can Gessner's 'herring-like' species be distinguished successfully based on the characteristics he lists? In modern biology, identifying species often still poses a problem. Related species are described and depicted with special attention to the characteristics that set them apart, and determination keys listing the morphological details that separate one species from another are composed as guidelines. Such determination keys are not perfect tools for identifying species; often they have to be adjusted in light of new insights, and in some cases they can only be applied successfully if one is familiar with all species being discussed. However, in practice they form a fairly efficient way to distinguish related species, and the most reliable method of doing so apart from genetic research. Do Gessner's descriptions of species he identifies and names as herring-like measure up to these modern standards, and can a determination key be created successfully based on the information he provides? Although this is difficult in the case of the wrongly identified atherina and the chalcis Rondeletii, which was mistakenly described as a freshwater fish, it is certainly possible for those species Gessner studied personally, the *Alosa alosa* (Allis shad), *Alosa fallax* (Twaite shad), *Clupea harengus* (herring), and *Sardina pilchardus* (sardine). In addition, Gessner's descriptions of the *Clupea harengus membras* (Baltic herring) and *Alosa macedonia* (Macedonian shad), although not based on original research, also provide sufficient information to identify the species. The main difference with a modern identification of species is that Gessner orders species by habitat. It is, however, possible to create a workable determination key.

Determination Key for Clupeidae in Gessner's Work

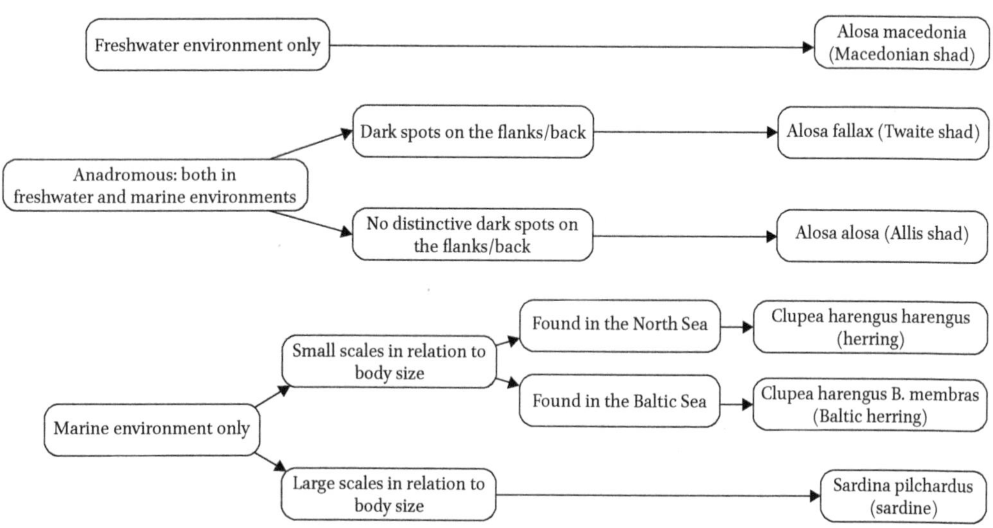

Gessner's attention to morphology and overall attention to detail made it possible to successfully group species together based on physical similarities, as well as to successfully distinguish between species within a group. In contrast, the only contemporary ichthyologist whose work could be compared to Gessner's, Guillaume Rondelet, only identified very limited groups of species. Even though seven of the eight species Gessner includes as clupeids are present in Rondelet's work, Rondelet only identified the herring, the sardine, and the Allis shad as belonging to the same group. Although he also connects these three species through nomenclature, calling each of them *thrissis*, he does not connect the other clupeid species he describes to this group, nor does he list sufficient morphological characteristics to distinguish these three species. Distinguishing features that Rondelet lists include relative size, the fact that the sharp scales on the herring's keel are relatively soft, and the fact that only sardines have dark spots on their backs. As Rondelet categorises all of these species as marine, habitat is not a distinguishing characteristic here. Distinguishing species based on the relative sharpness of the scales on the keel is nearly impossible, and since all fish start their lives small, relative size is also not a suitable characteristic to tell species apart. Consequently, only the black spots on the flanks and back of the sardine are a suitable distinguishing feature, and out of the three species only the sardine can be identified.

In contrast, Gessner's ichthyological work shows a clear conception of classification of species into groups comparable to modern taxonomical families, as well as an attention to the differences between the various members of one family that enabled him to successfully distinguish between them. This attention to the characteristics of species is also visible in Gessner's ichthyological illustrations, which emphasise the common and distinguishing characteristics of species and are comparable to the taxonomical depictions used in biology today. Gessner's nomenclature reflects his identification of species into groups; terms that delineate a group of species as effectively as Gessner's use of the terms *Hering* and *Hering Art* cannot be found in other ichthyological works of his time. Gessner's attention to morphology enabled him to present a coherent organisation of species and introduce a logical nomenclature that reflects the relation he recognised between species. In this Gessner's ichthyological work forms a distinct innovation compared with that of his contemporaries.

Selective Bibliography

Aristotle, *History of Animals*, trans. D'Arcy Wentworth Thompson (Oxford, Clarendon Press: 1910).

Belon Pierre, *De aquatilibus* [...] (Paris, Charles Estienne: 1553).

Belon Pierre, *La nature et diversité des poissons* [...] (Paris, Charles Estienne: 1555).

Cuvier G. – Valenciennes A., *Histoire naturelle des poissons* (Paris: 1847).

Cuvier G. *Dictionnaire des sciences naturelles* (Strasbourg: 1820).

Gessner Conrad, *Fischbuch* (Zurich, Christopher Froschauer: 1563).

Gessner Conrad, *Historiae animalium liber IV, qui est de piscium et aquatilium animantium natura* (Zurich, Christopher Froschauer: 1558).

Gessner Conrad, *Nomenclator aquatilium animantium* (Zurich, Christopher Froschauer: 1560).

Kentmann Johan, *Animalium aquatilium* [...] (Unknown publisher: 1549).

Rondelet Guillaume, *L'histoire entière des poisons* (Lyon, Matthieu Bonhomme: 1558).

Rondelet Guillaume, *Libri de piscibus marinis* (Lyon, Matthieu Bonhomme: 1554).

Rondelet Guillaume, *Universae aquatilium historiae* (Lyon, Matthieu Bonhomme: 1555).

Salviani Hippolito, *De historia aquatilium animalium* (Rome, Hippolito Salviani: 1558).

CHAPTER 4

Der Wal als Schauobjekt: Thomas Bartholin (1616–1680), die dänische Nation und das Ende der Einhörner *With an English Summary*

Bernd Roling

Einleitung

Daß Einhörner eine der erfolgreichsten Chiffren in der europäischen Kunst- und Literaturgeschichte begründen konnten, hat nicht zuletzt die gewaltige Arbeit von Jürgen Einhorn materialsatt dokumentieren können.[1] Seitdem Ktesias in seiner Ethnographie Indiens von einem Wesen berichtet hatte, das die Gestalt eines Pferdes besaß, mit einem rötlich-weißen Fell versehen, bewaffnet mit einem großen Horn auf seinem Haupt und das, wie man wußte, kaum zu erlegen war, war die Karriere dieser Kreatur nahezu beispiellos und ist bis in die Niederungen der Kitschposter und Zeichentrickserien verfolgt worden.[2] Zu einem aristokratischeren Aspekt dieses Ruhms hatte Ktesias selbst einen wichtigen Beitrag geleistet; Männer wie Aelian hatten für die Verbreitung seiner Behauptung gesorgt.[3] Gejagt worden war das Einhorn in Indien wegen seines Horns, so Ktesias, das nicht nur Krämpfe und Epilepsie

---

1 Ich danke der Staatsbibliothek Berlin und der Universitätsbibliothek Göttingen für ihre Unterstützung bei der Ausarbeitung dieser Studie.

2 Einhorn J.W., *Spiritalis unicornis. Das Einhorn als Bedeutungsträger in Literatur und Kunst des Mittelalters* (München: 1976) *passim*. Unter der jüngeren Literatur zum Einhorn seien außerdem zuvorderst genannt Shepard O., *The Lore of the Unicorn* (London: 1967, zuerst 1930) *passim*; daneben Beer R., *Einhorn. Fabelwelt und Wirklichkeit* (München: 1972), dort zur symbolischen Komponente des Einhorns und seiner antiken und mittelalterlichen Naturgeschichte vor allem 11–117; Lavers C., *The Natural History of Unicorns* (London: 2009) v.a. 44–111, und jetzt die sehr wertvolle Darstellung von Gerritsen W., *Het spoor van de eenhoorn. De geschiedenis van een dier dat niet bestaat* (Leiden: 2011). Für die Diskussion der Einhörner im 16. und 17. Jahrhundert ist grundlegend Garboe A., *Enhjørningen især i naturog lægevidenskabens historie* (Kopenhagen: 1915). Garboes exzellente Arbeit ist außerhalb Dänemarks unbekannt geblieben.

3 Ktesias, *La Perse—L'Inde—Autres fragments*, ed. D. Lenfant (Paris: 2004), *Indica* § 45, 182 f. und 203 f.

 | DOI 10.1163/9789004279179_006

kurieren konnte, sondern sich vor allem als ein nahezu unschlagbares Heilmittel gegen Vergiftungen aller Art erwiesen hatte.[4] Nachdem sich die Übersetzer der Septuaginta dafür entschieden hatten, in eine der vielen rätselhaften Kreaturen des Alten Testamentes, das *re'em*, das *monoceros* hineinzulesen und auch Hieronymus diesen Vorschlag dankbar annahm, hatte das Einhorn auch die Autorität der Offenbarung auf seiner Seite.[5]

Dieser Beitrag soll sich nicht mit den symbolischen Bedeutungen des Einhorns beschäftigen, die in ihrem Ursprung der *Physiologus* generiert hatte, seiner Rolle als Christussymbol oder seiner Funktion in den Liebesallegorien, sondern mit seiner Stellung in der frühneuzeitlichen Zoologie.[6] Seit dem Mittelalter verwahrte man in den Wunderkammern der Alten Welt die Hörner des Einhorns, stellte sie zur Schau, nutzte sie jedoch vor allem als nahezu universales, wenn auch nur für Fürsten erschwingliches Medikament, das, wie im Fall von Königin Elisabeth I. von England, für zehntausend Pfund gehandelt wurde. Von Interesse soll hier das Ende dieser Einhörner sein, das sich vor allem mit einer Figur verbindet, der Gestalt des dänischen Naturwissenschaftlers, Mediziners, Anatomen, Antiquars und Professors der Universität Kopenhagen, Thomas Bartholin, der zu diesem Ende den entscheidenden Beitrag geleistet hatte. Es war Bartholin, der ganz Europa offenbaren mußte, das die Einhörner in Wirklichkeit die Stoßzähne eines Narwals waren. Gezeigt werden hier im folgenden drei Dinge: Zum ersten, daß die Enttarnung der ebenso ästhetisch aufgeladenen wie pharmazeutischen Schauobjekte das Ergebnis einer allgemeinen Erschließung der polaren Fauna war und ihrer Meeresbewohner im Besonderen, die innerhalb von nur zwei Jahrzehnten entscheidend vorangetrieben wurde; zugleich jedoch auch ein Produkt einer akademischen Vernetzung des Nordens, die erst in diesen Jahren erreicht wurde. Zum zweiten soll deutlich werden, daß die Dekonstruktion der Hörner weiter ausgriff, als es Bartholin selbst beabsichtigt hatte, und er selbst, vielleicht auch aus kommerziellen Interessen, versuchte, den Status der Hörner aufrechtzuerhalten, obwohl sie ihren ursprünglichen Charakter verloren hatten. Zum dritten wird gezeigt, daß diese Bemühungen Bartholins ins Leere greifen mußten und er die Konsequenzen seiner Entdeckung nicht mehr bändigen konnte. Die gleiche

4 Aelian, *On the Characteristics of Animals*, ed. A.F. Schoolfield, 3 Bde. (Cambridge: 1958–1959), Bd. I, lib. III, cap. 41, S. 200–201; Bd. III, lib. XIII, cap. 25, S. 124 f.; lib. XV, S. 236 f.

5 In der Heiligen Schrift z.B. Ps. 91, 11 oder Ps. 21, 22, ebenso Num. 23, 22 und 24, 8 und Dt. 33, 17. Zum Einhorn in der Bibel auch Shepard, *The Lore* 41–43; Einhorn, *Spiritalis unicornis* 42–44; Beer, *Einhorn* 21 f., Lavers, *The Natural History* 44–52; Gerritsen, *Het Spoor* 30–37.

6 Zum Einhorn im ‚Physiologus' z.B. Shepard, *The Lore* 45–51; Einhorn, *Spiritalis unicornis* 47–81; Beer, *Einhorn* 45–66; Lavers, *The Natural History* 63–78; Gerritsen, *Het spoor* 38–45.

Erfahrungswissenschaft, die Bartholin für sich in Anspruch nahm und die seine Rettungsversuche des Einhorns zeitweilig unterstützen konnte, mußte sich gegen ihn wenden. Die Hörner wurden, was sie vielleicht von Anfang an hätten sein sollen, ein Gegenstand der Kunst und der Naturgeschichte, doch verloren sie den Rang eines Heilmittels.

I Die Zweifel an der Existenz der Einhörner und der Beginn der modernen Walkunde

In der zweiten Hälfte des 16. Jahrhunderts war das Einhorn zum ersten Mal in seiner zweitausendjährigen Geschichte in eine wirkliche Krise geraten. Gesehen wurde es durchaus: Berühmt geworden sind die Beschreibungen, die die Jerusalempilger Felix Fabri und Bernhard von Breytenbach zum Ende des 15. Jahrhunderts von ihm gegeben hatte,[7] oder Lodovico de Varthema, der eine wieder und wieder zitierte Schilderung eines Einhorns geliefert hatte, dem er im Inneren Arabiens begegnet sein wollte.[8] Der Jesuit Geronimo Lobo und ein spanischer Franziskaner mit Namen Luis de Urreta, die beide weitläufig Abessinien bereist hatten, hörten noch Anfang des 17. Jahrhunderts von Exemplaren, die alle Eigenschaften aufwiesen, die ihnen die Tradition zugeschrieben hatte. Groß wie ein Pferd, ein erhabenes, mehrere Ellen langes Horn auf dem Haupt und so behende, daß es nur schwer erlegt werden konnte.[9]

---

7 Unter den vielen Ausgaben dieser Pilgerberichte z.B. Fabri F., *Evagatorium in Terrae Sanctae, Egyptiae et Arabiae peregrinationem*, ed. K.D. Hässler, 2 Bde. (Stuttgart: 1843–1849), Bd. II, 441 f. (39b f.), und von Breydenbach B., *Peregrinatio in Terram sanctam*, ed. I. Mozer (Berlin – New York: 2010) 551 f. (fol. 139v). Eine aktuelle Monographie zu diesen zwischen Phantastik und Erbauung schillernden Texten liegt vor von Klußmann A., *In Gottes Namen fahren wir: die spätmittelalterlichen Pilgerberichte von Felix Fabri, Bernhard von Breydenbach und Konrad Grünemberg im Vergleich* (Saarbrücken: 2012), dort zum Einhorn in beiden Reiseberichten 98–99.

8 Varthema Ludovicus, *The Travels in Egypt, Syria, Arabia deserta, Arabia felix, in Persia, India and Ethiopia, A.D. 1503 to 1506*, hrsg. und übers. J. Winter Jones – G.P. Badger (London: 1853; ND Farnham: 2010) 46–49.

9 Zu Lobos Einhorn in der berühmten Übersetzung Samuel Johnsons, *A Voyage to Abyssinia by Father Jerome Lobo, a Portugese Missionary, containing the History, Natural, Civil and Ecclesiastical of that remote and unfrequented country* (London, Elliot and Kay: 1779), A description of Abyssinia, chapt. 2, 68 f.; ebenso Urreta Louis de, *Historia ecclesiastica, politica, natural y moral, de los grandes y remotos Reynos de la Etiopia, Monarchia del Emperador, llamado Preste Juan de las Indias* (Valencia, Pedro Patricio May: 1610), lib. I, cap. 25, S. 246–247.

Kritik an den Einhörnern kam von einer anderen Seite, von den Medizinern, die das seit der Spätantike etablierte Heilmittel des Einhorns nicht mehr widerspruchlos hinzunehmen bereit waren. Aus ihrem Kreise wurde der Mythos des Einhorns selbst in Frage gestellt. Odell Shephard hat in seinem fulminanten, in den dreißiger Jahren erschienenen *Lore of the Unicorn*, das Grundlage aller Arbeit am Einhorn bilden muß, mit großem Recht auf die Rolle hingewiesen, die der Traktat über die Einhörner, den der Italiener Andrea Marini im Jahre 1566 verantwortet hatte, in dieser ersten Debatte gespielt hatte.[10] Niemals war die in Mitteleuropa verbreitete Substanz, so Marini, in der Lage gewesen, vor Vergiftungen zu schützen, auch wenn die Marketingexperten der frühneuzeitlichen Pharmamafia das Gegenteil behauptet hatten. Zugleich so Marini, mußte die Herkunft der Hörner zur Gänze im Obskuren bleiben.[11] Trotz der Gegenschrift, die sein Landsmann Andrea Bacci schon wenig später folgen ließ und der sofortigen Reaktion der Einhorn-Freunde,[12] genannt sei nur Julius Caesar Baricelli,[13] waren Marinis Zweifel sofort auf fruchtbaren Boden gefallen. Auch Koryphäen der frühneuzeitlichen Heilkunst, darunter Hieronymus Mercurialis,[14] Gabriel Frascati oder Ambroise Paré,[15] aber auch Geologen wie Anselm de Boodt, schlossen sich seiner Auffassung an,[16] oder konnten sie wie Amatus Lusitanus sogar antizipieren.[17] Paré stellte die berechtigte Frage, warum weder Galen noch Hippokrates je das Einhorn als Therapeutikum ver-

---

10 Shepard, *The Lore* 155–168, außerdem auch Gerritsen, *Het Spoor* 160–165. Eine Zusammenfassung des Streites und einen erschlagenden Katalog an Autoritäten liefert Aldrovandi Ulisse, *De quadrupedibus solidipedibus volumen integrum* [...] (Frankfurt, Johannes Treudel: 1623), lib. I, cap. 6, S. 187–190.

11 Marini Andrea, *Discorso contra la falsa opinione dell'Alicorno* (Venedig, Paolo Manuzio: 1566), z.B. 31–36. Als Pestmedikament empfiehlt es fast zeitgleich Joubert Laurent, *De peste liber unus* (Lyon, Johannes Frellonius: 1567), cap. 18, S. 126–127.

12 Bacci Andrea, *L'Alicorno: discorso del'eccellente medico e filosofo M. Andrea Bacci, nel quale si tratta della natura dell'alicorno, e delle sue virtù* (Florenz, Giorgo Marescotti: 1573), terza parte, 57–80.

13 Baricelli Giulio Cesare, *Hortus genialis sive arcanorum valde admirabilium tam in arte medica quam reliqua philosophia compendium* (Köln, Matthias Schmitz: 1620) 221–222.

14 Mercurialis Hieronymus, *Consultationes et responsa medicinalia quatuor tomis comprehensa* (Venedig, Iunta: 1624), Bd. I, „Consultatio" 94, 121–123.

15 Frascati Gabriele, *De aquis Returbii Ticinensibus commentarii mineras, facultates et usum earum explicantes* (Ticini, Hieronymus Bartholus: 1575), lib. III, 137a.

16 De Boodt Anselmus, *Gemmarum et lapidum historia* (Hannover, Claudius Marnius: 1609), lib. II, cap. 244, S. 211–215.

17 Lusitanus Amatus, *In Dioscoridis Anazarbei de medica materia libros quinque enarrationes* (Venedig, Scotus: 1553), lib. I, „Ennaratio" 52, 195–197.

ordnet hatten.[18] War es je von Nutzen gewesen, so Paré, dann mit ihm auch jeder andere Hornwuchs eines gewöhnlichen Paarhufers. Doch wenn es nie Einhörner gegeben hatte, wie Marini alludierte, welche Exponate lagen dann in den Schatzkammern und Apotheken der europäischen Fürsten?[19] Schon Marini hatte die These aufgestellt, etliche der in Frankreich und Italien verbreiteten Hörner seien in Wirklichkeit Artefakte aus Elfenbein gewesen und schon aus diesem Grunde ohne jede Wirkung. Anfang des 17. Jahrhunderts stößt man auf die ersten Überlegungen, welche Substanzen statt des sagenhaften Tiers als Einhorn gehandelt wurden. Jacob Primerose, Verfasser der *Errores in medicina*, macht im Jahre 1638 einen bemerkenswerten Vorschlag. Keine Landtiere, sondern die Knochen und Zähne eines Meerestiers, oder besonderes Elfenbein seien das wahre Einhorn gewesen.[20] Der englische Arzt und Aufklärer bleibt noch weitgehend ungehört.

Will man die zweite Hälfte des 16. Jahrhunderts als eine Epoche der neuen Erfahrungswissenschaft werten, so kam im Falle der Einhörner eine zweite Komponente hinzu, die auf eine ähnliche Weise für eine Authentifizierung der Naturgeschichte sorgte. Mochte der Walfang auch über Jahrhunderte eine wichtige Einnahmequelle der atlantischen Küstenbewohner gewesen sein, das Bild des Wals und seiner diversen Varianten war dennoch bis in die Frühe Neuzeit verschwommen gewesen.[21] Allenfalls die kaum rezipierte Aufarbeitung des Walfangs in den *Libri de animalibus* des Albertus Magnus war hier eine Ausnahme gewesen,[22] doch die Angaben der hoch- und spätmittelalterlichen Enzyklopädisten, doch selbst der Katalog der Walfische,

---

18 Eine ausführlichere Betrachtung der Kritik Parés findet sich bei Gerritsen W., „De Eenhoorn en de Apothekers. Opvattingen omtrent de antitoxische werking van ‚eenhoornhoorn' in het laatste kwart van de 16e en het eerste kwart van de 17e eeuw", *Gewina* 30 (2007) (1–10) 3–8, und auch schon bei Garboe, *Enhjørningen* 16–17; Shepard, *The Lore* 168–172.

19 Paré Ambroise, *Les Œvvres* (Lyon, Claude Rigaud: 1641), *Chirurgia*, liv. 21, chapt. 47, S. 509. Parés medizinische Schriften erhielten zahllosen Auflagen.

20 Primerose James, *De vulgi erroribus in medicina* (Amsterdam, Johannes Jansson: 1644), lib. IV, cap. 38, S. 194–199.

21 Eine allgemeine Skizze der Walkunde des Mittelalters und der Frühen Neuzeit gibt die Arbeit von Szabo V.E., *Monstrous Fishes and the Mead-Dark Sea. Whaling in the Medieval North Atlantic* (Leiden – Boston: 2008). Zum Wal in der spätmittelalterlichen Enzyklopädik z.B. Megenberg Konrad von, *Das Buch der Natur*, ed. F. Pfeiffer (Stuttgart: 1861; ND Hildesheim: 1962), lib. III, cap. 7, S. 247–248; Thomas von Cantimpre, *Liber de natura rerum*, ed. H. Boese (Berlin – New York: 1973), lib. VI, cap. 6, S. 233–234; Vinzenz von Beauvais, *Speculum naturale* (Douai: 1624; ND Graz: 1964), lib. XXVII, cap. 43, Sp. 1275 f.

22 Magnus Albertus, *De animalibus libri XXVI*, ed. H. Stadler, 2 Bde. (Münster: 1920), lib. XXIV, cap. 1, Nr. 23, § 16, Bd. II, S. 1523–1524.

den Olaus Magnus in seiner *Geschichte der Nordvölker* offerierte, schillerte zwischen vagen Beschreibungen und rein monströsen Passagen.[23] Auch die in der Jahrhundertmitte entstandenen ersten Fischkunden Pierre Belons, Guillaume Rondelets und mit ihnen Conrad Gessners, so bedeutsam sie für die Genese der Ichthyologie auch gewesen sein mochten, hatten Mühe, zwischen mehr als drei oder vier Spielarten des Wals zu unterscheiden.[24] Diese Situation änderte sich in der letzten Dekade des 16. Jahrhunderts fast schlagartig, ja man darf sagen, der Wal hatte die europäische Bühne betreten. Ausführliche und detailreiche Beschreibungen der Walfische beruhten nun auf Autopsie. Für die Jahre 1595 bis 1600 verzeichnet die *Bibliotheca ichthyologica* Dirk Mulder-Bosgoeds allein in den Niederlanden ein Dutzend auf Augenschein beruhende Traktate, die sich gestrandete und öffentlich ausgeweidete Tiere zum Gegenstand gewählt hatten, die meisten von ihnen scheinbar Pot- und Grönlandwale.[25] Auch im Hinterland war binnen weniger Jahre auf diese Weise ein Bewußtsein für die Existenz der Meeressäuger entstanden. Hinzu kamen die ersten Reiseberichte, die über den Walfang vor Island, Grönland und Spitzbergen Auskunft erteilen konnten.[26] Nicht entgangen war einem Autor wie William Baffin, dessen Bericht Samuel Purchas 1625 gedruckt hatte, dabei auch, daß die Polarmeere einen bizarren Walfisch vorweisen konnten, der einem Einhorn zumindest in einem Detail sehr nahe kam, dem Horn, und der aus diesem Grunde zurecht Meereseinhorn, *Sea-Unicorn* genannt werden konnte. Bemerkenswert war darüber hinaus, so Baffin, daß die Einwohner

---

23 Magnus Olaus, *Historia de gentibus septentrionalibus* (Rom: 1555; ND Kopenhagen: 1972), lib. XXI, cap. 7, S. 736; cap. 9, S. 738; cap. 13, S. 741 f.; cap. 16, S. 746; cap. 25–26, 754 f. Englisch als Magnus Olaus, *Historia de gentibus septentrionalibus (Rom 1555)—Description of the Northern Peoples*, transl. P. Fisher – H. Higgens, 3 Bde. (London: 1996), Bd. III, book XXI, chapt. 7, S. 1089; 9, S. 1091; 13, S. 1095–1096; 16, S. 1099; 25–26, 1108–1109.

24 Gessner Conrad, *Historiae animalium Libri V* (Zürich: 1551–1587); *Liber IIII, qui est de piscium et aquatilium animantium natura cum iconibus singulorum fere omnium expressis* (Zürich, Christoph Froschauer: 1558), „De aquatilibus, Balaena", B, 132–143; als Grundlagen Rondelet Guillaume, *Libri de piscibus marinis, in quibus verae piscium effigies expressae sunt* (Lyon, Matthias Bonhomme: 1554), lib. XVI, cap. 11–14, S. 475–487; Belon Pierre, *Libri duo de aquatilibus* (Paris, Stephanus: 1553), cap. 2–3, S. 3–18.

25 Mulder-Bosgoed D., *Bibliotheca ichthyologica et piscatoria. Catalogus van boeken en geschriften over de naturlijke geschiedenis van de visschen en walvischen, de kunstmatige vischteelt, de visscherijen, de wetgeving op de vischerijen* (Haarlem: 1873), § 1, S. 157–176, dort z.B. Nr. 2759–2761 und 2771–2783.

26 Die letzte große Zusammenfassung der spitzbergischen Walfischerei liefert Zorgdrager Cornelis G., *Alte und neue Grönländische Fischerei und Walfischfang mit einer kurzen historischen Beschreibung von Grönland, Island, Spitzbergen, Nova Zembla, Jan Mayen Eiland, der Strasse Davis* (Leipzig, Peter Conrad Monath: 1723).

Grönlands die Auswüchse dieses Tiers auch zu verkaufen pflegten.[27] Schon vorher war dieses seltsame Wesen von Albertus Magnus, Thomas von Cantimpré, aber auch Olaus Magnus erwähnt worden, doch hatten diese Autoren sich kaum ein klares Bild von ihm machen können.[28]

II Die Familie Bartholin und das Einhorn

Die hier umrissene Vorgeschichte liefert die Rahmenbedingungen der Einhorndebatte, die Bartholin Zeit seines Lebens begleiten sollte, eine mit Blick auf die Hörner langsam um sich greifende Skepsis, und die Erschließung der Tierwelt der Polarregion, die sich den neuen akademischen Netzwerkbildungen der Zeit verdankte, gleichzeitig aber auch der Unwille, die alten Überlieferungen zur Gänze aufzugeben. Thomas Bartholins Entlarvung der in den Wunderkammern Europas gestapelten Pretiosen war im besten Sinne ein Familienunternehmen, das drei Generationen der bartholinschen Sippe in Anspruch nehmen konnte.[29] Vergessen werden sollte dabei nicht, daß es Bartholins Familie war, deren mehr als zwei Dutzend verwandte oder verschwägerte Mitglieder über fast hundert Jahre die Universität Kopenhagen nicht nur geprägt, sondern förmlich regiert hatten.

II.1 *Caspar Bartholin und Ole Worm*

Im Jahre 1628 veröffentlichte Thomas Bartholins früh verstorbener Vater, Caspar Bartholin, einen ersten Traktat zum Thema, der allen weiteren familieninternen Arbeiten die Grundlage liefern sollte. Bartholin senior war noch nicht in der Lage, die Angebote, die ihm die Naturordnung gemacht hatte, zu einer Synthese zu vereinigen, auch wenn ihre Bestandteile gleichsam in der

27 Baffin William, *A briefe and true Relation or Iournall, contayning such accidents as happened in the fift voyage, for the discouerie of a passage to the North-west*, in *Samuel Purchas his Pilgrimes: in five books*, 5 Bde. (London, Williams Stansby: 1624–25) 844–848, hier 846.

28 Magnus, *De animalibus libri XXVI*, Bd. II, lib. XXIV, § 84, S. 1540; Thomas von Cantimpre, *Liber de natura rerum*, lib. VI, cap. 35, S. 243. Zur Identifikation des albertschen Meereinhorns auch in der englischen Übersetzung Albert the Great, *Man and the Beasts—De animalibus (Books 22–26)*, transl. J.G. Scanlan (Binghamton: 1987) 363.

29 Zur Auseinandersetzung der Bartholins mit den Einhörnern Garboe, *Enhjørningen* 9–14; 35–37; 95–98, und außerdem kurz ders., *Thomas Bartholin. Et Bidrag til dansk natur- og lægevidenskabs historie i det 17. Aarhundrede*, 2 Bde. (Kopenhagen: 1949), Bd. I, 75–77; Shepard, *The Lore* 175–179; 268–269; 296; 301; auch Gerritsen, *Het Spoor* 170–175; Ders., „De Eenhoorn" 8–9.

Luft, oder vielmehr im Wasser lagen. Caspar Bartholin repetiert die Zweifel, die Andrea Marini oder Ambroise Paré an der Wirksamkeit der in Europa verkauften Hörner vorgebracht hatten,[30] auch wenn sich Männer wie Conrad Gessner vehement für sie stark gemacht hatten.[31] Caspar Bartholin schließt sich der Partei der Skeptiker an. Warum hatte Galen nie von der Wirkmächtigkeit des Horns gesprochen? *Opinio, quando in idiotas cadit, est indelibilis.* Man war einer Fabel der spätantiken Naturkunde und vor allem dem Einfallsreichtum des Aelian zum Opfer gefallen.[32] Wenn es keine Pferdehörner waren, was hatte man stattdessen in der Hand? In einem langen Kapitel mustert Bartholin die bekanntesten der Hörner, die in Europa zur Schau gestellt wurden,[33] ein wunderbares Exemplar, das in der Kathedrale von Saint-Denis aufgebahrt, und ein ebenso ansehnliches Stück, das im Gotteshaus von Metz bewahrt wurde, denen beiden Pierre Belon eine kurze Beschreibung gewidmet hatte.[34] Philipp II. von Spanien verwahrte in seinem Kronschatz, wie man dem flämischen Historiker Emanuel van Meteren entnehmen konnte, ein weiteres Prachtexemplar. Ein letztes erwähnenswertes Stück hatte der Arzt Heinrich Smet aus Breda beschrieben, dessen Echtheit gleich mehrere Gelehrte beglaubigt hatten.[35] Mit gleicher Konsequenz repetiert Bartholin auch die Alternativen, die die Zweifler des vorausgegangenen Jahrhunderts vorgebracht hatten. Waren es Elefantenknochen, denen man in Gestalt der genannten Gegenstände gegenüberstand?[36] Caspar Bartholin hatte den Schlüssel in der Hand, denn ein weiteres Kapitel seines Einhorn-Traktates gebührt der marinen Variante des Einhorns, das Bartolin aus den Enzyklopädien und Olaus Magnus nur verschwommen geläufig war.

---

30 Bartholin Caspar, *Opuscula IV singularia: I. De unicornu eiusque affinibus et succedaneis. II. De lapide nephritico et Amuletis praecipuis. III. De pygmaeis. IV. Consilium de studio medico inchoando, continuando et absolvendo* (Kopenhagen, GeorgHantzsch: 1628); *De unicornu*, cap. 6, fols. 31r–36v.

31 Gesner Conrad, *Historiae animalium Libri V*, lib. I, *De quadrupedibus viviparis* (Zürich, Christoph Froschauer: 1551), „De monocerote" 785–786.

32 Bartholin, *De unicornu*, cap. 4, fols. 18v–19r.

33 Ebd., cap. 4, fols. 19r–25v. Zu den von Bartholin besprochenen Hörnern auch Garboe, *Enhjørningen* 20–21.

34 Belon Pierre, *Plurimarum singularium et memorabilium rerum in Graecia, Asia, Aegypto, Judaea, Arabia, aliisque exteris provinciis ab ipso conspectarum Observationes* (Antwerpen, Christoph Plantin: 1589), lib. I, cap. 19, S. 36–37.

35 Smet Heinrich, *Miscellanea medica communicata et in libros XII digesta* (Frankfurt, Jonas Rhodius: 1611), lib. VIII, „Epistula 1", S. 388–389.

36 Bartholin, *De unicornu*, cap. 7, fols. 36v–39r.

Daß es sich um Wale handelte, mußte Bartholin noch verborgen bleiben.[37] 14 Jahre vor Abfassung seiner Abhandlung war Caspar selbst in den Besitz eines Horns gelangt, das ihm der Bischof von Island beschafft hatte, und hatte es gemeinsam mit dem berühmten Astronomen Christian Longomontanus in Stücke zerteilt und untersucht. Bekannt war Bartholin darüber hinaus, daß im Kronschatz des dänischen Monarchen weitere Hörner dieser Art gelagert waren.[38] Dennoch mußte die Autopsie des Horns Stückwerk bleiben, weil beiden Gelehrten weitere Hintergrundinformationen fehlten.

Der offenen Einhornfrage sollte sich nur wenige Jahre später Caspars Schwager, Ole Worm, annehmen, Worm war nicht nur Mitbegründer der dänischen Nordistik und sollte mit dem „Museum Wormianum" den Grundstock der wichtigsten naturkundlichen Sammlung Skandinaviens legen, sondern hatte nach dem Tod Caspars auch die Erziehung des früh verwaisten jungen Thomas Bartholin übernommen.[39] Im Jahre 1636 tritt der königliche Kanzler Frisius de Kragerup an den Großgelehrten und fordert ihn auf, die Gerüchte zu überprüfen, die über das Einhorn, das er gerade zu einem mittleren Vermögen erworben hatte, in Kopenhagen kursierten. Handelte es sich bei dem schweren Kleinod womöglich um das Relikt eines Wassertiers und nicht eines Paarhufers, wie man bisher geglaubt hatte?[40] Die nun angestellte Untersuchung war der Anfang vom Ende der alten Einhörner. Worms Bericht, den er 1640 abgeschlossen hatte, konnte Thomas Bartholin selbst als „Corollarium" in seine eigene Abhandlung einfügen.[41] Nach eingehender Prüfung und einem Abgleich mit anderen Exponaten war Worm zu dem Schluß gekommen, daß dem Gebilde, das ihm vorlag, alle Merkmale eines Horns fehlen mußten. Von seiner

37 Ebd., cap. 1, fols. 7v–8r. Beschrieben fand sich das *unicornu* fast zeitgleich auch bei den ersten isländischen Schriftstellern, dazu Skúlason Þ. *Responsio subitanea* in *Two Treatises on Iceland from the 17th Century*, ed. J. Benediktsson (Kopenhagen: 1943), § 9, 13.

38 Bartholin, *De unicornu*, cap. 1, fols. 8r–10v.

39 Allgemein zum „Museum Wormianum" Schepelern H.D., *Museum Wormianum. Dets Forudsætninger og Tilblivelse* (Odense: 1971), *passim*, und kurz auch ders. „The Museum Wormianum reconstructed", *Journal of the History of Collections* 2 (1990) 81–85.

40 Worm Ole, *Museum Wormianum, seu historia rerum rariorum tam naturalium, tam artificialium, tam domesticarum, quam exoticarum, quae Hafniae Danorum in aedibus authoris servantur* (Amsterdam, Johannes Elsevier: 1655), cap. 13, S. 282.

41 Bartholin Thomas, *De unicornu observationes novae, acesserunt de aureo cornu Olai Wormii eruditorum judicia* (Padua, Cribelli: 1645), cap. 15, S. 94–106; ders., *De unicornu observationes novae* (Amsterdam, Heinrich Wetstenius: 1678), cap. 15, S. 107–124. Ebenso Bartholin Thomas, *Cista medica Hafniensis, variis consiliis, curationibus, casibus rarioribus, vitis medicorum Hafniensis aliisque ad rem medicam, anatomicam, botanicam et chymicam spectantibus referta* (Kopenhagen, Peter Haubold: 1662), „Loculus 37", 394–398.

Konsistenz war es in Wirklichkeit ein Zahn, entstanden durch eine Verdickung des Zahnschmelzes, eine Gomphose, wie man sagte, und entnommen dem Oberkiefer eines großen Meerestiers. Es handelte sich, wie Worm erkannte, um den Stoßzahn jenes Wals, den die Islandreisenden schon als *unicornu marinus*, als marines Einhorn angesprochen hatten. Kragerup war darüber hinaus im Besitz eines Narwalschädels, der jedoch über keine Zähne mehr verfügte. Ein Vergleich mit diesen Überresten konnte Worms Hypothese noch weiter unterstützen.[42] Alle jene, die zuvor noch spekuliert hatten, die Stoßzähne eines Elefanten oder Walroßes vor sich zu haben, waren ebenfalls eines Besseren belehrt worden. Allerdings waren bei weitem noch nicht alle Fragen beantwortet worden. Was hatte es mit der wieder und wieder diagnostizierten Wunderkraft des Horns auf sich? Halfen seine Bestandteile gegen Vergiftungen oder andere Malaisen, wie man in den alten Texten behauptet hatte, oder sollte Paré recht behalten? Waren also die astronomischen Summen berechtigt, für die man womöglich nur Walzähne erworben hatte? Worm unternimmt in der Apotheke seines Freundes Johann Woldenberg, wie sein Neffe später in einem Brief berichtet, einen ersten Versuch, um dem Mysterium des Horns beizukommen. Eine Gruppe Tauben erhält Arsen und daraufhin 15 Körner des geriebenen Narwalstoßzahns; die Vögel speien die Körner wieder aus und verenden. Zwei Katzen verabreicht Worm im Anschluß ebenfalls Gift und daraufhin geriebenen Stoßzahn; die eine stirbt sofort, während die zweite immerhin bis zur Nacht am Leben bleibt. Worm kann die Versuche nicht zum Anlaß nehmen, dem Einhorn jede Wirkung abzusprechen. Daß Stoßzähne anstelle von faktischen Einhörnern vorgelegen hatten, mußte noch keine Bedeutung haben.[43]

II.2 *Thomas Bartholin und das* unicornu groenlandicus

Thomas Bartholin war seinen Onkel zeitlebens eng verbunden, ja er hatte ihn als seinen zweiten Vater verehrt. Im Museum Worms hatte er die Relikte des Einhorns schon als Kind bewundern dürfen. Es verwundert so kaum, daß er sich berufen fühlte, der These Ole Worms ausgreifend Gehör zu verschaffen und sie restlos abzusichern. Drei Dinge sollten ihn dabei besonders umtreiben. 1. Es galt die Zugehörigkeit der in Europa kursierenden Hörner zu den Narwalen widerspruchsfrei zu belegen. 2. Die mirakulösen Fähigkeiten des Horns mußten noch immer überprüft werden. 3. Welche Bedeutung sollte die Aufdeckung der Narwahlstoßzähne für die vor allem in Indien und Äthiopien

42 Worm, *Museum Wormianum*, cap. 13, S. 282–283.

43 Bartholin, *Cista medica Hafniensis*, „Loculus 37“, S. 393–394. Zu den mit Woldenberg durchgeführten Versuchen Garboe, *Enhjørningen* 58–60.

beobachteten pferdeartigen Einhörner besitzen? Waren sie damit als bloße Ausgeburten der Phantasie abgetan?

In der weiteren Bestimmung des Narwals konnten Thomas seine eigenen antiquarischen Arbeiten zugute kommen. Aus Island erhält er von einem Pfarrer ein Manuskript, das sich später als Auszug des altnordischen *Königsspiegels* entpuppen sollte, eine Liste von 22 Spezies der Gattung Wal und ein Musterbeispiel der Ethnobiologie, das bestätigen konnte, wie komplex die alten Isländer schon im 13. Jahrhundert zwischen diversen Walarten unterscheiden konnten.[44] Neben den *Andhualur*, den *Svinhualur* und den *Raudkemlingur* erschienen hier als elfte Spezies die *Narhualur*, die Nar- oder Leichenwale. Ihren Namen hatten sie erhalten, weil Kadaver, wie der Verfasser des *Königsspiegels* geglaubt hatte, ihre wichtigste Nahrung bildeten.[45] Narwale besaßen einen Stoßzahn von mehr als zwei Ellen, gemasert nach Art einer Muschel und von schneeigem Weiß; ihr Fleisch war ungenießbar. Die Isländer hatten also von Anfang an um die Bedeutung dieser Tiere gewußt.

Seine Bemühung um die Absicherung der Wormschen Hypothese schildert Bartholin in seinen 1645 zum ersten Mal gedruckten und während seiner Zeit in Padua geschriebenen *Observationes novae de unicornu* und in den zahlreichen Briefen, die er mit seinem Onkel über den Gegenstand austauschen sollte. Bartholins *Observationes* liefern eine mehr als 250 Seiten starke Enzyklopädie des Einhorns, die nicht nur Kapitel über gehörnte Menschen, Vögel, Reptilien, Insekten und Fische enthielt, exotische Tiere wie die Oryxantilope, Bisone, Auerochsen und Nashörner behandelt, sondern vor allem auch alle Traditionen zusammenführt, derer er zum Einhorn habhaft werden konnte. Auch die Bibel, die dem *R'em*, wie es in der Schrift hieß, reiche Aufmerksamkeit geschenkt

---

44 Bartholin Thomas, *Historiarum anatomicarum rariorum centuria I–IV* (Kopenhagen, Peter Haubold: 1654–57), „Centuria" IV, „Historia" 24, S. 167–176. Ebenfalls auf diesen Text rekurriert schon unabhängig von den Bartholins Schonefeld Stephan von, *Ichthyologia et nomenclaturae animalium marinorum, fluviatilium, lacustrium, quae in florentissimis ducatibus Slesvici et Holsatiae et celeberrimo Emporio Hamburgo occurrunt triviales* (Hamburg, Heringianus: 1624) 27–28, doch kann er den *Nahvval* noch nicht weiter mit Bedeutung füllen. Nach Bartholin druckt die Wal-Liste Torfason Þormóður, *Gronlandia antiqua seu veteris Gronlandiae descriptio* (Kopenhagen, Typographicus regius: 1706), cap. 12, S. 89–97.

45 *Konungs Skuggsjá—Speculum regale, udgivet efter håndskrifterne af det Kongelige Nordiske Oldskriftselskab* (Kopenhagen: 1920) 33–38. Englisch als *The King's Mirror* (*Speculum regale—Konungs Skuggsjá*), transl. L.M. Larson (New York: 1917; ND New York: 1972), chapt. 12, S. 119–126.

hatte, fehlt bei Bartholin nicht.[46] Im Zentrum des Werkes stehen der Narwal und seine Bedeutung. Über Generationen hatte man die Stoßzähne dieser Tiere an den Küsten Islands und Grönlands gesammelt und hatten Kaufleute dieser Inseln sie als Einhorn-Hörner in ganzen Schiffsladungen an den Höfen Europas feilgeboten, auch wenn ihnen nun, so Bartholin gleich zu Beginn, die Larve abgezogen worden war.[47] Möglich geworden war diese Enttarnung nun durch den Fund vollständiger Schädel, die noch über Zähne verfügten. Auch Bartholin gelangt mit Unterstützung Worms in den Besitz eines Narwalkopfes.[48] Zugleich spielt ihm der isländische Bischof Thorlak Skulason die Skizze eines vollständigen gestrandeten Exemplars zu und weitere Skelettteile, die jeden Zweifel ausräumen konnten.[49] Die Aufzeichnungen seines Vaters und Onkels, aber vor allem seine eigene ausgedehnte Reiseerfahrung geben dem dänischen Professor die Möglichkeit, die in Europa vorhandenen Exemplare ein weiteres Mal zu mustern. Das Schaustück in Saint Denis, das sein Vater schon beschrieben hatte, offenbart sich unzweifelhaft als Narwalstoßzahn, ebenso das Horn aus dem Besitz Philipp II.[50] Bartholin beschränkt sich nicht auf diese Exponate. Eines der schönsten Hörner wußte die Kathedrale von San Marco in ihrem Bestand, das, wie man stolz zu berichten wußte, Seeleute schon im Jahre 1202 aus Byzanz herbeigeschafft hatten. Es war ein Narwalzahn.[51] Ein weiteres Horn verwahrte der Domschatz von Utrecht, in seiner Größe der Pretiose von Frankreich vergleichbar. Auf der Rückseite kann Bartholin eine Runen-Inschrift ausfindig machen, eine Art Waschzettel, die das Horn als Geschenk

46 Eine eigenes Kapitel zum Einhorn als Christus-Symbol findet sich außerdem bei Bartholin Thomas, *De latere Christi aperto dissertatio* (Leiden, Johannes Maire: 1646), cap. 9, S. 117–128.

47 Bartholin, *De unicornu observationes novae* (1645), cap. 15, S. 94–106; 27, S. 195–206; 29, S. 213–231. Zitiert wird im Folgenden nach der Neuausgabe Bartholin, *De unicornu observationes novae* (1678), dort cap. 15, S. 107–124; 27, S. 251–267; 29, S. 275–304.

48 Bartholin Thomas, *Epistolarum medicinalium a doctis vel ad doctos scriptarum centuria I–IV* (Den Haag, Pierre Gosse: 1740; zuerst Kopenhagen: 1663–1668), Centuria I, Ep. 21, S. 99–100; Ep. 25, S. 109–110.

49 Bartholin, *Historiarum anatomicarum rariorum centuria I–IV*, Centuria III, „Historia“ 4, S. 149–152.

50 Bartholin, *De unicornu observationes novae* (1678), cap. 27, S. 250–253. Das Horn von Saint-Denis findet sich z.B. auch beschrieben bei de Renou Jean de, *Dispensatorium medicum* (Frankfurt, Theobald Schönwetter: 1609), *De materia medica*, lib. III, cap. 21, S. 593.

51 Bartholin, *De unicornu observationes novae* (1678), cap. 27, S. 253–255. Zu diesem Horn, einem Klassiker, auch Pona Francesco, *Trattato de' veneni e lor cura* (Verona, Bortolamio Merlo: 1643), cap. 15, S. 57–58; Doglioni Nicolo, *Le cose notabili e maravigliose di citta de Venetia* (Venedig, Battista Cestari: 1666), lib. II, S. 225–226.

oder Eigentum des norwegischen Königs Snorre auswies. Schon aufgrund seiner skandinavischen Herkunft lag die wahre Natur des Horns auf der Hand; es war ein Walstoßzahn.[52] Bartholin autopsiert noch weitere Hörner, darunter in Brügge und Danzig. Auch für sie läßt sich die gleiche Diagnose stellen.[53] In der Schatzkammer des dänischen Regenten Friedrichs III. fand sich ein ganzer Thron, den sich der Monarch aus Einhorn hatte zimmern lassen—er steht heute auf Schloß Rosenborg—, ebenso im Domschatz des Bischofs von Roskilde. Sie alle waren aus Narwalen gefertigt worden. Es war also, wie Bartholin noch einmal summiert, nie eine andere Substanz gewesen, die man allerortens förmlich in Gold aufgewogen hatte, als profaner Walstoßzahn.[54]

Die zweite Frage war, ob diese Erkenntnis dem Horn oder vielmehr Zahn jeden Wert geraubt hatte. Waren die Erfolge, die man mit dem geriebenen Einhorn über Epochen erzielt hatte, bloße Täuschung gewesen? Nach seinem Vater und Onkel macht sich Bartholin diese Schwierigkeit ein drittes Mal zur Aufgabe, gestützt auf Material, das ihm Worm zuspielt, und auf eigene Versuche. Auf bemerkenswerte Weise verfolgt Thomas Bartholin nun einen weit weniger kritischen Ansatz, was die Wirksamkeit der Hörner betraf, als scheinbar noch sein Vater wenige Jahrzehnte zuvor. Auf den ersten Blick sprach die Empirie für die Stoßzähne, wie zuvor für die Einhörner. Noch ein anderer Grund mochte dazugekommen sein. Hatte Bartholin nicht teuer erworbene Objekte neu taxiert, deren Schätzwert zu einem nicht unwesentlichen Teil im Interesse der Isländer liegen mußte? Waren es nicht die gleichen Isländer, deren Geschichte als Teil der dänischen Nation von Männern wie Worm und Resenius aufgearbeitet wurde? Konnte Dänemark daran gelegen sein, die Hörner zu nutzlosem Beiwerk herabzustufen? Bartholin mußte sich über die Bedeutung im Klaren geworden sein, die der Vertrieb der Hörner für die isländische und damit zugleich dänische Fischerei besaß. Die zweite Auflage der *Observationes*, erscheint im Jahre 1678, zwei Jahre vor dem Tod des großen Gelehrten und wird von seinem Sohn Caspar Bartholin verantwortet. Eingefügt finden wir hier eine ausgreifende Analyse der medizinischen

52 Bartholin, *De unicornu observationes novae* (1678), cap. 27, S. 256–258. Zu diesem Horn und seiner inzwischen berühmten Inschrift van Vlierden M., „De eenhoorns van Sint-Marie", *Bulletin van het Rijksmuseum* 37 (1989) 5–13, und vorher schon Quack A., „Eine nordische Runeninschrift in Utrecht" *Amsterdamer Beiträge zur älteren Germanistik* 21 (1984) 73–83, die jedoch beide keine bessere Lesart als Bartholin bieten können.

53 Bartholin, *De unicornu observationes novae* (1678), cap. 27, S. 258–265.

54 Ebd., cap. 27, S. 265–267. Eine Liste der Einhornartefakte in Dänemark gibt Garboe, *Enhjørningen* 26–27; 71–76.

Funktion der Stoßzähne.[55] Thomas erinnert daran, daß schon Ole Worm als junger Mann, als er noch in Basel bei Felix Platter zur Ausbildung weilte, erste Berichte über die Wirksamkeit der Hörner gesammelt hatte.[56] Ende des 16. Jahrhunderts hatte sich in Augsburg ein Ärzteteam eingefunden, um einen Hund erst mit Arsen, dann mit zu Einhorn traktieren. Tatsächlich war das Experiment zufriedenstellend verlaufen und das Tier kam mit dem Leben davon.[57] Philipp Höchstetter, ein anderer Mediziner, hatte, wie er selbst berichtete, bei der Therapie von Durchfallerkrankungen mit Einhorn gute Ergebnisse erzielen können.[58] Sollten sich die Kritiker der Einhornkur etwa geirrt haben? Ab den fünfziger Jahren greifen Bartholin und Worm selbst wiederholt zu Narwalstoßzähnen und empfehlen sie, wie man aus der überarbeiteten Fassung der *Observationes* erfährt, zur Probe auch ihren Kollegen.[59] 1655 gelingt es, wie man in einem Brief Bartholin lesen kann, seinem Leidener Kollegen Albert Kyper, das Einhorn während einer Epidemie mit Erfolg zur Anwendung zu bringen.[60] Bartholin selbst verwendet es, wie er berichtet, während einer Fieberepidemie in Kopenhagen.[61] Wenn es in warmes Bier getunkt wird, konnte es offenkundig Wassersucht kurieren; in Gelatine gelegt, half es, den Geburtsvorgang zu beschleunigen.[62] Sogar als Amulette konnten die Fragmente des Narwal-Einhorns von Nutzen sein.[63] Was bei seinem Vater bloße Abwehr war und dreißig Jahre zuvor beim jungen Bartholin noch vorsichtige Zurückhaltung, hatte sich auf diese Weise im Jahre 1678 in eine Werbeschrift

55 Bartholin, *De unicornu observationes novae* (1678), cap. 29, S. 275–304. Zur medizinischen Funktion der Einhörner bei Bartholin auch Garboe, *Enhjørningen* 60–66, und Garboe, *Thomas Bartholin*, Bd. I, 76–77.

56 Worm Ole, *Selecta controversiarum medicarum centuria* (Basel, Jacob Genathius: 1611), dort laut Thomas Bartholin, „Controversia II, Exercitatio 18". Dieser Text war mir nicht zugängig.

57 Bartholin, *De unicornu observationes novae* (1678), cap. 29, S. 299–302.

58 Höchstetter Philipp, *Rararum observationum medicinalium*, 2 Bde. (Augsburg, Sebastian Mylius: 1627), Bd. II, decas V, casus 9, S. 397–398.

59 Bartholin, *De unicornu observationes novae* (1678), cap. 29, S. 302–303.

60 Bartholin, *Epistolarum medicinalium centuria II*, Ep. 52–53, S. 577–580; Ep. 81, S. 661–665.

61 Bartholin, *De unicornu observationes novae* (1678), cap. 29, S. 303–304.

62 Ähnlich zum medikamentösen Gebrauch auch Bartholin Thomas, *De medica Danorum domestica dissertationes* (Kopenhagen, Peter Haubold: 1666), „Dissertatio" I, 29; IV, 145; V, 214 f.; „Additiones" 517.

63 Daß die Hörner in Europa zu Amuletten verarbeitet wurden, betont auch schon kurz Bartholin Caspar, *De aere pestilenti corrigendo consilium* (Kopenhagen, Sartorius: 1619), cap. 6, fol. B4r; doch scheint er auch diesem Gebrauch wenig Vertrauen zu schenken.

verwandelt, die Einhörner wie eine frühe Form des Aspirin anpreisen konnte: *princeps est medicamentum quo utimur Hafniae!*

Auch wenn Bartholin also die Einhörner in Stoßzähne verwandelt hatte, war ihr Nimbus einfach auf das marine Einhorn übertragen worden. Es blieb noch die dritte Frage zu beantworten, die Frage nach den gesichteten Tieren. Gab es trotz allem auch pferdeartige Einhörner? Tatsächlich sieht Bartholin keinen Grund, ihre Existenz zu bezweifeln, denn zu häufig waren sie seit der Zeit eines Ktesias oder Aelian in freier Wildbahn gesichtet worden, auch wenn es sicher keiner Jungfrau bedurfte, um sie einzufangen.[64] Wie sehr Bartholin bereit war, sich noch immer auf die alten Einhörner einzulassen, offenbart eine Episode, die in er in seinen *Anatomischen Centurien* verzeichnet.[65] 1652 erscheint beim Herzog von Kurland ein Legat einer Afrikaexpedition und berichtet von der Begegnung mit einem Tier in Guinea, das fortan mit seinem afrikanischen Namen ‚Tire Bina', ‚gehörntes Tier', als weiteren Variante des Einhorns die Literatur bereichern sollte, eine wilde, äußerlich den Pferden vergleichbare Kreatur, die ein Horn von drei Ellen Länge besaß, ein Einhorn, mit anderen Worten. Bartholin konnte es als Bestätigung seiner Annahme verbuchen, daß auch die Berichte über Einhornsichtungen auf eine realen Grundlage beruhen mußten.[66] Daß ihre Hörner die europäischen Wunderkammern nicht erreicht hatten, mußte dabei auf einem anderen Blatt stehen.

III Bartholins Rezeption: Placebos oder heilkräftige Stoßzähne?

Wie aber nahm die Wissenschaft der Zeit, die Medizin und Zoologie, die Decamouflage der Hörner auf und wie reagierte sie auf die Kompromißlösung, die Bartholin vorgebracht hatte, die Effizienz der Walstoßzähne? Es gab durchaus kritische Stimmen. Caspar de Reyes Franco, einer jener Ärzte, die sich in diesen Jahren mit dem Einhorn beschäftigt hatten, wehrt sich mit Nachdruck gegen mögliche Zweifel an der Authentizität der Hörner. Waren nicht auch andere Tiere, wie der antike Satyr, den de Reyes wie andere Zeitgenossen mit dem Orang-Utang identifiziert, oder der gewaltige Vogel Roc im Nachhinein

64 Bartholin, *De unicornu observationes novae* (1678), cap. 19, S. 151–153; cap. 28, S. 268–275. Zum letzten Argument wird für Bartholin die Erwähnung eines Tiers, das sich am Hofe Philipp II. befunden haben soll, dazu Nieremberg Juan Eusebio, *Historia naturae, maxime peregrinae libri XVI* (Antwerpen, Balthasar Moret: 1635), lib. VII, cap. 2, fol. 111r.

65 Zu Bartholins Glauben an die Pferdeeinhörner Garboe, *Enhjørningen* 87–89.

66 Bartholin, *Historiarum anatomicarum rariorum centuria I–IV*, Centuria II, „Historia" 61, S. 251–254.

doch noch aus der Wirklichkeit bestätigt worden? Warum sollten sich unter den in Europa verbreiteten Hörnern nicht auch echte Einhorn-Hörner befunden haben?[67] Auch wenn viele der Schaustücke vielleicht einen anderen Ursprung hatten, sollte daher niemand mutwillig, wie de Reyes unterstreicht, allen Hörnern ihre besondere Herkunft absprechen.[68]

Andere große Figuren der europäischen Gelehrtenwelt akzeptieren Bartholins Dekonstruktion der Hörner sofort. Als die medizinische Fakultät der Universität Leipzig im Jahre 1656 von Seiten der Obrigkeit befragt wird, ob die erworbenen Hörner wahrhaftig Einhörnern zuzuschreiben waren, liefert ihr Vorstand Paul Ammann in seiner *Medicina critica* eine eindeutige Expertise; sie waren es nicht, wie Bartholin gezeigt hatte.[69] Isaac de La Peyrère fügt die Abbildungen eines Narwals in seine 1663 neu erschienene *Beschreibung Grönlands* ein, druckt lange Auszüge aus den Briefen Worms und Bartholins ab, und läßt sich die Hörner, wie er berichtet, im Anschluß sogar vor Ort von Worm in Kopenhagen zeigen. Daß die Narwalstoßzähne die Grundlage der europäischen Einhörner gelegt hatten, lag für den Gelehrten auf der Hand.[70] In England beruft sich Thomas Browne in seiner *Pseudodoxia epidemica*, der vielleicht berühmtesten kritischen Aufarbeitung überkommener Ideen, die in dieser Zeit entsteht, ausdrücklich auf Bartholin. Den Hörnern mußte fortan mit Vorsicht begegnet werden, denn ihre Varietät mußte auch ihren Nutzen in Frage stellen.[71] Auch Nicolaas de Tulp schließlich, jener berühmte niederländische Anatom, den Rembrandt einst bei der Arbeit porträtiert hatte, räumt den neuen Erkenntnissen in der Neuauflage seiner *Observationes* einen eigenen Platz ein, doch macht sich, was die Wirksamkeit der Hörner betrifft, die Zurückhaltung des älteren Bartholin zu eigen.[72]

67 De Reyes Franco Gaspar, *Elysius iucundarum quaestionum campus omnium literarum aemonissima varietate refertus, medicis imprimis* (Frankfurt, Johannes Beyer: 1670), „Quaestio" 67, S. 897–898.

68 Ebd., „Quaestio" 67, S. 900–901.

69 Ammann Paul, *Medicina critica sive decisoria, centuria casuum medicinalium in concilio Facultatis Medicinae Lipsiensis antehac resolutorum comprehensa* (Erfurt, Johann Georg Hertzl: 1670), „Casus et Discursus" 67, S. 343–345.

70 De la Peyrère I., *Histoire du Groenland*, in *A Collection of Documents on Spitzbergen and Greenland*, ed. A. White (London: 1855; ND New York: 1970) 197–205. Die erste Ausgabe aus dem Jahre 1647 erwähnt das Einhorn noch nicht.

71 Browne Thomas, *Pseudodoxia epidemica* (The Works II), ed. G. Keynes (London: 1928), book III, chapt. 23, S. 237–242, bes. 238–239.

72 Tulpius Nikolaus, *Observationes medicae* (Amsterdam, Ludovicus Elzevirius: 1652), lib IV, cap. 58, S. 394–400.

Wichtiger als diese allgemeinen Einlassungen, deren Zahl sich leicht vermehren ließe, sind die Arbeiten, die im Anschluß an Bartholin unmittelbar zu den Einhörnern Stellung nehmen und die sich den neuen Erkenntnissen aus Kopenhagen zu stellen hatten. Franz Christoph Berens, Zoologe aus Leipzig, oder Simon Frenzel aus Wittenberg, jeweils 1667 und 1679, schließen sich in ihren Disputationen direkt an Bartholin an und referieren die in der Literatur überlieferten einhornigen Huftiere ebenso wie ihre marinen Antagonisten. Die alleinige Anwesenheit der letzteren in den Schatzkammern war kein ausreichendes Ausschlußkriterium, wie Berens und sein Disputant betonen, um die alten Berichte zur Gänze zu den Akten zu legen.[73] Auch Frenzel sieht die Bedenken, die Marini und Paré vorgebracht hatten, nicht bestätigt. Wale konnten ebenso für sich existieren, wie die Einhörner in den Weiten Äthiopiens.[74] Daß die Hörner, welcher Abkunft sie auch immer waren, allerdings in der Lage waren, das Fieber zu senken, Giftattacken zu abwehren, aber auch Epilepsie kurieren konnten, hatte die Medizin hinreichend bewiesen.[75]

Am weitesten im Versuch, den Nutzen der alten und neuen Hörner unter Beweis zu stellen, geht ein Mann aus Hamburg, Paul Ludwig Sachse, dessen zu diesem Zweck durchgeführte Testreihen heute vielleicht nicht vor jeder Ethikkommission Bestand hätten. Sachses mehr als zweihundert Seiten starke *Monocerologia* trägt den konventionellen Einhörnern ebenso Rechnung wie ihren Freunden im Wasser, wendet sich jedoch vor allem letzteren zu.[76] Sachse selbst sieht sich mit einem Narwal konfrontiert, als ein von Hamburgern bei Grönland harpuniertes Exemplar in den Hafen geschafft und vom Rat der Stadt begutachtet wird. Einige Jahre später gelingt es einem hanseatischen Kaufmann, in Island elf Stoßzähne zu erwerben, wie Sachse berichtet.[77] Daß Bartholin in seiner Vereinnahmung der Narwale recht behalten hatte, stand

---

73 Berens F. G. – Homilius J.H. (resp.), *Dissertatio zoologica de monocerote* (Leipzig, Bauer: 1667), cap. 1, § 1 und § 6, fols. A2r–v; cap. 2, § 5, fols. A4r–v.

74 Frenzel Simon Fredericus – Vater Christianus (resp.), *Disquisitio naturalis de unicornu* (Wittenberg, Wilcke: 1679), § 5, fols. A4r–v. Ähnlich auch mit Bezug auf Bartholin: Densovius Christoph – von Somnitz Friedrich Albert, *Disputatio physica de monocerote—Einhorn* (Stargard, Johannes Nicolaus Ernst: 1701), Thesis I, fol. 3v, oder Kirchmeier Georg Caspar. *Disputationes zoologicae de Basilisco, Unicornu, Phoenice, Behemot, et Leviathan, Dracone ac Aranea* (Jena, Johann Friedrich Ritter: 1736; zuerst Wittenberg 1660), „Disputatio" II, mit Friedrich Hubrigk als Respondenten, cap. 1, § 5, S. 25–26; cap. 2, § 4–9, S. 35–39; zum Heilmittel cap. 2, § 11, S. 40–41.

75 Frenzel – Vater, *Disquisitio naturalis de unicornu*, §§ 21–22, fols. C3v–C4v.

76 Kurz zu Sachses Beschäftigung mit den Einhörnern auch Shepard, *The Lore* 263–264.

77 Sachse Paul Ludwig, *Monocerologia, seu de genuinis unicornibus dissertatio* (Ratzeburg, Nicolai Nissen: 1676), cap. 5–6, S. 55–96.

außer Zweifel. Wieder stellte sich jedoch die Frage nach der fast magischen Heilkraft der teueren Importe. Paul Sachse beschließt, der Sache gemeinsam mit seinem Vater auf den Grund zu gehen und Bartholins Behauptung weiter abzusichern. Die Methode der beiden Mediziner war einfach; sie bestand in der Verabreichung von geriebenem Einhorn und einem starken Extrakt der *nux vomica*, der noch heute bekannten Brechnuß, die eine hohe Dosis Strychnin enthielt. Bei einer Gruppe von Kaninchen, der Vater und Sohn beide Stoffe einflößen, zeigt der Stoßzahn noch keine Wirkung. Die aufgrund des Gifts zappelnden Tiere werden nach der Einnahme des Einhorns zwar kurzzeitig ruhiger, verenden aber nichtsdestoweniger nach kurzer Zeit. Eine zweite Versuchskette nimmt sich Hunde vor, die nun in der Tat, wie Sachse notiert, auf die Doppelung von Narwal und Strychnin anschlagen, wenn auch zunächst nur verhalten. Beim ersten vergifteten Tier kann das Horn den Tod immerhin um fünf Stunden hinauszögern, bevor die Welpe mit Schaum vor dem Mund zusammenbricht. Beim einem zweiten, größeren Hund, dem mit dem Gift nun 25 Gramm des Zahns verabreicht werden, läßt sich der Effekt noch verlängern und der Hund scheint die Tortur zu überleben. Erst als Sachse die Dosierung des Strychnins auf 15 Gramm erhöht, beginnt der Hund Blut zu spucken und verendet. Eine abermalige Aufstockung des Einhorns beim nächsten Hund, der die gleiche Menge Gift erhält, führt bei dem bedauernswerten Tier zwar zunächst zu Konvulsionen, doch scheint das Einhorn den Hund tatsächlich resistent zu machen. Eine Gegenprobe mit Spiritus und Strychnin läßt auch dieses Tier verenden. Mit Bedauern stellt Sachse fest, daß die einbrechende Nacht weitere Experimente verhindert hatte.[78] Auch wenn offen bleibt, wieviele Hunde dem Forscherdrang des Hamburger Gelehrten in den folgenden Tagen noch zum Opfer fallen mußten, steht für den Mediziner zum Ende seiner Experimente die Schlußfolgerung fest: die Stoßzähne hatten sich als die wahren Einhörner der Tradition erwiesen. Die Wissenschaft mußte Gott dankbar sein, so Sachse, daß die Natur dem Menschen ein solches Heilmittel zur Verfügung gestellt hatte.[79]

IV Das Ende des Medikamentes

Auf den ersten Blick war Bartholins Strategie also aufgegangen. Unter den Medizinern hatte sich vordergründig ein Konsens durchgesetzt, der auch den Geschäftsinteressen der isländischen Kaufleute entgegenkommen mußte.

78 Ebd., cap. 10, S. 158–175.

79 Ebd., „Epilogus" 175–182.

Noch im Jahre 1706 konnte Bartholins Landsmann und Schüler Tycho Lassen Tychonius die Geschichte der grönländischen Einhörner, deren Enttarnung und Anatomie er mit Hilfe der Schriften der Familie Bartholin noch einmal zusammenfaßt, mit einem glanzvollen Eulogium auf die Bartholins verbinden, die Zierde der Universität Kopenhagen.[80] Die Einhörner waren zwar ihrer Aura beraubt worden, doch war an ihre Stelle ein neues Medikament in Gestalt des alten getreten, das die bisherige Funktion ebenso ausfüllen konnte.[81] Es war nur natürlich, daß dieses vage Gleichgewicht der Interessen irgendwann implodieren mußte, denn nicht allen Parteien, vor allem nicht den Kritikern Marini, Paré und ihren Anhängern hatte man auf diese Weise genüge getan. Die Erfahrungswissenschaft mußte sich ein weiteres Mal Gehör verschaffen. Anton Deusing, Mediziner in Groningen, scheint in seiner Abhandlung zu den Einhörnern im Jahre 1659 vor allem die Ansichten des dänischen Großgelehrten und seiner Familie zu referieren. Auch Deusing leugnet zunächst nicht, daß immer wieder Einhörner gesehen worden waren, vor allem in Zentralafrika, und kann mit weiteren Sichtungen aufwarten.[82] Daß die Hörner, die in Mitteleuropa vermarktet wurden, Narwalzähne waren, kann Deusing wie Sachse vor Ort nachvollziehen, als er im Jahre 1645 ein Narwalskelett in Amsterdam einer Autopsie unterzieht und ein Jahr später noch ein zweites.[83] Etliche Kollegen seiner Heimatstadt bestätigen ihm sogar, daß die geriebenen Stoßzähne gegen Fieber eine sehr heilsame Wirkung entfalten konnten.[84] Anders als Sachse oder Bartholin bleibt Deusing jedoch mißtrauisch. War man

80 Tychonius Tycho Lassen – Weichhart Justus Heinrich (resp.) – Gyrsting Johann Peter (resp.), *Monoceros piscis haud monoceros, ad veram formam nuperi e mari Gronlandico hospitis depictus et descriptus* (Kopenhagen, Typographeus Universitatis: 1706–1707), „Prologus" 4–9. Ein Skelett des Tiers gelangt in die königliche Sammlung, dazu Thomas Bartholins Schwiegersohn Jacobaeus Oligerus, *Museum Regium seu catalogus rerum tam naturalium, quam artificialium* (Kopenhagen, Joachim Schmetgen: 1696), sectio III, S. 14.

81 Abschreibetraditionen konnten den Narwalstoßzahn als mögliches fiebersenkendes Mittel bis in die Mitte des 18. Jahrhunderts weitertragen, doch war hiermit kaum ein Wort über reale Therapien verloren worden, hierzu z.B. Alpinus Prosper, *De medicina methodica libri XIII* (Leiden, Boutestein: 1719), lib. V, S. 309–310; Hoffmann Friedrich, *Medicina rationalis systematica*, 3 Bde. (Halle, Renger: 1727), Bd. III, sectio II, cap. 6, § 2, S. 535–536; Piquer Andrés, *Medicina vetus et nova* (Valencia, Joseph Garcia: 1735), Tractatus IV, § 49, S. 252, oder als besondere Spezialität der dänischen Heilkunst noch immer Roncalli Parolino Francisco Roncalli, *Europae medicina a sapientibus illustrata* (Brixen, Marcus Vendramenius: 1747), „Danorum medicina" 43.

82 Deusing Anton, *Dissertationes de unicornu et lapide Bezaar* (Groningen, Cöllenius: 1659), „De unicornu", § 18–21, S. 14–16; § 25, S. 19–20.

83 Ebd., *De unicornu*, § 29, S. 24; § 32, S. 27–28.

84 Ebd., *De unicornu*, § 34, S. 29–30.

sich in der Vergangenheit über die Homogenität der verabreichten Stoffe wirklich im Klaren gewesen? Hatte man auch früher immer den Zahnschmelz des Narwals verabreicht, oder war gewöhnliches Horn zur Anwendung gelangt? Bartholins Ergebnisse mußten dieser Vermutung entgegenkommen.[85] Besser war es daher, das Einhorn von der Liste der Medikamente zu streichen, so Deusing, schon um sich als Arzt nicht des Betruges schuldig zu machen und niemanden in Gefahr zu bringen.[86] Anton Deusing bleibt mit seiner neuen Haltung nicht allein. Ein weiteres Mal geht der Leidener Arzt Cornelis Stalpart van der Wiel im Jahre 1687 auf die Angelegenheit ein. Pflichtbewußt verzeichnet auch Stalpart neue Einhornsichtungen an der kalifornischen Küste und in Indien,[87] doch sieht er in den Tieren eher eine *splendida fabula*, ein elegantes Märchen, wie er selbst betont.[88] War es nicht seltsam, daß noch nie ein lebendes Tier gefangen worden war, obwohl Kreaturen aller Art inzwischen nach Europa gelangt waren? Warum hatten die Römer auf ihren Triumphzügen nie ein Einhorn präsentieren können?[89] Hatten die Narwale nicht von dem enormen Kredit gelebt, den die Pferdeeinhörner genossen hatten?[90] Bartholins Hörner, die Stoßzähne, waren also kein Heilmittel mit einem Sonderstatus mehr, sie konnten auf Grund ihrer Konsistenz nicht mehr oder weniger therapeutische Kraft beanspruchen als die Stoßzähne eines gewöhnlichen Elefanten oder eines Walroßes.[91]

Beiden Medizinern schließt sich auch Robert Sibbald an, mit dessen *Phalainologia* die moderne Walkunde ihren Anfang nimmt.[92] Endgültig hatte sich damit getrennt, was Bartholin noch hatte verbinden wollen, die Attribute des alten Horns und seine neue Standortbestimmung im Naturreich. Bartholin war es trotz erheblicher Mühen nicht gelungen, den Nimbus des Horns länger als drei Jahrzehnte über seine verhängnisvolle Enttarnung hinaus aufrecht zu erhalten. Narwal und Einhorn durften fortan getrennte Wege gehen. In den großen Zoologien des frühen 18. Jahrhunderts konnte der Narwal problemlos als *unicornu marinus* oder *monoceros* angesprochen werden. Fundberichte

85 Ebd., *De unicornu*, § 47–48, S. 44–46.

86 Ebd., *De unicornu*, § 50, S. 48–49.

87 Stalpart van der Wiel Cornelis, *Observationum rariorum medicorum, anatomicorum, chirurgicarum Centuria prior, accedit De unicornu dissertatio* (Leiden, Kerkhem: 1727; zuerst Leiden: 1687) 480–481; 485–486.

88 Ebd., 491–492.

89 Ebd., 497–498.

90 Ebd., 499.

91 Ebd., 502–503; 506–507.

92 Sibbald Robert, *Phalainologia sive observationes de rarioribus quibusdam balaenis in Scotiae littus nuper eiectis* (Edinburgh, Johannes Redius: 1698) sectio II, cap. 6, 24.

aus dem Hinterland, wie sie in der Royal Society erscheinen konnten—ein Narwalkadaver aus dem Jahre 1735, angetrieben in der Weser bei Bremen—sprachen vom *unicorn fish*, ohne über die Funktion des Stoßzahns oder die urprüngliche Bedeutung des Terminus Einhorn noch ein Wort zu verlieren.[93] Als *monoceros* oder *monodon* nehmen Petrus Artedus und Carl von Linné den Narwal in ihr Artensystem auf,[94] deutsche Fassungen des *Systema naturae*, wie sie Philipp Statius Müller oder Johann Erxleben liefern, sprechen bewußt vom *Meer-* oder *See-Einhorn* als eigener Spezies.[95] Auch Johann Blumenbach in seiner *Naturgeschichte* wählt den Artbegriff des *See-Einhorns*, um den Narwal zu klassifizieren.[96]

Ein Platz im Artensystem war dem Landeinhorn nicht vergönnt gewesen. Dennoch hatten Bartholins Einlassungen nicht erreichen können, daß sie von der Bildfläche der globalen Marginalen verschwanden, im Gegenteil, gerade Bartholins Bereitschaft, die eigentlichen Einhörner noch immer in einem empirischen Zwischenreich der Semiverifikation zuverorten, mußte einen erheblichen Beitrag dazu geleistet haben, sie auch weiterhin mit einem gewissen Eigenrecht zu versehen. Gesichtet werden sie auch weiterhin, vor allem in ihren angestammten Habitaten, in Äthiopien, wo sie Hiob Ludolf im Jahre 1680 ausmachen möchte,[97] oder in Südafrika, wo sie noch Linnés Apostel Anders Sparrmann in der Mitte des 18. Jahrhunderts ein weiteres Mal verortet, als er im Auftrag seines Meisters die Artenwelt am Kap der Guten Hoffnung kartographiert.[98] Von den letzten lebenden Exemplaren in Bhutan oder Tibet

93 Steigerthal J.G., „Letter to Sir Hans Sloane, giving an Account of a Narhvaal or Unicorn Fish, lately found in the River Ost, Dutchy of Bremen, dated at Hannover May 1. 1736“, *Philosophical Transactions of the Royal Society* 40 (1739) 147–149. Im Anschluß, ebd. 149–150, folgt Hampe J.H., „A Description of the Same Narhvaal“.

94 Artedus Petrus, *Ichthyologia sive opera omnia de piscibus*, 4 Bde. (Leiden, Conrad Wishoff: 1738), Bd. I, Plagiuri, Genus 49, Nr. 1, 78. Unter vielen Auflagen des *Systema naturae* z.B. Linné Carl von, *Systema naturae sistens regna tria naturae in classes et ordines, genera et species* (Leiden, Theodor Haak: 1756), Pisces, Plagiuri, Ordo 8, Nr. 98, 39.

95 Müller Philipp Ludwig Statius, *Des Ritters Carl von Linné vollständiges Natursystem: Erster Theil: Von den säugenden Thieren* (Nürnberg, Nicolaus Raspe: 1773) 37. Erxleben Johann Christian Polykarp, *Systema regni animalis per classes, ordines, genera, species, varietates: Classis I: Mammalia*, (Leipzig, Weygand: 1778), Nr. 51, § 1, S. 626–628.

96 Blumenbach Johann Friedrich, *Handbuch der Naturgeschichte*, 2 Bde. (Göttingen, Johann Christian Dieterich: 1799), Bd. I, *Von den Säugethieren*, X. „Cetacea“, Nr. 45, § 1, S. 124–125.

97 Ludolf Hiob, *Historia aethiopica sive brevis et sucincta descriptio Habessinorum* (Frankfurt, Johann David Zuner: 1681), cap. 10, § 78–87.

98 Sparrman Anders, *Resa till Goda Hopps-Udden, Södra Pol-kretsen och Omkring Jordklotet, samt till Hottentott- och Caffer-Landen, Åren 1772–1776: Förste Delen* (Stockholm, Nordström: 1783), Januar 1776, 550–553, in deutscher Übersetzung Sparrman Anders, *Reise nach dem*

ist noch im 19. Jahrhundert die Rede.[99] Vergleichbare Schilderungen mochten auch Zoologen wie Eberhard Zimmermann oder Johann Christian Schreber noch zum Ende des 18. Jahrhunderts veranlaßt haben, das Landeinhorn zumindest noch als obskure Spezies in ihren Handbüchern zu erwähnen.[100] Als Medikament konnte auch dieses Horn keine Rolle mehr spielen.

Vorgebirge der guten Hoffnung, den südlichen Polarländern und um die Welt, hauptsächlich aber in den Ländern der Hottentotten und Kaffern in den Jahren 1772 bis 1776, aus dem Schwedischen frey übersetzt von Christian Heinrich Großkurt, ed. Georg Forster (Berlin, Haude und Spener: 1784), XI. Abschnitt, 453–457.

99 Turner Samuel, *An Account of an Embassy to the Court of the Teshoo Lama in Tibet* (London, W. Bulmer & Co: 1800), Part I, chapt. 8, S. 157.

100 Von Zimmermann Eberhard August Wilhelm, *Geographische Geschichte der Menschheit und der vierfüßigen Thiere*, 3 Bde. (Leipzig, Weygand: 1780), Bd. II, 157–160, und kurz und skeptischer z.B. Schreber Johann Christian von, *Die Säugethiere in Abbildungen nach der Natur mit Beschreibungen*, 3 Bde. (Erlangen, Palm: 1775), Bd. II, 237–238, oder Erxleben, *Systema regni animalis*, Nr. 22, 201. Auch Uhse Erdmann, *Der gelehrte Criticus*, 3 Bde. (Leipzig, Johann Gleditsch: 1704–07), Bd. I, 78. Frage, S. 876–877., beantwortet die Frage nach der Existenz der Einhörner noch immer positiv.

English Summary

In the early modern period, the unicorn was a creature whose existence was guaranteed just as firmly by the ancient authorities on the natural world, such as Ctesias and Aelian, as by Scripture and contemporary travel reports. Its presence in the natural order was not a matter of any doubt. The near legendary healing powers of its horn, which was especially effective against poisoning, had also been stressed even in the ancient sources. The prices people were prepared to pay for the miraculous horns in the 16th century were correspondingly high. But when did belief in unicorns come to an end and the myths about their horn start to erode?

Responsibility for demystifying the unicorn should be assigned, as this paper aims to show, above all to a dynasty of Scandinavian scholars, the Bartholin family in Copenhagen. Already in the 16th century, doctors in Italy had questioned the efficacy of the horn, but their debate in the universities had had no effect. The unmasking of the horn was a northern European phenomenon and was closely linked to the advances achieved in these regions by zoology and pre-industrial whaling. At the end of the 16th century, William Baffin provided the first description of a narwhal. Further public dissections of a bowhead whale or sperm whale had followed. When Caspar Bartholin, the father of Thomas Bartholin, for the first time systematically studied the unicorn horns preserved in the cabinets of curiosities, cathedrals and apothecaries of Europe, he was forced to conclude that these objects were not the horns of an ungulate, but he was not yet prepared to identify the exhibits with the tusks of the narwhal, a familiar creature in Denmark. In 1636 Ole Worm turned his attention to the unicorns and completed what his brother-in-law Caspar had begun: from comparison with skeletons that Worm had acquired from Iceland, it was clear that the horns touted in Europe had always been the tusks of the polar marine mammal. Worm's nephew Thomas Bartholin, too, subsequently chose unicorns as an object of study: did all the unicorn artefacts circulating in Europe really come from Scandinavian narwhals? Could their tusks perhaps nonetheless have the powers to neutralise poison that medical science had attributed to them until now? And what about the horse-like unicorns that had been observed repeatedly since ancient times, especially in Asia and Africa? In his work *Observationes novae de unicornu*, published in 1645 and revised in 1678 by his son Caspar, Thomas surveyed the horns one more time and demonstrated, with reference to Old Icelandic sources like the *King's Mirror*, that the ancient Icelanders, well aware of the true nature of the horn, had pursued a busy trade in the horns. A series of experiments then demonstrated to Bartholin's satisfaction that their therapeutic role had been ascribed to them quite correctly;

Thomas even thought he was able to halt a fever epidemic in Copenhagen with unicorn rubs. Finally, up-to-date travel accounts from Ethiopia could show, so Thomas believed, that the horse-like unicorns, too, had their counterpart in the natural world, even if they had no part in the horns themselves.

The curious synthesis presented by Thomas Bartholin, which aimed to do justice both to contemporary science and Iceland's economic interests, had great success in the following years, as will be shown in the paper. Above all in *disputationes* held on the subject of the horns from 1660 onwards, and in various medical handbooks, the Danish anatomist's theories are encountered repeatedly. The Hamburg doctor Paul Sachse thought he could demonstrate, by a series of experiments in which he tested on dogs the interactions of a unicorn rub and strychnine, that unicorn must be a useful medical remedy against poison. Nonetheless, Bartholin's compromise would eventually fail. Medical scholars like Anton Deusing and Cornelis Stalpart van der Wiel, who undertook further investigations into the efficacy of the horn, were able to falsify Bartholin's results and the unicorn disappeared from the apothecaries of the early 18th century. Though the narwhal retained the name *Unicornu groenlandicus*, the taxonomists following Linnaeus in the mid-18th century struck off the unicorn from the system of species.

Bibliographie

Bartholin Thomas, *Cista medica Hafniensis, variis consiliis, curationibus, casibus rarioribus, vitis medicorum Hafniensis aliisque ad rem medicam, anatomicam, botanicam et chymicam spectantibus referta* (Kopenhagen, Peter Haubold: 1662).

Bartholin Thomas, *De latere Christi aperto dissertatio* (Leiden, Johannes Maire: 1646).

Bartholin Thomas, *Epistolarum medicinalium a doctis vel ad doctos scriptarum centuria I–IV* (Den Haag, Pierre Gosse: 1740; zuerst Kopenhagen: 1663–1668).

Bartholin Thomas, *Historiarum anatomicarum rariorum centuria I–IV* (Kopenhagen, Peter Haubold: 1654–1657).

Beer R., *Einhorn. Fabelwelt und Wirklichkeit* (München: 1972).

Einhorn J.W., *Spiritalis unicornis. Das Einhorn als Bedeutungsträger in Literatur und Kunst des Mittelalters* (München: 1976).

Garboe A., *Enhjørningen især i natur- og lægevidenskabens historie* (Kopenhagen: 1915).

Garboe A., *Thomas Bartholin. Et Bidrag til dansk natur- og lægevidenskabs historie i det 17. Aarhundrede*, 2 Bde. (Kopenhagen: 1949).

Gerritsen W., „De Eenhoorn en de Apothekers. Opvattingen omtrent de antitoxische werking van ‚eenhoornhoorn' in het laatste kwart van de 16e en het eerste kwart van de 17e eeuw", *Gewina* 30 (2007) 1–10.

Gerritsen W., *Het spoor van de eenhoorn. De geschiedenis van een dier dat niet bestaat* (Leiden: 2011).

Klußmann A., *In Gottes Namen fahren wir: die spätmittelalterlichen Pilgerberichte von Felix Fabri, Bernhard von Breydenbach und Konrad Grünemberg im Vergleich* (Saarbrücken: 2012).

Lavers C., *The Natural History of Unicorns* (London: 2009).

Quack A., „Eine nordische Runeninschrift in Utrecht" *Amsterdamer Beiträge zur älteren Germanistik* 21 (1984) 73–83.

Schepelern H.D., „The Museum Wormianum reconstructed", *Journal of the History of Collections* 2 (1990) 81–85.

Shepard O., *The Lore of the Unicorn* (London: 1967, zuerst 1930).

Szabo V.E., *Monstrous Fishes and the Mead-Dark Sea. Whaling in the Medieval North Atlantic* (Leiden – Boston: 2008).

Vlierden M., „De eenhoorns van Sint-Marie", *Bulletin van het Rijksmuseum* 37 (1989) 5–13.

CHAPTER 5

Snakes, Fungi and Insects. Otto Marseus van Schrieck, Johannes Swammerdam and the Theory of Spontaneous Generation[1]

Eric Jorink

In the cold winter of 1667–68, the young Florentine Prince Cosimo de' Medici (1642–1723) started a journey through Northern Europe, not—as a persistent rumor would have it—to escape the horrors of a dynastic marriage, but with the explicit aim of learning about the blossoming intellectual and artistic culture in England and—especially—the Dutch Republic.[2] On Monday 19th December 1667, Cosimo and his modest retinue arrived in Amsterdam, the city which at that time was the undisputed global hub of trade, knowledge and an emerging creative industry.[3] Guided by his agents Francesco Feroni (1614–1696) and Pieter Blaeu (1637–1706), Cosimo visited the new town hall; the headquarters of the East and West Indian Trading Companies; churches and the Jewish synagogues; publishing houses (including, of course, the famous one owned by Pieter Blaeu's brother Johannes) and many cabinets

1 Many thanks to Esther van Gelder and Daniël Margócsy for their comments on an earlier version of this paper, as well as to Anthony Ossa-Richardson for correcting my English and additional remarks.

2 A detailed account of Cosimo's travels through the Netherlands in: Hoogewerff G.J., *De twee reizen van Cosimo de' Medici, prins van Toscane door de Nederlanden (1667–1669). Journalen en documenten* (Amsterdam: 1919); by far the best analysis of Cosimo's journey: Veen H. Th. van, "Cosimo de' Medici's reis naar de Republiek in een nieuw perspectief", *Bijdragen en mededelingen betreffende de geschiedenis der Nederlanden* 102 (1987) 44–52.

3 Cf. Cook H.J., *Matters of Exchange: Commerce, Medicine, and Science in the Dutch Golden Age* (New Haven: 2007); Berkel K. van, "The Dutch Republic: Laboratory of the Scientific Revolution", in Berkel K. van – Goei L. de (eds.), *The International Relevance of Dutch History* (= Special issue *Bijdragen en medelingen betreffende de geschiedenis der Nederlanden* 125 (2010) 3–367; Dupré S. – Lüthy C. (eds.), *Silent Messengers: The World of Goods and the Circulation of Knowledge in the Early Modern Netherlands* (Berlin: 2011); Jorink E. – Ramakers B., "Undivided territory. 'Art' and 'Science' in the Early Modern Netherlands", in Jorink E. – Ramakers B. (eds.), *Art and Science in the Early Modern Low Countries*. Netherlands Yearbook for History of Art/Nederlands Kunsthistorisch Jaarboek 61 (Zwolle 2011) 6–33.

 | DOI 10.1163/9789004279179_007

FIGURE 5.1 *Otto Marseus van Schrieck. Oil on canvas, 38 × 47.8 cm, 1655. Florence, Galleria degli Uffizi, Inv. 5268.*

of curiosities.[4] Cosimo took a special interest in paintings; he visited Rembrandt's workshop (who, much to his regret, did not have anything suitable in stock) and saw the *Nightwatch* in the Doelen, the homebase of the *Schutterij*, the civic guards. However, the first painter whom the prince met was Otto Marseus van Schrieck (ca. 1620–1678), the creator of a strange genre of paintings in which snakes, insects and toads were depicted in mutual combat, against a dark background of thistles, fungi and moss [Fig. 5.1].[5] This genre,

4 Cf. Cools H., "Francesco Feroni (1614/16–1696): Broker in Cereals, Slaves and Works of Art", in Keblusek M. – Cools H. – Noldus B. (eds.), *Your Humble Servant: Agents in Early Modern Europe* (Hilversum: 2006) 39–50; Mirto A. – Veen H. Th. van (eds.), *Pieter Blaeu: lettere ai Fiorentini: Antonio Magliabechi, Leopoldo e Cosimo III de' Medici, e altri, 1660–1705: edizione con commento e saggio introduttivo in Italiano e Inglese / Pieter Blaeu: letters to Florentines: Antonio Magliabechi, Leopoldo and Cosimo III de' Medici, and others, 1660–1705: edited and annotated with an introductory essay in Italian and English* (Florence – Amsterdam: 1992).

5 Hoogewerff, *Twee reizen* 47. On Marseus see Steensma S., *Otto Marseus van Schrieck. Leben und Werk* (Hildesheim: 1999) and the excellent but as yet unpublished dissertation by Hildebrecht D.R., *Otto Marseus van Schrieck (1619/20–1678) and the Nature Piece. Art, Science,*

known since the nineteenth century as *sottobosco*, is now largely forgotten and in retrospect seems to have been a dead end in the evolving history of art.[6] However, in his own time Marseus was highly successful: he gained a small but notable circle of followers; high prices were paid for his work; and his paintings were appreciated in the world of the learned, including the Medici Court. During this visit, Cosimo bought three paintings of 'monsù Otto pittore' for the large sum of 500 guilders—in addition to the works by the Dutch painter which at that time were already in the Court's possession and on display in the Palazzo Pitti.[7]

Far from being marginal, Marseus was a central figure in Amsterdam in the 1660s. Besides the accounts of Cosimo's travels, many other sources testify to his fame, and show his connections not only with the artistic milieu, but with the wider world of learning in general. His paintings sold for high prices, among such eminent figures as Feroni, the Amsterdam burgomaster and collector of curiosities Roetert Ernst (1616–1685) and the famous anatomist Frederik Ruysch (1638–1731). As I will demonstrate in this contribution, Marseus consciously operated at the borders between art and science, *ars* and *scientia* to be more precise, and his work (and the obviously high regard in which it was held) can only be fully understood if we take a deeper look at the subjects and themes he depicted, and the responses they evoked in the eyes of the contemporary beholders. By depicting snakes, fungi and insects, Marseus chose to depict some of the lowest and most despised subjects anyone could imagine. But, much more than being a mere provocation to good taste and art theory, his work was also intended to communicate the knowledge of nature in a non-verbal way. To fully appreciate Marseus' work, a contemporary had to be not only a *connaisseur* of art and a man of letters, but actively involved in the cutting-edge scientific research of the age.

Marseus, and the circle of artists and other students of nature in which he operated, most notably Francesco Redi (1626–1697) and Johannes Swammerdam (1637–1680), were instrumental in putting hitherto highly neglected creatures at the center of attention, and arguing for the uniformity and ontological stability of nature. According to them, there was no essential distinction between man, the higher animals, and the lowest forms of life. The basic distinction was that man possessed a soul. But all beings were created by God, and worthy of being observed, collected, studied and depicted. It is no

---

Religion and the Seventeenth-Century Pursuit of Natural Knowledge (diss., University of Michigan, Ann Arbor: 2004).

6 Vries A.D., "Otto Marseus", *Oud-Holland* 1 (1883) 166–168.

7 Hoogewerff, *Twee reizen* 47.

coincidence that the Medici court, where under the auspices of the Accademia del Cimento groundbreaking research was done into the generation of animals, put such a high value on Marseus' work. Nor was it by chance that, some weeks after his encounter with the painter, Cosimo made the acquaintance of Swammerdam, the author of a pioneering work on the generation of insects.

Marginal Creatures

Before focusing on Marseus, Swammerdam and their network, it is worth taking a closer look at the objects of their observations. According to a deeply rooted tradition going back to antiquity, snakes, toads, mushrooms, insects, and also mice, reptiles, worms, and damp and slimy beings such as shellfish and snails, had their origin in spontaneous generation, i.e. in rotting flesh, decaying plants, putrefying wood etc.[8] Essentially, they came into being by chance, without descent from a similar organism. Spontaneous generation was not the result of sexual reproduction nor of some kind of teleology or entelechy, but of the random interaction of material, moist and heat. A decaying fungus could, for example, result in the rise of an amphibian, or the rotting remains of a horse could either be the breeding ground for worms or for lizards, or bees could have their origin in oxen. Different types of insects could be also generated spontaneously in different situations, for example out of mud, dung, wood, animal hair, flesh, or within a living animal. All such creatures were generally assumed to have no, or no complex, internal structure and they were consequently seen as the lowest link in the 'great chain of being'.[9] Aristotle, Pliny the Elder and others had elaborated in depth on the nature of these creatures, and the Bible incorporated like conceptions into the Christian tradition, providing an impressive list of unclean animals, including insects, crayfish, moles and snails.

The low nature of such creatures also implied that they were unworthy of serious study. For example, until well into the seventeenth century, countless species of insect were completely ignored by both artists and naturalists. There were a few exceptions, such as the bee, the butterfly, the spider and the grasshopper, which became symbols beloved of the artists who created

8 Lennox J., "Teleology, Chance and Aristotle's Theory of Spontaneous Generation", *Journal of the History of Philosophy* 20 (1983) 219–238; Cobb M., *The Egg and Sperm Race. The Seventeenth-Century Scientists who unravelled the Secrets of Sex, Life and Growth* (London: 2006) 9–30 and *passim.*

9 Lovejoy A., *The Great Chain of Being. A Study of the History of an Idea* (Cambridge MA: 1936).

illuminated manuscripts, paintings and emblems, partly because of their biblical connotations.[10] For centuries the butterfly, assumed to arise out of the corpse of a dead caterpillar, symbolized the Resurrection.

Partly under the influence of Pliny's dictum, 'rerum natura nusquam magis quam in minimis tota est', scholars like Conrad Gessner (1516–1565) and Ulisse Aldrovandi (1522–1605) began to turn their attention to a far broader range of creatures. Aldrovandi's *De insectis*, illustrated with fairly unsophisticated woodcuts, appeared in 1602, while Gessner's work would be included in Thomas Moffet's *Theatrum insectorum* (1634). They worked along the same lines as Joris Hoefnagel (1542–ca. 1600), who depicted countless insect species, spiders, snails, shells and other small creatures in beautiful watercolors and in the famous *Archetypa* series of engravings (1592).[11] Hoefnagel's example was followed by Jacques de Gheyn II (1565–1629).[12]

It has often been claimed that Hoefnagel and De Gheyn helped to create the specifically Dutch genre of the still life with flowers, shells and small living creatures, one exponent being Johannes Goedaert (1620–1668).[13] This pious painter, born around the same time as Marseus, earned his living with still lifes and was also active as an observer and describer of insects [Fig. 5.2].[14] In about 1635, Goedaert became one of the first scholars in Europe to begin a systematic study of the life and genesis of insects. He was an enthusiastic advocate of the theory of spontaneous generation, and described watching with his own eyes as two identical caterpillars died and different insects then emerged from their remains. From one little corpse rose a beautiful butterfly, while out of the other came a swarm of 'flies'. Goedaert noted that he was 'extremely astonished'

---

10 Freeman R., "llustrations of insect anatomy from the beginning to the time of Cuvier", *Medical and Biological Illustration* 12 (1962) 174–183; Lehman-Haupt H., "The Microscope and the Book", in Benzing J. *et al.* (eds.), *Festschrift für Claus Nissen. Zum siebzigsten Geburtstag 2. September* 1971 (Wiesbaden: 1973) 471–502.

11 See Vignau-Wilberg T., *Archetypa studiaque patris Georgii Hoefnagelii 1592. Natur, Dichtung und Wissenschaft in der Kunst um 1600* (Munich: 1994).

12 Swan C., *Art, Science, and Witchcraft in Early Modern Holland: Jacques de Gheyn II (1565–1629)* (Cambridge: 2005).

13 See for example Bergström I., *Dutch still-life painting in the seventeenth century* (London: 1956); idem, "On Georg Hoefnagel's manner of working with notes on the influence of the *Archetypa* series of 1592", in Cavalli-Björkman G. (ed.), *Netherlandish Mannerism* (Stockholm: 1985) 177–188.

14 Bol L.J., *'Goede onbekenden'. Hedendaagse verkenning en waardering van verscholen, voorbijgezien en onbekend talent* (Utrecht: 1982); Jorink E., *Reading the Book of Nature in the Dutch Golden Age, 1575–1715* (Leiden: 2010) 240–260.

FIGURE 5.2 *Johannes Goedaert,* A boucquet of roses in a glass vase. *Oil on canvas, date unknown. Middelburg, Zeeuws Museum, inv.nr. M96-031.*

on seeing this second transformation.[15] It must indeed have been a remarkable sight, although the contemporary reader will detect in it not so much the hand of God as the activities of the parasitic ichneumon wasp, using a living caterpillar as the breeding place for its own eggs. The general process of metamorphosis was interpreted by Goedaert as the coming into existence of a life form (caterpillar, grub), which then died and from whose remains new life (butterfly, fly) emerged by spontaneous generation. After a while these died too, 'until a new resurrection'.[16]

Equally symbolically laden, although much rarer in the Dutch Republic, were snakes. They were mostly depicted in paintings representing the story of the Fall, or the myth of Hercules killing the two snakes. Snakes were also a beloved theme in the extremely popular genre of emblem books, and moreover described and depicted in vivid detail—including their myriad hidden meanings and occult properties—by both Gessner and Aldrovandi.[17] Besides their mysterious origins and highly negative symbolic meanings, snakes had also medical connotations, being both highly poisonous and (partly as a result of that) providing an antidote and a basis for medicine. They shared these properties with a form of life equally ambiguous in value, namely fungi.[18] While some sorts were edible and considered a *delicatesse*, other, virtually identical sorts, were extremely poisonous. Again, they were considered the result of spontaneous generation, which is not very surprising given their moist and slimy nature and their preference for decaying matter as a habitat. Only very few naturalists took an interest in them, notable exceptions being Hadrianus Junius (1511–1575) and Carolus Clusius (1526–1609), both of whom published substantial treatises in the late sixteenth century [Fig. 5.3].[19]

---

15 Goedaert J., *Metamorphosis naturalis, ofte historische beschrijvinge van den oirspronk, aerdt, eygenschappen ende vreemde veranderingen der wormen, rupsen, maeden, vliegen, witjens, byen, motten en diergelijke dierkens meer; niet uit eenige boeken, maar aleenlyck door eygen ervaretntheyt uytgevonden, beschreven ende na dekonst afgeteykent* (3 vols.; Middelburg: Jacob Fierens: s.d. [1660]–1669) I, 4, 45.

16 Goedaert, *Metamorphosis naturalis* II, 140. See also ibidem I, 7–8.

17 Gessner Conrad, *Historiae animalium liber V qui est de serpentum natura: ex variis schedis et collectaneis eiusdem compositus per Iacobum Carronum* [...] *adiecta est ad calcem, scorpionis insecti historia* (Zurich, Christoph Froschauer: 1587); Aldrovandi Ulisse, *Serpentum, et draconum historiae libri II. Bartholomaeus Ambrosinus* [...] *summo labore opus concinnavit* [...] (Bologna, Antonio Bernia, Clemente Ferronio: 1640).

18 Ainsworth G.C., *Introduction to the history of mycology* (Cambridge: 1975).

19 Leonhard K., "Pictura's fertile field. Otto Marseus van Schrieck and the Genre of *sottobosco* Painting", *Simiolus* 34 (2010) 95–118; Gelder E. van, *Tussen hof en keizerskroon. Carolus*

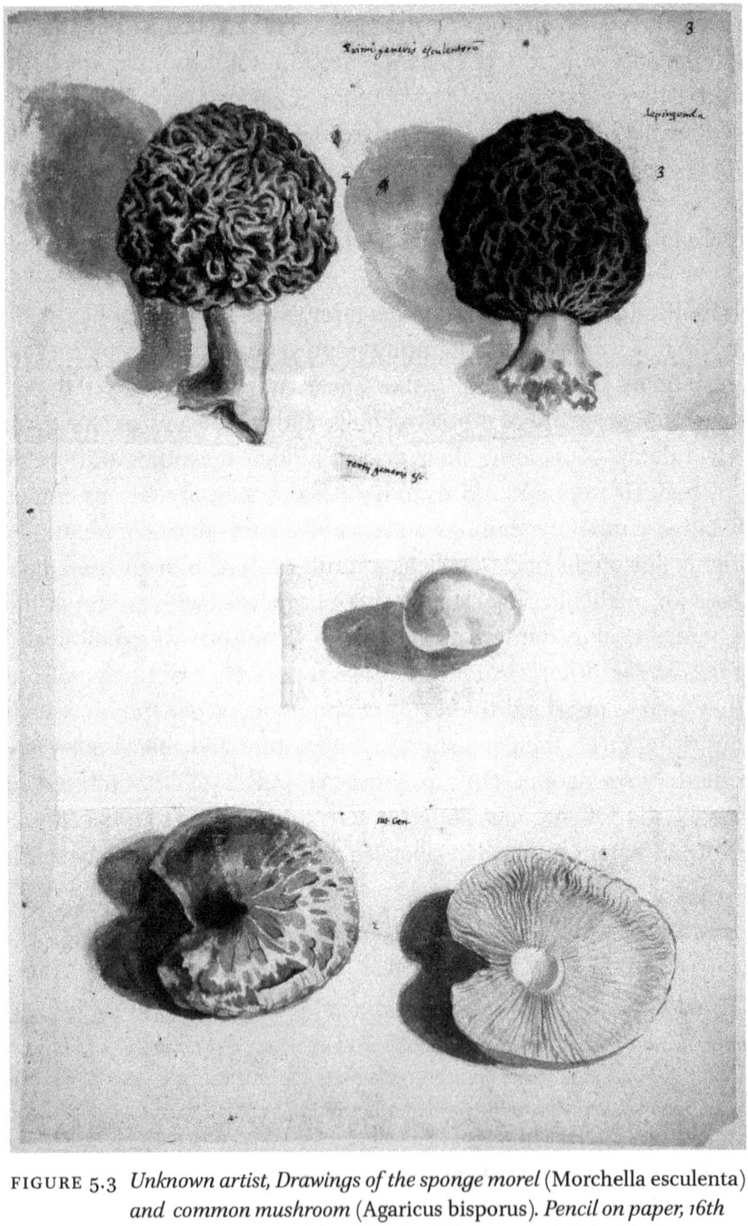

FIGURE 5.3 *Unknown artist, Drawings of the sponge morel* (Morchella esculenta) *and common mushroom* (Agaricus bisporus). *Pencil on paper, 16th century, University Library Leiden, Ms.* BPL *303, fol. 18r. The image was sent to Clusius, who published it in his* Fungorum in Pannoniis observatorum historia, *in Carolus Clusius,* Rariorum plantarum historia *(Antwerp, Plantin: 1601)* *(cclxi–ccxcv).*

Thus, in the lowest regions of the hierarchy of beings, below fishes, plants and stones, there existed an largely undefined, unstable and contingent category of creatures, which seemed to be beyond the order of nature. Seventeenth-century books on natural history were abundant with descriptions of hybrids, but also of slime changing into toads, putrefying meat changing into snakes, showers of frogs falling from the air, and recipes for creating mice. Authors on these and similar stories included famous scientists and philosophers like Jan Baptista van Helmont (ca. 1580–1644), William Harvey (1578–1657) author of pioneering works on the theory of the circulation of blood and on the generation of animals, and Robert Hooke (1635–1703), amanuensis of the Royal Society of London and author of the paradigmatic *Micrographia* (1664).[20]

At this time, René Descartes (1596–1650) provided the philosophical basis for a staunch and, as it turned out, devastating attack on this theory of spontaneous generation. The Frenchman, who lived in the Dutch Republic from 1628 to 1649, was a man of tremendous influence in his second fatherland.[21] By contrast to scholastic Aristotelianism, in which the explanatory factors were the different elements and qualities and their diverse aims and hidden potentials, Descartes regarded the universe as a vast mechanism in which the only factors that truly mattered were matter and motion. Planets, clouds, and animals consisted purely of moving particles [Fig. 5.4]. The universe was basically an enormous clockwork, obeying fixed laws. Nature was stable and uniform. The body was essentially a machine, a complex piece of engineering, comparable to an hydraulic construction or a laboratory. Anatomically speaking, there was no difference between men and animals: the only distinction being the fact that man had a thinking soul, with its material seat in the pineal gland. Thus, animals were basically automata.[22]

The relevance of Descartes' claim is the following. First, The mechanistic model gave an enormous boost to anatomical research. The idea that the body was comparable to a machine, mechanism or laboratory created a revolutionary new interpretative framework that, for example, allowed Harvey's theory of the circulation of the blood (*De motu cordis*, 1628) to be given a philosophical basis,

---

Clusius en de ontwikkeling van de botanie aan de Midden-Europese hoven (1573–1593) (Leiden: 2011).

20 Cobb, *Egg and Sperm Race.*

21 Verbeek Th., *Descartes and the Dutch. Early Reactions to Cartesian Philosophy, 1637–1650* (Carbondale: 1992); Ruler J.A. van, *The Crisis of Causality. Voetius and Descartes on God, Nature and Change* (Leiden: 1995); Bunge W. van, *From Stevin to Spinoza. An Essay on Philosophy in the Seventeenth-Century Dutch Republic* (Leiden: 2001).

22 Des Chene D., *Spirits and Clocks: Machines and Organism in Descartes* (Ithaca: 2004).

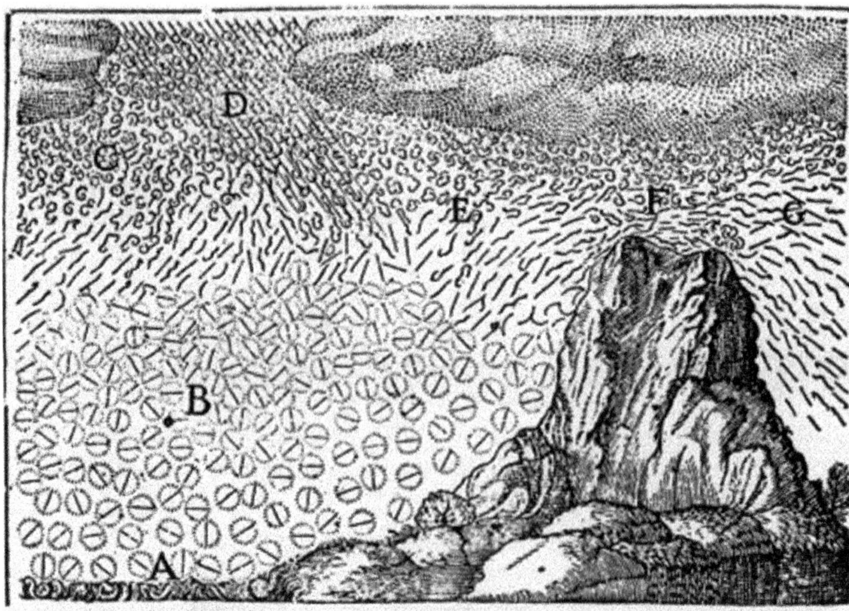

FIGURE 5.4 *Frans van Schooten jr., Woodcut depicting formation of clouds. From: René Descartes,* Discours de la méthode *(Leiden, Maire: 1637), "Les météores".*
COURTESY UNIVERSITY LIBRARY LEIDEN.

and encouraged countless other studies of anatomy and physiology.[23] The idea that animals were just ingenious automata provided an agenda for very elaborate experiments, mostly on living dogs and frogs, to study their anatomy and physiology. Secondly and even more importantly, from Descartes' point of view, nature was stable, orderly and uniform, governed by laws.[24] Those same laws applied at all times and in all places, and there was no ontological difference between the various forms of life. The traditional distinction between high and low, large and small, beautiful and ugly, rare and common, or between man and woman, was dispensed with. Everything was merely matter-in-motion; the cosmos was essentially a meaningless mechanism.

The Cartesian concept of nature as a mechanism raised a number of questions. For example: If people and animals are machines, where do baby machines come from? If the laws of nature are the same always and everywhere, can spon-

23 Lindeboom G.A., *Descartes and Medicine* (Amsterdam: 1974).

24 Vermij R., "Een nieuw concept. De wetten der natuur", in Egmond F. – Jorink E. – Vermij R. (eds.), *Kometen, monsters en muilezels. Het veranderende natuurbeeld en de natuurwetenschap in de zeventiende eeuw* (Haarlem: 1999) 105–119; Jorink, *Reading the Book of Nature* 85–88.

taneous generation be ruled out? If the body consists of tiny particles, can these also be observed under the lens of that wonderful new invention, the microscope?

Descartes gave a new impulse to the existing debate on procreation. The philosopher himself struggled heavily with the problem without bringing it to a satisfactory conclusion, and composed a *Traité de l'homme* and a *De la formation de l'animal*, which both remained unpublished during his life. At one point he even wrote: 'so little is required to make an animal, it is really not surprising that we see so many animals, so many worms, so many insects spontaneously forming in all putrefied matter'.[25] In other words, the founding father of what has been called the mechanistic worldview adhered to the idea that matter-in-motion, in some random way, could coagulate to a living being. In essence, this was the theory of spontaneous generation.

Others were puzzled by this theme as well. The Utrecht Cartesian Lambertus van Velthuysen issued a—rather conventional—*Tractatus de generatione* in 1657. He sent a copy to his friend and kindred spirit Johannes Hudde. Hudde, a brilliant mathematician and staunch defender of Descartes, would later become one of the most powerful politicians of the Dutch Republic, amongst other things burgomaster of Amsterdam and Director of the East India Trade Company (VOC).[26] As such, he became an influential patron of the arts and sciences. In 1657, he was a still unknown but highly ambitious figure. In his thank-you letter to Velthuysen, he wrote of his intention

> [...] to start, as soon as I have mastered the basics of medicine, to investigate by means of microscopes whether one cannot find and demonstrate *ad oculum* the generation of many things. I have reason to be very optimistic about that, because of various experiments I have already undertaken. So that presently, moved by that, I am working to determine the best microscope.[27]

---

25 Descartes R., *Œuvres*, ed. C. Adam – P. Tannery (Paris: 1897–1913) 1-I, 506.

26 Vermij R., "Bijdrage tot de bio-bibliografie van Johannes Hudde", *Gewina* 18 (1995) 25–35.

27 Hudde to Velthuysen, 13 October 1657, University Library Amsterdam Ms D 23: '[...] alzo ik van meening ben, zo haast als ik de fondamenten van de medicijnen zal geleert hebben, door vergrootglazen te onderzoeken, of men *ad oculum* de generatie van veel dingen niet zal konnen vinden en demonstreren; en hier toe zie ik door verscheyden experimenten, die ik alreede gedaan heb groote hoop, zulx ook dat ik bezigh ben de beste vergrootglazen te determineren'. Here quoted after Vermij R., "Instruments and the Making of a Philosopher: Spinoza's Career in Otics", *Intellectual history review* 22 (2013) 74.

It was not Hudde himself who would embark on this voyage of discovery. But he would provide the agenda, as well as the optical instruments to a small but influential group of friends in Amsterdam in the early 1660s. Two of them were Otto Marseus and Johannes Swammerdam.

Marseus and His Inventions

Although it is impossible to pinpoint an exact date, or even a year in which Marseus and Swammerdam first met, evidence suggests that they knew each other since the latter settled near Amsterdam, ca. 1660. They would cooperate until the former's death in 1678.

Otto Marseus was born around 1620 near Nijmegen, presumably in a Catholic family.[28] Like his elder brother Evert (ca. 1612–1681), who was a painter of landscapes, Otto chose a career as an artist. His earliest known work (ca. 1645; present whereabouts unknown) is a rather conventional still life painting with flowers and butterflies in a glass vase.[29] According to contemporary art theory, this genre was the lowest in the hierarchy of paintings: the most esteemed (and most difficult) being huge pieces depicting biblical and historical scenes.[30] Nevertheless, still life paintings were quite popular, and Dutch artists produced hundreds, if not thousands of them. Although the delicate way in which these ensembles of flowers, insects, shells and other objects were composed and depicted is still the subject of debate among historians of art, it seems safe to assume that, in the eye of the contemporary beholder, they were much more than expressions of *vanitas* or the result of a presumed objective 'art of describing'.[31] It can be argued that they were intended to evoke a strong

---

28 I follow here mainly Hildebrecht, *Marseus*, who provides by far the best data on Marseus' life and work.

29 The work was auctioned in 1999: an image can be found at the website of the Rijksbureau voor Kunsthistorische Documentatie: http://www.rkd.nl/rkddb/(S(53ljsemrpcueeqvi52xwulaf))/detail.aspx?parentpriref=.

30 On still life painting see a.o. Bergström, *Dutch still-life Painting*: Bol, *'Goede onbekenden'*; Taylor P., *Dutch Flower Painting 1600–1720* (New Haven – London: 1995); Weststeijn T., *The Visible World: Samuel van Hoogstraten's Art Theory and the Legitimation of Painting in the Dutch Golden Age* (Amsterdam: 2008).

31 On the interpretation of still life paintings see, besides the previous note, also Alpers, A., *The Art of Describing. Dutch Art in the Seventeenth Century* (Chicago: 1983); Franits W. (ed.), *Looking at Seventeenth-Century Dutch Art: Realism Reconsidered* (Cambridge: 1997); Hochstrasser J., *Still Life and Trade in the Dutch Golden Age* (New Haven: 2007); Woodall J., "Laying the Table. The Procedures of Still Life", *Art History* 35 (2012) 976–1003.

sense of wonder, confronting the viewer with the beauty and ingenuity of the Creator.

Leaving this issue aside, we return to young Otto, who in or around 1648 decided to travel to Italy. Early in the eighteenth century, Arnold Houbraken (1660–1719) wrote in his book on the lives of Dutch painters that Marseus at some time was working in Florence for the Grand Duke Ferdinando II—although he did not mention a source. Be this as it may, around 1650 Marseus had settled in Rome, where he joined the *Bentveugels*, a group of northern artists notorious for their initiation ceremonies, drunken festivities and provocative artistic attitudes [Fig. 5.5]. Here, Marseus transformed his earlier style into his own unique genre of painting. As Hildebrecht put it: Marseus could have continued in the style of Dutch still life paintings, but

> [f]aced with numerous possibilities of how to set himself apart from others, Marseus made one crucial decision: to begin to paint a selection of natural specimens that had never before been the subject of easel painting.[32]

As a member of the Bent, he became known as the *Snuffelaer* (ferret) because, as Houbraken wrote, 'he was always sniffing about everywhere for strange coloured and speckled snakes, lizards, caterpillars, spiders, butterflies and strange plants and herbs'.[33] The Italian countryside offered many opportunities for bodily engagement with these lowest of creatures.[34] In addition, Marseus also became fascinated by fungi, and started to use them as a motif in his work. However, Marseus not only took obvious delight in collecting these creatures, he also very soon acquired fame for the excellent way in which he depicted these and other unusual *naturalia*. For example, after receiving a *cedrato* (a rare kind of lemon) from the De Medici gardens in Florence, the Roman collector Cassiano Dal Pozzo (1588–1657), famous for his 'paper museum', commissioned Otto Marseus to draw a detailed image on parchment of this strange

32 Hildebrecht, *Marseus*, 89.

33 Houbraken A., *De groote schouburgh der Nederlantsche konstschilders en schilderessen* (The Hague: Swart – Boucquet – Gaillard: 1718–1721) I, 358 as quoted by Hildebrecht, *Marseus*, 34: 'In de 'Bent' doopten zy hem den bynaam Snuffelaer; om dat hy allerwegen naar vremd gekleurde of gespikkelde slangen, hagedissen,rupsen, spinnen, flinterjes, en vremde gewassen en kruiden rondsnuffelde'.

34 Cf. Smith P.H. *The Body of the Artisan: Art and Experience in the Scientific Revolution* (Chicago: 2004).

FIGURE 5.5 *Anonymous,* Initiation of a new member of the Bentveughels in Rome, *ca. 1660. Oil on canvas, 95.5 cm × 134.5 cm. Rijksmuseum Amsterdam, inv.nr. SK A 4672.*

fruit ('[...] non più visto ramo del cedrato unico [...]').[35] It is interesting to note that Dal Pozzo at that time owned a large collection of drawings of fungi, a subject which certainly would have interested Marseus, and another indication that these creatures were gaining respect in the world of the learned. Another visitor of Marseus, the painter and writer Samuel van Hoogstraten (1627–1678), later presented Marseus' methods of representing nature as an illuminating example for others:

> But certainly Otto Marseus (alias Snuffelaer) has made clear his special aptitude and the part of art toward which he is inclined. For when I was with him in Rome in 1652, I was amazed at the number of monstrous creatures he kept and fed, and how his thorough understanding of their nature was just as wondrous as his lifelike representation of their forms.[36]

---

35 Rutgers J., 'Otto Marseus van Schrieck in Rome: een opdracht van Cassiano Del Pozzo', *Oud Holland* 125 (2012) 59–64.

36 Hoogstraten, S. van, *Inleyding tot de Hooge Schoole der Schilderkonst* (Rotterdam: F. van Hoogstraeten: 1678) 69 as quoted by Hildebrecht, *Marseus*, 34: 'Maer zeker onzen Otho Marseus (alias Snuffelaer) heeft zijn bequaemheit in de kunst, en tot wat deel hy neygde,

Others were struck by Marseus' intimate knowledge of these creatures as well, and much later, Houbraken even claimed that snakes would strike a pose for the painter! The astonishment at Marseus' ability to depict these 'monstrous creatures' was completely justified. Not only were they a rather unusual subject for an artist; the way in which Marseus depicted them 'naer het leeven' (after life) was remarkable as well. Because of the texture of their skin and their twisting locomotion, snakes are notoriously difficult to paint. Marseus not only managed to evoke the illusion of real snakes ready to strike with a venomous bite, he also painted species that had never been depicted before. In the same way, he painted other rare and exotic creatures, like chameleons.

During his stay in Italy, Marseus transformed the peaceful and quiet still life paintings into a new genre, in which action and interaction between all kinds of creatures takes place. The subjects were moved from the tranquil glass vases, out of the comfortable rooms, into nature itself. The setting now was usually a dark, gloomy place in a forest, strongly suggesting moisture and decay. Marseus consciously challenged artistic propriety by putting marginal creatures like snakes, toads, insects and mushrooms at the center of his paintings. It was a completely new genre of painting. The extremely complex paintings offered multiple layers of meaning, including biblical references, emblematic connotations, notions of sympathy and antipathy and, above all, factual evidence tested against textual knowledge. In a way which obviously was meant to remind the viewer of biblical and historical stories and figures, snakes, butterflies and lizards were caught in action and mutual interaction, waiting to start a deadly attack, or to escape in a miraculous way. For a couple of decades, this genre—for which no contemporary term was coined—was very much *en vogue*. Marseus' style was copied or emulated by, amongst others, Matthias Windhoos, Melchior d'Hondecoeter, Johan Teyler, Rachel Ruysch and, in Italy, Paolo Porpora (1617–1673) [Figs. 5.6–5.7].[37]

Most of the creatures taking center stage in Marseus' paintings were traditionally considered the very lowest forms of life. Snakes, fungi, insects were, like shellfish, frogs etcetera, considered the random outcome of spontaneous generation. At the same time, they were the focus of scientific attention; just as Marseus was starting his new genre, natural philosophers like Van Helmont, Descartes, Henricus Regius (1598–1679) and William Harvey (1578–1657) were

---

genoeg laten blijken, want als ik in 't jaer 1652 by hem tot Romen was, verwonderde ik my over zoo veel gedrochten, asl hy onderhielt en voede: welkers natuaer hy ook zoo wonderlijk deurgronde, als hy hare gedaentens levendich heeft uitgebeelt'.

37 Hildebrecht, *Marseus* 59–83; Rikken M., *Melchior d'Hondecoeter, Bird Painter* (Amsterdam: 2008).

FIGURE 5.6 *Johan Teyler,* Snake and butterfly, *ca. 1670. Engraving and etching a là poupée in blue, red, green brown and black, hand-colored, 17.5 cm × 43.8 cm. Rijksmuseum Amsterdam, inv.nr. RP-P-1939-92.*

FIGURE 5.7 *Melchior d'Hondecoeter,* Animals and plants, *ca. 1668. Oil on canvas, 66.5 cm × 52.5 cm. Rijksmuseum Amsterdam, inv.nr. SK A 169.*

showing a renewed interest in the problem of procreation. Harvey's *De generatione animalium* appeared in 1651 (two Latin editions were then published in London; two simultaneously in Amsterdam) and became for a short time the new paradigm in the scientific debate.[38] Basing himself on experiments and observations, Harvey mainly focused on the problem of the generation of higher animals, claiming that all of these creatures came out of an egg. Unfortunately, his notes on insects were not included in the book, but from his digression on other low creatures, it is very clear that he endorsed the theory of spontaneous generation. By depicting and showing the magnificence of those low creatures, Marseus elevated them from their traditional status, mirroring one of the main scientific themes of the age. He would not remain entrenched in the paradigm of spontaneous generation, but became actively engaged in the new research on procreation that took place in the 1660s.

But beside the relation between the creatures in Marseus' 'moving life' paintings and contemporary debates on procreation, the life forms he painted had yet another intriguing connotation. Most of them—snakes, many sorts of mushrooms and fungi—were highly poisonous, and others—like toads—were believed to be so as well.[39] As such, these creatures were also a focus of scientific attention, most notably at the Medici court. Since antiquity, courts had been the stage for intrigues, betrayal and murder—often with poison. Besides other factors, this had led to a lively interest in medicine and antidotes in mid-seventeenth century Florence. Under the patronage of Ferdinando II and Leopoldo, scholars put ancient recipes to the test, and started new observations and experiments, especially with vipers.[40] These were considered one of nature's most deadly creations; thought to provide an antidote to their own poison; and were an essential component for the production of theriac, the highly complex and costly panacea which was produced once a year during a solemn ritual at the court. In the mid-1650s, the court's openness to the prospects of the New Science gave rise to the Accademia del Cimento, a rather informal scientific society that operated under the patronage and watchful eye

38 Cobb, *Egg and Sperm Race, passim.*

39 Hildebrecht, *Marseus* 109–133, 142–192; Leonhard K., "Painted Poison. Venomous Beasts, Herbs, Gems, and Baroque Colour Theory", in Jorink – Ramakers (eds.), *Art and Science* 116–147.

40 Knoefel P. (ed.), *Fransesco Redi on Vipers* (Leiden: 1988); Findlen P., *Possessing Nature: Museums, Collecting, and Scientific Culture in Early Modern Italy* (Berkeley: 1994); Tribby J., "Club Medici: Natural Experiment and the Imagineering of 'Tuscany'", *Configurations* 2 (1994) 215–235.

of the Grand Duke.[41] It was within this context that Franscesco Redi started his groundbreaking experimental research on vipers, resulting in the publication of his *Osservazioni intorno alle vipere* in 1664. During these experiments, thousands of animals were killed, dissected, observed or used in trials.

It is also in this context that the Medici court, that great family of patrons of the arts and sciences, started to take an interest in Otto Marseus. Although it is still unclear—as Houbraken reported—if he ever entered into the Grand Duke's service, in the mid-1650s the court started to buy his paintings, collecting at least twelve of them during the subsequent decades. As Hildebrecht put it: 'This made the Medici, members of the court, and visitors to the court one of Marseus' most important audiences and established him as one of the most visible northern artists in the Grandducal collections'.[42]

Moreover, Marseus started to explore the boundaries of the art of painting in yet another way. After first having challenged the traditional hierarchy between history paintings and still life paintings rather successfully, he now started to subvert contemporary conceptions of representation. Although it is a much more problematic term than usually acknowledged, we could say that Marseus was very successful in painting 'naer het leeven' (after the life), in the sense that the creatures depicted in his images very much resembled the actual creatures in living nature.[43] But to determine if snakes really attack butterflies, or toads indeed feed by ballistically projecting their long tongues from their mouths to capture flies, was a rather different story. And was that fungus, depicted in an early stage of its development, the deadly poisonous Death Cap (Amanita phalloides) or the virtually identical field mushroom (Agaricus campestris)? Marseus took obvious delight in scrambling and decoding knowledge, thereby inviting the viewers of his work to test and display their erudition. But apart from playing with these more theoretical dimensions of the idea of representation, Marseus also toyed with the material aspects of it. Picking up a technique occasionally used by Joris Hoefnagel, Marseus started to stick real butterfly wings in the still wet paint in order to evoke nature.[44] Moreover, in many of the paintings in a dark, mossy setting, the paint of the brown-

41 Beretta M. (ed)., *The Accademia del Cimento and its European Context* (Sagamore Beach: 2009).

42 Hildebrecht, *Marseus* 251.

43 Cf. Swan C., "Ad vivum, naer het leven, from the life. Defining a mode of representation", *Word & Image* 2 (1995) 353–372; Bakker B., "Au vif—naar 't leven—ad vivum. The Medieval Origin of a Humanist concept", in Boschloo A. e.a. (eds.), *Aemulatio. Imitation, Emulation and Invention in Netherlandish Art from 1500 to 1800. Essays in Honour of Eric Jan Sluijter* (Zwolle: 2011).

44 Berthier S. – Boulenguez J. – Mottin B., "Butterfly Inclusions in Van Schrieck Masterpieces. Techniques and Optical Properties", *Applied Physics A* 92 (2008) 51–57.

green foreground appears to have been applied with the aid of a sponge—or a piece of moss?—so as to create an almost three-dimensional texture. Moreover, there are many indications—yet to be confirmed by technical analysis—that Marseus used pigments extracted from the objects he painted: painting rock with rock pigments; earth with earth, and herbs with herbal dyes as Karin Leonhard has suggested.[45]

Thus, operating in the twilight zone between *ars* and *scientia*, Marseus was no unimportant figure, and his work was the focus of a small and erudite community of discourse. Within the narrative, multilayered composition of his fusion paintings, he depicted rare and exotic creatures, some of them for the very first time in history, thereby transcending the métier of *pictor et inventor*. From different sources we know that Marseus had the ambition to be more than an innovative and successful painter.[46] His observational and representational skills, as well as his great knowledge of some nature's less esteemed creatures, fostered the ambition to publish an illustrated treatise on natural history—a project that never materialized.

Marseus, Swammerdam and the Amsterdam Circle of curiosi

No signed paintings of Mareus are known between 1656 and 1660, and this episode in his life is something of a mystery. Houbraken wrote that Marseus spent some time in France; a contemporary source much closer to Marseus gives the same impression.[47] It adds that the painter had suffered from a severe tumor in one of his eyes which took away his vision, and was successfully cured by the notorious mystic and alchemist Guiseppe Franscesco Borri (1627–1695), who at that time was residing in Amsterdam. Be that as it may, we are on safe ground if we assume that shortly before 1660 Marseus was back in the Dutch Republic. True to his somewhat eccentric character, he did not settle in one of the more fashionable quarters of Amsterdam, but instead took lodgings in a small countryhouse, 'Waterijk', in the marshy area near Diemen, just outside the city

45 Leonard, "Painted Poison".

46 De Monconys B., *Journal des voyages* (Lyon, Horace Boissat – George Remeus: 1666), vol. II, 178; Tolmer L., "Une page d'histoire des sciences 1661–1669. Vingt-deux lettres inédites d'André de Graindorge à P.D. Huet", *Mémoires de l'Académie des sciences, arts et belles-lettres de Caen* 10 (1941) 245–330: 'L'on m'a dit qu'il y avait un Flamand à Rome qui nourrissait des lézards, des crapauds et mille vilaines bètes qui faisaient ses délices, dont il examinait la nature et qu'il peignant par excellence, mais peu savant pour en faire la relation'. From the context of this source, it is clear that this is a statement by Swammerdam, and that he is referring to Marseus.

47 Monconys, *Journal*, vol. II, 178.

and known as 'the land of snakes'. Here, he maintained his gathering of living animal models, picked up painting again, and further developed his unique style, producing many works which sold at high prices. He seems to have supplemented his income by acting as an art dealer. As in Italy, Marseus moved in the shared world of artists and scientists. He turned out to be on excellent terms with the local elite of regents and collectors, including Johannes Hudde. The paintings produced by Marseus in 1660 and later continued to elaborate on the theme of procreation; taking their inspiration no longer from the experimental culture in Florence, but from Amsterdam.

Partly as a result of his respect for Descartes, Hudde was fascinated by optics and the microscope, the new instrument constructed around 1620 which, along with the telescope and the air pump, would emerge as an icon of the scientific revolution.[48] In the early 1660s Hudde succeeded in manufacturing simple microscopes using drops of molten glass, with the help of which Swammerdam and later Antonie van Leeuwenhoek made their revolutionary discoveries. Hudde generously shared his invention with a wide circle of likeminded enthusiasts, and by about 1662 both Otto Marseus and Johannes Swammerdam were in possession of Hudde's microscopes.[49] As individuals, but also often in cooperation, they started to study and represent subjects related to the theme of procreation: not only snakes, frogs, insects but also fungi, galls and ferns.

The two appear to have met for the first time in Amsterdam, in the early 1660s, and would often meet, cooperate and exchange observations over the next two decades.[50] Both were active in a network that, besides Hudde, also included the physician Mattheus Sladus (1628–1689), the later burgomaster and Director of the VOC Nicolaas Witsen (1641–1717), the later Leiden professor of physics Burchardus de Volder (1641–1709), the anatomist (and later bishop) Nicolaus Steno and—*bien étonné*—Benedictus Spinoza (1632–1677).[51] They

48 Vermij R. – Atzema E., "*Specilla circularia*. An unknown work by Johannes Hudde", *Studia Leibnitiana* 27 (1995) 104–121; Ruestow E.G., *The Microscope in the Dutch Republic. The Shaping of Discovery* (Cambridge: 1996) 6–60.

49 Monconys, *Journal*, II, 152–164; Swammerdam Johannes, *Historia generalis insectorum ofte Algemeene verhandeling van de bloedeloose dierkens* (Utrecht: Meindardus van Dreunen: 1669) 81.

50 That the two met before 1664 can be deduced from the letter of André Graindorge to Huet, xx April 1665, published in Tolmer L., "Une page d'histoire des sciences", in which an unnamed person—who can only be Swammerdam—discusses Marseus. Swammerdam matriculated at Leiden University, and was on a Grand Tour to France 1664–1665.

51 On this network see Jorink E., '"Outside God, there is Nothing'. Swammerdam, Spinoza and the Janus-face of the early Dutch Enlightenment", in Bunge W. van (ed.), *The Early Enlightenment in the Dutch Republic, 1650–1750. Selected papers of a conference held at*

were all students of nature fascinated by the microscope and the order of creation. All were deeply religious, although their ideas were mostly unorthodox from the point of view of Dutch reformed church. Much more could be said about this network—mostly educated at the Leiden medical faculty and living in Amsterdam—but I have to limit myself to Marseus, Swammerdam and, in the background, Steno and Spinoza.

Swammerdam was born on 12 February 1637, as the son of the pharmacist Jan Jacobszoon Swammerdam (1606–1678) of Amsterdam. In his house the apothecary kept a famous cabinet of curiosities, including exotic *naturalia* and *artificialia*, antiquities, artworks and scientific instruments.[52] Cabinets like these had an important part to play in the seventeenth-century information explosion, as the centerpoint of a community of discourse.[53] As with Marseus' paintings, perceptive visitors examining such cabinets learned never to accept any story as true but to look at all the evidence with their own eyes and analyze it with their own intellects. Johannes would later become a fervent propagator of this lesson. From his early years on, he was fascinated by the wonders of God's creation, but whereas his father had gathered exotic objects and species, young Swammerdam was especially fascinated by insects and other 'bloodless creatures'. Swammerdam must have got to know Hudde in about 1660, before going on to study in Leiden from 1661 and his subsequent tour of France in 1664–65. Back in the Dutch Republic, he took his MD in 1667 with a dissertation on respiration, causally announcing, both in words and images, forthcoming work on insects and the question of procreation. The frontispiece of the work shows two snails mating [Fig. 5.8].[54]

---

the Herzog August Bibliothek, Wolfenbüttel, 22–23 March 2001 (Leiden: 2003) 81–108; idem "Beyond the Lines of Apelles. Johannes Swammerdam, Dutch Scientific Culture, and the Representation of Insect Anatomy", in Jorink – Ramakers (eds.), *Art and Science* 148–184; Jorink E., "In the Twilight Zone. Isaac Vossius and the Scientific Communities in France, England and the Dutch Republic", in Jorink E. – Miert D. van (eds.), *Isaac Vossius (1618–1689) between Science and Scholarship* (Leiden: 2012) 119–156.

52 Bergvelt E. – Kistemaker R. (eds.), *De wereld binnen handbereik. Nederlandse kunst- en rariteitenverzamelingen 1585–1735* (Zwolle: 1992); Roemer G.M. van de, *De geschikte natuur. Theorieën over natuur en kunst in de verzameling van zeldzaamheden van Simon Schijnvoet (1652–1727)* (unpub. diss., University of Amsterdam: 2005); Cook, *Matters of Exchange*; Jorink, *Reading the Book of Nature* 257–346. For the international context see Findlen, *Possessing Nature* 1996; Daston L. – Park K., *Wonders and the Order of Nature 1150–1750* (New York: 1998).

53 Findlen, *Possessing Nature* 97–154; Jorink, *Reading the Book of Nature* 257–346.

54 Swammerdam Johannes, *Tractatus physico-anatomico-medicus de respiratione usuque pulmonum* (Leiden: Gaasbeeck: 1667).

FIGURE 5.8 *Frontispiece of Johannes Swammerdam,* Tractatus de respiratione *(Leiden, Gaasbeek: 1667). The design was presumably by Swammerdam himself; the engraving was made by Herman Wingendorp.*
COURTESY UNIVERSITY LIBRARY LEIDEN.

In the 1660s, Leiden University was one of the most exciting centers of scholarship and scientific research in Europe, especially with regard to medicine and natural philosophy. Under the leadership of the professors Franciscus de le Boë Sylvius (1614–1672) and Johannes Van Horne (1621–1670), Swammerdam excelled at anatomical dissection.[55] Along with three other students—Niels Stensen (better known as Steno, 1637–1689), Reiner de Graaf (1642–1673) and Frederik Ruysch—he was part of a brilliant quartet that made innumerable discoveries. The focus of the research these professors and their students was on two related fields: the conception of the body as a laboratory or machine, and the problem of procreation. Both subjects were to an high degree inspired by Cartesian philosophy. During his student years, Swammerdam would become more and more fascinated by the latter and, with the use of the microscope, would write a book, the *Historia Generalis Insectorum* (1669) in which he described the generation of insects in astonishing detail and dealt a devastating blow to the conception of spontaneous generation. Swammerdam was a gifted experimentalist and a keen observer. But at the heart of his quest for nature's secrets was a clear and distinct axiom: that nature was stable and uniform. God's laws of nature were inmutable and applied throughout the whole of creation. From a philosophical and ontological point of view, there was no distinction between the 'higher' and 'lower' animals. Echoing Pliny, Swammerdam stated that the tiniest louse was worth as much attention as the biggest lion or elephant.

Swammerdam expressed an even more explicit preference for the 'least esteemed creatures' than his friend Otto Marseus. He too studied amphibians (especially frogs), reptiles, fungi and, above all, insects. Time and again he stressed that they had an admirable anatomy whose minuscule size actually made it all the more admirable. The assumption that they were produced by spontaneous generation was not only untrue, it was blasphemous. After all, such a theory implied that part of nature had exempted itself from the order and laws laid down by God. Belief in spontaneous generation was a 'direct route to Atheism', Swammerdam kept repeating on nearly every page of his writings.[56] Nature was orderly; there could be no such thing as chance. The

55 Lindeboom G.A., *Descartes and medicine* (Amsterdam: 1974); Kooijmans L., *De doodskunstenaar. De anatomische lessen van Frederik Ruysch* (Amsterdam: 2004); Margócsy D., "Advertising cadavers in the Republic of Letters: Anatomical publications in the Early Modern Netherlands", *British Journal for the History of Science* 42 (2009) 187–210; Huisman T., *The Finger of God. Anatomical Practice in 17th-century Leiden* (Leiden: 2009).

56 See for example Swammerdam Johannes, *Bybel der natuure of historie der insecten. Biblia naturae, sive historia insectorum* (Leiden, Isaac Severinus – Bodewijn van der Aa – Pieter van der Aa: 1737–1738) 669.

same rules apply everywhere at any time. Humans could only give birth to humans; sheep could only give birth to sheep; and lice could only bring forth lice. Even borderline cases like hybrids obeyed underlying rules. Nature was stable and uniform; God's laws were fixed. With explicit reference to Descartes, Swammerdam translated this belief into natural history.

In 1669, Swammerdam published the tentative results of his research into *insecta* (in the broad early modern meaning of the word of divided animals; spiders and scorpions were also included). At the heart of *Historia generalis insectorum* is the axiom that insects do not arise from spontaneous generation or by means of a mystical metamorphosis but, like other species of animal, from a female egg, or ovum. The process of metamorphosis in insects was purely a matter of change in form and scale. A caterpillar became a butterfly not as a consequence of a puzzling 'inadvertent transformation, abandonment of form, known as death and resurrection',[57] but rather of a process of growth that was precisely the same as that of 'a chick, which does not *change* into a hen but grows in its limbs and so *becomes* a hen'.[58] One of Swammerdam's finest hours was when he demonstrated to his visitor Cosimo de Medici how parts of the future butterfly already can be discerned in the caterpillar.[59]

There was no such thing as chance or spontaneous generation. Like produces like. Procreation and generation are not the result of chance, but obey fixed laws. Every animal, regardless of its size, has its origin in an egg and then grows out, stage by stage, into a mature being.

Illustrations were of great importance to Swammerdam's book and supported the verbal argument made by its author.[60] Like several of his friends and colleagues—Witsen, Christiaan Huygens and Spinoza for example—Swammerdam was exceptionally good at drawing. In his engravings in *Historia generalis insectorum* Swammerdam shows step by step how the various insect species change their shape and size, starting with a drawing of an egg [Fig. 5.9]. The realistic effect is enhanced by the fact that the entire process is shown life-sized. The reader sees not just the egg from which the caterpillar emerges and the caterpillar that results but the parts of the butterfly that are already

57 Swammerdam, *Historia Generalis Insectorum* 27.

58 Swammerdam, *Historia Generalis Insectorum* 9, emphasis added.

59 Swammerdam, *Historia Generalis Insectorum* 27.

60 Cobb M., "Malpighi, Swammerdam and the Colourful Silkworm. Replication and Visual Representation in Early Modern Science", *Annals of Science* 59 (2002) 111–147; Meli D.B., "The Representation of Insects in the Seventeenth Century. A Comparative Approach", *Annals of Science* 67 (2010) 405–429; Jorink E., "Beyond the Lines of Apelles. Johannes Swammerdam, Dutch Scientific Culture, and the Representation of Insect Anatomy", in Jorink – Ramakers (eds.), *Art and Science* 148–184.

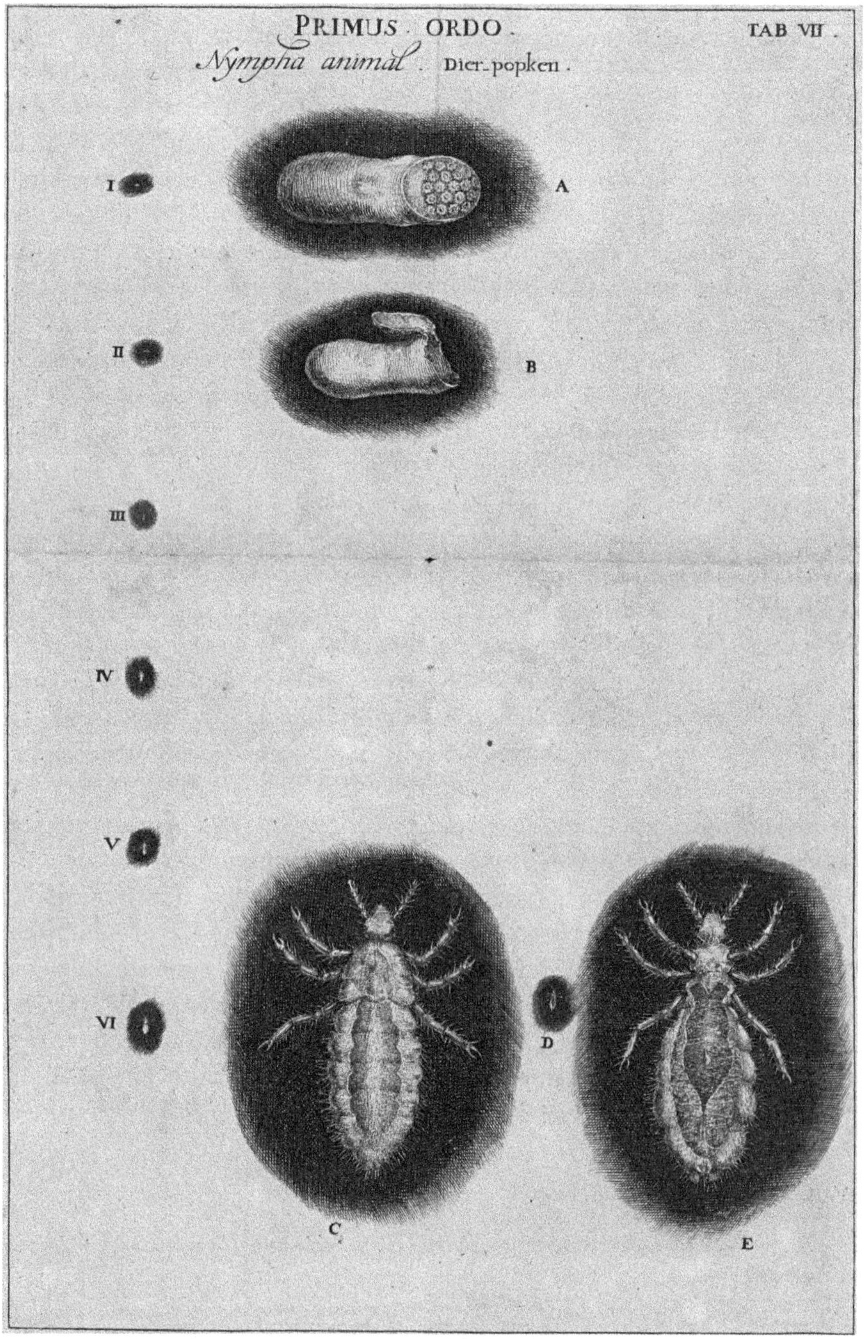

FIGURE 5.9 *The development of the louse. Engraving taken from: Johannes Swammerdam,* Historia generalis insectorum *(Utrecht, Van Dreunen: 1669).*
COURTESY UNIVERSITY LIBRARY LEIDEN.

present in the caterpillar, along with the pupa and the butterfly itself [Fig. 5.10]. In order to underline the uniformity of nature, Swammerdam included an engraving showing the stage-by-stage development of a frog and a carnation, pointing out the similarities in the process of growth [Fig. 5.11]. Every one of Swammerdam's engravings can be read as a narrative, each depiction within it representing a chapter. It was a visual technique that Mario Biagioli recently has called 'visual sequencing' and in older German art history is known as *kontinuierende Darstellung*.[61] The narrative technique resembles the one that was used in medieval and Renaissance visual culture to depict biblical stories in one image. It was also very suitable to illustrate the process of generation, and had already been used by Marseus to illustrate the grow and decay of the Death Cap in his famous 1662 painting of insects and a giant toad, now in the Herzog Anton Ulrich-Museum, Braunschweig [Fig. 5.12].[62]

Swammerdam was very much aware of his abilities as a draftsman. Like Marseus, he consciously tried to cross disciplinary borders. Whereas the painter had had the ambition to publish a book on natural history, Swammerdam occasionally signed his elaborate illustrations with a proud 'auctor delineavit' [Fig. 5.13]. His descriptions used the idiom of a *connaisseur* of art, and the descriptions he gave of anatomical details consciously evoked comparisons with still life paintings. Obliquely revealing his friendship with Marseus, the *Historia Generalis Insectorum* contained a chapter on how 'one could ornament a painting with the wings of a butterfly'.[63]

There are many more indications that Swammerdam and Marseus cooperated very closely in these years, and were investigating nature on a cutting-edge level. Both observed and made images of the mating of dragonflies. In April 1671 Swammerdam dissected a chameleon—the rare animal that was one of the pet subjects of Marseus.[64] Fascinated by the anatomy of frogs, Swammerdam discovered that they have a particular form of lungs [Fig. 5.14]. This observation, he wrote to the secretary of the Royal Society, Henry Oldenburg (1618–1677), made him ask if similar creatures had the same structure, and underlined the importance of comparative anatomy:

61 Biagioli M., *Galileo's Instruments of Credit. Telescopes, Instruments, Secrecy* (Chicago: 2006) 135–218.

62 I would like to thank Jan Hengstmengel, Naturalis Biodiversity Centre Leiden, for identifying the Amanita phalloides for me.

63 Swammerdam, *Historia Insectorum Generalis* 129.

64 Swammerdam to Thévenot, 16 April 1671, in Lindeboom G.A. (ed.), *The letters of Jan Swammerdam to Melchisedec Thévenot* (Amsterdam: 1975) 60.

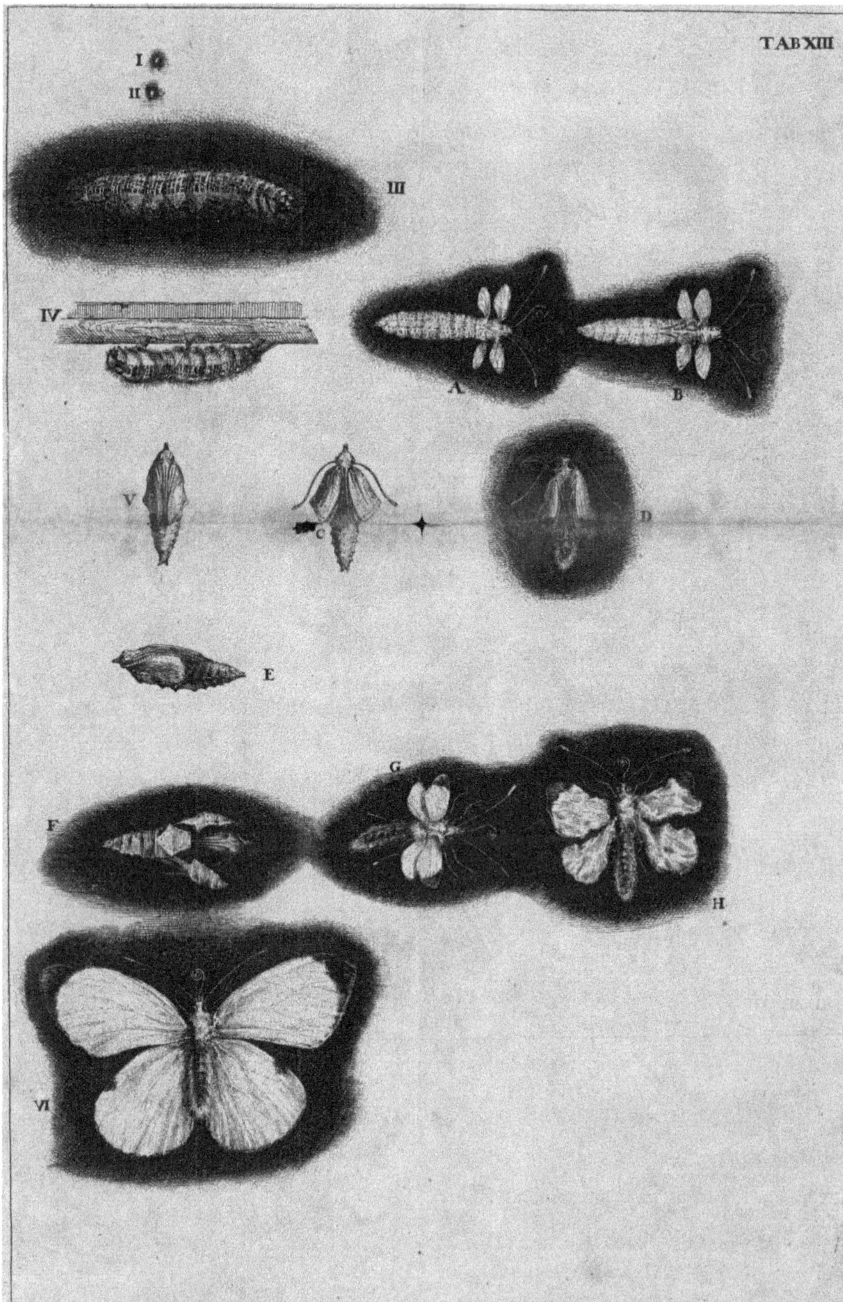

FIGURE 5.10 *The development of the caterpillar into the butterfly. Engraving taken from: Johannes Swammerdam,* Historia generalis insectorum *(Utrecht, Van Dreunen: 1669).*

COURTESY UNIVERSITY LIBRARY LEIDEN.

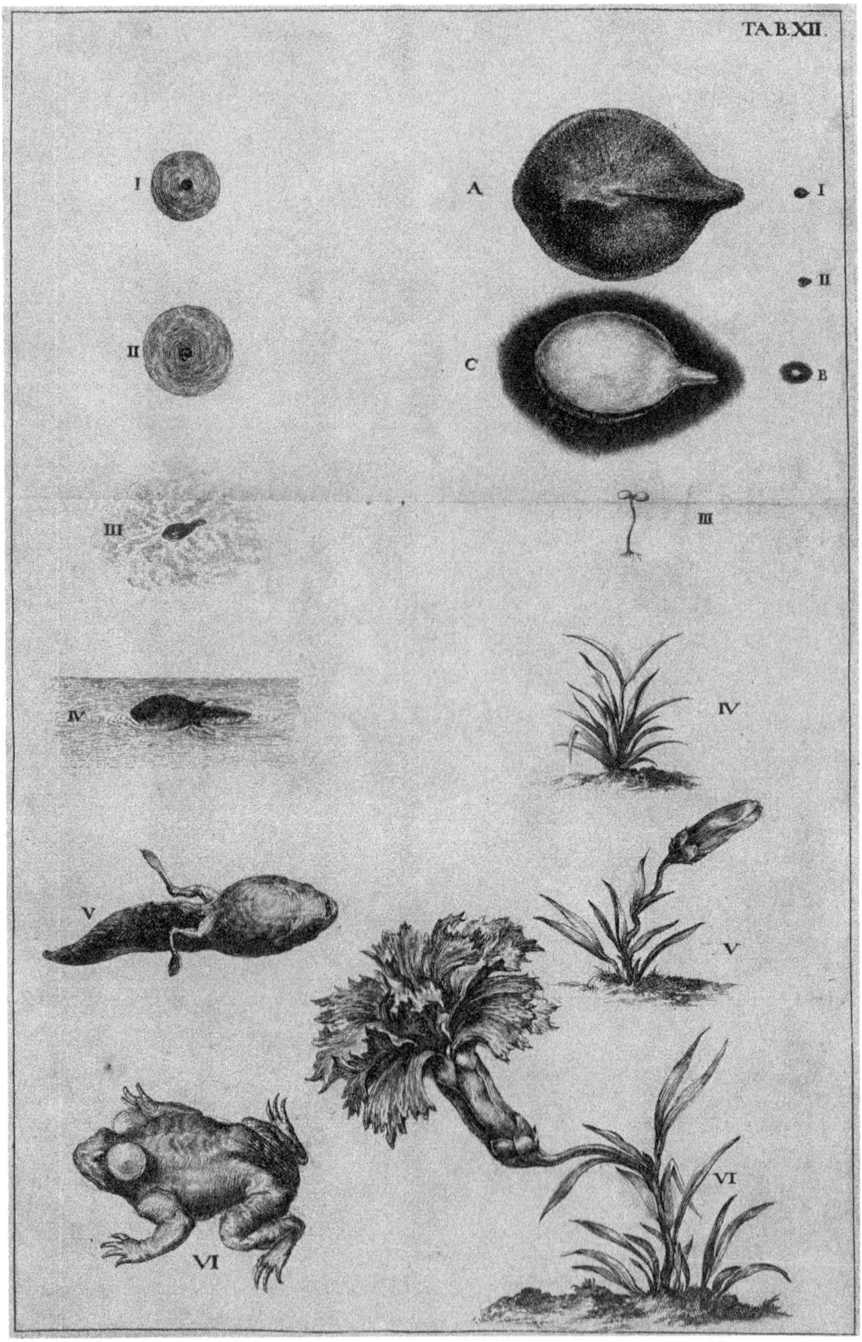

FIGURE 5.11 *The stage-by-stage development of a frog and a carnation, stressing the order and uniformity of nature. Engraving taken from: Johannes Swammerdam,* Historia generalis insectorum *(Utrecht, Van Dreunen: 1669).*
COURTESY UNIVERSITY LIBRARY LEIDEN.

FIGURE 5.12 *Otto Marseus van Schrieck,* Still life with Insects and Amphibians, *1662. Oil on canvas, 50.7 cm × 68.5 cm. Herzog Anton Ulrich Museum, Braunschweig, inv. no. 431.*

> The animals in which, I imagine, the same structure of the lung obtains are toads, lizards, snakes, chameleons, tortoises, aquatic salamanders, and any others provided with membranaceous lungs whose structure I have not been able to investigate yet since I first happened on a first idea of the observation this present winter.[65]

One could imagine how Swammerdam later that year, during the summer, would walk to Waterrijk and there, together with Marseus, would dissect some of the painter's models to check the hypothesis. Both were constantly looking for underlying patterns in the process of procreation, and from Swammerdam's own account it becomes evident that Marseus' talent and knowledge certainly were not inferior to his own. Facing the problem already encountered by Goedaert of a sort of caterpillar bringing forth two different sorts of insects, Swammerdam described in 1678 or 1679:

---

65 Swammerdam to Oldenburg, 4 March 1672, in *The Correspondence of Henry Oldenburg* ed. A.R. Hall – M.B. Hall (1965–1973), vol. IX, 524.

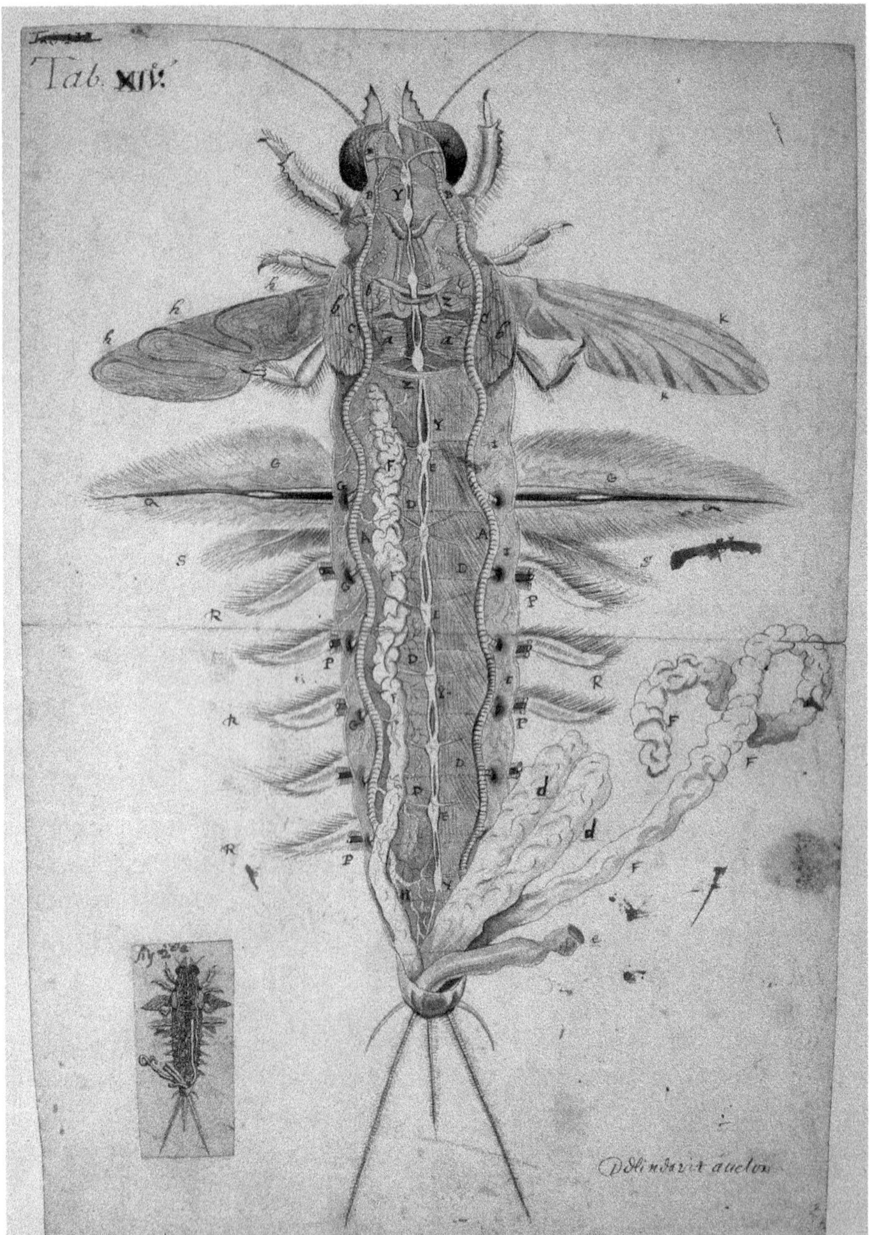

FIGURE 5.13 *Johannes Swammerdam, drawing of the intestines of the mayfly* (Ephemeron), *1673. Red pencil, chalk and ink on paper. University Library Leiden Ms* BPL *126 B.*

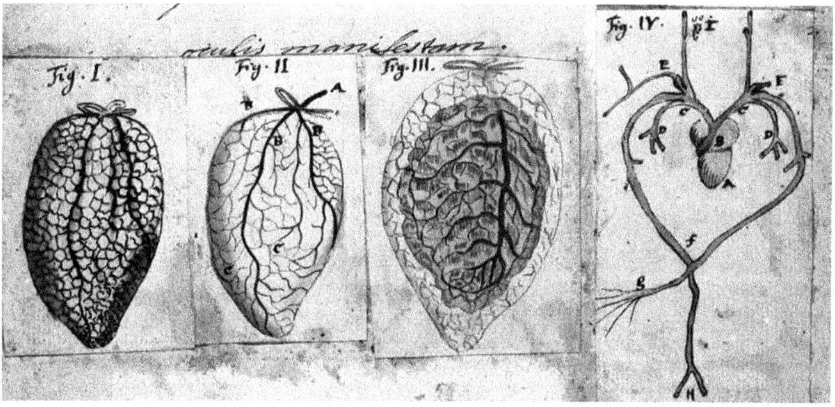

FIGURE 5.14 *Johannes Swammerdam, drawing of the lungs of a frog injected with colored wax. Ink on paper, taken from a letter from Swammerdam to Henry Oldenburg, April 1672.*
ARCHIVES ROYAL SOCIETY LONDON.

I must not omit a circumstance which I heard form Otto Marsilius, the late painter of flower and insects. It was this that he had seen, at the time when caterpillars were busy in divesting themselves of their skins, or when they had just performed the operation, and were grown faint and weak of it, a great many flies, some bigger and others less, piercing the bodies of such caterpillars, and depositing in the wounds, so made, quantities of little eggs; from which proceed the worms, that are every year so commonly found in caterpillars. As yet I must own, I never saw this myself, but allowing it to be a fact, it would perhaps greatly help us in explaining the generation of worms found in the viscera, or bowels of larger animals; provided especially such worms were found afterwards to turn to flies, or other winged insects, which I as yet have not been able to determine experimentally, notwithstanding the pains and attention I must have bestowed on the observations of changes of this kind [...] It must be owned, after all, that these things are as yet buried in a cloud of darkness, and awaiting to be illuminated by a bright light.[66]

66 Swammerdam, *Bybel der Natuure* 709: 'Alleen is my van Otto Marsilius, vermaart bloem en Insecten Schilder in syn leeven; verhaalt, dat hy gesien heeft, wanneer de Rupsen op haar vervellen laagen, of dat se effen haar vel uyt getrokken hadden, en magteloos waaren, dat se dan van sommige kleene, en ook grooter Vliegen doorboort wiereden, die haare Eyeren daar in brogten, en waar uy hy seyde, dat de Wurmene voort quamen, die men gemeenelyk, en alle jaaren, in de selve vint. Ik moet meede bekenne dit nog niet

Swammerdam's description of Marseus' discovery is one of earliest on parasites.[67]

For other important discoveries by Swammerdam, no direct evidence of Marseus' involvement can be found, but from the context it becomes clear that they were inspired by the subjects of his dark, gloomy paintings. In 1673, Swammerdam closely studied ferns – until then their procreation was shrouded in mystery as well. Swammerdam soon discovered the little boxes, *sporangia*, containing the seeds of the plant, which he described in a letter to the Leiden professor of botany Arnoldus van Seyen (1640–1678).[68] In these years, Swammerdam often cooperated with Van Seyen, and both made observations with the aid of a complex microscope made by the famous Musschenbroek workshop. Interestingly enough, Van Seyen was in the possession of the original manuscript and drawings of Clusius' treatise on fungi mentioned earlier.[69] Again, here was the question of procreation: it was generally assumed that fungi were freaks of nature that had their origin in spontaneous generation as well. In the literature on mycology, Pier'Antonio Michelli (1679–1737), Cosimo III's court botanist, is usually credited with the discovery that fungi have reproductive bodies or spores, as he described in his *Nova plantarum genera iuxta Tournefortii methodum disposita* (1729). In a passage that until now has been overlooked, it becomes clear that Swammerdam was on the trail of this important discovery, recently described as the 'last bastion in abiogenesis'.[70] On 21 June 1679 Swammerdam wrote, in a letter published only in 1975:

> As I firmly trust, I have also found the seed of the fungi which is still 20 times smaller than that of the filices (ferns). It is placed as the seed of the poppy, inside and at both sides, on the membranous lamellae of the fungus, upon which it is present in an indescribable quantity. The colour is for the major part black, and they give this colour to the water wherein

gesien te hebben; maar indien het waar was, soo was mogelyk de voortteeling van alle die Dieren, die men in de Ingewanden der Dieren vint, seer ligt te expliceren, indien sy haar in Vliegen of andere Dieren weer quaamen te vergroeyen, dat ik meede tot nog toe niet geobserveert hebbe: en naar welke manier van verandering ik lang gesogt, maar nooit gevonden heb [...] Maar dit syn alle donkere dingen, die nog een groot ligt verwagten'.

67 Lenteren J.C. van, "Early Entomology and the Discovery of Insect Parasitoids", *Biological Control. Theory and Application in Pest Management* 32 (2005) 2–7.

68 Swammerdam, *Bybel* 906.

69 Gelder, *Tussen Hof en Keizerskroon* 348.

70 Leonhard, "Pictura's Fertile Field" 117.

> the seed is washed. Till now I did not yet sow it, so I do not properly know how it arises and brings forth its kind.[71]

Swammerdam also added some sketches of what he saw under the microscope. He finished his letter by saying that the fact he had discovered was not irrefutable, but at least highly probable: 'It should be confirmed by time and by a further examination'. Swammerdam did not live to see it confirmed: he died half a year later, on 15 February 1680; nearly two year after his fellow-traveler to the heart of darkness, Otto Marseus.

Concluding Remarks

Via Swammerdam, fungi and Michelli we return to the person we started with: Cosimo de Medici. The two visits he paid to the Netherlands, and the interest he kept in the developments in religion, politics, art and science that took place there after he became Grand Duke in 1670, are illustrative of the central role the Republic now played in European culture. Cosimo's visit set an example, later famously to be followed by Peter the Great. More particularly, the interest Cosimo showed in the work of Otto Marseus and Johannes Swammerdam is revealing. He visited both on his Tour, and after returning to Italy he stayed in touch with them, purchasing some new paintings by Marseus, and receiving a copy of Swammerdam's *Historia Insectorum Generalis* as well as original drawings made by the latter. When Swammerdam experienced his much debated religious crisis in 1674–5, Cosimo even considered buying his collection of natural specimens and anatomical preparations and offered him a position at the Medici court. Steno acted as a go-between.

The Italian interest in these students of snakes, fungi and insects is a good indication of the sudden European fascination with the problem of procreation in general, and with the question of the hierarchy of beings in particular. Within a decade or so, traditional notions of the order of things were severely attacked, and the philosophical basis as well as empirical evidence for the distinction between the higher and lower creatures, big and small, beautiful and ugly, and male and female was seriously called into question. Generation became a major issue, catching the attention of scientists such as Swammerdam, De Graaf, Ruysch, Steno and Redi. One factor stimulating this process was the intellectual climate in the Dutch Republic, more notably the rise of rationalism and the mechanical worldview. As we saw, Johannes Hudde, highly influenced

71 Swammerdam to Thévenot, 21 June 1979, in Lindeboom, *Letters* 153.

by Descartes, both put the theme of procreation on the agenda, and provided a toolkit for research. Swammerdam was very explicit about his debts to both Hudde and Descartes, writing that nature is stable and uniform, and that the theory of spontaneous generation essentially was atheistic.

This presupposition brings us right to one of the most hotly debated issues of the age: the power of God and the order of nature. There is a striking parallel in the arguments used by Swammerdam in his attack on spontaneous generation in the *Historia Generalis Insectorum*, published in September 1669, and the logic of Spinoza in the notorious chapter 6, "On Miracles", in the *Tractatus Theologico-Politicus*, that was published half a year later. Nature's laws were fixed and immutable, both maintained, and to believe in contingency and chance, in miracles and spontaneous generation, was to deny the power of God. Although the exact relation between the two men falls outside the scope of this essay, it is of relevance to note that Hudde and Steno knew them both very well.[72] In 1677, Steno would write how, when he was a student in Leiden, Spinoza 'paid me daily visits to see the anatomical investigations of the brain that I carried out on several animals in order to discover the place where motion begins and sensation ends'.[73] The resonance of the debate started by the materialist philosophy of nature as put forward by Descartes and brought to its logical consequences by Spinoza was heard in Italy as well. Far from being marginal, the use of the microscope and fascination with the origins, structure and visual representation of the lowest creatures were at the center of European culture in the second half of the seventeenth century. The paintings of Otto Marseus, now largely forgotten, are the silent witnesses of this field of enquiry.

Bibliography

Ainsworth G.C., *Introduction to the History of Mycology* (Cambridge: 1975).

Aldrovandi Ulisse, *Serpentum et draconum historiae libri II. Bartholomaeus Ambrosinus* [...] *summo labore opus concinnavit* [...] (Bologna, Antonio Bernia, Clemente Ferronio: 1640).

Alpers A., *The Art of Describing. Dutch Art in the Seventeenth Century* (Chicago: 1983).

Bakker B., "Au vif—naar 't leven—ad vivum. The Medieval Origin of a Humanist concept", in Boschloo A. e.a. (eds.), *Aemulatio. Imitation, Emulation and Invention in Netherlandish Art from 1500 to 1800. Essays in Honour of Eric Jan Sluijter* (Zwolle: 2011).

72 Jorink, '"Outside God, There is Nothing"'.

73 Spruijt L. – Totaro P. (eds.), *The Vatican Manuscript of Spinoza's* Ethica (Leiden: 2011) 41.

Beretta M. (ed.)., *The Accademia del Cimento and its European context* (Sagamore Beach: 2009).

Bergström, I., *Dutch still-life Painting in the Seventeenth Century* (London: 1956).

Bergström, I., "On Georg Hoefnagel's manner of working with notes on the influence of the *Archetypa* series of 1592", in Cavalli-Björkman G. (ed.), *Netherlandish Mannerism* (Stockholm: 1985) 177–188.

Bergvelt E. – Kistemaker R. (eds.), *De wereld binnen handbereik. Nederlandse kunst- en rariteitenverzamelingen 1585–1735* (Zwolle: 1992).

Berkel K. van, "The Dutch Republic: Laboratory of the Scientific Revolution", in Berkel K. van – Goei L. de (eds.), *The International Relevance of Dutch History* (= Special issue: *Bijdragen en medelingen betreffende de geschiedenis der Nederlanden* 125 (2010) 3–367).

Berthier S. – Boulenguez, J. – Mottin, B., "Butterfly Inclusions in Van Schrieck Masterpieces. Techniques and Optical Properties", *Applied Physics A* 92 (2008) 51–57.

Biagioli M., *Galileo's Instruments of Credit. Telescopes, Instruments, Secrecy* (Chicago: 2006).

Bol L.J., *'Goede onbekenden'. Hedendaagse verkenning en waardering van verscholen, voorbijgezien en onbekend talent* (Utrecht: 1982).

Bunge W. van, *From Stevin to Spinoza. An Essay on Philosophy in the Seventeenth-Century Dutch Republic* (Leiden: 2001).

Cobb M., "Malpighi, Swammerdam and the Colourful Silkworm. Replication and Visual Representation in Early Modern Science", *Annals of Science* 59 (2002) 111–147.

Cobb, M., *The Egg and Sperm Race. The Seventeenth-Century Scientists who Unravelled the Secrets of Sex, Life and Growth* (London: 2006).

Cook H.J., *Matters of Exchange: Commerce, Medicine, and Science in the Dutch Golden Age* (New Haven: 2007).

Cools H., "Francesco Feroni (1614/16–1696): Broker in Cereals, Slaves and Works of art", in Keblusek M. – Cools H.- Noldus B. (eds.), *Your Humble Servant: Agents in Early Modern Europe* (Hilversum: 2006) 39–50.

Daston L. – Park K., *Wonders and the Order of Nature 1150–1750* (New York: 1998).

De Monconys B., *Journal des voyages* (Lyon, Horace Boissat – George Remeus: 1666).

Dupré S. – Lüthy C. (eds.), *Silent Messengers: The World of Goods and the Circulation of Knowledge in the Early Modern Netherlands* (Berlin: 2011).

Des Chene D., *Spirits and Clocks: Machines and Organism in Descartes* (Ithaca: 2004).

Findlen P., *Possessing Nature: Museums, Collecting, and Scientific Culture in Early Modern Italy* (Berkeley: 1994).

Franits W., ed., *Looking at Seventeenth-Century Dutch Art: Realism Reconsidered* (Cambridge: 1997).

Freeman R., "llustrations of Insect Anatomy from the Beginning to the Time of Cuvier", *Medical and Biological Illustration* 12 (1962) 174–183.

Gelder E. van, *Tussen hof en keizerskroon. Carolus Clusius en de ontwikkeling van de botanie aan de Midden-Europese hoven (1573–1593)* (Leiden: 2011).

Gessner Conrad, *Historiae animalium liber V qui est de serpentum natura: ex variis schedis et collectaneis eiusdem compositus per Iacobum Carronum* [...] *adiecta est ad calcem, scorpionis insecti historia* (Zurich, Christoph Froschauer: 1587).

Goedaert Johannes, *Metamorphosis naturalis, ofte historische beschrijvinge van den oirspronk, aerdt, eygenschappen ende vreemde veranderingen der wormen, rupsen, maeden, vliegen, witjens, byen, motten en diergelijke dierkens meer; niet uit eenige boeken, maar aleenlyck door eygen ervaretntheyt uytgevonden, beschreven ende na dekonst afgeteykent* (Middelburg, Jacob Fierens: s.d. [1660]–1669) 3 vols.

Hildebrecht, D.R., *Otto Marseus van Schrieck (1619/20–1678) and the Nature Piece. Art, Science, Religion and the Seventeenth-Century Pursuit of Natural Knowledge* (diss. University of Michigan, Ann Arbor: 2004).

Hochstrasser, J., *Still Life and Trade in the Dutch Golden Age* (New Haven: 2007).

Hoogstraten Samuel van, *Inleyding tot de Hooge Schoole der Schilderkonst* (Rotterdam, F. van Hoogstraeten: 1678).

Houbraken Arnoldus, *De groote schouburgh der Nederlantsche konstschilders en schilderessen* (The Hague, Swart – Boucquet – Gaillard: 1718–1721).

Huisman T., *The Finger of God. Anatomical Practice in 17th-century Leiden* (Leiden: 2009).

Jorink E., '"Outside God, there is Nothing'. Swammerdam, Spinoza and the Janus-face of the early Dutch Enlightenment", in Bunge W. van (ed.), *The Early Enlightenment in the Dutch Republic, 1650–1750. Selected Papers of a Conference held at the Herzog August Bibliothek, Wolfenbüttel, 22–23 March 2001* (Leiden: 2003) 81–108.

Jorink E., *Reading the Book of Nature in the Dutch Golden Age, 1575–1715* (Leiden: 2010).

Jorink E., "In the Twilight Zone. Isaac Vossius and the Scientific Communities in France, England and the Dutch Republic", in Jorink E. and Miert D. van (eds.), *Isaac Vossius (1618–1689) between Science and Scholarship* (Leiden: 2012) 119–156.

Jorink E., "Beyond the Lines of Apelles. Johannes Swammerdam, Dutch Scientific Culture, and the Representation of Insect Anatomy", in Jorink – Ramakers (eds.), *Art and Science* 148–184.

Jorink E. – Ramakers B., "Undivided territory. 'Art' and 'Science' in the Early Modern Netherlands", in Jorink E. – Ramakers B. (eds.), *Art and Science* 6–33.

Jorink E. – Ramakers B. (eds.), *Art and Science in the Early Modern Low Countries*. Netherlands/Yearbook for History of Art/Nederlands Kunsthistorisch Jaarboek 61 (Zwolle 2011).

Jorink E., "Between emblematics and the 'Argument from Design'. The representation of Insects in the Dutch Republic", in Enenkel K.A.E. – Smith P.J. (eds.), *Early Modern Zoology: The Construction of Animals in Science, Literature and the Visual Arts*, Intersections: Yearbook for Early Modern Studies 7 (2007) (Leiden – Boston) 147–175.

Knoefel P. (ed.), *Fransesco Redi on Vipers* (Leiden: 1988).

Kooijmans, L. *De doodskunstenaar. De anatomische lessen van Frederik Ruysch* (Amsterdam: 2004).

Lehman-Haupt H., "The Microscope and the Book", in J. Benzing et al. (eds.), *Festschrift für Claus Nissen. Zum siebzigsten Geburtstag 2. September 1971* (Wiesbaden: 1973) 471–502.

Lennox J., "Teleology, Chance and Aristotle's Theory of Spontaneous Generation", *Journal of the History of Philosophy* 20 (1983) 219–238.

Lenteren J.C. van, "Early Entomology and the Discovery of Insect Parasitoids", *Biological Control. Theory and Application in Pest Management* 32 (2005) 2–7.

Leonhard K., "Pictura's Fertile Field. Otto Marseus van Schrieck and the Genre of Sottobosco Painting", *Simiolus* 34 (2010) 95–118.

Leonhard K., "Painted Poison. Venomous Beasts, Herbs, Gems, and Baroque Colour Theory", in Jorink – Ramakers (eds.), *Art and Science* 116–147.

Lindeboom G.A., *Descartes and medicine* (Amsterdam: 1974).

Lindeboom G.A. (ed.), *The letters of Jan Swammerdam to Melchisedec Thévenot* (Amsterdam: 1975).

Lovejoy A., *The Great Chain of Being. A Study of the History of an Idea* (Cambridge MA: 1936).

Margócsy D., "Advertising cadavers in the Republic of Letters: Anatomical publications in the Early Modern Netherlands", *British Journal for the History of Science* 42 (2009) 187–210.

Meli D.B., "The Representation of Insects in the Seventeenth Century. A Comparative Approach", *Annals of Science* 67 (2010) 405–429.

Mirto A. – Veen H. Th. van (eds.), *Pieter Blaeu: lettere ai Fiorentini: Antonio Magliabechi, Leopoldo e Cosimo III de' Medici, e altri, 1660–1705: edizione con commento e saggio introduttivo in Italiano e Inglese / Pieter Blaeu: letters to Florentines: Antonio Magliabechi, Leopoldo and Cosimo III de' Medici, and others, 1660–1705: edited and annotated with an introductory essay in Italian and English* (Florence – Amsterdam: 1992).

Rikken M., *Melchior d'Hondecoeter, Bird Painter* (Amsterdam: 2008).

Roemer G.M. van de, "Neat nature. The relation between art and nature in a Dutch cabinet of curiosities from the early eighteenth century", *History of Science* 42 (2004) 47–84.

Roemer G.M. van de, *De geschikte natuur. Theorieën over natuur en kunst in de verzameling van zeldzaamheden van Simon Schijnvoet (1652–1727)* (unpub. diss., University of Amsterdam: 2005).

Ruestow E.G., *The Microscope in the Dutch Republic. The Shaping of Discovery* (Cambridge: 1996).

Ruler J.A. van, *The Crisis of Causality. Voetius and Descartes on God, Nature and Change* (Leiden: 1995).

Smith P.H., *The Body of the Artisan: Art and Experience in the Scientific Revolution* (Chicago: 2004).

Steensma S., *Otto Marseus van Schrieck. Leben und Werk* (Hildesheim: 1999).

Swammerdam Johannes, *Tractatus physico-anatomico-medicus de respiratione usuque pulmonum* (Leiden, Gaasbeeck: 1667).

Swammerdam Johannes, *Historia generalis insectorum ofte Algemeene verhandeling van de bloedeloose dierkens* (Utrecht, Meindardus van Dreunen: 1669).

Swammerdam Johannes, *Bybel der natuure of historie der insecten. Biblia naturae, sive historia insectorum* (Leiden, Isaac Severinus – Bodewijn van der Aa – Pieter van der Aa: 1737–1738).

Swan C., "Ad vivum, naer het leven, from the life. Defining a mode of representation", *Word & Image* 2 (1995) 353–372.

Swan C., *Art, Science, and Witchcraft in Early Modern Holland: Jacques de Gheyn II (1565–1629)* (Cambridge: 2005).

Taylor P., *Dutch Flower Painting 1600–1720* (New Haven – London: 1995).

Tribby J., "Club Medici: Natural Experiment and the Imagineering of 'Tuscany' ", *Configurations* 2 (1994) 215–235.

Tolmer L., "Une page d'histoire des sciences 1661–1669. Vingt-deux lettres inédites d'André de Graindorge à P.D. Huet", *Mémoires de l'Académie des sciences, arts et belles-lettres de Caen* 10 (1941) 245–330.

Verbeek Th., *Descartes and the Dutch. Early Reactions to Cartesian Philosophy, 1637–1650* (Carbondale: 1992).

Vermij R., "Bijdrage tot de bio-bibliografie van Johannes Hudde", *Gewina* 18 (1995) 25–35.

Vermij R., "Een nieuw concept. De wetten der natuur", in Egmond F. – Jorink E. – Vermij R. (eds.), *Kometen, monsters en muilezels. Het veranderende natuurbeeld en de natuurwetenschap in de zeventiende eeuw* (Haarlem: 1999) 105–119.

Vermij R., "Instruments and the making of a philosopher: Spinoza's career in optics", *Intellectual history review* 22 (2013).

Vermij R. – Atzema E., "*Specilla circularia.* An unknown work by Johannes Hudde", *Studia Leibnitiana* 27 (1995) 104–121.

Vignau-Wilberg T., *Archetypa studiaque patris Georgii Hoefnagelii 1592. Natur, Dichtung und Wissenschaft in der Kunst um 1600* (Munich: 1994).

Vries A.D., "Otto Marseus", *Oud-Holland* 1 (1883) 166–168.

Weststeijn T., *The Visible World: Samuel van Hoogstraten's Art Theory and the Legitimation of Painting in the Dutch Golden Age* (Amsterdam: 2008).

Woodall J., "Laying the Table. The Procedures of Still Life", *Art History* 35 (2012) 976–1003.

CHAPTER 6

Insects in John Ray's Natural History and Natural Theology

Brian W. Ogilvie

Introduction

Early modern physico-theology has a strong panglossian streak.[1] Theodicy was a serious problem for many seventeenth-century theologians and philosophers, but physico-theological treatises at best gave it the slightest of nods.[2] Reading John Ray's *The Wisdom of God Manifested in the Works of Creation*, William Derham's *Physico-Theology*, or Friedrich Christian Lesser's *Insecto-Theologia*, it is indeed difficult to avoid the impression that, for their authors, all was for the best in this best of all possible worlds. Not only was 'each partial evil, universal good'; there was little evil at all in their pages.[3]

Some of the panglossian overtones in such works was due to the fact that they were not solely apologetic texts; rather, they also served to encourage Christian devotion. Though Ray insisted that his *Wisdom of God* would prove God's existence beyond the ability of 'atheistical persons' to doubt, and that it would reveal the Creator's 'infinite power and wisdom', he also emphasized that his examples would 'stir up and increase in us the affections and habits of admiration, humility and gratitude,' just as the Psalmist had done.[4] Even William Derham, at the end of his *Physico-Theology*, turned to the devotional import of his lengthy catalogue of God's marvelous contrivances.[5] As Sara

---

1 Research for this paper was supported by the University of Massachusetts Amherst and the Institut d'études avancées—Paris.

2 For an introduction to the vast literature on early modern theodicy, see Nadler S.M., *The best of all possible worlds: A story of philosophers, God, and evil* (New York: 2008).

3 See Knight D.M., "Science and beliefs", in Knight D.M. – Eddy M.D. (eds.), *Science and beliefs: From natural philosophy to natural science, 1700–1900* (Aldershot, Hants, England, and Burlington, VT: 2005) 6.

4 Ray John, *The wisdom of God manifested in the works of the creation*, 3rd ed. (London, Samuel Smith and Benjamin Walford: 1701) fol. [A7]r–[A8]r. In quotations from early modern English sources I have modernized capitalization but retained contemporary spellings.

5 Derham William, *Physico-theology: or, a Demonstration of the being and attributes of God, from his works of creation: Being the substance of sixteen sermons preached in St. Mary le*

 | DOI 10.1163/9789004279179_008

Stebbins showed long ago, the devotional content increased and the argumentative content decreased in German Protestant physico-theology over the course of the eighteenth century.[6] It is a mistake to consider physico-theology strictly as a branch of apologetics and to overlook its intended use for stirring Christian piety.

However, physico-theologians did make empirical claims in support of their devotional and apologetic ends. This paper examines a specific aspect of those empirical claims: whether they changed in response to new empirical discoveries in natural history. In other words, did physico-theologians take their empiricism seriously? Was their panglossianism up to date? I approach this question through the works of John Ray, focusing on Ray's treatment of insects. Ray's *Wisdom of God* appeared in 1691, just before Ray resumed the study of insects that he had set aside over a decade earlier.[7] He produced three revised, expanded editions over the next decade and a half, as he was compiling material for his *History of Insects*, which appeared posthumously in 1710. Hence, Ray provides an interesting test case for assessing the response of physico-theology to empirical research. By examining what he added and changed between editions of *The Wisdom of God*, we can see how, and to what extent, his investigations into insects shaped his theological interpretation of them.

Ray's changes to successive editions of *The Wisdom of God* have received little systematic attention. Richard Yeo has noted that Ray's original subtitle referred to the theological lectures he had delivered at Cambridge on which the book was based, but that that reference was dropped in the second edition.[8] Neal Gillespie has argued that Ray's readers encountered no new arguments in his book; rather, his contribution to physico-theology was to support the arguments with his 'extensive knowledge and field experience' and present them in a form that was amenable to 'the common reader'.[9] And while Gillespie noted that subsequent editions of the *Wisdom of God* added material, making it increasingly a textbook of 'natural history *and* natural theology,' he did not examine what Ray added and why.[10]

Bow-Church, London, at the Hon.ble Mr. Boyle's Lectures, in the years 1711 and 1712: With large notes, and many curious observations, 2nd ed. (London, William Innys: 1714) 436 ff.

6 Stebbins S., *Maxima in minimis: Zum Empirie- und Autoritätsverständnis in der physikotheologischen Literatur der Frühaufklärung* (Frankfurt a.M.: 1980).

7 Raven C.E., *John Ray, naturalist: His life and works* (Cambridge: 1950) 389.

8 Yeo R., "Between memory and paperbooks: Baconianism and natural history in seventeenth-century England", *History of Science* 45 (2007) 32 n. 7.

9 Gillespie N.C., "Natural history, natural theology, and social order: John Ray and the 'Newtonian ideology'", *Journal of the History of Biology* 20 (1987) 39.

10 Gillespie, "Natural history" 46.

Charles Raven, Ray's biographer, summarized the changes Ray made to *The Wisdom of God*, emphasizing the range of printed books on which Ray drew as he revised.[11] But Raven was content to note the broad areas in which Ray added material and to list his sources. Moreover, Raven assessed the *Wisdom of God* by what it revealed about Ray's scientific interests and because its theology adumbrated Raven's own: it was 'the first example' of a theology 'capable of giving appropriate expression to the Christian faith in a scientific age'.[12] By treating Ray as a forerunner of twentieth-century Anglicanism, Raven wrenched his physico-theology out of its late seventeenth-century context.

The fullest analysis of the changes Ray introduced in successive editions is by Lisa Zeitz.[13] Zeitz emphasizes the rhetorical strategy and effect of Ray's successive revisions. She identifies nine distinct types of material that Ray added between the 1691 and the 1704 edition, from 'elaboration and restatement of the design argument' to 'the inclusion of more reports of experiments, dissections, and common observations—we might call this strategy the multiplication of witnesses'.[14] Of Zeitz's categories, those that relate to my analysis are the third—'responses to objections, or to contemporary *scientific* controversies' (her emphasis), the fourth—'further illustrations of 'design' in nature', the eighth—'modification of knowledge', and the final, the multiplication of witnesses. Zeitz identifies several instances of each of these kinds of additions, without attempting an exhaustive account. But her analysis emphasizes the rhetoric of Ray's additions, which serve to 'generate a sense of shared experience and consensus' in the reader through multiplying Ray's own observations and those of his correspondents.[15] Zeitz offers an important examination of the rhetoric of revision in *The Wisdom of God*; her work provides a substantial complement to my analysis.

My purpose here focuses not on Ray's rhetoric but, rather, on how he related his empirical investigations (including reading the letters and publications of other naturalists) and his physico-theological exegesis. As I have already noted, Ray's chief concern in natural history during the period in which he was revising successive editions of *The Wisdom of God* was the history of insects; for that reason, my analysis focuses on additions involving insects. After a brief overview of Ray's research on insects, I turn to how Ray integrated new material

11 Raven, *John Ray* 469–476.

12 Raven, *John Ray* 478.

13 Zeitz L.M., "Natural theology, rhetoric, and revolution: John Ray's Wisdom of God, 1691–1704", *Eighteenth-Century Life* 18.1 (1994) 120–133.

14 Zeitz, "Natural theology" 124.

15 Zeitz, "Natural theology" 125–127.

into successive editions of *The Wisdom of God*. I end by discussing the limits to empiricism in Ray's *Wisdom* and note the generic distinction in his work between natural history and physico-theology, despite their intersections. Physico-theology built on the edifice of sound natural history, but Ray's natural history of insects had little place for physico-theology in its pages.

Ray's *History of Insects* (1710) and Its Antecedents

Ray's interest in insects appears to date back to his years in Cambridge and his collaboration with the young gentleman Francis Willughby.[16] Willughby matriculated at Trinity College in 1652, and between then and his departure in 1660, he and Ray frequently discussed natural history. Ray reported one of Willughby's observations on the metamorphosis of caterpillars in his 1660 *Catalogue of Plants that Grow near Cambridge*.[17] Insects were not a new subject for naturalists; from the middle of the sixteenth century, naturalists such as Conrad Gessner, Thomas Penny, Edward Wotton, and Ulisse Aldrovandi had taken an interest in them, an interest reflected in Aldrovandi's 1602 *Seven Books on Insect Animals* and the 1634 *Theater of Insects* edited from Gessner's, Penny's, and Wotton's notes by Thomas Moffett.[18] But Willughby and Ray were

16 This section is condensed from Ogilvie B.W., "Attending to insects: Francis Willughby and John Ray", *Notes and Records of the Royal Society* 66, 4 (2012) 357–372, which has a fuller discussion of Ray's and Willughby's empirical studies of insects.

17 Ray John, *Catalogus plantarum circa Cantabrigiam nascentium* […] (London, John Martin, James Allestry, Thomas Dicas: 1660) 134–138.

18 Aldrovandi Ulisse, *De animalibus insectis libri septem* […] (Bologna, Giovanni Battista Bellagamba: 1602); Moffett Thomas, *Insectorum sive minimorum animalium theatrum* […] (London, Thomas Cotes: 1634). On the early modern study of insects, see Bodenheimer F.S., *Materialien zur Geschichte der Entomologie bis Linné*, 2 vols. (Berlin: 1928); Dicke M., "Insects in western art", *American Entomologist* 46.4 (2000) 228–237; Cobb M., "Malpighi, Swammerdam and the colourful silkworm: Replication and visual representation in early modern science", *Annals of Science* 59 (2002) 111–147; Cambefort Y., "Artistes, médecins et curieux aux origines de l'entomologie moderne (1450–1650)", *Bulletin d'Histoire et d'Épistémologie des Sciences de la Vie* 11, 1 (2004) 3–29; Jorink E., *Het Boeck der Natuere: Nederlandse geleerden en de wonderen van Gods schepping, 1575–1715* (Leiden: 2006), translated as Jorink E., *Reading the Book of Nature in the Dutch Golden Age, 1575–1715*, tr. P. Mason, Brill's studies in intellectual history, 191 (Leiden: 2010); Vignau-Wilberg T., "*In minimis maxima conspicua*. Insektendarstellungen um 1600 und die Anfänge der Entomologie", in Enenkel K.A.E. – Smith P.J. (eds.), *Early modern zoology: The construction of animals in science, literature and the visual arts*, 2 vols. (Leiden – Boston: 2007), vol. I, 217–243; Ogilvie B.W., "Nature's Bible: Insects in seventeenth-century European

at the forefront of a new wave of interest in the nature of generation and what insects could reveal about it.[19]

The interest appears to have been primarily Willughby's. On several occasions, Ray noted Willughby's devotion to the subject. In October 1668, Ray explained in a letter to Martin Lister that he had a lingering childhood fear of spiders and had avoided them. 'I have not wholly neglected', he added, 'other genera of insects, both coleopterous and without elytrae. But since Mr. Willughby had painstakingly devoted himself for many years to acquiring, examining, describing, and comparing them, I have studied them only occasionally, as a pastime'.[20] After Willughby's untimely death in 1672, Ray set about the work of organizing and publishing Willughby's notes on natural history, a task that involved substantial work. Many of those notes involved insects. In 1674, Ray reported to Henry Oldenburg that 'The History of Insects is that wherein Mr. Willughby did chiefly labour and most considerably advance'.[21] And in the preface to the *Ornithology*, the first of Ray's publications based on Willughby's material, Ray remarked that on his deathbed his friend spoke most highly of his 'new and pretty observations about Insects'.[22]

However, Ray returned to insects only in 1690, at which point his heath was increasingly poor; he informed Edward Lhwyd that summer that he was reviewing his notes on insects, and in the summer of 1691 he began collecting in earnest.[23] For the remaining fourteen years of his life he diligently pursued their study, with the aim of publishing a history of British insects and exotics

art and science", *Tidsskrift for kulturforskning* 7.3 (2008) 5–21; Bertoloni Meli D., "The representation of insects in the seventeenth century: A comparative approach", *Annals of Science* 67.3 (2010) 405–429; and Neri J., *The insect and the image* (Minneapolis: 2011).

19 On the new interest in generation after 1650, see Cobb M., *Generation: The seventeenth-century scientists who unraveled the secrets of sex, life, and growth* (New York: 2006).

20 Ray J., *The correspondence of John Ray: Consisting of selections from the Philosophical Letters published by Dr. Derham, and original letters of John Ray, in the collection of the British Museum*, ed. E. Lankester (London: 1848) 29–30: 'Alia Insectorum genera non omnino neglexi κουλεόπτερα et ἀνέλυτρα: at vero cum Ds. Willughby iis conquirendis, examinandis, describendis, conferendis sedulam a multis retro annis navavit operam, ego obiter tantum et animi causa hac in parte versatus sum'.

21 Ray to Oldenburg, November 30, 1674, in Ray J., *Further correspondence of John Ray*, ed. R.W.T. Gunther (London: 1928) 69.

22 Willughby Francis, *The ornithology of Francis Willughby of Middleton in the county of Warwick Esq, fellow of the Royal Society in three books* (London, John Martyn: 1678) fol. A3r–v.

23 Raven, *John Ray* 392–393. Ray's health began declining in the 1680s: Raven, *John Ray* 216, 228.

in British collections.[24] Initially he pursued his studies without referring to Willughby's manuscript, which he had long since returned to Willughby's family.[25] He immersed himself in new publications on the subject, including the *General History of Insects* by Jan Swammerdam, and collected, bred, and described insects, emphasizing butterflies and moths. As he weakened, he shifted his attention to specimens collected by his family and friends or shipped from London and Oxford. Finally, in 1703, with death drawing ever closer, he decided to acquire Willughby's manuscript and integrate its material with his own. By August 1704 he had it in his hands.[26]

When Ray died in January 1705 (new style), his *Historia insectorum* was unfinished, and the material from Willughby only partially integrated with his own. Eventually a committee of the Royal Society took on the task of editing the work; it was finally published in 1710.[27] It contains four heterogeneous parts:

1. A *Method of insects*, which Ray had composed in 1704 and published separately in 1705, drawing in part on Willughby's classification.[28]
2. A systematic description of insects, more or less but not entirely following the order in the *Method*, but with some extraneous material inserted.
3. A series of chronologically organized observations of caterpillars and adult butterflies and moths, based on Ray's investigations in the 1690s.
4. As an appendix, a series of descriptions of British beetles by Martin Lister.

Despite its imperfect form, the *History of insects* reveals that Ray was intensely focused on insects during this period. His correspondence, too, shows his concern for the project as well as his increasingly pessimistic remarks on whether he could ever complete it, given his health and the sheer number of different species to describe. In sum, Ray was actively investigating insects at the very time that he was revising *The Wisdom of God*. And as we shall see, his physico-theological work fattened on the fruits of that labor.

---

24 Ray, *Correspondence* 410–411, 414–415, 417–418, 430, 432–433.

25 Ray, Further correspondence 98–100.

26 Ray, *Correspondence* 414–417, 430–431, 448–449, 452.

27 Ray John, *Historia insectorum:* [...] *opus posthumum jussu Regia Societatis Londinensis editum; cui subjungitur appendix De scarabais Britannicis, autore M. Lister* (London, A. and J. Churchill: 1710).

28 Ray John, *Methodus insectorum seu Insecta in methodum aliqualem digesta* [...] (London, Samuel Smith and Benjamin Walford: 1705).

Ray's *Wisdom of God* and English Physico-Theology

Ray returned to natural theology and insects around the same time. As the title page of the first edition of *The Wisdom of God* indicated, he had preached 'some common Places', that is, morning sermons, in the Trinity College chapel during his years as a Fellow. But his refusal to take the oath required by the Act of Uniformity led to his resignation of his Fellowship in 1662 and his suspension of activities as a divine, including preaching and writing on theology. With regime change in 1688, and an offer of ecclesiastical preferment in 1690, Ray felt that he could once more take up the public exercise of divinity, which he had considered his profession even during the decades when he could not exercise it. A number of physico-theological and devotional treatises soon followed, including *The Wisdom of God*.[29]

This proved to be one of Ray's most popular books. The first edition of 1691 was soon followed by a second in 1692, a third in 1701, and a fourth, the last published during Ray's lifetime, in 1704. Each new edition proclaimed itself to be 'very much enlarg'd' compared to the previous. The work continued to be popular long after Ray's death early in 1705, with English editions continuing until the 1840s and translations appearing in French (*L'Existence et la sagesse de Dieu*, 1714), German (*Gloria Dei oder Spiegel der Weißheit und Allmacht Gottes*, 1717), and Dutch (*Gods wysheid geopenbaard in de werken der schepping*, 1732).[30] While the *Three Physico-Theological Discourses* of 1692 went into more depth on specific subjects,[31] *The Wisdom of God* offered, in Scott Mandelbrote's words, 'a coherent natural theology in which the evidence of the heavens, geology, botany, zoology, and human anatomy suggested the providential action of a benevolent deity who was responsible for the creation of all things'.[32] Ray was particularly concerned to counter the arguments of 'pretended theists', like the Cartesians, and 'atheists' like Spinoza; in this regard his work had clear apologetic aims.

29 Mandelbrote S., "Ray, John (1627–1705)", *Oxford Dictionary of National Biography*, http://www.oxforddnb.com/view/article/23203 (Oxford: 2004); Zeitz, "Natural theology" 128–129.

30 OCLC WorldCat union catalogue, http://www.worldcat.org (consulted 30 April 2012).

31 The works were individually published in 1692, and gathered into one volume the following year: Ray John, *Three physico-theological discourses, concerning I. the primitive chaos, and creation of the world; II. the general deluge, its causes and effects; III. the dissolution of the world, and future conflagration*, 2nd ed. (London, Samuel Smith: 1693).

32 Mandelbrote, "Ray, John (1627–1705)".

The Wisdom of God was a key contribution to late seventeenth-century English physico-theology.[33] The association of natural history and natural theology in England has long been noted. In a seminal article, Neal Gillespie analyzed this association, noting that earlier historians had often conflated two distinct forms of natural theology in the seventeenth century: a 'cosmic' theology that appealed to the general order of the world, and physico-theology, which used examples of 'adaptative design' drawn from the living world to demonstrate the existence of a divine creator.[34] As Gillespie showed, Robert Boyle and other Christian virtuosi concluded that the cosmic argument had been weakened by materialist accounts of the universe, such as 'Hobbism' and Spinozism; simple material objects could plausibly be the result of chance.[35] And though Descartes admitted the existence and benevolence of God, his diffidence regarding final causes and his emphasis on efficient causation was considered to be a real problem by late seventeenth-century physico-theologians.[36]

Descartes had argued that God's purposes were so subtle, and the human understanding so weak, that questions of final causes should simply be set aside as insoluble.[37] Robert Boyle's response to Descartes was to admit his premisses but reject his conclusions. Human understanding was limited, and God's purposes were indeed subtle. Yet for Boyle, this did not mean that the natural philosopher should set aside questions of final causality; rather, he should approach them diffidently, and be sure of matters of fact before proceeding to their interpretation. When the proper mechanical account of a phe-

33 Important and recent studies of English physico-theology include Gillespie, "Natural history, natural theology, and social order"; Harrison P., *The Bible, Protestantism and the rise of natural science* (Cambridge: 1998); Mandelbrote S., "The uses of natural theology in seventeenth-century England", *Science in Context* 20.3 (2007) 451–480; and Berry R.J., "John Ray, physico-theology and afterwards", *Archives of Natural History* 38.2 (2011) 328–348; for the broader context, see Stebbins, *Maxima in minimis*; Brooke J.H., "Science and the fortunes of natural theology: Some historical perspectives", *Zygon* 24.1 (1989) 3–22; Trepp A.-C., *Von der Glückseligkeit alles zu wissen: Die Erforschung der Natur als religiöse Praxis in der Frühen Neuzeit* (Frankfurt am Main: 2009); and Jorink, *Reading the Book of Nature*.

34 Gillespie, "Natural history" 4–9.

35 Gillespie, "Natural history" 27.

36 Ogilvie B.W., "Natural history, ethics, and physico-theology", in Pomata G. – Siraisi N. (eds.), *Historia: Empiricism and erudition in early modern Europe* (Cambridge, MA: 2005) 75–103.

37 Osler M.J., "Whose ends? Teleology in early modern natural philosophy", *Osiris* 16 (2001) 155–156.

nomenon had been established, and only then, the natural philosopher could turn to a consideration of divine purpose.[38]

Boyle developed this line of thought most extensively in his *Disquisition about the Final Causes of Natural Things* (1688).[39] After dismissing the Epicurean position and responding to Cartesian arguments against considering final causes in philosophy, Boyle offered a range of 'observations' from the natural world that warranted the conclusion that many natural objects had a purpose. His analysis of final causes distinguished four distinct areas: 'grand and general ends of the whole world', more specific purposes in the arrangement of parts of the world, the ways in which parts of the animal serve 'the welfare of the whole animal himself', and the framing of natural productions 'for the use of man'.[40] Boyle was confident that the universe as a whole had been framed to manifest God's glory and to convey his benefits to humanity.[41] But when it came to its constituent parts, he was far more willing to draw teleological conclusions from the animate creation than from the system of the world:

> I cannot but think, that the situations of the coelestial bodies, do not afford by far so clear and cogent arguments, of the wisdom and design of the author of the world, as do the bodies of animals and plants. And for my part I am apt to think, there is more of admirable contrivance in a mans muscles, than in (what we yet know of) the celestial orbs; and that the eye of a fly is (at least as far as appears to us,) a more curious piece of workmanship, than the body of the sun.[42]

---

38 Osler, "Whose ends" 161–164; see also Lennox J.G., "Robert Boyle's defense of teleological inference in experimental science", *Isis* 74.1 (1983) 38–52; Hunter M. – Davis E.B., "The making of Robert Boyle's 'Free enquiry into the vulgarly receiv'd notion of nature' (1686)", *Early Science and Medicine* 1.2 (1996) 235, 237–239; Carlin L., "The importance of teleology to Boyle's natural philosophy", *British Journal for the History of Philosophy* 19 (2011) 663–680; and Carlin L., "Boyle's teleological mechanism and the myth of immanent teleology", *Studies in History and Philosophy of Science* 43 (2012) 54–63.

39 Boyle Robert, *A disquisition about the final causes of natural things: Wherein it is inquir'd, whether, and (if at all) with what cautions, a naturalist should admit them. To which are subjoyn'd, by way of appendix, some uncommon observations about vitiated sight* (London, John Taylor: 1688).

40 Boyle, *Disquisition* 6–9; see Carlin, "Importance of teleology" 668 ff.

41 Carlin, "Importance of teleology" 671–672; cf. Gillespie, "Natural history" 25–27, who argues that Boyle entertained doubts about the cogency of natural theological arguments resting on the harmony or order of the cosmos, leading him to emphasize the evident purpose in living nature.

42 Boyle, *Disquisition* 43–44.

Boyle based this conclusion on the evident connection between the parts of animals' bodies and their use by the creatures. The eye was his chief example, but he introduced others as well, comparing them favorably with even the finest products of human ingenuity: 'There is incomparably more art express'd in the structure of a doggs foot, than in that of the famous clock at Strasburg'.[43] The naturalist must be diffident, aware of the limits of human understanding and the multiplicity of divine purpose, when drawing conclusions about the purpose of bodily structures. And he must not be distracted by final causes from careful investigation of efficient causes, which is the proper task of natural philosophy and leads to firmer knowledge of final causes.[44]

Boyle was aware of the difficulty even of identifying final causes in animals and their organs. Since man was the most perfect being, one might wonder why God made other, less perfect beings in the first place, and why even in animals of the same kind, the safety of some should be so much better provided for than that of others; as we see, that some ants, and some glow-worms, are furnish'd with wings; and some not'. This was a reminder that 'cosmical, and therefore primary and over-ruling ends', as well as a creature's particular needs, governed its organization.[45] Nonetheless, the naturalist should start with the latter.

Above all, Boyle insisted, the naturalist must establish the facts on the firmest possible ground. This was the approach that Ray adopted in *The Wisdom of God*. In the work's Preface he insisted on its factual veracity: 'I have been careful to admit nothing for matter of Fact or Experiment but what is undoubtedly true, lest I should build upon a Sandy and Ruinous Foundation; and by the admixture of what is False, render that which is True, suspicious' (1691, fol. [A6] r–v). In good Protestant fashion, Ray insisted on literal truth as the basis for his interpretation of the Book of Nature.[46] And he insisted further on the need to add new truths to this stock.

> Let it not suffice us to be Book-learned, to read what others have written, and to take upon trust more Falshood than Truth: but let us our selves examine things as we have opportunity, and converse with Nature as well as Books [...] Let us not think that the bounds of Science are fixed like Hercules his Pillars, and inscribed with a Ne plus ultra. Let us not think

43 Boyle, *Disquisition* 47.

44 Boyle, *Disquisition* 214–234.

45 Boyle, *Disquisition* 219–220.

46 See Harrison, *Bible*, for further discussion.

> we have done when we have learnt what they have delivered to us. The Treasures of Nature are inexhaustible.[47]

Ray was familiar with Boyle's work, which he cited at three reprises: first, Boyle's refutation of Descartes's 'confident assertion' that we can know nothing of God's ends in nature; second, his remarks about the beaver, which constructs 'houses for shelter and security in winter-time'; and finally, his observations on the exquisite design of the teeth.[48] The range of material drawn from Boyle demonstrates that Ray had carefully read and digested the *Disquisition*. And as John Hedley Brooke has shown, Ray followed Boyle in the latter's anthropocentric interpretation of nature: 'Wise men now knew that all things were not made for man alone; but all things were still of some use to man'.[49] And one of those uses, at least on a secondary level, was to demonstrate God's power and wisdom.

Insects in the First Edition of *The Wisdom of God* (1691)

In the manner of a sermon, Ray's text was ostensibly an explication of the 104th Psalm, verse 24: 'How Manifold are Thy Works, O Lord? In Wisdom hast thou made them all'.[50] In keeping with the verse's two claims, Ray began with the diversity of the Creation, then turned to demonstrations of the wise hand that it reveals. I should note that insects were not the only focus of his analysis. Much of Ray's most detailed evidence of divine craftsmanship came from vertebrate anatomy, especially human anatomy.[51] Nonetheless insects had a significant place within Ray's book.

> If man ought to reflect upon his Creator the glory of all his works, then ought he to take notice of them all, and not to think any thing unworthy

47 Ray John, *The wisdom of God manifested in the works of the creation* (London, Samuel Smith: 1691) 124–125.

48 Ray, *Wisdom* (1691) 21, 99, 187.

49 Brooke J.H., "'Wise men nowadays think otherwise': John Ray, natural theology and the meanings of anthropocentrism", *Notes and Records of the Royal Society* 54, 2 (2000) 199–213, esp. 207, 210.

50 Ray had also used this verse as the epigraph to the English version of Willughby's *Ornithology*, 1678, along with the concluding phrase: 'the earth is full of thy riches'. Willughby, *Ornithology* t.p.

51 In this he was following in the footsteps of predecessors going back to antiquity, including non-Christian writers such as Galen.

> of his cognizance. And truly the wisdom, art and power of almighty God, shines forth as visibly in the structure of the body of the minutest insect, as in that of a horse or elephant: Therefore God is said to be, *maximus in minimis*. We men, esteeming it a more difficult matter, and of greater art and curiosity to frame a small watch, than a large clock: and no man blames him who spent his whole time in the consideration of the nature and works of a bee, or thinks his subject was too narrow.[52]

Ray's last sentence may be a reference to Jan Swammerdam, though Swammerdam's detailed drawings and notes on the anatomy of the bee would not be published until 1737, long after Ray's death.[53] It might also refer to Moses Rusden, beekeeper to Charles II and author of a treatise on bees and beekeeping, or even to Samuel Hartlib, who compiled an epistolary treatise on bees and silkworms.[54] Ray may not have had a specific individual in mind; the bee, identified by Aldrovandi and Moffett as the noblest insect, was the ideal example of an insect worthy of attention.[55]

For Ray, one evidence of God's creative activity was the sheer variety of living species in the world. 'Animate bodies', he wrote, 'are divided into four great genera or orders, beasts, birds, fishes and insects'. Of beasts, including 'serpents', barely 150 had been described. There were perhaps 500 birds, and a thousand fishes and shellfishes. Ray recognized that not every species was known, but he estimated that at most a quarter of beasts and birds remained to be found, and half of the fishes and shellfishes.[56]

Insects, on the other hand, were far more numerous: Ray estimated there were at least 150 species each of butterflies (including moths) and beetles in England, as well as flies, bees, ants, and other insects, winged and wingless. There were perhaps a thousand total in the island. If the ratio of English to

---

52 Ray, *Wisdom* (1691) 130.

53 Swammerdam Jan, *Bybel der natuure* [...] *of Historie der insecten tot zeekere zoorten gebracht*, ed. H. Boerhaave, 2 vols. (Leiden, Isaak Severinus: 1737).

54 Rusden Moses, *A further discovery of bees: Treating of the nature, government, generation & preservation of the bee: With the experiments and improvements arising from the keeping them in transparent boxes, instead of straw-hives: Also proper directions* (*to all such as keep bees*) *as well to prevent their robbing in straw-hives, as their killing in the colonies* (London, the author: 1679); Hartlib Samuel (ed.), *The reformed common-wealth of Bees, presented in several letters and observations to Samuel Hartlib Esq.: With the reformed Virginian silk-worm: Containing many excellent and choice secrets, experiments, and discoveries for attaining of national and private profits and riches* (London, Giles Calvert: 1655).

55 Aldrovandi, *De animalibus insectis* 20; Moffett, *Insectorum theatrum* 1.

56 Ray, *Wisdom* (1691) 5.

foreign species was one to nine, as Ray thought it was in plants (his special area of expertise), then there must be no fewer than ten thousand species of insects: 'And I do believe they rather exceed than fall short of that sum'.[57] In that, he was certainly correct! Ray argued that the vast number of creatures offered 'a demonstrative proof of the unlimited extent of the Creators skill, and the foecundity of his wisdom and power'. And since the less perfect a genus was, the more numerous its species, there were 'more Insects than of any of the rest' of the animals.[58]

But the bulk of Ray's treatise was devoted to the second part of the Psalm verse: 'In Wisdom hath thou made them all'. This wisdom was revealed both in the exquisite workmanship of insects' bodies and in the finely-tuned instincts that governed their behavior. Looking through the microscope, Ray said (borrowing the words of John Wilkins), 'There are such inimitable gildings in the smallest seeds of plants, but especially in the parts of animals, in the head or eye of a small fly; such accuracy, order, and symmetry in the frame of the most minute creatures, a louse, for example, or a mite, as no man were able to conceive without seeing of them'.[59]

Insects' instincts were a further proof of divine craftsmanship. The bee knows how to use her sting for defense without being taught. Bees were also nature's geometers *par excellence*: 'The bee, a creature of the lowest form of animals, so that no man can suspect it to have any considerable measure of understanding, or to have knowledge of, much less to aim at any end, yet makes her combs and cells with that geometrical accuracy, that she must needs be acted by an instinct implanted in her by the wise Author of nature'.[60] Not only honeybees, but other species as well, make perfect cells despite completely lacking any understanding of the principles by which they acted.

In keeping with his unwillingness to admit unconfirmed matters of fact, Ray rejected stories of insects' apparent instinctual prudence if they lacked empirical confirmation. He dismissed the story that ants bite off the germ of seeds before hoarding them as 'a mere fiction', and added that without Scriptural support he would even have rejected the belief that ants set aside stores for the winter, 'because I could never observe any such storing up of grain by our country-ants'.[61] In this sense, his physico-theology was empirical in a negative sense: received stories needed to be tested (Pierre Lyonet would later employ

57 Ray, *Wisdom* (1691) 5–7.
58 Ray, *Wisdom* (1691) 1, 4.
59 Ray, *Wisdom* (1691) 41.
60 Ray, *Wisdom* (1691) 96.
61 Ray, *Wisdom* (1691) 99.

the same critical sifting of fact from fiction in his notes on Friedrich Christian Lesser's *Insect Theology*).[62] This is one of the crucial ways in which physico-theology differed from medieval and Renaissance natural theology.

Another negative point Ray made, to counter the Epicurean account of how animals originated, was his 'observation and affirmation [...] that there is no such thing in nature, as aequivocal or spontaneous generation, but that all animals, as well small as great, not excluding the vilest and most contemptible insect, are generated by animal parents of the same species with themselves'.[63] In support, Ray cited Francesco Redi's experiments and others made by members of the Royal Society.[64] Ray remained puzzled by the origin of intestinal worms, considering it the greatest challenge to his opinion: 'But seeing the round Worms do manifestly generate, and probably the other kinds too; it's likely they come originally from Seed, which how it was brought into the Guts, may afterwards possibly be discovered'.[65] This brief discussion of spontaneous generation may date back to Ray's Cambridge common-place sermon, because he and Willughby had made more extensive observations on generation during Willughby's time in Cambridge, and Ray had returned to the question later.[66] As Ray continued his study of insects, he would give far more attention to the problem of spontaneous generation.

Insects in the Second Edition (1692)

When he was preparing *The Wisdom of God* for the press, Ray had only just resumed his study of insects. The three subsequent editions demonstrate that Ray continued to enrich his physico-theological arguments with material about insects, drawing upon his own observations as well as on knowledge communicated by his correspondents and published by other naturalists. The

62 Lesser Friedrich Christian, *Théologie des insectes, ou demonstration des perfections de Dieu dans tout ce qui concerne les insectes*, ed. P. Lyonet, 2 vols. (The Hague, Jean Swart: 1742).

63 Ray, *Wisdom* (1691) 221.

64 On the complex history of spontaneous generation, see Farley J., *The spontaneous generation controversy from Descartes to Oparin* (Baltimore: 1977); Findlen P., "Controlling the experiment: Rhetoric, court patronage and the experimental method of Francesco Redi", *History of Science* 31 (1993) 35–64, esp. 41–44; and Goodrum M.R., "Atomism, atheism, and the spontaneous generation of human beings: The debate over a natural origin of the first humans in seventeenth-century Britain", *Journal of the History of Ideas* 63.2 (2002) 207–224, esp. 217–219.

65 Ray, *Wisdom* (1691) 221–222.

66 Ogilvie, "Attending to insects" 359, 364–365.

second edition, which appeared only a year after the first, contained numerous additions. Ray had produced his initial estimate of the number of insect species before his investigations of 'our English insects', especially butterflies and moths, in the summer of 1691. That autumn he concluded that he had underestimated the number of English species by a factor of at least two, and that therefore the earth contained no fewer than 20,000 species of insects.[67]

Ray noted that flying insects' bodies were covered with scales or other hard parts, which protected them from enemies and prevented their internal fluids from evaporating. He reported having found insects everywhere, even on the highest mountains. And he reported experiments suggesting that ants' venom was the same as 'oil of vitriol' (i.e., sulfuric acid).[68] But the most significant additions about insects to this edition involved the question of generation: in particular, Ray's opposition to spontaneous generation.

In a lengthy passage, Ray answered many of the questions he had left open in the first edition, drawing this time not only on Redi's researches but also on the publications of Swammerdam, Malpighi, Lister, and Leeuwenhoek, careful and circumspect observers whom Ray contrasted to 'the common herd of philosophers.'[69] Given the importance of the subject for Ray's physico-theology, since Ray was intent on denying that matter could organize itself, the length and fervor of the additions strongly suggest that Ray had returned seriously to the question of the generation of insects only after completing the first edition of *The Wisdom of God*, at the same time that he began collecting and describing insect species.

Swammerdam, Ray noted, had observed the origins of those insects found in galls, which even Redi had conceded might have been engendered by the plant. Swammerdam insisted this was impossible:

> They all come of the seed of animalcules of their own kind, that were before laid there. For these insects do thrust their seed or eggs so deep into the plants, that they come to be afterwards as it were united with them, and the aperture or orifice by which they entred, quite closed up, and obliterated; the eggs being hatched and nourished within. We have

67 Ray John, *The wisdom of God manifested in the works of the creation*, 2nd ed. (London, Samuel Smith: 1692), vol. I, 8–9; Ray's reference to 'this summer, An. 1691', suggests he wrote that passage in the autumn.

68 Ray, *Wisdom* (2nd ed.: 1692), vol. I, 148; 204; vol. II, 130.

69 Ray, *Wisdom* (2nd ed.: 1692), vol. II, 75 ff.

> often found the eggs of insects so deeply sunk into the tender buds of trees, that without hurting of them it was impossible to draw them out.[70]

Malpighi, too, asserted the same and had observed insect larvae boring inside plants, as had Martin Lister.

Ray remained puzzled about the origin of intestinal worms, but 'the constancy to their species', the way successive generations continued to resemble one another, was to him strong evidence that they were engendered from one another, probably entering their hosts along with food, since 'children in their first infancy, and as long as they are constantly confined to a milk diet, are seldom troubled with them'.[71] The puzzle he had identified in the first edition was still unresolved, but he had identified a possible explanation that excluded spontaneous generation.

An even more puzzling subject was the origin 'of such insects as are found and seem to be bred in the bodies of others of different kinds'. Ray had observed two species of caterpillar produce 'small maggots, to the number sometimes of threescore, or more'. Still other underwent a 'dubious metamorphosis, sometimes changed into the aurelia of a butterfly, sometimes into a fly-case'.[72] Ray imagined an interlocutor insisting that here one must 'necessarily have recourse to a spontaneous generation'. He replied negatively:

> The most that can be inferred from hence is, a transmutation of species; one insect may, instead of generating another of its own kind, beget one or more of a different. But I can by no means grant this. I do believe that these flies do either cast their eggs upon the very bodies of the forementioned caterpillars, or upon the leaves on which they feed, all in a string; which there hatching, eat their way into the body, where they are nourished till they be come to their full growth. Or it may be, the fly may with the hollow and sharp tube of her womb punch and perforate the very skin of the eruca, and cast her eggs into its body. So the ichneumon will convey her eggs into caterpillars.[73]

In this passage, Ray drew on the investigations into insect parasitism that had been made by his friends Willughby and Lister (and the subject of one of Willughby's few publications, a letter in the *Philosophical Transactions*),

70 Ray, *Wisdom* (2nd ed.: 1692), vol. II, 76–77.

71 Ray, *Wisdom* (2nd ed.: 1692), vol. II, 87–88.

72 Ray, *Wisdom* (2nd ed.: 1692), vol. II, 90–91.

73 Ray, *Wisdom* (2nd ed.: 1692), vol. II, 91–92.

as well as the work of Swammerdam, who bitterly criticized his predecessor Johannes Goedaert for failing to distinguish between normal transmutations and the results of parasitism. He was, of course, aware of Willughby's investigations, but Willughby had not ruled out spontaneous generation. The earliest unequivocal evidence that Ray had read Swammerdam's *General History of Insects* dates from 1692—the very year he cited the book in the second edition of *The Wisdom of God*.[74] In this case, as in the case of the number of species, Ray's new research almost immediately added additional empirical support for his physico-theological objections to spontaneous generation.

Ray took a final stab at the theory's supporters: 'And whereas the Assertors of equivocal generation were wont to pretend the imperfection of their animals, as a ground to facilitate the belief of their spontaneous generation; I do affirm, that they are as perfect in their kind, and as much art shewn in the formation of them, as of the greatest; nay more too, in the judgment of that great wit and natural historian Pliny'.[75] Here, too, Ray was reiterating a point made by Swammerdam: insects were not imperfect creatures but possessed their own form of perfection.[76]

Insects in the Third Edition (1701)

Nonetheless, Ray remained doubtful about whether the problem of spontaneous generation had been completely resolved. In the third edition of *The Wisdom of God*, he encouraged further research into the subject, due to its great importance:

> For if this point be but cleared, and it be demonstrated that all creatures are generated univocally by parents of their own kind, and that there is no such thing as spontaneous generation in the world, one main prop and support of atheism is taken away, and their strongest hold demolished:

74 On 10 September 1683, Tancred Robinson wrote to Ray that he had seen scorpions in Montpellier that resembled those described by Swammerdam in his history of insects: Ray, *Correspondence* 135–138. However, the earliest reference I have found in Ray's own letters is in one to Edward Lhwyd, 5 April 1692: Ray, *Further correspondence* 226–228.

75 Ray, *Wisdom* (2nd ed.: 1692), vol. II, 93.

76 Swammerdam Jan, *Histoire generale des insectes: Ou l'on expose clairement la maniere lente & presqu'insensible de l'accroissement de leurs membres, & ou l'on decouvre evidemment l'Erreur ou l'on tombe d'ordinaire au sujet de leur prétendué* [sic] *transformation*, trans. G. de Walcheren (Utrecht, Guillaume de Walcheren: 1682) 1–4. Ray used one of the French editions of Swammerdam's work (1682 and 1685), originally published in Dutch in 1669.

> They cannot then exemplify their foolish hypothesis of the generation of man and other animals at first, by the like of frogs and insects at this present day.[77]

Ray added a passing remark about insects and the laws of motion to this edition.[78] But his longest addition concerning insects involved a problem that they posed for physico-theology: 'A wise agent acts for ends. Now what end can there be of creating such a vast multitude of insects, as the world is fill'd with; most of which seems to be useless, and some also noxious and pernicious to man, and other creatures?'.[79]

Ray offered two responses, addressing both the 'multitude of species or Kinds' and the 'number of individuals in each kind'. The abundance of species served 'to manifest and display the riches of the power and wisdom of God'. Here Ray referred to his revised estimate of the number of insect species. Though 'this sublunary world' is finite, God's creations are, practically speaking, infinite: 'And therefore, I believe, never any man yet did, never any man shall, so long as the world endures, by his utmost industry, attain to the knowledge of all the species of nature'.[80]

Nonetheless, this should not stop humans from exercising their 'contemplative faculty' on them: the human mind quickly grows weary unless presented with variety, and

> New objects afford us great delight, especially if found out by our own industry. I remember Clusius saith of himself, 'That upon the discovery of a new plant, he did not less rejoice, than if he had found a rich treasure.' Thus God is pleased, by reserving things to be found out by our pains and industry, to provide us employment most delightful and agreeable to our natures and inclinations.[81]

This response was theological in nature, though it drew on his empirical estimate of the number of insect species. But Ray returned to his empirical

---

77 Ray, *Wisdom* (3rd ed.: 1701) 328–329; Ray John, *The wisdom of God manifested in the works of the creation*, 4th ed. (London, Samuel Smith: 1704) 372.

78 Ray, *Wisdom* (3rd ed.: 1701) 370.

79 Ray, *Wisdom* (3rd ed.: 1701) 380 ff.

80 Ray, *Wisdom* (3rd ed.: 1701) 380. Here Ray adumbrates the principle of plenitude discussed by Lovejoy: Lovejoy A.O., *The great chain of being: A study of the history of an idea* (Cambridge, MA – London: 1964).

81 Ray, *Wisdom* (3rd ed.: 1701) 381.

investigations when he turned to the question of why there were so many 'individuals in each kind of insect'. Here he noted, if only in qualitative terms, how many of them were the prey of others: 'It is designed to secure the continuance and perpetuity of the several species; which, if they did not multiply exceedingly, scarce any of them could escape the ravine of so many enemies as continually assault and prey upon them, but would endanger to be quite destroyed and lost out of the world'.[82]

Many birds, for instance, survive wholly or primarily by eating insects: 'Now insects supply land-birds the chiefest part of their sustenance: some, as the entire genus of swallows, living wholly upon them', while others made insects a large part of their diet, 'as appears by dissecting their stomachs'. Even birds that feed on other substances as adults are often fed insects while still in the nest.

> Neither do birds alone, but many sorts of fishes feed upon insects, as is well known to anglers, who bait their hooks with them. Nay, which is more strange, divers quadrupeds feed upon insects, and some live wholly upon them, as two sorts of Tamunduus upon ants, which therefore are called in English ant-bears; the camelion upon flies; the mole upon earth-worms: the badger also lives chiefly upon beetles, worms, and other insects. Here we may take notice by the way, that because so many creatures live upon ants and their eggs, Providence hath so ordered it, that they should be the most numerous of any tribe of insects that we know.[83]

Even noxious insects had a purpose: in part theological, to serve as scourges of the wicked, but also quasi-ecological, as food for birds and other creatures, as Ray and others had discovered. In his response to this objection, therefore, Ray employed knowledge of insects that was not merely based on their individual natural histories but that also drew upon his qualitative yet empirical understanding of their place in the broader scheme of life on Earth that God had created. Yet we should bear in mind that Ray's interpretation did not involve the dynamic processes of modern ecology: rather, it was an instance of how, in Boyle's terms, 'cosmical' natural theology—the general order of nature—could modify the utility of individual creatures.[84]

82 Ray, *Wisdom* (3rd ed.: 1701) 382.

83 Ray, *Wisdom* (3rd ed.: 1701) 383–385.

84 See Boyle, *Disquisition* 219–220.

Insects in the Fourth Edition (1704)

Ray made few additions on insects to the fourth edition of *The Wisdom of God*. By that point he was suffering considerable pain in his legs, especially in the cold, damp winters; his energies were going rather toward his ongoing project of describing insect species. Nonetheless he added material. He inserted a note from Derham about the vast numbers of aquatic insects.[85] He commented on the providential mission of the louse, which was 'designed to deterr men and women from sluttishness and sordidness, and to provoke them to cleanliness and neatness'.[86]

The most interesting insertion, however, was a correction:

> Having lately had an opportunity more curiously to view and examine the great flesh-colour'd, thin-hair'd English caterpillar, which is so like that sent me by Dr. Sloane, that it differs little but in magnitude, which may be owing to the climate,) I observed that it had a power of drawing its eight hind legs or stumps so far up in its body, that they did altogether disappear, so that the creature seemed to want them, and of thrusting them out again at pleasure.

Ray concluded from this observation that a Jamaican insect larva that Sloane had sent to him was also a caterpillar, though Ray had taken it to be a 'hexapod', that is a beetle grub. 'Thus much I thought fit to add to Dr. Lister, and do the Truth right, by retracting my former conjecture concerning the cossi'.[87] Even at this point, Ray was unwilling to let an error remain in his text, even an error that by itself had nothing to do with his physico-theological argument. The reader of *The Wisdom of God* deserved nothing less than the natural truth.

Conclusion

This brief overview permits two conclusions about the empirical basis of Ray's physico-theology.

First, insects provided significant support for physico-theology. The number of insect species and the vast quantity of individual insects demonstrated the

85 Ray, *Wisdom* (4th ed.: 1704) 430–431.

86 Ray, *Wisdom* (4th ed.: 1704) 357.

87 Ray, *Wisdom* (4th ed.: 1704) 355–356.

magnitude of God's creative power. Through their remarkable structure and finely honed instincts, insects also revealed the quality and immense subtlety of divine craftsmanship. This use of insects would continue in eighteenth-century Protestant physico-theology, culminating in Lesser's *Insecto-Theologia*, as well as in Catholic natural theology, for instance that of the Abbé Noël Pluche in the first volume of his *Spectacle de la nature*.[88]

Second, as Ray revised successive editions of *The Wisdom of God*, he integrated new empirical information into his natural theology. As he did so, he proceeded by accretion: that is, he generally added new material without substantially revising it.[89] However, both the second and the third editions drew upon insects to resolve specific challenges to natural theology.

One was the challenge posed by spontaneous or equivocal generation. Despite Ray's insistence that a 'plastick nature' was responsible for development and growth, rather than direct divine intervention, he rejected the Renaissance notion of *lusus naturae*.[90] Nature, for Ray, proceeded always in the same fashion except when some problem arose to produce abortions or monsters. His careful study of insects revealed that they reproduced in the same fashion as larger creatures.[91] Moreover, they were not imperfect creatures; rather, each attained its proper stage of perfection, even though it was less perfect than quadrupeds, birds, or fishes.

Ray's approach to the specific problem of how insects (in the early modern sense) generated within other bodies reveals both the importance of empirical data to his natural theology and its limits. The investigations that Willughby, Ray, and Lister had carried out on galls and on parasitic wasps and flies, along with further investigations in published works, gave Ray a clear understanding

---

88 Lesser Friedrich Christian, *Insecto-Theologia, oder: Vernunfft- und Schriftmäßiger Versuch, wie ein Mensch durch aufmercksame Betrachtung derer sonst wenig geachteten Insecten zu lebendiger Erkänntniß und Bewundering der Allmacht, Weißheit, der Güte und Gerechtigkeit des grossen GOttes gelangen könne* (Frankfurt – Leipzig, Michael Blochberger: 1738); Pluche Noël-Antoine, *Le Spectacle de la nature, ou Entretiens sur les particularités de l'histoire naturelle qui ont paru les plus propres à rendre les jeunes gens curieux et à leur former l'esprit*, 8 parts in 9 vols. (Paris, La veuve Estienne: 1732–52); see Jorink, *Reading the Book of Nature*, ch. 4–5.

89 This point is emphasized by Zeitz, "Natural theology".

90 On this concept, see Findlen P., "Jokes of nature and jokes of knowledge: The playfulness of scientific discourse in early modern Europe", *Renaissance Quarterly* 43 (1990) 292–331.

91 Swammerdam, too, emphasized this point. See Ogilvie B.W., "Order of insects: Insect species and metamorphosis between Renaissance and Enlightenment", in Nachtomy O. – Smith J.E.H. (eds.), *The life sciences in early modern philosophy* (Oxford: 2014) 222–245.

of how these insects could apparently be generated inside vegetable or animal matter. Without analogous investigations into intestinal worms, however, Ray was reduced to speculation, based on his opposition to equivocal generation and circumstantial evidence, such as the fact that nurslings generally did not suffer from intestinal parasites. In this case, though, even in his physico-theological work he noted that further investigation was required.

Another challenge, ironically, was posed by the very variety and number of insects that Ray had offered as evidence of the manifold nature of God's creation. This number and variety posed a challenge to an anthropocentric natural theology, for it raised the question of *why* God would create so many things that did not benefit man, and often caused him harm. Here Ray's response was, in good part, empirical: he noted that insects fed many creatures, especially birds, that were themselves useful to man.

Of course, *The Wisdom of God* also displays non-empirical, theological, ways of thinking about insects. Despite Ray's insistence that insects were not imperfect creatures, he still drew upon the notion of degrees of perfection, and implicitly on the chain of being, when he discussed the number of species. And his anti-anthropocentric argument drew upon Puritan theology: a variety of creatures existed not so they could serve man but, instead, that they might, in their own limited way, praise God.[92] Nonetheless the noxious species could also serve as a scourge for humans. And finally, Ray did not rule out the possibility of discovering some utility in the future for creatures that seemed to have no human use.[93]

I do not wish to imply that Ray's natural theology was somehow a mix of old and new elements, as if a new layer of empirical data had been superimposed on an older theological framework. Rather, theology provided an organizing principle for making sense of empirical data. That is why the perspective that Ray adopted on the number and function of individuals was not yet a dynamic, ecological perspective.[94] Despite his attention to how different sorts of insects

92 Cf. Goodman Godfrey, *The creatures praysing God: or, The religion of dumbe creatures, an example and argument, for the stirring up of our devotion, and for the confusion of atheisme* (London, Felix Kingston: 1622).

93 A point emphasized by Brooke, "Wise men".

94 Natalie Zemon Davis has characterized Maria Sibylla Merian's approach to insects in the context of their food plants as 'ecological', but Merian, like Ray, did not have a dynamic perspective on this relationship: Davis N.Z., *Women on the margins: Three seventeenth-century lives* (Cambridge, MA: 1995) 151.

served as food for one another and for other creatures, Ray envisioned a static order.

It is worth noting that the transfer of empirical data from Ray's insect studies into his physico-theology was one way. In the whole *Historia insectorum* there is only one passage that refers to physico-theology, and that indirectly: after describing the elaborate behavior of an *Ammophila* wasp that had paralyzed a caterpillar and then buried it, Ray challenges his reader: 'Who could attribute this behavior to a mere machine?'.[95] His implicit target was, of course, the Cartesian view of animals as automata.[96] Ray saw authentic natural history as the foundation of his physico-theological discourse, but even if physico-theology motivated Ray's pursuit of natural history (at least in part), he saw no need to regularly introduce it into his works in the natural history genre.

To close, I will return to my opening question. Ray's physico-theology in *The Wisdom of God* is deeply panglossian. But its significant empirical component was enriched and updated in successive editions as Ray learned more about insects and their generation. Of course, Ray is an exceptional case, a naturalist of the first rank who was also a leading physico-theologian. But his epigones—Derham, Lesser, Lyonet, Paley, and other—continued to insist on the empirical foundation of their theological demonstrations. This empirical physico-theology would ultimately bear scientific fruit, but only later, when its panglossian elements had been stripped away, and when Ray's and his successors' insistence on the perfect adaptation of creatures to their lives and environments had been succeeded by the imperfect, only apparent design produced by natural selection.

---

95 Ray, *Historia insectorum* 254.

96 Descartes's actual position was more complicated, but many seventeenth-century readers interpreted Descartes and his followers as holding not only that animals were machines but that they had no feelings: Harrison P., "Descartes on animals", *Philosophical Quarterly* 42, 167 (1992) 219–227; Harrison P., "The virtues of animals in seventeenth-century thought", *Journal of the History of Ideas* 59.3 (1998) 463–484; Wolloch N., "Dead animals and the beast-machine: Seventeenth-century Netherlandish paintings of dead animals, as anti-Cartesian statements", *Art History* 22, 5 (1999) 705–727.

Bibliography

Aldrovandi Ulisse, *De animalibus insectis libri septem* [...] (Bologna, Giovanni Battista Bellagamba: 1602).

Berry R.J., "John Ray, physico-theology and afterwards", *Archives of Natural History* 38.2 (2011) 328–348.

Bertoloni Meli D., "The representation of insects in the seventeenth century: A comparative approach", *Annals of Science* 67.3 (2010) 405–429.

Bodenheimer F.S., *Materialien zur Geschichte der Entomologie bis Linné*, 2 vols. (Berlin: 1928).

Boyle Robert, *A disquisition about the final causes of natural things: Wherein it is inquir'd, whether, and (if at all) with what cautions, a naturalist should admit them. To which are subjoyn'd, by way of appendix, some uncommon observations about vitiated sight* (London, John Taylor: 1688).

Brooke J.H., "Science and the fortunes of natural theology: Some historical perspectives", *Zygon* 24.1 (1989) 3–22.

Brooke J.H., " 'Wise men nowadays think otherwise': John Ray, natural theology and the meanings of anthropocentrism", *Notes and Records of the Royal Society* 54.2 (2000) 199–213.

Cambefort Y., "Artistes, médecins et curieux aux origines de l'entomologie moderne (1450–1650)", *Bulletin d'Histoire et d'Épistémologie des Sciences de la Vie* 11.1 (2004) 3–29.

Carlin L., "The importance of teleology to Boyle's natural philosophy", *British Journal for the History of Philosophy* 19 (2011) 663–680.

Carlin L., "Boyle's teleological mechanism and the myth of immanent teleology", *Studies in History and Philosophy of Science* 43 (2012) 54–63.

Cobb M., "Malpighi, Swammerdam and the colourful silkworm: Replication and visual representation in early modern science", *Annals of Science* 59 (2002) 111–147.

Cobb M., *Generation: The seventeenth-century scientists who unraveled the secrets of sex, life, and growth* (New York: 2006).

Davis N.Z., *Women on the margins: Three seventeenth-century lives* (Cambridge, MA: 1995).

Derham William, *Physico-theology: or, a Demonstration of the being and attributes of God, from his works of creation: Being the substance of sixteen sermons preached in St. Mary le Bow-Church, London, at the Hon.ble Mr. Boyle's Lectures, in the years 1711 and 1712: With large notes, and many curious observations*, 2nd ed. (London, William Innys: 1714).

Dicke M., "Insects in western art", *American Entomologist* 46.4 (2000) 228–237.

Farley J., *The spontaneous generation controversy from Descartes to Oparin* (Baltimore: 1977).

Findlen P., "Jokes of nature and jokes of knowledge: The playfulness of scientific discourse in early modern Europe", *Renaissance Quarterly* 43 (1990) 292–331.

Findlen P., "Controlling the experiment: Rhetoric, court patronage and the experimental method of Francesco Redi", *History of Science* 31 (1993) 35–64.

Gillespie N.C., "Natural history, natural theology, and social order: John Ray and the 'Newtonian ideology' ", *Journal of the History of Biology* 20 (1987) 1–49.

Goodman Godfrey, *The creatures praysing God: or, The religion of dumbe creatures, an example and argument, for the stirring up of our devotion, and for the confusion of atheisme* (London, Felix Kingston: 1622).

Goodrum M.R., "Atomism, atheism, and the spontaneous generation of human beings: The debate over a natural origin of the first humans in seventeenth-century Britain", *Journal of the History of Ideas* 63.2 (2002) 207–224.

Harrison P., "Descartes on animals", *Philosophical Quarterly* 42.167 (1992) 219–227.

Harrison P., *The Bible, Protestantism and the rise of natural science* (Cambridge: 1998).

Harrison P., "The virtues of animals in seventeenth-century thought", *Journal of the History of Ideas* 59.3 (1998) 463–484.

Hartlib Samuel (ed.), *The reformed common-wealth of Bees, presented in several letters and observations to Samuel Hartlib Esq.: With the reformed Virginian silk-worm: Containing many excellent and choice secrets, experiments, and discoveries for attaining of national and private profits and riches* (London, Giles Calvert: 1655).

Hunter M. – Davis E.B., "The making of Robert Boyle's 'Free enquiry into the vulgarly receiv'd notion of nature' (1686)", *Early Science and Medicine* 1.2 (1996) 204–271.

Jorink E., *Het Boeck der Natuere: Nederlandse geleerden en de wonderen van Gods schepping, 1575–1715* (Leiden: 2006).

Jorink E., *Reading the Book of Nature in the Dutch Golden Age, 1575–1715*, tr. P. Mason, Brill's studies in intellectual history, 191 (Leiden: 2010).

Knight D.M., "Science and beliefs", in Knight D.M. – Eddy M.D. (eds.), *Science and beliefs: From natural philosophy to natural science, 1700–1900* (Aldershot etc.: 2005) 3–13.

Lennox J.G., "Robert Boyle's defense of teleological inference in experimental science", *Isis* 74.1 (1983) 38–52.

Lesser Friedrich Christian, *Insecto-Theologia, oder: Vernunfft- und Schriftmäßiger Versuch, wie ein Mensch durch aufmercksame Betrachtung derer sonst wenig geachteten Insecten zu lebendiger Erkänntniß und Bewundering der Allmacht, Weißheit, der Güte und Gerechtigkeit des grossen GOttes gelangen könne* (Frankfurt – Leipzig, Michael Blochberger: 1738).

Lesser Friedrich Christian, *Théologie des insectes, ou demonstration des perfections de Dieu dans tout ce qui concerne les insectes*, ed. P. Lyonet, 2 vols. (The Hague, Jean Swart: 1742).

Lovejoy A.O., *The great chain of being: A study of the history of an idea* (Cambridge, MA, and London: 1964).

Mandelbrote S., "Ray, John (1627–1705)", *Oxford Dictionary of National Biography*, http://www.oxforddnb.com/view/article/23203 (Oxford: 2004).

Mandelbrote S., "The uses of natural theology in seventeenth-century England", *Science in Context* 20, 3 (2007) 451–480.

Moffett Thomas, *Insectorum sive minimorum animalium theatrum: olim ab Edoardo Wottono, Conrado Gesnero, Thomaque Pennio inchoatum, tandem Tho<mae> Moufeti Londinatis opera sumptibusque maximis concinnatum, auctum, perfectum* (London, Thomas Cotes: 1634).

Nadler S.M., *The best of all possible worlds: A story of philosophers, God, and evil* (New York: 2008).

Neri J., *The insect and the image* (Minneapolis: 2011).

Ogilvie B.W., "Natural history, ethics, and physico-theology", in Pomata G. – Siraisi N. (eds.), *Historia: Empiricism and erudition in early modern Europe* (Cambridge, MA: 2005) 75–103.

Ogilvie B.W., "Nature's Bible: Insects in seventeenth-century European art and science", *Tidsskrift for kulturforskning* 7.3 (2008) 5–21.

Ogilvie B.W., "Attending to insects: Francis Willughby and John Ray", *Notes and Records of the Royal Society* 66.4 (2012) 357–372.

Ogilvie B.W., "Order of insects: Insect species and metamorphosis between Renaissance and Enlightenment", in Nachtomy O. – Smith J.E.H. (eds.), *The life sciences in early modern philosophy* (Oxford: 2014) 222–245.

Osler M.J., "Whose ends? Teleology in early modern natural philosophy", *Osiris* 16 (2001) 151–168.

Pluche Noël-Antoine, *Le Spectacle de la nature, ou Entretiens sur les particularités de l'histoire naturelle qui ont paru les plus propres à rendre les jeunes gens curieux et à leur former l'esprit*, 8 parts in 9 vols. (Paris, la veuve Estienne: 1732–1752).

Raven C.E., *John Ray, naturalist: His life and works* (Cambridge: 1950).

Ray John, *Catalogus plantarum circa Cantabrigiam nascentium: In quo exhibentur quotquot hactenus inventae sunt, quae vel sponte proveniunt vel in agris seruntur; una cum synonymis selectioribus, locis natalibus et observationibus quibusdam oppido raris* [...] (London, John Martin – James Allestry – Thomas Dicas: 1660).

Ray John, *The wisdom of God manifested in the works of the creation* (London, Samuel Smith: 1691).

Ray John, *The wisdom of God manifested in the works of the creation*, 2nd ed. (London, Samuel Smith: 1692).

Ray John, *Three physico-theological discourses, concerning I. the primitive chaos, and creation of the world; II. the general deluge, its causes and effects; III. the dissolution of the world, and future conflagration*, 2nd ed. (London, Samuel Smith: 1693).

Ray John, *The wisdom of God manifested in the works of the creation*, 3rd ed. (London, Samuel Smith – Benjamin Walford: 1701).

Ray John, *The wisdom of God manifested in the works of the creation*, 4th ed. (London, Samuel Smith: 1704).

Ray John, *Methodus insectorum seu Insecta in methodum aliqualem digesta* (London, Samuel Smith and Benjamin Walford: 1705).

Ray John, *Historia insectorum: [...] opus posthumum jussu Regia Societatis Londinensis editum; cui subjungitur appendix De scarabais Britannicis, autore M. Lister* (London, A. & J. Churchill: 1710).

Ray John, *The correspondence of John Ray: Consisting of selections from the Philosophical Letters published by Dr. Derham, and original letters of John Ray, in the collection of the British Museum*, ed. E. Lankester (London: 1848).

Ray John, *Further correspondence of John Ray*, ed. R.W.T. Gunther (London: 1928).

Rusden Moses, *A further discovery of bees: Treating of the nature, government, generation & preservation of the bee: With the experiments and improvements arising from the keeping them in transparent boxes, instead of straw-hives: Also proper directions (to all such as keep bees) as well to prevent their robbing in straw-hives, as their killing in the colonies* (London, the author: 1679).

Stebbins S., *Maxima in minimis: Zum Empirie- und Autoritätsverständnis in der physikotheologischen Literatur der Frühaufklärung* (Frankfurt a.M.: 1980).

Swammerdam Jan, *Histoire generale des insectes: Ou l'on expose clairement la maniere lente & presqu'insensible de l'accroissement de leurs membres, & ou l'on decouvvre evidemment l'Erreur ou l'on tombe d'ordinaire au sujet de leur prétendué* [sic] *transformation*, trans. G. de Walcheren (Utrecht, Guillaume de Walcheren: 1682).

Swammerdam Jan, *Bybel der natuure [...] of Historie der insecten tot zeekere zoorten gebracht*, ed. H. Boerhaave, 2 vols. (Leiden, Isaak Severinus: 1737).

Trepp A.-C., *Von der Glückseligkeit alles zu wissen: Die Erforschung der Natur als religiöse Praxis in der Frühen Neuzeit* (Frankfurt am Main: 2009).

Vignau-Wilberg T., "*In minimis maxima conspicua*. Insektendarstellungen um 1600 und die Anfänge der Entomologie", in Enenkel K.A.E. – Smith P.J. (eds.), *Early modern zoology: The construction of animals in science, literature and the visual arts*, 2 vols. (Leiden – Boston: 2007), vol. I, 217–243.

Willughby Francis, *The ornithology of Francis Willughby of Middleton in the county of Warwick Esq, fellow of the Royal Society in three books* (London, John Martyn: 1678).

Wolloch N., "Dead animals and the beast-machine: Seventeenth-century Netherlandish paintings of dead animals, as anti-Cartesian statements", *Art History* 22.5 (1999) 705–727.

Yeo R., "Between memory and paperbooks: Baconianism and natural history in seventeenth-century England", *History of Science* 45 (2007) 1–46.
Zeitz L.M., "Natural theology, rhetoric, and revolution: John Ray's Wisdom of God, 1691–1704", *Eighteenth-Century Life* 18.1 (1994) 120–133.

CHAPTER 7

Exkurs ins Pflanzenreich: Die Rose des Paracelsus. Die Idee der Palingenesie und die Debatte um die natürliche Auferstehung zwischen Mittelalter und Neuzeit *With an English Summary*

Bernd Roling

Einleitung

In einer seiner Parabeln mit dem schönen Titel „Die Rose des Paracelsus" läßt der argentinische Schriftsteller Jorge Luis Borges einen Schüler die Studierstube des berühmten Naturphilosophen und Alchemisten Paracelsus betreten.[1] Der junge Johannes Griesebach, so war sein Name, legt eine große Summe von Goldmünzen auf den Tisch, in der anderen Hand hält er eine Rose. Nach einem kurzen Wortwechsel, der schon ahnen lässt, dass der Schüler die Geheimnisse der Philosophie nicht durchdringen wird, fordert der junge Mann Paracelsus auf, einen Beweis zu erbringen, um den Wert seiner Lehren zu beglaubigen. Paracelsus soll die Rose verbrennen und wiederauferstehen lassen. Borges lässt den Weisen eine mehrdeutige oder vielleicht auch eindeutige Antwort geben: Glaubst Du, jemand könnte diese Rose vernichten? Hatte Adam im Paradies nur einen Grashalm vernichten können? Sind es nicht nur die Erscheinungen, die sich in Wirklichkeit verändern? Griesebach wirft die Rose in die Flammen, doch Paracelsus macht keine Anstalten, die Rose wieder aus ihrem Staub erstehen zu lassen, weder durch ein magisches Wort, wie man hätte erwarten können, noch durch den Brennkoben seines Laboratoriums und besondere Instrumente. War Paracelsus also ein Scharlatan und Phantast,

---

1 Ich danke der Alten Abteilung der Staats- und Universitätsbibliothek Göttingen, der Staatsbibliothek Berlin und dem Thomas-Institut der Universität Köln für ihre Hilfe bei der Erstellung dieser Studie. Ein Teil ihrer Thesen wurden auf Vorträgen in Göttingen, Innsbruck und Prag diskutiert; allen Diskutanten sei ebenfalls herzlich gedankt. Gewidmet ist diese Arbeit Magdalena Biela-Nastase (Riihimäki/Finnland).

 | DOI 10.1163/9789004279179_009

wie es den Schüler dünkt, der einen unendlichen Augenblick lang auf die neue Rose wartet?[2]

Wie so viele der Geschichten des argentinischen Erzählers erscheint auch diese auf den ersten Blick wie ein Spiel mit Metaphern, eine Geschichte über den Glauben und den Unterschied zwischen wahrer und falscher Erkenntnis. Gerade heute mag sie auch ein Lehrstücksein sein über die Ökonomisierbarkeit von Wissen, das in Module gepresst lieber evaluierbare Kompetenzen vermitteln möchte, als bis zur wirklichen Erkenntnis vorzudringen. Wie so oft in den Erzählungen Borges erkennt man jedoch auch, daß die geschilderten Begebenheiten nicht nur der Metaphysik als Bühne dienen sollten, sondern mit tiefem Fachkenntnis durchsetzt waren, Fachwissen, auf das Borges für seine Leser nur noch anspielen musste. In den großen Forschungsbibliotheken der Welt konnte sich der Dichter seiner Mitverschworenen sicher sein; die anderen hatten ihn vielleicht nie interessiert.

Die Rose war keine Erfindung des Romanciers, sondern sollte ebenso eine Theorie bezeugen, wie sie ihre praktischen Umsetzung unter Beweis stellte. Sie dokumentierte die Möglichkeit einer natürlichen Auferstehung, die nicht nur Pflanzen betraf, sondern auch den Menschen und seinen Körper miteinschließen musste. Diese Möglichkeit einer natürlichen Wiederstellung, die man vor allem im 17. Jahrhundert als Palingenesie bezeichnen konnte, soll hier im folgenden Thema sein. Drei Dinge sollen zu diesen Zweck geleistet werden. Zunächst werden die für die christlichen Konfessionen verbindliche Lesart der Auferstehung rekonstruiert, wie sie die Theologie des Hochmittelalters etwa ab dem 13. Jahrhundert hatte etablieren können. In einem zweiten Schritt sollen Autoren vorgestellt werden, die im Mittelalter tatsächlich die Möglichkeit einer natürlichen Auferstehung gelehrt haben. Der dritte Teil wird dann den Gelehrten gebühren, die in die Tat umsetzen konnten, was die Theologie zuvor nur als Option proklamiert hatte, die Auferstehung. Wir werden sehen, daß sich diese Praktiker der leiblichen Wiederherstellung in ihren Versuchsanordnungen tatsächlich bis zum Menschen emporarbeiten konnten.

2 Borges Jorge Luis, „La rosa de Paracelso", in *Obras completas*, 4 vols. (Barcelona: 1996), vol. III, 387–389. Deutsch als Borges Jorge Luis, „Die Rose des Paracelsus", in *Erzählungen 1975–77* (Gesammelte Werke, vol. IV), übersetzt von D.E. Zimmer (München: 1982) 105–108.

I Hylemorphismus und Auferstehung

I.1 *Das Mirakel der Wiederherstellung*

Auf den ersten Blick hatte Borges allen Grund, den penetranten Bittsteller im Laboratorium des großen Meisters am Mysterium der auferstandenen Rose zweifeln zu lassen. Für die Schulphilosophie der frühen Neuzeit, aber auch für die Mehrheit ihrer mittelalterlichen Stichwortgeber mußte die Restitution einer Pflanze ein Skandalon sein. Verantwortlich war die aristotelische Ontologie, deren Hylemorphismus, die Zweiteilung aller Dinge in Form und Materie, auch die Auferstehungsdebatte nicht unberührt gelassen hatte. Menschen, Tiere und Pflanzen hatten eine Seele, wie Aristoteles glaubt, die ihren Körper formte und aktualisierte. Trennte sich die Form von ihrem Leib, musste die Materie in ihre Möglichkeit zurückfallen oder sich einer neuen, anderen Form unterwerfen. Mit dem Tod einer Pflanze oder eines Tiers verließ die Körperform ihr Gegenüber und löste sich auf; eine Wiederkehr war nicht vorgesehen.[3] Nur die menschliche Seele konnte für sich selbst existieren; ihr musste am Tage des Jüngsten Gerichtes ein neuer Körper zugeführt werden. Liest man die Einlassungen der Aristoteliker zur Frage der Wiederherstellung eines Körpers, so fielen sie, was die Auferstehung einer Pflanze oder eines Tiers betraf, mehr als eindeutig aus. Nur mit Blick auf den Menschen verhielt sich die Angelegenheit aus bekannten theologischen Gründen komplexer. Die Auferstehung des Fleisches, die ihm die Schrift in Aussicht stellte, verlangte ihre Ausnahme.

Angenommen, das Zeitkonto einer Rose war überschritten und sie verfaulte und starb, gab es für das Gewächs eine Aussicht, wieder zu jener einen und besonderen Rose zu werden, als welche sie zuvor hatte blühen dürfen? Die Schulphilosophie hatte ihre Antwort in den Kommentaren zur aristotelischen Schrift *Über das Werden und Vergehen* gegeben. Um die numerische Identität einer Substanz zu gewähren, also das Fortbestehen eines Individuums in seiner besonderen Unverwechselbarkeit, mussten Form und Materie, die beiden bestimmenden Faktoren der Wirklichkeit, zusammenbleiben oder in gleicher Ausprägung wieder zueinanderfinden. Akzidentien, Farben, äußere Gestalt oder Ort, mochten sich verändern, doch musste die Substanz ihre beiden Bestandteile bewahren. Auch wenn eine Form in sich erhalten blieb, verging ihre Materie und musste erneut zu einer Möglichkeit werden. Verband sich diese Form mit einer neuen Materie, hatte sie als Substanz ihre numerische

3 Ähnlich wiederholen es in der Frühen Neuzeit die Standardphysiken, z.B. Sperling Johann, *Institutiones physicae* (Wittenberg, Berger: 1672), lib. VII, cap. 2, S. 990–1002, oder Kyper Albert, *Institutiones physicae*, 2 vols. (Leiden, Adrian Wyngaerden: 1646), vol. II, lib. VI, cap. 3, S. 231–235.

Identität verloren, wie man glaubte; sie hatte sich lediglich in ihrer Spezies, ihrer Art, wiederhergestellt.[4] Konkret gesprochen: Selbst wenn die Form einer Rose als Idee für sich allein hätte existieren können, konnte sie nicht wieder zur selben Rose werden. Sobald sich das Gewächs wieder materialisiert hatte, war eine neue Rose entstanden, doch nicht die vergangene wiederhergestellt worden. Es war für die Philosophen des Mittelalters eine Frage der Logik. Wenn ein Mensch am Abend einschlief, so der Aristoteles-Kommentator Francesco von Toledo, und am Morgen wieder aufwachte, hatte er mit etwas Glück sein Sehvermögen bewahrt, doch das Sehen selbst, die konkrete *visio*, war eine numerisch andere geworden.[5] Wenn die Form des Sokrates, so sagt es Johannes Philoponos, nach dem Ablauf des großen Weltenjahres wieder in die Materie entlassen wurde, war ein neuer Sokrates geboren worden. Jener Mann jedoch, der auf dem Marktplatz die Athener belästigt hatte, war verschwunden. Die Identität eines Individuums konnte nur gewahrt bleiben, wenn Form und Materie ineinander verflochten blieben.[6] Das Wesen der Materie jedoch war Veränderung und Chaos; ihr beharrlicher Fluss machte es unmöglich, dass die Seele wieder in dieselbe Körperlichkeit entlassen wurde, die sie im Moment des Todes aufgegeben hatte. Eine Auferstehung eines Individuums war daher zur Gänze ausgeschlossen, wie alle Aristoteliker von Marsilius von Inghen bis zu Agostino Nifo in ihren Kommentaren unterstreichen konnten.[7]

Dass die Theologie diesen Satz nicht unterschrieb, lag auf der Hand, denn die Auferstehung des Fleisches zählte seit den Kirchenvätern zu den Kernbeständen der Dogmatik. Schon beim Propheten Ezechiel war von den Knochen die Rede

---

4 Aristoteles, *De generatione et corruptione*, ed. M. Rashed (Paris: 2005), lib. II, 337b–338b (S. 82–84), und ders., *Catégories*, ed. R. Bodéus (Paris: 2001), cap. 10, 12b–13a, S. 56–58.

5 Toletanus Franciscus, *Commentaria una cum quaestionibus in universam logicam Aristotelis* (Köln, Arnold Mylius: 1639), In Categorias, cap. 11, 179–181, ders., *Commentaria cum quaestionibus in libros De generatione et corruptione* (Köln, Hermann Mylius: 1615), lib. II, q. 13, S. 338–340.

6 Philoponus Johannes, *In Aristotelis libros De generatione et corruptione*, ed. H. Vitelli (Berlin: 1897), lib. II, S. 313, und in der lateinischen Übersetzung ders., *Commentaria in libros De generatione et corruptione*, übersetzt von Hieronymus Bagolinus (Venedig, Hieronymus Scotus: 1558; ND Stuttgart – Bad Cannstatt: 2004), lib. II, S. 136 f.

7 Nifo Agostino, *In libros Aristotelis De Generatione et Corruptione interpretationes et commentaria* (Venedig, Hieronymus Scotus: 1557), lib. II, fol. 102r–104r; Aegidius Romanus, *In Aristotelis libros De generatione et corruptione commentaria* (Venedig, Hieronymus Scotus: 1567), lib. II, S. 102–105, und ibid., Marsilius von Inghen, *Quaestiones super libris de generatione et corruptione*, lib. II, q. 20, S. 279–281. Thomas von Aquin, *In libros De generatione et corruptione*, in *Opera omnia*, vol. XXIII (Paris: 1875), lib. II, Lectio XII, geht auf diese Frage an dieser Stelle nicht weiter ein.

gewesen, die aus den Gräbern erstehen sollten,[8] das Evangelium hatte diese Verse aufgegriffen, Kirchenväter wie Tertullian, Augustinus, Johannes Chrysostomos oder Gregor von Nyssa in ihren Predigten, Trakaten und Bibelkommentaren hatten die Auferstehung des Leibes als unabdingbaren Gehalt des Glaubens gegen alle Platoniker und Gnostiker verteidigt.[9] Glücklicherweise war Gott allmächtig und nicht an die Gesetze der Natur gebunden, geschweige denn an die Vorgaben des Aristoteles. Für die aristotelisch geprägte, thomistische Theologie musste nach der Lektüre des großen Philosophen jedoch feststehen, dass die Auferstehung des Menschen kein natürlicher Vorgang sein konnte, sondern die Gesetze der Physik weit hinter sich lassen mußte. Nur Gott in seiner Gnade konnte die Wiederherstellung eines einzelnen Körpers ermöglichen.

Wie hatte man sich als Katholik des Mittelalters oder der Frühen Neuzeit, aber auch als Lutheraner des 16. oder 17. Jahrhunderts diese Auferstehung des Fleisches vorzustellen?[10] Hier hilft die Konsultation der großen Summenkommentare, allen voran des Francesco Suarez, die sich in diesen Fragen vergleichsweise einig waren und die gesamten vorausgegangenen Diskussionen summieren konnten. Ihre Argumentation war trocken, doch so stichhaltig, dass sie auch von Lutheranern, wenn auch gern ohne Nennung der

---

8 Ezechiel 37, 1–14.

9 Tertullian, *De resurrectione mortuorum*, ed. J.G.Ph. Barleffs (CCSL 2) (Turnhout: 1954), cap. 30–32, S. 959–963; Gregor von Nyssa, *De anima et resurrctione dialogus quo inscribitur Macrinia* (MPG 46) (Turnhout: o. J.), Sp. 137–160; Johannes Chrysostomos, *Homilia de resurrectione mortuorum* (MPG 50) (Paris: 1862), Sp. 427–432; Augustinus, *De Genesi in litteram*, ed. Joseph Zycha (CSEL 28), 2 vols. (Prag – Wien: 1894), lib. XII, § 35, 432 f.; ders., *De civitate Dei libri XXII*, ed B. Dombart – A. Kalb, 2 vols. (CCSL 47/48) (Turnhout: 1955), lib. XX, cap. 19–21, S. 837–842.

10 Zu den Fragen der postmortalen Identität und Personalität und ihrer Diskussion im Hochmittelalter die Überblicke von Weber H.J., *Die Lehre von der Auferstehung der Toten in den Hauptraktaten der scholastischen Theologie von Alexander von Hales zu Duns Scotus* (Freiburg: 1973) 125–158; 217–253; Heinzmann R., *Die Unsterblichkeit der Seele und die Auferstehung des Leibes. Eine problemgeschichtliche Untersuchung der frühscholastischen Sentenzen- und Summenliteratur von Anselm von Laon bis Wilhelm von Auxerre* (Münster: 1965) 6–146; Walker Bynum C., *The Resurrection of the Body in Western Christianity, 200–1338* (New York: 1995) 229–278 oder Solère J.-L., „Was the eye in the tomb? On the metaphysical and historical interest of some strange quodlibetal questions“, in Schabel C. (ed.), *Theological Quodlibeta in the Middle Ages. The Thirteenth Century* (Leiden – Boston: 2006) (507–558) 510–526. Eine Zusammenschau der frühneuzeitlichen Debatten existiert noch nicht. Als sehr wertvolle Diskussion eines Teilproblems jedoch bald Sander C. „For Christ's Sake: Pious Notions of the Human and Animal Body in Early Jesuit Philosophy and Theology", in: Lo Presti R. – Buchenau S. (eds.), From *‚Animal' to ‚Man'? Human and Animal Perception in Early Modern Philosophical and Medical Anthropology (16th–18th century)* (Pittsburgh: 2015, im Druck).

Quelle repetiert wurde. Es war die thomistische Position, die hier überwiegend zu lesen war.

Die menschliche Seele konnte als Substanz nicht vergehen, doch bedurfte sie, so Suarez, im Augenblick der Auferstehung derselben Materie, um als Person im Jüngsten Gericht wiederhergestellt zu werden.[11] Ein Kritiker des Thomas von Aquin, Durandus von San Porciano, hatte Ende des 13. Jahrhunderts zwar behauptet, dass die bloße Form der Seele als Garant der Identität ausreichen musste, weil die Materie nie mehr als eine Möglichkeit sein konnte, doch hatte sich diese Lehre im späteren Mittelalter niemand mehr zueigen machen wollen.[12] Thomas von Aquin in seiner *Summa contra gentiles*, aber auch Thomisten wie Johannes Capreolus oder Petrus Paludanus hatten das Verhältnis von Seele und Körpermaterie eindeutig bestimmen können.[13] Die Seele hatte ihren Leib, den sie seit dem Moment der Zeugung ausbilden durfte, geformt und belebt; er war ihr ebenso zum Instrument der Sünde wie des Verdienstes geworden. Sie blieb auch nach dem Tod in ihrem ganzen Verlangen auf diesen einen Körper hingeordnet und musste ihn daher, wie Thomas betont, wieder zurückerhalten. Hätte man ihr ein neues Fleisch gewährt, wäre die Person des Menschen verschwunden und eine Seelenwanderung hätte stattgefunden.[14] Identität und Verantwortlichkeit jedes Einzelnen, der

---

11 Suarez Francisco, *Commentaria ac disputationes in Tertiam partem Divi Thomae*, in *Opera omnia*, ed. C. Berton, 26 vols. (Paris: 1856–1878), vol. XIX, Disputatio 44, Sectio I, §§ 1–4, S. 744 f.

12 Durandus a Sancto Porciano, *Scriptum super IV libros Sententiarum. Buch IV, Dist. 43–50*, ed. T. Jeschke (Leuven – Paris: 2012), Distinctio 44, q. 1, S. 67–79, und auch in der alten Ausgabe Durandus a Sancto Porciano, *In Sententias theologicas Comentariorum libri IIII* (Venedig: 1571) (Reprint Farnborough: 1964) 2 vols., vol. II, In librum quartum, Dist. 44, q. 1, fol. 395r–396r.

13 Thomas von Aquin, *Commentum in libros Sententiarum*, in *Opera omnia*, vol. VII–XI (Paris: 1873), lib. IV, Distinctio 44, a. 1, q. 1–3; ders., *Quodlibetales*, in *Opera omnia*, vol. XV (Paris: 1876), Quodlibet IX, q. 6, a. 6, S. 589 f.; ders., *In Aristotelis librum De anima commentarium*, ed. A.M. Pirotta (Turin: 1959), lib. II, Lectio 1, §§ 224–225; ders., *Summa contra gentiles* (*Opera omnia: Editio leonina*, vol. XIII–XIV) (Rom: 1918–1930), lib. IV, cap. 81; Petrus de Palude, *Scriptum in quartum Sententiarum* (Köln, Jean Petit: 1514), Distinctio 43, q. 1, fol. 196r–197v; auch schon Albertus Magnus, *De anima*, in *Opera omnia*, vol. VII, ed. C. Stroick (Münster: 1968), lib. III, Tract. 2, cap. 12, S. 193. Die ausführlichste Rechtfertigung der thomasischen Position gegen ihre Kritiker liefert im Spätmittelalter Johannes Capreolus, *Defensiones theologicae Divi Thomae Aquinatis*, ed. C. Paban – T. Pègues, 7 vols. (Tours: 1908; ND Frankfurt: 1967), vol. VII, lib. IV, Distinctio 43, q. 1, a. 1–2, 1–7; a. 3, S. 14–19; Distinctio 44, q. 1, a. 1–3, S. 51–61.

14 Suarez, *Commentaria ac disputationes in Tertiam partem*, Disputatio 44, Sectio II, §§ 1–6, S. 745–749. Zur philosophischen Widerlegung der Seelenwanderung, die durch Thomas

Unterschied zwischen dem Sünder und dem Heiligen, damit aber auch der Gegenstand des Gerichtes am Jüngsten Tag waren nur durch diese Identität auch des Leibes gerechtfertigt. Wie weit sollte diese Regel reichen? Sollte der Körper eins zu eins und vollkommen deckungsgleich wiederhergestellt werden? Was geschah mit den Deformierten, den siamesischen Zwillingen oder den Menschen, die Kannibalen zum Opfer gefallen war, nach denen schon Augustinus in seinem *Enchiridion* gefragt hatte?[15] Für Thomas bestand kein Zweifel, daß auch Kinder als Erwachsene auferstehen würden und die körperlichen Mängel während der Wiederherstellung des Leibes keine Rolle mehr spielen konnten. Zum Objekt der Auferstehung musste ein Körper werden, der wie ein arithmetisches Mittel die Vollkommenheit der menschlichen Spezies abbildete, zugleich jedoch noch immer die Individualität des Verstorbenen spiegeln konnte.[16] Auch ein zerstückelter oder verbrannter Leichnam konnte die göttliche Allmacht nicht in Verlegenheit bringen.

Tatsächlich wusste man es noch genauer. Das Gesetz der hylemorphistischen Identität verlangte nach einem Kernbestand an Materie, die der Seele als Gegenüber dienen konnte. Gott könnte dieses stoffliche Fundament, die wenigen Brocken, die die Geschichte vielleicht von der Leiche eines Menschen noch bewahrt hatte, wie Suarez im Gefolge seiner mittelalterlichen Autoritäten weiter ausführt, durch die Ausdehnung und Veränderung ihrer Dichte in einen vollständigen Körper zurückverwandeln und gleichsam aufblähen, wie er einst aus der Rippe Adams eine komplette Frau hatte heranbilden können. Wahrscheinlicher war es jedoch, so Suarez, daß der Kernbestand, der individuierende Bausatz, durch zusätzliche Materie wiederaufgefüllt wurde.[17] Die Identität des Leibes konnte nicht nur auf einigen Partikel beruhen, die die Kontinuität des Stoffes herstellten, sie musste sich auch, wie die meisten der mittelalterlichen Philosophen glaubten, in der proportionalen Anordnung des Leibes zum Ausdruck bringen. Um eine vollkommene Auferstehung herbeizuführen benötigte man,

---

kanonisch wurde, George M.I., „Aquinas on Reincarnation", *The Thomist* 60 (1996) 33–52, als Beispiele Thomas von Aquin, *Summa contra gentiles*, lib. IV, cap. 82, oder auch Thomas von Straßburg, *Commentaria in IIII libros Sententiarum* (Venedig: 1564) (ND Farnborough: 1964), lib. IV, Distinctio 44, q. 1, fol. 170va f.

15 Augustinus, *Enchiridion ad Laurentium de fide et spe et caritate*, ed. M.P.J. van den Hout (CCSL 46) (Turnhout: 1969), cap. 23, §§ 84–93, 95–99. Hierzu vor allem auch Durandus a Sancto Porciano, *Scriptum super IV libros Sententiarum*, lib. IV, Distinctio 44, q. 3, S. 86–92.

16 Thomas von Aquin, *Commentum in libros Sententiarum*, lib. IV, Distinctio 44, a. 3, q. 1–4, und Suarez, *Commentaria ac disputationes in Tertiam partem*, Disputatio 44, Sectio II, §§ 8–10, S. 750.

17 Suarez, *Commentaria ac disputationes in Tertiam partem*, Disputatio 44, Sectio II, § 7, S. 749 f.

aus allen Schlüsselbestandteilen des vergangenen Körpers elementare Materialpartikel, aus denen der neue Leib zusammengestellt wurde, auch aus den Armen, Beinen und den Sinnesorganen. Auch das Auge, mit denen die Tränen der gnadengesättigten Zerknirschung geweint worden waren, musste im Idealfall wiederauferstehen und mit neuem Leben gefüllt werden.[18] Suarez selbst gibt zu, dass diese Forderung nach der übergreifenden Identität des Leibes nicht überbeansprucht werden durfte. Eine komplette Rekonstruktion des alten Körpers in seiner gesamten Dauer und Geschichte konnte nicht im Interesse Gottes liegen. Dennoch war ein Kernbestand an Akzidentien erforderlich, der das alte Leben mit dem neuen verbinden konnte und die Seele im Moment ihrer Vereinigung mit dem alten Leib mit Bekanntem konfrontierte. Selbst wenn diese Körperteile, also das Gesicht, Nase, Augen, oder die Extremitäten nicht mehr numerisch identisch waren, also die gleiche Materie hatten, wie vor dem Tod, mussten sie in der Spezies ihren Vorgängern gleichen.[19] Ausgeschlossen war es zum Beispiel, dass ein Mann sein neues Leben in einem Frauenkörper in Angriff nehmen musste, auch wenn niemand mehr nach dem Tode für seine Geschlechtsteile noch Verwendung hatte, ja alles Begehren zur Gänze zum Erliegen gekommen war. Die Vollkommenheit der menschlichen Natur als ganzer verlangte es, dass die Varietät des Menschen im Moment der Auferstehung neu zur Geltung gebracht wurde.[20] Allen platonisierenden Ideen, einem Urmenschen, der wieder beide Geschlechter besaß, oder der Aufhebung der geschlechtlichen Unterschiede, die noch Johannes Scotus Eriugena gelehrt hatte, war damit eine eindeutige Absage erteilt worden.[21]

Vor dem Jüngsten Gericht sorgten die Heerscharen der Engel in einer gewaltigen konzertierten Aktion, wie man glaubte, für die Bereitstellung der Aschepartikel und der Knochenreste, die noch vorhanden waren, um die Menschen angemessen zu präparieren. Die scholastischen Philosophen waren auch in der Lage gewesen, den Modus der Auferstehung näher zu bestimmen. Geschah sie in einem einzigen Augenblick oder handelte es sich um einen zeitlich ausgedehnten Vorgang? Sollte sie eine, wenn auch nur kurze Zeitspanne in

18 Thomas von Aquin, *Commentum in libros Sententiarum*, lib. IV, distinctio 44, a. 2, q. 1–5; Petrus de Palude, *Scriptum in quartum Sententiarum*, Distinctio 44, q. 1, fol. 203–204r, und Wilhelm von Ockham, *Quaestiones in librum quartum Sententiarum* (*Reportatio*), ed. Rega Wood – Gedeon Gál (St. Bonaventure: 1984), Distinctio 44, q. 13, S. 257–277.

19 Suarez, *Commentaria ac disputationes in Tertiam partem*, Disputatio 44, Sectio III, §§ 1–2, S. 752 f.

20 Ebd., *Commentaria ac disputationes in Tertiam partem*, Disputatio 44, Sectio III, §§ 3–4, S. 753 f.

21 Johannes Scotus Eriugena, *Periphyseon. De divisione naturae*, ed. I.P. Sheldon Williams – É.A. Jeauneau, 4 vols. (Dublin: 1978–1995), vol. II, lib. II, lateinisch und englisch, S. 20–33.

Anspruch nehmen?[22] Gott war für die Auferstehung verantwortlich. Wer den Urmenschen in einem einzigen Moment hatte schaffen können, hätte sicher auch, wie Suarez konzediert, in der Lage sein müssen, ihm in einem einzigen Moment wieder zu einer vollständigen Existenz zu verhelfen; alle Ingredientien dieser Transformation hätten die Engel zudem schon vorher bereitlegen können. Die Ausdehnung und räumliche Modifikation der Ausgangsmaterie, also der noch vorhandenen Knochen, Eingeweide, Knorpel oder Hautfetzen, war dennoch nur als zeitgebundener Vorgang denkbar. Zumindest mittelbar musste die Heranbildung des neuen Leibes also für den Menschen, der wieder zu Bewusstsein gekommen war, auch ein Erlebnispotential besitzen. Seine Ausformung würde sich zwar in ihrer Geschwindigkeit, nicht jedoch in ihrem Charakter, wie Suarez zu wissen meint, von der natürlichen Generierung eines menschlichen Körpers unterscheiden. Auch der neue Leib musste wachsen, an Größe und Gestalt gewinnen, doch war es Gott, der dieses kaum fassbare Wachstum herbeiführte.[23]

Mit dieser Feststellung verband sich für die Theologen des Mittelalters und der Frühen Neuzeit eine entscheidende Frage, die, wie schon gesehen, die Mehrzahl der Scholastiker abschlägig beantwortet hatte. Handelte es sich bei der Auferstehung um einen natürlichen Vorgang? Daß der neue Körper in seinem Wesen der alte sein musste und der geschaffene Status der Seele diesen Körper einforderte, stand außer Zweifel. Auch dass die Seele ihren alten Körper wieder auf natürliche Beweise bewegen und auch die alten Körperfunktionen, bis zu einem gewissen Grad, wieder mit Leben erfüllen konnte, musste zunächst zu denken geben.[24] Dennoch existierte keine natürliche Ursache, wie Thomas oder Thomisten wie Johannes von Neapel deutlich machen, die eine Neubelebung des Leibes veranlassen und den alten Zusammenhang von Form und Materie wieder hätte herbeiführen können; geschweige denn, daß irgendeine natürliche Ursache die Materie selbst in ihrem numerischen Fortbestand garantiert hätte.[25] Gott selbst allenfalls konnte dafür sorgen, dass

22 Suarez, *Commentaria ac disputationes in Tertiam partem*, Disputatio 44, Sectio IV, §§ 1–2, S. 755. Ausführlich zur Frage, ob die Auferstehung *in instanti* erfolgte oder eine *mutatio successiva* war, vor allem Petrus de Palude, *Scriptum in quartum Sententiarum*, Distinctio 43, q. 4, fol. 202r–203v, und Johannes Capreolus, *Defensiones theologicae*, vol. VII, Distinctio 43, q. 2, a 1–3, S. 29–51.

23 Suarez, *Commentaria ac disputationes in Tertiam partem*, Disputatio 44, Sectio IV, §§ 3–6, S. 755 f.

24 Ebd., *Commentaria ac disputationes in Tertiam partem*, Disputatio 44, Sectio V, §§ 1–4, S. 756–758.

25 Thomas von Aquin, *Commentum in libros Sententiarum*, lib. IV, distinctio 44, a. 1, q. 3; Suarez, *Commentaria ac disputationes in Tertiam partem*, Disputatio 44, Sectio V, § 5, S. 758, und die in Fußnote 13 genannten Autoritäten.

ein vollkommen auf seine Seele abgestimmter Körper über alle Epochen hinweg auf sein neues Leben wartete, doch hätte diese Bereitschaft eine Fülle von Folgewundern nach sich gezogen. Eine verfaulte Leiche wiederum hatte die Merkmale des alten Leibes verloren und konnte ihre Seele nicht mehr auf natürlichem Wege einfordern, wie es einst beim Embryo im Mutterleib noch der Fall gewesen war. Körper und Seele, Wirklichkeit und Möglichkeit, konnten nicht mehr miteinander in Einklang gebracht werden.[26] Die Schlussfolgerung war evident: Der Zusammenhang von Leib und Seele wurde erst im Moment des Endgerichtes wiederhergestellt; ausschließlich Gott und seine Gnade und Wunderkraft waren für diese Vereinigung verantwortlich.[27]

1.2 *Optionen einer natürlichen Auferstehung in der scholastischen Philosophie*

In den Ausführungen des Aquinaten hatte der scheinbar gesunde theologische Menschenverstand gesprochen. Kein Weg jedoch führt uns von diesen Beweisführungen zur auferstandenen Rose, geschweige denn zu einer natürlichen Wiederherstellung des Menschen. Tatsächlich jedoch gab es mittelalterliche Philosophen, die das Verhältnis von Seele und Körper im postmortalen Status anders definiert hatten als die genuinen Anhänger des Aristoteles. Wir finden sie vor allem im Kreis der Franziskaner, unter jenen Denkern, die sich fast gewohnheitsmäßig darum bemühten, der thomistischen Mehrheitsmeinung Widerstand zu leisten. Welche Gründe hatte man in Wirklichkeit, so fragt Duns Scotus in seiner Diskussion der Auferstehung, um die Vereinigung von Leib und Seele nach dem Tod als wunderbares Ereignis zu beschreiben? Im Leib des auferstandenen Menschen erfüllte die Seele die Aufgabe, die sie auch vor dem Tod erfüllt hatte; sie formte ihren Körper und diente ihm als Lebensprinzip.[28] Traf sie auf ihr natürliches Gegenüber, den auf sie vorbereiteten Leib, wie auch Marsilius von Inghen, Heinrich von Gent

26 Suarez, *Commentaria ac disputationes in Tertiam partem*, Disputatio 44, Sectio VI, §§ 5–9, S. 760–764.

27 Ebd., *Commentaria ac disputationes in Tertiam partem*, Disputatio 44, Sectio VII, §§ 1–5, S. 764–767, Sectio VIII, §§ 1–2, S. 768 f.

28 Duns Scotus Johannes, *Quaestiones in quartum librum Sententiarum a distinctione 43 usque ad 48*, in *Opera omnia*, ed. L. Wadding, 24 vols. (Paris: 1891–94), vol. XX, lib. IV, Distinctio 43, q. 3, S. 103 f., S. 106–108; q. 4, S. 112–115, ders., *Reportatio parisiensia. lib. IV a distinctione 7 usque ad 49*, in *Opera omnia*, vol. XXIV, lib. IV, Distinctio 43, q. 3–4, S. 508–523, und aus der Perspektive der Naturphilosophie ders., *Quaestiones super libros Metaphysicorum Aristotelis* (Opera philosophica 3–4), ed. R. Andrews – G. Etzkorn, 2 vols. (St. Bonaventure 1997), lib. VII, q. 12, S. 195–214; ders., *In libros physicorum commentaria*, (Opera omnia, vols. II–III), vol. II, lib. II, q. 4, S. 521–525.

oder Richard von Middleton deutlich machen, so bewegte sie den Leichnam als Ergebnis eines *impetus naturalis*, eines zur Gänze natürlichen Triebs, der keiner göttlichen Hilfe bedurfte. Warum sollte diese Situation nicht wiederholbar sein?[29] Wenn Akteur und Empfänger identisch waren, musste dasselbe Ergebnis die Folge sein. Nicht anders als die Seele würde sich der Same des Petrus verhalten, wenn er wieder auf dasselbe *menstruum*, den weiblichen Lebenssaft stieß. Wieder müsste derselbe Petrus ins Leben treten, wieder läge das dasselbe Individuum, und nicht nur der gleiche Mensch vor.[30]

Selbst wenn ein Engel also der Seele die entsprechende Körpermaterie beschaffte und sie, solange sie noch für sich existierte, mit ihrem alten auf sie abgestimmten Leib konfrontierte, war die Selbstverständlichkeit, mit der die Seele ihr altes Tagewerk wiederaufnahm und sich ihren Körper wieder zu eigen machte, ein Naturereignis, wie Scotus und seine Anhänger deutlich machen, wenn es auch ein in seiner Exklusivität unglaublicher Vorgang sein mußte. Individuen waren also, wie Scotus gezeigt hatte, der Natur nach zumindest theoretisch einer Auferstehung fähig, ohne dass Gott ein Wunder bereitstellen musste. Der Schöpfer hatte lediglich die entsprechenden Rahmenbedingungen zu liefern, um dann der natürlichen Kausalität ihren Lauf zu lassen. Leib und Seele fanden von alleine zueinander. Dass die Verbindung von Form und Materie dabei zeitweilig unterbrochen war, musste die numerische Identität, also die exakte Wiederherstellung des Einzelwesens, nicht in Frage stellen. Was immer in einem Medium ohne Unterbrechung produziert werden konnte, veränderte sich in seiner Identität nicht, wie Scotus betont, wenn seine Hervorbringung eine Zeitlang unterbrochen wurde. Ein Sonnenstrahl blieb der derselbe Sonnenstrahl, unabhängig davon, ob die Sonne durchgehend schien, oder eine Wolke den Strahl für einen Moment verschwinden ließ.[31] Die Zeit war nicht mehr als ein Akzidenz, eine nichtnotwen-

---

29 Bonaventura, *Commentaria in quatuor libros Sententiarum* (Opera omnia, vols. I–IV) (Quaracchi: 1889), vol. IV, Dist. 43, a. 1, q. 4, S. 887–890; Richardus de Mediavilla, *Commentum in libros Sententiarum*, 4 vols. (Brixen: 1591; ND Frankfurt: 1963), vol. IV, Dist. 43, a. 1, q. 2, S. 555–556; Marsilius von Inghen, *Quaestiones super quattuor libros sententiarum* (Straßburg, Martinus Flach: 1501; ND Frankfurt: 1966), lib. IV, q. 13, a. 1, fol. 588r–589v; Heinrich von Gent, *Quodlibet VII*, ed. G.A. Wilson (Leuven: 1991), q. 16, S. 109–115; Suarez, *Commentaria ac disputationes in Tertiam partem*, Disputatio 44, Sectio V, § 7, S. 759.

30 Ausführlich hierzu Wilhelm von Ockham, *Quaestiones in librum quartum Sententiarum*, Distinctio 44, q. 12, S. 239–256, bes. 248–250, und auch Suarez, *Commentaria ac disputationes in Tertiam partem*, Disputatio 44, Sectio VI, § 3, S. 760 f.

31 Duns Scotus Johannes, *Quaestiones in quartum librum Sententiarum*, lib. IV, Distinctio 43, q. 3, S. 71–74; Suarez, *Commentaria ac disputationes in Tertiam partem*, Disputatio 44, Sectio VI, § 8, S. 762 f.

dige Begleiterscheinung. Scotisten der Frühen Neuzeit wie Johannes von Rada repetierten diese These ihres Meisters und arbeiten sie weiter aus, auch wenn sie dafür von ihren Gegnern unter den Thomisten aufs heftigste gescholten wurden und man sie der üblen Phantasterei oder schlimmer noch der mangelhaften Logik bezichtigte.[32]

Gab es also ein inhärentes organisches Gesetz, eine Schwerkraft von Leib und Seele, die am Ende gleichsam physikalisch die Wiederherstellung der Individuen zur Folge haben konnte? Wohl kaum zufällig hatten sich die scholastischen Philosophen über die letzten konkreten körperlichen Details der Auferstehung weitgehend ausgeschwiegen. Von den physiko-theologischen Spekulationen der mittelalterlichen Scotisten war es dennoch nur ein kleiner Schritt zur nächsten, entscheidenden Hypothese. Was wäre, wenn niemand den dynamischen Zusammenhalt von Form und Materie, damit aber auch von Leib und Seele, nach dem Tode unterbrechen könnte, sondern beide miteinander verbunden blieben? Welche Schlußfolgerungen ergäben sich, wenn sich Leib und Seele lediglich in einem temporären Ruhezustand befänden? Was wäre also, wenn mit anderen Worten nicht nur Pflanzen nach diversen Veränderungen ihre Seele nicht vollständig verlieren würden, sondern auch die menschliche Seele ihren Leib bis zum Endgericht noch auf eine eingeschränkte Art und Weise besitzen würde? Wenn also die Identität eines Individuums über den Tod hinaus dadurch gesichert wäre, daß beide Komplementäre, Form und Materie, durchgehend eine Einheit formen würden?

II Paracelsus und die Palingenesie

Wenden wir uns jetzt von der Theorie ab und der Anwendung zu. Schon Denker der Schulphilosophie des 16. und 17. Jahrhunderts, die sich mit der Frage der Palingenesie, der Wiederherstellung von Individuen beschäftigten, hatten erkannt, daß die Anhänger des Paracelsus ein Modell auf praktische Weise zuende denken sollten, das ihnen die spekulative Theologie des Mittelalters

32 de Rada Johannes, *Controversiarum theologicarum inter S. Thomam et Scotum libri quatuor*, 4 vols. (Köln, Johannes Crithius: 1619–1620), vol. IV, Super quartum Sententiarum, Controversia XV, a. 3–6, S. 359–373; oder Tartaretus Petrus, *Lucidissima commentaria, sive (ut vocant) Reportata in libros Sententiarum Scoti*, 4 vols. (Venedig, Simon de Gagliani de Carera: 1633), vol. IV, In quartum librum, Distinctio 43, q. 3, S. 389–394. Eine ähnliche, scotistische Position vertritt z.B. auch der Jesuit Pellegrini Martino, *Resurrectio corporum humanorum probata rationibus naturalibus* (Rom, Jacob Dragondelli: 1674), cap. 21–22, S. 125–132.

als Modell angeboten hatte. Gelehrten wie Caspar Posner, der zu dieser Frage in Jena im Jahre 1686 hatte disputieren lassen, aber auch anderen mitteldeutsche Physikern, die mit alten wie neuen Denkstilen vertraut waren, war nicht entgangen, daß die Grenze zwischen den Epochen weniger klar verliefen, als wir sie vielleicht heute ziehen würden.[33] Es wird Zeit für Paracelsus, selbst die Bühne zu betreten. Sehen wir, was die experimentelle Alchemie der frühen Neuzeit aus den Vorgaben des Spätmittelalters hatte machen können.[34]

II.1 *Phoenix botanicus: die Pflanze aus der Asche*

Auch Paracelsus hatte seinen Aristoteles gelesen und sicher auch Thomas, doch war er, was die Gesetze der Materie betraf, zu anderen Schlussfolgerungen gelangt. Es kann hier nicht sinnvoll sein, alle Kernsätze der paracelsischen Physik zu referieren, so dass sich die weitere Darstellung auf einige Schlüsselgedanken beschränken muss. In seiner Schrift *Von den natürlichen Dingen* hatte Paracelsus im sechsten Kapitel eine folgenschwere These aufgestellt. In den natürlichen Substanzen, vor allem in den Pflanzen, blieb die Form in ihrer Materie, auch wenn die Pflanze längst in einen anderen Aggregatzustand übergegangen war; sie verharrte auch in jenen Resten, die auf den ersten Blick keine Eigenschaft mehr mit der alten Pflanze teilen konnten. Selbst eine Verbrennung konnte die völlige Abtrennung des Lebensprinzips nicht herbeiführen; vor allem wenn der Tod vor dem vorbestimmten Todesdatum der Kreatur und durch Gewalt herbeigeführt worden war. Im Holz befanden sich, wie Paracelsus glaubte, Phlegma, *Feiste*, also Fett, und Asche, die den drei Elementen Quecksilber,

33 Posner Caspar – B.D. Meyring (resp.), *Dissertatio physica de Palingenesia sive reditu corruptorum et speciatim quoque resurrectione mortuorum naturae impossibli, secundum sententiam Aristotelis* (Jena, Johannes Zacharias Nisus: 1686), §§ 18–38, dort zu Scotus und den Paracelsisten fol. C2v–Fv.

34 Den bisher einzigen wirklichen Beitrag zur Idee der chemischen Wiederherstellung im 17. Jahrhundert liefert Telle J., „Chymische Pflanzen in der deutschen Literatur", *Medizinhistorisches Journal* 8 (1973) 1–34. Zur Idee der Palingenesie in der Alchemie, vor allem mit Blick auf die späte Rezeption im 19. Jahrhundert Marx J., „Alchimie et Palingénésie", *Isis* 62 (1971) 274–289. Einen anderen Aspekt, die Verlängerung des menschlichen Lebens durch die Alchemie, behandelt der Beitrag von Secret F., „Palingenesis, Alchemy and Metempsychosis in Renaissance Medicine", *Ambix* 26 (1979) 81–92. Lediglich Leibniz, Herder und Bonnet, die außerhalb dieser Studie liegen, referiert Unger R., „Zur Geschichte des Palingenesiegedankens im 18. Jahrhundert", *Deutsche Vierteljahrsschrift für Literaturwissenschaft und Geistesgeschichte* 2 (1924) 257–274. Verflechtungen mit dem Mittelalter sind zur Gänze unarbeitet geblieben.

Schwefel und Salz entsprachen.[35] Verbrannte man eine Pflanze, blieb ihre Asche, das Salz, zurück. Gab man *spiritus* und andere Ingredienzien zum Salz und versah es mit Licht, konnte das Holz durch die Form, die ihm immer noch innewohnte, wieder zum Leben erweckt werden.[36] Die Ausgangsbedingungen der paracelsischen Physik mochten absurd anmuten, doch konnten sie die Antwort auf die Frage nach der natürlichen Auferstehung geben, die man in weiten Teilen gesucht hatte. Anders als die Aristoteliker behauptet hatten, war das Leben nie aus den Organen eines Körpers verschwunden. Es war, wie man stattdessen glauben sollte, lediglich in einen Wartemodus versetzt worden, der mit einigen Kunstgriffen wieder aktiviert werden konnte.

Eine ganze Generation von Gelehrten machte sich im beginnenden 17. Jahrhundert daran, die natürliche Auferstehung, die Palingenesie, durch Experimente zu untermauern und die Fehlerhaftigkeit der aristotelischen Ontologie unter Beweis zu stellen. Die Mehrzahl dieser Wissenschaftler gab sich mit Pflanzen zufrieden. Joseph DuChesne, Pariser Leibarzt des Henri Quatre, schenkt uns in seiner *Verteidigung der hermetischen Philosophie* jenes berühmte Exempel, das Borges direkt dem Paracelsus zugeschrieben hatte.[37] Ein polnischer Kollege war in der Lage gewesen, wie DuChesne zu berichten weiß, die Asche einer Rose, die er in einer Phiole aufbewahrt und mit einer Reihe von Extrakten versehen hatte, im Licht einer besonderen Lampe ins Leben zurückzubringen. Zitternd und gleich einem Bienenschwarm geballt, hatte sich die Aschewolke langsam, wie es hieß und wie es zugleich viele

35 Zur paracelsischen Dreiteilung des Menschen und der Rolle des *spiritus* als Medium unter vielen Arbeiten z.B. Blümlein K., *Naturerfahrung und Welterkenntnis. Der Beitrag des Paracelsus zur Entwicklung des neuzeitlichen, naturwissenschaftlichen Denkens* (Frankfurt – Bern: 1992) 99–107; 256–259; Weeks A., *Paracelsus. Speculative Theory and the Crisis of the Early Reformation* (New York: 1997) 109–111; 117 f.; 157 f.; Pagel W., *Das medizinische Weltbild des Paracelsus, seine Zusammenhänge mit Neuplatonismus und Gnosis* (Wiesbaden: 1962) 46–52; 105–108; Daems W.F., „‚Sal-Merkur-Sulfur' bei Paracelsus und das ‚Buch von der Heiligen Dreifaltigkeit'", in *Nova Acta Paracelsica* (1982) 189–205; Kämmerer E.W., *Das Leib-Seele-Problem bei Paracelsus und einigen Autoren des 17. Jahrhunderts* (Wiesbaden: 1971) 3–41.

36 Philippus Theophrastus von Hohenheim, *De natura rerum: Neun Bücher* (Straßburg, Bernhard Jobin: 1584), lib. VI, fol. 36v–43r. Kurz geht Paracelsus auch auf die Unvergänglichkeit der Rose ein, z.B. ders., *Archidoxa* (Straßburg, Theodor Rihel: 1570), lib. IV, fol. Iv f., und lateinisch ders., *Achodoxae libri X* (Krakau, Wirzbieta: 1569), lib. IV, S. 89 f. Eine Zusammenfassung gibt auch Dorn Gérard, *Fasciculus Paracelsicae medicinae veteris et non novae per flosculos chymicos et medicos* (Frankfurt am Main, Sigismund Feyerabend: 1581), cap. 14, S. 9 f.

37 Kurz zu den Experimenten DuChesnes auch Telle, „Chymische Pflanzen" 5–7.

Zeugen gesehen haben wollten, zu ihrer alten Größe emporgehoben und sich erneut zur Gestalt einer Rose verdichtet, wenn auch ohne ihre einstige rote Farbe. Nahm man das äscherne Gewächs aus dem Lichtkegel, fiel es wieder in sich zusammen.[38] Andere europäische Gelehrte imitieren, wie man lesen kann, den Mann aus Krakau, darunter Petrus Servius oder Ole Borch, die am Hofe des dänischen Regenten ihre eigenen Rosen generieren.[39] Einem Gelehrten aus DuChesnes Bekanntenkreis gelingt ein vergleichbarer Kunstgriff mit einem Strauß verbrannter Nesseln, die er durch besonderes Licht und die Hinzugabe von *spiritus* wieder abbildhaft in seinen alten Zustands zurückversetzt.[40] Andreas Libavius, ein weiterer prominenter Paracelsist, schildert, wie der Zwickauer Mediziner Johannes Cornarius aus Kulmbach aus den Überresten eines Rosmarin-Krautes ein Extrakt destilliert, aus dem er fast ein halbes Jahr später dasselbe Pflänzchen, wenn auch nur in Miniaturform, zum Erblühen bringt. Es hatten hier sicher keine *praestigiae* vorgelegen, wie Libavius unterstreicht, sondern der im *sal plantalis* virulente Same war wieder wirksam geworden.[41] Der englische Anatom Nathaniel Highmore beschreibt, wie er aus den Aschepartikeln, den *atomi spirituales*, eines Veilchens und den Resten einer Pappel, nachdem er ihre Asche eingefroren hatte, im Eise wieder neue blütenartige Strukturen hatte hervorbringen können. Keine Saat wurde hier ausgegeben, kein Same eines neuen Exemplars zum Wachsen gebracht, sondern ein und derselben Pflanze, wie Highmore betont, hatte man neues Leben

38 DuChesne Joseph, *Ad veritatem hermeticae medicinae ex Hippocraticis veterumque decretis ac therapeusi necnon vivae rerum anatomiae exegesi responsio* (Frankfurt, Wolfgang Richter: 1605), cap. 23, S. 231 f. Kurz zu ähnlichen Praktiken auch DuChesne Joseph, *Tetras gravissimorum totius capitis affectuum* (Marburg, Paul Egenolph: 1606), cap. 29, S. 350 f.

39 Bartholin Thomas, *Acta Medica et Philosophica Hafniensia*, 5 vols. (Kopenhagen, Peter Haubold: 1673–1680), vol. I, § 42, S. 78 f.; Servius Petrus, *Dissertatio de unguento armario sive de naturae artisque miraculis* (Rom, Dominicus Marcianus: 1643), § 67, S. 70. Ein vergleichbares Experiment, wenn auch nur in einem Glas, schildert noch im Jahre 1726 Bauer J.F. „Regeneratio rosarum rubrarum spontanea in aceto rosarum", *Acta physico-medica Academiae Caesareae Leopoldino-Carolino naturae curiosorum exhibentia Ephemerides* 1 (1727), Observatio 219, S. 484–490.

40 DuChesne, *Ad veritatem hermeticae medicinae responsio*, cap. 23, S. 233f., und ders., *Pestis Alexicus, sive Luis pestiferae fuga, auxiliaribus selectorum utriusque medicinae remediorum copiis procurata* (Leipzig, Thomas Schurer und Bartholomaeus Voigt: 1615), cap. 5, S. 300–302.

41 Libavius Andreas, *Syntagma selectorum undiquaque et perspicue traditorum Alchymiae Arcanorum*, 2 vols. (Frankfurt, Petrus Kopff: 1613–1615), vol. I, lib. I, cap. 22, S. 47 f. Gleiches gelingt einem Kollegen mit einer Quitte, dazu Höfer Wolfgang, *Hercules medicus sive locorum communium liber* (Nürnberg, Endter: 1675), lib. VIII, cap. 2, S. 461 f.

abgerungen.[42] Die Form der Pflanze war nicht mit in ihrer Materie vergangen oder hatte sich von ihr getrennt, wie es Thomas Browne in seiner berühmten *Religio medici* formuliert, sondern war in den verbrannten Partikeln geblieben und reaktiviert worden.[43]

Hatten hier nicht Männer wie Joseph DuChesne und Libavius, wie Ezechiel de Castro, ein spanischer Mediziner, offen zur Diskussion stellt, die Lehren des Aristoteles durch die Erfahrungswissenschaft widerlegt? War nicht die *anima vegetativa* von unsterblicher Natur und blieb ihrem Körper als Materie immer verhaftet?[44] Blieb nicht zumindest das *principium generationis* in Gestalt des *spiritus*, so der Wittenberger Paracelsist Daniel Sennert etwas vorsichtiger, in den Pflanzen bestehen, und konnte, wie die Versuche DuChesnes und Libavius gezeigt hatten, zumindest als eine abbildliche Variante des Lebens wieder abgerufen werden?[45] Der Kieler Paracelsist Johann Ludwig Hannemann bejaht diese Fragen mit Nachdruck: Das Leben war unzerstörbar, denn es ruhte als Form in den Elementarteilchen, den *particulae seminificae*, die nur Gott wieder vernichten konnte und die ihre *virtus resuscitatrix* nie verloren.[46] Paracelsus, DuChesnes, aber auch schleswig-holsteinische Lokalhelden wie Jodocus de Haen oder Joel Langelott hatten es in ihren Studien bewiesen,[47] Hannemann

---

42 Highmore Nathaniel, *Corporis humani disquisitio anatomica, in qua sanguinis circulationem prosequutus est* (Den Haag, Samuel Brown: 1651), lib. I, Pars IV, S. 105 f. Laut Mergenius J.C., *Democritus reviviscens, sive vita et philosophia Democriti* (Leiden, Adriaan Wygaerden: 1648), Disputatio II, cap. 2, S. 182–185, lag in vergleichbaren Versuchen zugleich ein Beleg vor, um den Atomismus Demokrits empirisch abzusichern.

43 Browne Thomas, *Religio medici and other Works*, ed. L.C. Martin (Oxford: 1964), lib. I, Sectio 47–48, S. 45 f., und auch lateinisch ders., *Religio medici cum annotationibus* (Straßburg, Johann Friedrich Spoor: 1665), Sectio 47, S. 272 f., mit Nicolaus Moltrenius in seinem Kommentar, Annotata ad Sect. 47, S. 275–277.

44 De Castro Ezechiel, *Ignis lambens historia medica, prolusio physica* (Verona, Franciscus Rubeus: 1642), cap. 4, S. 155–158.

45 Sennert Daniel, *De chymicorum cum Aristotelicis et Galenicis consensu et dissensu liber I* (Wittenberg, Zacharias Schurer: 1619), cap. 10, S. 262 f.

46 Hannemann Johann Ludwig, *Phoenix botanicus ceu diatriba physica curiosa de plantarum ex suis cineribus resuscitatione* (Kiel, Samuel Spörel: 1678), §§ 5–6, fol. A3v–A4v; § 7, fol. A4v f.; § 9, fol. B2r f.; ders., *Nova et accurata Methodus cognoscendi simplicia vegetabilia iuxta triplicem cognitionem 1. grammaticam 2. philosophicam 3. medicam neotericorum philosophorum et medicorum propriisque principiis superstructa* (Kiel, Johann Beumann: 1677), §§ 30–31, S. 58–60; ders. – Eccardus Joannes Michael (resp.), *Triumphus naturae et artis seu dissertatio physica Fridericiana naturae Phoenicem sistens* (Kiel, Barthold Reuther: 1710), Thesis II, 7–9; Thesis 6, S. 28–34.

47 Hannemann, *Phoenix botanicus*, §§ 13–15, fol. B3v–Cv, §§ 18–19, fol. C2r–C4v.

selbst bewerkstelligt die Wiederherstellung eines Rosmarin-Pflänzchens.[48] Athanasius Kircher, der letzte wahre Universalgelehrte, und sein Schüler Caspar Schott liefern im *Mundus subterraneus* und anderen Werken einen ganzen Katalog an weiteren Experimenten.[49] Schon Albertus Magnus war mit ähnlich gelagerten Wiederbelebungsversuchen erfolgreich gewesen, wie Kircher dabei nicht vergisst zu erwähnen, auch wenn sich in den bestehenden Werken des Dominikaners kein Hinweis mehr auf seine Errungenschaften finden ließ.[50] In der Alchemie waren es vor allem der berühmte *Mayenthau*, den Johannes Tacke, einer der vielen weiteren Paracelsisten, die den Autoritäten des frühen 17. Jahrhunderts folgten, in einem eigenen *Eulogium* feiert,[51] oder der Weinstein, das *sal Tartarii*, mit dem sich, wenn man Phillip Sachse in seiner *Ampelographia* Glauben schenkt, die Palingenese herbeiführen ließ.[52] Auch andere Zusätze hatten sich in den Laboratorien des frühen 17. Jahrhunderts

48 Hannemann Johann Ludwig, „Rosmarini resucitatio" in *Misellanea curiosa, sive Ephemeridum medico-physicarum Germanicarum Academiae naturae curiosorum, Decuriae II, Annus I (1682), continens observationes medicas, physicas et chemicas necnon mathematicas* (Nürnberg, Wolfgang Moritz Endter: 1683), Observatio 135, S. 337.

49 Kircher Athanasius, *Mundus subterraneus in XII libros digestus*, 2 vols. (Amsterdam, Johannes Janssonius: 1665–68), vol. II, lib. XII, Sectio 5, cap. 6, S. 431–433; ähnlich auch ders., *Magnes sive De Arte Magnetica opus tripartitum: quo Universa Magnetis Natura, eiusque in omnibus Scientiis et Artibus usus, nova methodo explicatur* (Rom, Blasius Deversin: 1654), lib. III, Pars V, cap. 3, S. 497 f.; Schott Caspar, *Mechanica hydraulico-pneumatica opus bipartitum* (Frankfurt, Johannes Gottfried Schönwetter: 1658), Pars II, Classis I, cap. 6, machina 8, S. 313–315; ders., *Technica curiosa sive mirabilia artis libris XII comprehensa* (Nürnberg, Johannes Andreas Endter: 1664), lib. XI, cap. 16, S. 885–888. Gleiches war auch Jacobus Johannes Wenceslaus Dobrzensky de Negroponte gelungen, dazu die *Nova et amoenior de admirando fontium genio philosophia nova* (Ferrara, Alphonsus de Marestis: 1658), Pars III, Propositio II, Problema II, 83–85. Kurz zu Kircher auch Telle, „Chymische Pflanzen" 11–13.

50 Ausführlich referiert Kirchers Experimente auch Mögling Johann Ludwig – Daßdorf Johann Adam (resp.), *Palingenesia sive resurrectio plantarum, eiusque ad resurrectionem corporum nostram futuram applicatio* (Tübingen, Martin Rommey: 1683), §§ 5–13, 7–13. Mögling verbittet sich zum Ende jeden Vergleich mit der menschlichen Auferstehung.

51 Tacke Johann, *Mysterium resurrectionis rerum sive Phasis III, spei mortalium ad immortalitatem et incoruptibilitatem consecratus* (Frankfurt, Zubrodt: 1673) 60 f.; ders. – Susemihl Paul (resp.) *Ros benedictio coeli, et menstruum mundi, hoc est disputatio physico-chemica de rore* (Giessen, Utzius: 1661), *passim.*

52 Sachse von Lewenhaimb Philipp Jakob, *Ampelographia sive Vitis viniferae eiusque partium consideratio physico-philologico-historico-medico-chymica* (Leipzig, Jacob Trescher: 1661), Pars II, Sectio 9, cap. 5, S. 635–637. Zum *sal Tartarii* zum Beispiel auch Grembs Franz Oswald, *Arbor integra ruinosa hominis, id est: Tractatus medicus theorico-practicus in tres libros divisos* (Frankfurt, Wilhelm Richard Stock: 1671), lib. III, cap. 3, S. 510 f.

bewährt: *Wermuth* konnte wahre Wunder vollbringen, wie die Arbeiten Philipp Müllers und Johannes Horsts gezeigt hatten.[53] Entscheidend für die alchemische Praxis der Frühen Neuzeit war, dass man das noch immer in der Asche einer Pflanze virulente Salz veranlassen konnte, seine Lebensenergie neu zu entfalten, wie Matthias Untzer in seinem entsprechenden Traktat betont.[54]

Durchläuft man diese und zahllose andere Berichte und auch die Disputationen, die der Palingenese vor allem in Mitteldeutschland, in Jena oder Wittenberg, geschenkt werden,[55] erkennt man rasch, wie sehr im schillernden Oberbegriff der natürlichen Auferstehung denkbar heterogene Phänomene in einem neuen Rahmensystem gebündelt werden konnten. Ausdrücklich suchten ihre Anhänger, aber auch ihre Gegner den Anschluss an die Sprache der mittelalterlichen Scholastik und der Schulphilosophie. Sie galt es empirisch herauszufordern, zu verteidigen, oder ihre Maßgaben zu überwinden. Die wenig präzise Terminologie und der orakelnde Tonfall der Paracelsisten gereichten den eigenen Zielen dabei eher zum Vorteil. Komplexe Hydrokulturen, die bei Athanasius Kircher oder Caspar Schott mit raffinierten hydrostatischen Techniken arbeiten sollten, ließen sich unter dem Lemma einer *resurrectio naturalis* ebenso abbuchen, wie schlichte Eisblumen, die nur auf einer guten Mischung aus Frost, Fäulnis und Geduld beruhten. Schließlich fanden sich hier auch jene völlig obskuren Prozeduren, die Gesellenstücke der wahren Weisheit, wie sie uns in Gestalt der wiederhergestellten Rose vorliegen. Niemand war in der Lage, den Funken des Lebens zur Gänze zu tilgen, wie man einhellig glaubte. Schon zum Ende des 16. Jahrhunderts stoßen wir auf erste lutheranische Theologen wie Georg Calixt, die bereit waren, aus den neuen, wie auch immer gearteten wissenschaftlichen Erkenntnissen ihre Schlüsse zu ziehen. Hatte man hier, und Calixt nennt ausdrücklich die ascherne Rose DuChesnes, womöglich einen erfahrungsgesättigten Beweis dafür gefunden, dass auch die Auferstehung des menschlichen Leibes mehr war als nur ein Mirakel und ein Geschenk der göttlichen Gnade?[56] Handelte es sich

53 Müller Philipp, *Miracula et mysteria chymico-medica libris quinque enucleata* (Amsterdam, Janssonius: 1655), cap. 4, S. 68–73; Horst Johann Daniel, *Epistolarum medicinalium decas* (Frankfurt, Serlin und Fickwirth: 1656) 59–61.

54 Untzer Matthias, *Opus chymico-medicum, in quo Anatomia spagirica trium principiorum, nec non corporis humani affectus breviter et perspicue explicantur* (Halle, Melchior Oelschlege: 1635), Tractatus I, cap. 21, Sp. 134–138. Kurz z.B. auch Beguin Jean, *Tyrocinium chymicum* (Genf, Blasius LeMelais: 1659), lib. I, cap. 1, S. 19 f.

55 Als weiteres Beispiel Loescher Martin – Edzardi Esdras Heinrich – Martini Samuel Gottfried (resp.), *Disputatio de corporis humani palingenesia* (Wittenberg, Gerdes: 1722), Disputatio prior, dort 13–15.

56 Calixt Georgius, *De immortalitate animae et resurrectione carnis liber unus* (Helmstedt, Typographeum Calixtinum: 1649), cap. 8, §§ 17–18, S. 92 f.

um ein Naturereignis, das den neuen Körper des Auferstandenen kraft chemischer Gesetze aus dem alten hervorbringen konnte, wie der Glaser den Kolben aus der Schlacke?[57] Der Helmstedter Professor bleibt noch zurückhaltend und wagt es nicht zu beantworten.[58]

II.2 *Der Traum von der natürlichen Wiederherstellung des Fleisches*

Andere Physiker der Zeit gingen noch einen Schritt weiter und ihre Ergebnisse beginnen bizarrer zu werden. Jan Marek Marci von Kronland, ein enger Freund des Athanasius Kircher, möchte im gekochten Extrakt einer Schlange noch immer das Abbild des Reptils, das inhärente Lebensprinzip, erkannt haben,[59] Phillip Sachse das Gleiche im zerkochten Herzen eines Rehs,[60] Daniel Sennert schließlich meint, im Urin eines Mannes, der von einem tollwütigen Hund gebissen wurde, noch immer die organisch-fleischlichen *imagines* des aggressiven Tiers sichten zu können,[61] Warum sollte nicht auch der Phönix der Mythologie ein Vogel gewesen sein, so Petrus Faber, der sich aus dem Seminalprinzip seiner eigenen Asche wieder zum Leben erweckte?[62] Es lässt sich kaum noch bestimmen, was diese Gelehrten in ihren Glaskolben wirklich vor Augen hatten. Martin Kerger, ein Professor aus Wittenberg, nimmt diese und andere Begebenheiten in seinem Traktat über die Hefe zum Anlaß, von zwei Formen des Lebens zu sprechen, einer *vita prima*, die mit dem gewohnten Leben zusammenfiel, und einer *vita media*, einem weitgehend inaktiven Restleben, das allein auf der Fortexistenz der Form beruhte und auch in einem Tierkadaver noch existieren konnte.[63] Für viele Mediziner war diese

---

57 Calixt entnimmt den Glaservergleich aus Wilhelm von Auvergne, *De universo*, in *Opera omnia*, 2 vols. (Paris: 1674; ND Frankfurt: 1963), vol. I, Pars II, cap. 25, S. 730b f. Wilhelm, Pars II, cap. 21, 720a f., läßt allerdings keinen Zweifel daran, daß weder separate Intelligenzen noch die Natur eine Auferstehung herbeiführen können, sondern nur Gott allein.

58 Calixt, *De immortalitate animae*, cap. 8, §§ 22–23, S. 96 f.

59 Kircher, *Mundus subterraneus*, vol. II, lib. XII, Sectio 2, cap. 8, S. 373 f., außerdem Marek Marci von Kronland J. *Philosophia vetus restituta partibus V comprehensa* (Frankfurt, Christian Weidmann: 1676), Pars IV, Sectio I, S. 389–392. Eine ausführliche Beschreibung der Schlangenversuche findet sich bei Francisci Erasmus, *Ost- und West-Indischer, wie auch Sinesischer Lust- und Stats-Garten* (Nürnberg, Endter: 1668), Erster Theil, S. 760–766.

60 Sachse von Lewenhaimb, *Ampelographia*, Pars II, Sectio 9, cap. 5, S. 637.

61 Sennert Daniel, *Institutionum medicinae libri V* (Paris: 1632), lib. II, Pars III, Sectio 2, cap. 4, S. 388 f.

62 Faber Pierre Jean, *Panchymicum seu anatomia totius universi, in quo omnibus, quae in et sub caelo sunt spagyricae tractantur*, in *Opera omnia*, 2 vols. (Frankfurt, Johannes Beyer: 1652), vol. I, lib. III, 324–326.

63 Kerger Martin, *De fermentatione Liber Physico-medicus, cui de inseparabilitate formarum materialium et Vita singularia sunt innexa, omnia perpetuis experimentis firmata* (Wittenberg, Moevius: 1663), Sectio I, cap. 6, S. 48 f., 54–56.

Information von großer Bedeutung: Wenn jedem animalischen Körper noch immer die *flamma vitalis* des organischen Lebens innewohnte und sie auch einem Leichnam noch abgerungen werden konnte, ließen sich aus vielen fleischlichen Überresten eines Tiers belebende Medikamente gewinnen, die dem Menschen unmittelbar zu Gute kamen. Die Flamme des Lebens aus der Asche zu destillieren, so Johann Schröder in seiner *Pharmocopeia*, mußte daher zu den vordringlichen Pflichten des Arztes gezählt werden.[64] Extrakte aus dem Fleisch der Schlange konnten das Leben des Menschen entscheidend verlängern, wie Giovanni Spontone behauptete.[65] Nach den gleichen Maßgaben ließen sich vielleicht, wie andere Kollegen proklamierten, auch aus Würmern und anderen revitalisierbaren Kreaturen wirksame Medikamente gegen den Skorbut und andere Fäulniskrankheiten erstellen.[66]

Kenelm Digby, der immerhin Mitglied der Royal Society war und den man mit großem Recht für einen der einflussreichsten Naturwissenschaftler des 17. Jahrhunderts halten darf, läßt in seinem *Discourse concerning the vegetation of plants* eine ganze Reihe dieser Phänomene Revue passieren und leistet seinen eigenen Beitrag zu ihrer Durchdringung,[67] er kocht eine Gruppe von Hummern in ihrem Sud, destilliert das erforderliche Salz, läßt es faulen und erkennt im Destillat erneut kleine Lebewesen. Digby kann sie sofort als kleine Hummer identifizieren, die aus dem gekochten Hummer wieder ins Leben getreten waren.[68] Für Digby nehmen die Schlussfolgerungen, die sich auch für den Theologen aus diesen Beobachtungen ergeben mussten, fast zwin-

---

64 Schröder Johann, *Pharmocopoeia medico-chymica, sive Thesaurus pharmacologicus, quo composita quaeque celebriora; hinc mineralia, vegetabilia et animalia chymico-medice describuntur* (Ulm, Görlinus: 1655), lib. VI, 265 f. Ähnlich auch Horst Johann Daniel, *Physica Hippocratea, Tackenii, Helmontii, Cartesii, Espagnet, Boylei* (Frankfurt, Friedrich Knoch: 1682), Dissertatio VI, § 4, S. 40 f.

65 Spontone Giovanni Battista, *Conechidnelogia, hoc est pulveris viperini discursus* (Rom: Manelphius, 1648), cap. 13, S. 90–92. Zum Schlangenpulver und seiner heilenden Funktion selbst noch der Comte de LaGaraye Claude Toussaint Marot, *Chymie hydraulique pour extraire les sels exxentiels des végétaux, des animaux et des minéraux, par le moyen de l'eau pure* (Paris, Didot: 1777) 343–350.

66 Moellenbrock Valentin Andreas, *De varis seu arthritide vaga scorbutica tractatu* (Leipzig, Johannes Gross: 1622), cap. 15, S. 275 f.

67 Zu den Experimenten Digbys auch Marx, „Alchimie et Palingenesie" 282 f.

68 Digby Kenelm, *A Discourse, concerning the Vegetation of Plants* (London, John Dakins: 1661) 76 f. Ähnlich zur *restitutio* von Krebsen auch Sachse von Lewenhaimb Philipp Jakob, *Gammarologia sive Gammarorum vulgo cancrorum consideratio physico-philologico-historico-medico-chymica* (Frankfurt, Esaias Fellgibel: 1655), cap. 13, § 6, S. 262 f., und vor allem Pezold Adamus Friedericus, „De Palingenesia", in *Academiae Caesaro-Leopoldinae naturae curiosorum ephemerides sive observationum medico physicarum centuria VII et*

genden Charakter an. Anders als die Aristoteliker und Thomisten geglaubt hatten, musste das Fleisch durchgehend, *in substance and reality,* mit sich selbst identisch bleiben und konnte nicht vergehen. Nur scheinbar löste sich der Leib des Menschen nach dem Tod zur Gänze in seine Bestandteile auf, weil sich die Form der Seele von ihm getrennt hatte. In Wirklichkeit blieben beide Einheiten miteinander verbunden. *By force of nature,* so Digby, konnte sich der Körper aus dem Lebensprinzip wiederherstellen lassen, so wie Wasser und Mehl immer wieder dasselbe Brot ergeben mussten, solange sie zusammenblieben. Durch den Konnex von Form und Materie war auch die Identität des auferstandenen Menschen garantiert; der Tod spielte nur eine sekundäre Rolle.[69] Digby wählt einen Vergleich, den auch Duns Scotus hätte wählen können, einen Springbrunnen, den er im Park in Saint-Germain hatte beobachten können. Gleichgültig, welche Form die Fontäne annahm, sie blieb immer mit ihren Ursprung, dem Wasserrohr, verbunden. Auch wenn das Wasser seine Gestalt veränderte oder die Fontäne fast zur Gänze abgedreht wurde, sie blieb mit sich selbst identisch und als Individuum erhalten.[70]

Es fehlte nicht an Kritik an der Palingenesie-Idee und vor allem die scheinbaren Experimente stießen bei Gelehrten, die nicht dem paracelsischen, sondern dem aristotelischen, oder strenger cartesianischen oder mechanistischen Lager angehörten, auf wenig Gegenliebe. Ein Mann wie Robert Boyle hatte für die Wiederbelebungsphilosophien seiner englischen wie deutschen Kollegen allenfalls Spott übrig; sie alle konnten nur auf Sinnestäuschung und kreativer Einbildung beruhen.[71] Anders als DuChesne behauptet hatte, musste die *forma substantialis* einer Pflanze im Moment ihrer Zerstörung vernichtet worden sein; eine identische Restituierung des Gewächses aus dieser Form war daher, so betonen es Aristoteliker und Kritiker der paracelsischen Medizin wie Anton Billich oder Johannes Freitag, völlig ausgeschlossen und widersprach jeder Naturordnung. Die Essenz einer Pflanze konnte sich nicht in der Asche oder im Salz *sicut in capsulis* verbergen, nur um sich bei Gelegenheit durch ein wenig Wärme wieder zu Tage fördern zu lassen; bei den Experimenten der Paracelsisten konnte es sich also nur um *fabulae* handeln.[72] Auch weniger

VIII (Nürnberg, Melchior Gottfried Hein: 1719), Observatio 12 (S. 31–37) S. 35–37, der die Experimente Digbys, wie er selbst behauptet, noch einmal erfolgreich nachvollzieht.

69 Digby, *A Discourse, concerning the Vegetation of Plants* 85–93.

70 Ebd., 97 f.

71 Boyle Robert, *Tentamina quaedam physiologica diversis temporibus et occasionibus conscripta* (Genf, Samuel de Tournes: 1670), 43 f. Boyle hat vor allem Kircher im Blick.

72 Billich Anton Günter, *Thesssalus in chymicis redivivus, id est, De vanitate medicinae chymicae, hermeticae seu spagyricae dissertatio* (Frankfurt, Johannes Beyer: 1643), cap. 7,

skeptische Geister wie der Wittenberger Gelehrte Johannes David Maior bemängelten, daß die vorgebrachten Beispiele sich in ihrem Charakter kaum in Einklang bringen ließen, und man viele Belege bei genauer Betrachtung als simple Fortpflanzung, als Totenstarre oder auf andere Weise erklären konnte.[73] In etlichen Handbüchern der Naturphilosophie des 17. Jahrhunderts zählte man die Palingenesie aus dem gleichen Grund zu den sogenannten *non-entia chemica*, den Hirngespinsten, die allenfalls zu den Gegenständen der halbwissenschaftlichen Mythologie gehören konnten.[74] Was als Experiment vielleicht im Reich der Mineralien noch mit Erfolg gekrönt sein konnte, musste im Reich der Pflanzen zu einem absurden Kunststück mutieren,[75] denn weitaus leichter fiel es, so Georg Michaelis, Werke der Natur zu erfinden, als die Natur nur zu beobachten.[76] Um der Kritik, die den Paracelsismus als ganzen zu diskreditieren drohte, entgegenzukommen, beeilten sich manche Anhänger des Paracelsus, den Anspruch der eigenen Versuche herunterzuschrauben: Allenfalls Schattenbilder oder *simulacra* des ursprünglichen Lebens, so Ole Borch, wurden, wie Sennert schon nahegelegt hatte, in Gestalt der *resuscitatio plantarum* heraufbeschworen, doch nicht das Leben selbst, das allein Gott oblag.[77] DuChesne hatte eine *repraesentatio spiritualis* seiner

---

S. 44–46; Freitag Johann, *Noctes medicae sive de abusu medicinae tractatus* (Frankfurt, Johannes Berner: 1616), cap. 64, S. 272 f. Ebenso z.B. Schoock Marten, *Physica generalis* (Groningen, Johannes Collenius: 1660), Disputatio XV, § 24, S. 290 f.; oder Voigt Gottfried, *Curiositates physicae de rebus iucundis raris et novis* (Leipzig, Johann Adam Plener: 1698), cap. 1, Sectio III, S. 7–13; cap. 2, Sectio III, S. 42–47.

73 Maior Johann Daniel, *Dissertatio epistolica de cancris et serpentibus petrefactis* (Jena, Esaias Fellgiebel: 1664), §§ 38–39, 30–32.

74 Rolfinck Werner, *Chymia in artis formam redacta sex libris comprehensa* (Frankfurt, Georg Erhard Martius: 1686), lib. VI, cap. 3, 422–424; Kunckel Johannes, *Philosophia chemica experimentis confirmata* (Amsterdam, Johannes Wolters: 1697), Perspicilium chemicum, 300–302. Als *phantasmata* bezeichnen sie auch Hofmann Friedrich, *Clavis pharmaceutica Schroederiana seu animadversiones cum annotationibus in Pharmacopeiam Schroederianam* (Halle, Vidua et haeredes Myliani: 1681), lib. V, S. 639 f., und Schmidt Johann Andreas, *Dissertatio de corporum naturalium Palingenesia*, in *Miscellaneorum physicorum Fasciculus XVI disserationes variorum argumentorum continens* (Jena, Vidua Wertheriana: 1685, zuerst 1680), Thesis III, o. S.

75 Ruland Martin, *Progymnasmata Alchemiae sive problemata chymica nonaginta et una quaestionibus dilucidata* (Frankfurt, Palthenianus: 1607), q. 16, S. 52 f.

76 Michaelis Gregorius, *Notae in Jacobi Gaffarelli Curiositates* (Hamburg, Gottfried Schulze: 1676), Nota 55 (249–253) 253. Kritisch auch Faber Petrus Johannes, *Palladium spagyricum* (Straßburg, Lazarus Zetzner: 1632), cap. 1, S. 86 f.

77 Borch Olaus, *Hermetis Aegyptiorum et Chemicorum Sapientia ab Hermanni Conringii animadversionibus vindicata* (Kopenhagen, Peter Haubold: 1674), cap. 5, S. 359–365.

Pflanzen erreicht, doch keine reale Auferstehung.[78] Auch Georg Francke von Franckenau, der dem Palingenesie-Problem zu Beginn des 18. Jahrhunderts ein monumentales Werk schenkt, zieht es ähnlich wie sein lateinischer Kommentator Johann Christian Nehring am Ende vor, die vielen von seinen Kollegen vorgebrachten Phänomene auf sich beruhen zu lassen. Ein Beweis der natürlichen Auferstehung waren sie, wie Franckenau und andere frühneuzeitliche Mediziner glaubten, nicht.[79] Francke von Franckenau schließt sich Christian Garmann an, der mit seinen *Miracula mortuorum* eine der größten und bizarrsten Stoffsammlungen zu dieser und zu verwandten Fragen geliefert haben dürfte.[80]

Kaum ein Gelehrter des 16. und 17. Jahrhunderts, der sich noch auf dem Boden des Christentums befand, hätte Gott jede Verantwortung für die Auferstehung des Fleisches abgesprochen. Man ahnt dennoch, dass für die Anhänger einer natürlichen Wiederherstellung in der Abfolge ihrer Versuchsanordnungen nach Pflanzen und Tieren irgendwann auch die Reihe an den Menschen kommen musste, der schon vorher das eigentliche Ziel ihrer Argumentationen gewesen war. Ließ sich auch in seinem Fall schon vor dem Jüngsten Gericht eine Palingenese herbeiführen? Konnte man das Leben vielleicht auch aus einem Haufen menschlicher Knochen noch wieder herauskitzeln und mit den Kräften der Natur wiederauferstehen lassen? Wir betreten nun das Reich des Bizarren und ein Gebiet, das auch von vielen Anhängern des Paracelsus mit Argwohn bedacht wurde. Johannes Tacke, ein weiterer Denker aus dem Kreis der Paracelsisten, erinnert sich in seinem *Mysterium resurrectionis*, dem *Geheimnis der Auferstehung*, an den im jüdischen Midrash beschriebenen 18. Knochen der Wirbelsäule, den eine angeblich unzerstörbare Natur auszeichnete und der auch den Anatomen der eigenen Zeit nicht verborgen geblieben

78 Tachenius Otto, *Hippocrates chimicus per ignem et aquam methodum inaudita* (Venedig, Combi und LaNou: 1697), cap. 20, S. 180–182.

79 Francke von Franckenau Georg, *Palingenesia Francica, oder: Tractätlein von der Künstlichen Auferweckung derer Pflanzen, Menschen und Thiere aus ihrer Asche* (Leipzig, Johann Christian Martini: 1716), §§ 38–44, S. 63–70, und Johann Christian Nehring in seinem monumentalen Kommentar der lateinischen Ausgabe Francke von Franckenau Georg, *De Palingenesia sive resuscitatione artificiali plantarum, hominum et animalium e sui cineribus liber singularis, commentario illustratus* (Halle, Felix du Serre: 1717), §§ 38–44, S. 237–258. Eine weitere Zusammenfassung einiger der hier genannten Autoritäten liefert de Vallemont Pierre, *Curiositez de la nature et de l'art sur la vegetation: ou l'agriculture et le jardinage dans leur perfection* (Paris, Claude Cellier: 1707), chapt. 16, S. 654–708. Zu Francke von Franckenau Telle, „Chymische Pflanzen" 24–26.

80 Garmann Christian Friedrich, *De miraculis mortuorum libri tres*, 2 vols. (Dresden, Johann Christoph Zimmermann: 1709), lib. III, Titulus IV, §§ 17–23, S. 1214–1220.

war. Lag man in diesem Gelenk vielleicht die Entsprechung des Salzes vor, die zentrale Materiepartikel, aus der der Körper des Menschen im Augenblick der Auferstehung rekonstruiert werden konnte? Tacke wagt es nicht beantworten, doch mußte sich, wie er betont, in den Überresten des Menschen das entscheidende Salz der Wiederherstellung finden lassen.[81] Vielleicht war es schon dem Propheten Elischa gelungen, so Johann Ludwig Hannemann, den Toten, die er selbst als Leichnam im zweiten Buch der Könige wiederauferweckt hatte,[82] mit Hilfe einer chemischen Palingenesie ins Leben zurückzubefördern. Warum sollte Gott seinem Propheten nicht die Möglichkeit gegeben haben, aus dem verbliebenen Blut, der Asche und vielleicht auch dem immanenten Phosphor die *vita innata* eines Verstorbenen neu zu aktivieren?[83]

Andere Gelehrte wollten das universale Lebensprinzip tatsächlich gefunden haben, wie Tacke glaubt, ja sie sollten sogar Scheinauferstehungen herbeigeführt haben. Johannes Karl Rosenberg berichtet in seiner Enzyklopädie der Rose, der *Rhodologia*, die sich nicht zufällig auch ihrer Auferstehung widmet, im Jahre 1628 von einem befreundeten Arzt aus Wittenberg. Erst war es jenem Mediziner gelungen, wie Rosenberg weiß, eine Tulpe aus der Asche wieder zum Leben zu erwecken, dann jedoch erscheinen ihm im Destillat der Nachgeburt einer Frau, wie es heißt, Phantasmen, die auf ihn den Eindruck menschliche Köpfe machten, ja den Häuptern von Engeln glichen. Die Seele hatte ihren Ort überall im Leib, wie man wusste. Hatte der Wittenberger Gelehrte, ein Doktor Frey, schon das Tor zur Seelenform öffnen können?[84] Robert Fludd, Englands berühmtester Anhänger des Paracelsus, beschreibt in seiner Anatomie, wie ein französischer Kollege etwa zur gleichen Zeit mit dem Blut Verstorbener zu experimentieren beginnt. Der Lebenssaft wird in Reagenzgläsern wiederholt erhitzt und diversen Fäulnisprozessen unterzogen. Nach einigen Stunden macht der Arzt eine Entdeckung, die ihm, wie Fludd berichtet, den kalten Schweiß auf die Stirn treibt. Aus einer Phiole erklingt ein markerschütternder Schrei, das dem Brüllen eines verwundeten Löwen oder Ochsen vergleichbar war; zugleich erscheint eine Nebelwolke mit menschlichen Zügen,

---

81 Tacke, *Mysterium resurrectionis rerum sive Phasis III*, 25 f. Zu ähnlichen Überlegungen zum Salz als Lebensträger aus dem Bereich der Cabala christiana z.B. de Vigènere Blaise de, *Traicté du feu et du sel. Excellent et rare opuscule* (Paris, Abel L'Angelier: 1618), Seconde partie, 219–267. Der gleiche Text konnte kaum zufällig noch im Jahre 1782 in deutscher Sprache erscheinen, als *Des Herrn Blaise Vigeneres Abhandlung von Feuer und Salz* (Breslau, Korn: 1782), dort Zweiter Theil, 279–336.

82 2 Kön 13, 21.

83 Hannemann – Eccardus, *Triumphus naturae et artis*, Thesis 6, S. 35–37.

84 Rosenberg Johann Karl, *Rhodologia seu philosophico-medica generosae rosae descriptio* (Straßburg, Marcus ab Heyden: 1628), Pars II, cap. 33, S. 311.

der Schattenriß eines Menschen, der sich scheinbar auf das offene Fenster im Laboratorium hinbewegte. Mehrere Zeugen hatten die Erscheinung gesehen, wie Fludd gelobt, und sie auch in den oberen Stockwerken des Gebäudes noch hören können. In einer Schale gefüllt mit erhitztem Blut, das vom selben Toten stammte, erkennt der gleiche Arzt wenig später die Gesichtszüge des Verstorbenen, die sich in der pulsierenden Brühe auf seinem Seziertisch abzeichnen. Wiederum fehlt es nicht an Augenzeugen, die das schaurige Ereignis, wie Fludd betont, bestätigen können. Hatte man hier eine Seele gesehen, die vielleicht schon für die Verdammnis vorgesehen war? Hatte sie aus Verzweiflung geschrien, weil sie in den Resten ihres Leibes nur noch auf das letzte Urteil warten musste?[85]

In eine ähnliche Richtung wie die Blutexperimente in der Studierstube Fludds gehen die Erfahrungen, die Pierre Borel, auch er ein Leibarzt des französischen Königs, wenn auch Ludwig des XIV, im Jahre 1656 in seinen *Observationes* zu Papier bringt. Borel dürfte der größte Euphoriker der Palingenesie gewesen sein. Nicht nur wollte er selbst diverse Pflanzen aus der Asche wiederhergestellt haben und hatte zahlreiche ähnliche Phänomene dieser Arbeit katalogisiert, ja er schwärmte von gläsernen Gärten, in dessen Kolben und Phiolen Pflanzen aller Art wie in einer Arche Noah in ihrem lebenserhaltenden Licht schillern konnten.[86] Wenn die menschliche Seele in den Kernbeständen des Leibes immer noch zu finden war, sollte es nicht auch möglich sein, diese Glaskolben mit menschlichem Blut zu füllen und auf diese Weise alle aus dem Leben geschiedenen Freunde als reale Seelenbilder wieder herbeizuholen? Ja war es nicht sogar denkbar, ganze Generationen der eigenen Familie in einer statthaften Form der Nekromantie um sich herum zu versammeln, eine Ahnengalerie, die auf der Gewalt des Faktischen beruhen konnte?[87] Auch die Schilderungen Fludds waren Borel vertraut. Wie Fludd kennt er französische Mediziner, denen es in einem mit Blut gefüllten Glaskolben gelungen war, weitere Seelenbilder zu generieren. Wieder andere Gelehrte hatten, wie Borel zu berichten weiß, ein Destillat aus Knochenresten gewonnen, die sie auf dem Stadtfriedhof von Paris ausgegraben hatten, und waren mit dem gleichen

85 Fludd Robert, *Utriusque cosmi maioris scilicet et minoris metaphysica, physica atque technica historia*, 2 vols. (Frankfurt, De Bry: 1617–1626), vol. II, Tractatus II, Sectio I, Portio 3: De anatomia triplici, Pars II: De mystica sanguinis anatomia, cap. 6, S. 233.

86 Borel Pierre, *Historiarum et observationum medicophysicarum Centuria IV, in quibus non solum multa utilia, sed etiam rara et stupenda ac inaudita continentur* (Paris, Jean Billaine: 1656), Centuria I, Observatio 21, S. 121–123; Centuria III, 252–254,;Centuria IV, Observatio 34, S. 303 f.

87 Ebd., Centuria IV, Observatio 62, S. 323–325; 326–328.

Ergebnis konfrontiert worden. Die wabernden Erscheinungsformen der Toten waren furchterregend gewesen und hatten alle früheren Erfahrungen weit hinter sich gelassen. Nicht der Teufel jedoch hatte bei diesen Erscheinungen seine Hände im Spiel gehabt, wie Borel mit Nachdruck deutlich macht, sondern die Wissenschaft, die nun mit ihren eigenen Mitteln bewiesen hatte, dass die Seele bis zur endgültigen Auferstehung im Leichnam verharren musste und während dieser Zeit sichtbar gemacht werden konnte. Die Auferstehung musste also ein in weiten Teilen natürliches Ereignis sein.[88]

Für die Kritiker der paracelsischen Medizin wie Johannes Freitag stand fest, dass Borel die Geister von Toten heraufbeschworen und sich in den Bannzirkel der verbotenen Magie begeben hatte. Der Franzose musste der Gewalt der Dämonen zum Opfer gefallen sein.[89] Zumindest in gewissen Kreisen war man jedoch von den Experimenten Borels überzeugt: Der Mindener Physiker David van der Becke sieht keinen Grund, Borel Umnachtung oder Phantasmen zu unterstellen: DuChesnes und Borels Versuche hatten gezeigt, daß auch in den sterblichen Resten der Menschen eine *idea* anwesend blieb, die zum Ausgangspunkt der Restituierung des Körpers wurde. Es war die gleiche *idea seminalis*, die als Lebensprinzip zuvor schon den Embyro im Mutterleib ausgeformt hatte, deren Wirkung sich im Naturreich an widerbelebten Fröschen oder Schlangen beobachten ließ und mit deren Hilfe die in der Offenbarung verheißene Auferstehung des Menschen bewerkstelligt wurde.[90] Die *scintilla* des Lebens konnte von der organischen Materie nicht gelöst werden; sie blieb als Seminalprinzip bestehen, wie Borel gezeigt hatte. Dass am Ende dennoch Gott zum Motor der erneuten Verkörperlichung und Ausfaltung dieses Prinzips, das in der Restmasse des Menschen noch tätig war, werden musste, hatte jedoch auch die Physik anzuerkennen, wie Christian Schmid in einer eigenen Arbeit zur *Resurrectio artificialis* unterstreicht.[91]

88 Ebd., Centuria IV, Observatio 62, S. 325 f.

89 Als Nekromantie apostrophieren die Auferstehungspraktiken z.B. Conring Hermann, *De Hermetica Aegyptiorum vetere et Paracelsicorum nova medicina liber unus* (Helmstedt, Henning Müller: 1648), cap. 17, S. 211 f.; Freitag Johann, *Detectio et solida refutatio novae sectae Sennerto-Paracelsicae* (Amsterdam, Wilhelm Blaeu: 1637), Disputatio II, dort vor allem der Brief des Adam Luchtenius an Johannes Freitag, 269–272; Praetorius Johannes, *Anthropodemus plutonicus, das isst Neue Welt-Beschreibung von allerley wunderbaren Menschen*, 2 vols. (Magdeburg: Johann Lüdewald, 1666–1677), cap. 3, S. 161–163.

90 Van der Becke David, *Experimenta et meditationes circa naturaliu rerum principia* (Hamburg, Gottfried Schulze: 1674) 241–249.

91 Schmid Christian, *Resurrectio rerum artificialis* (Leipzig, Samuel Spörel: 1677), Sectio II, §§ 6–8, fol. B4r f. Borel schaffte es sogar in ein zeitgenössisches Studienbuch, nämlich

II.3 *Die Welt aus der Asche*

Gothic fiction war im Fall Borels auf den hehren Anspruch der empirischen Erkenntnis gestoßen, doch mit einer Konsequenz, die vielleicht nur der inneren Logik des eigenen Systems folge leisten musste. Borels Entdeckungen verstörten auch die Anhänger des Paracelsus; in den anderen Lagern, sei es bei orthodoxen Lutheranern, Cartesianern oder einfachen Skeptikern waren seine Thesen kaum mehrheitsfähig oder mussten bestenfalls Heiterkeit hervorrufen. Leuwenhoecks oder Swammerdamms Entdeckungen, die Erfolge der Mikroskopie, mit dessen Hilfe sich die meisten der Experimente Kirchers oder Digbys auf einfache Weise falsifizieren ließen, aber auch der Bedeutungsverlust der Alchemie an den Universitäten der Frühaufklärung drängten die Debatte um die Palingenesie und die natürliche Auferstehung in die Randbereiche der Wissenschaft.[92] Die ersten Geschichten der Chemie, die im 18. Jahrhundert entstanden waren, behandeln sie nur noch wie einen grotesken Irrtum, eine kollektive Einbildung oder traktieren sie als Neurose einiger Exzentriker, wie Joachim Telle gezeigt hat.[93]

Attraktiv blieb die Palingenesie Mitte des 18. Jahrhunderts trotz allem im Kreise jener Gelehrter, die sich der Physiko-theologie verpflichtet sahen, der naturwissenschaftlichen Demonstration der christlichen Dogmatik, und die auf diese Weise die quasi-scotistischen Alternativkonzepte einer natürlichen Auferstehung und einer physikalischen Eschatologie am Leben erhielten.[94] Arbeiten zur Palingenesie entstehen auch im 18. Jahrhundert noch zuhauf.[95]

Bechmann Friedemann, *Systema physicum mandato Ernesti Saxon. Jul. Cliviae et Mont. Ducis ad usum Gymnasii Gothani* (Jena, Georg Sengenwald: 1664), lib. V, § 28, S. 498 f.

92 Zur Rolle, die das Mikroskop im allgemeinen und die Arbeiten Leeuwenhoecks im Besonderen in der Widerlegung der Urzeugung und anderer Theorien spielten, Wilson C., *Early Modern Philosophy and the Invention of the Microscope* (Princeton: 1995) 198–214, und ausführlich Ruestow E.G., „Leeuwenhoeck and the Campaing against Spontaneous Generation", in *Journal of the History of Biology* 17 (1984) 225–248; ders., *The Microscope in the Dutch Republic. The Shaping of Discovery* (Cambridge: 1996) 201–259.

93 Telle, „Chymische Pflanzen" 28 f. Als Beispiel Wegner Georg Wilhelm, *Adeptus ineptus, oder Entdeckung der falsch berühmten Kunst, Alchemie genannt* (Berlin, Ambrosius Haude: 1744) 197 f.

94 Zur Bewegung der Physiko-theologen allgemein die Überblicke von Michel P., *Physikotheologie: Ursprünge, Leistung und Niedergang einer Denkform* (Zürich: 2008); Stebbins S., *Maxima in minimis: Zum Empirie- und Autoritätsverständnis in der physikotheologischen Literatur der Frühaufklärung* (Frankfurt: 1980); Bots J., *Tussen Descartes en Darwin. Geloof en natuurwetenschap in de achttiende eeuw in Nederland* (Assen: 1972).

95 Vollständig kopiert die Angaben zur Palingenesie aus Francke von Franckenau z.B. noch der anonyme Traktat *Künstliche Auferweckung der Pflanzen, Menschen und Thiere*

Als junger Mann entwirft der Frankfurter Gelehrte Johann Michael von Loen, der einst sein Leben in Lingen an der Ems beschließen sollte, eine ausgreifende chemico-theologische Abhandlung mit dem Titel *Das Geheimnis der Verwesung*. Das 1729 zum ersten Mal gedruckte Werk sollte noch etliche Auflagen erhalten können.[96] Wie Loen glaubte, fand sich im körpereigenen Salz, das dem lichthaften Phosphor und seiner energetischen Aufladung vergleichbar war, die entscheidende Kraft der Auferstehung. Seine besondere Funktion in der Naturordnung ließ sich mit einer großen Zahl von chemischen Experimenten untermauern.[97] Sie alle konnten belegen, dass sich auch der menschliche Leib während seiner Verwesung von allen vergänglichen Ölen und Schlacken trennen musste, um seinen gereinigten salz- und lichtgetränkten Überrest, seine Asche, zum Ausgangspunkt seiner organischen Wiederstellung zu machen.[98] Loen lässt dabei kaum einen der bekannten Anhänger der natürlichen Auferstehung aus, und vielen der hier genannten Autoritäten können wir bei ihm noch einmal begegnen.[99] Noch 1770 beschenkt Friedrich Christoph Oetinger seine Leser mit einer paracelsistischen Abhandlung zum Salz als dem universalen Lebensprinzip, auch wenn Oetinger selbst keine Experimente mehr beisteuern konnte.[100]

Entscheidend für die Autoren dieser Generation war die Ausweitung der Perspektive. Der Transformationsvorgang, die Prozedur, die der Weise an der Rose vornehmen konnte und die sie mit dem Leib des Auferstandenen verband, hatte das ganze Universum zu erfüllen. Der Astronom und Naturphilosoph Eberhard Christian Kindermann, der auch den unfreiwilligen Emsländer Loen gründlich studiert, um nicht zu sagen abgeschrieben hatte, kann diese Überlegungen im Jahre 1744 noch einmal zusammenfassen und liefert damit

---

aus ihrer Asche: nebst einem kurzen Unterricht allerhand Farben auf das Glas zu brennen (Frankfurt: 1785), dort 1–75.

96 Von Loen Johann Michael, *Das Geheimnuß der Verwesung und Verbrennung aller Dinge: Nach seinen Wundern im Reich der Natur und Gnade, Macro- und Micro-Cosmice als die Schlüssel, dadurch der Weg zur Verbesserung eröffnet, das Verborgene der Creaturen entdecket, und die Verklärung des sterblichen Leibes gründtlich erkandt wird* (Frankfurt, Johann Friedrich Fleischer: 1742). Die Schrift erschien zuerst anonym 1729, dann noch einmal 1733 und 1759.

97 Ebd., 42–46.

98 Ebd., 15 f.; 34–42; 55–58.

99 Ebd., 61–68.

100 Oetinger Friedrich Christoph (Elias Artista Hermetica), *Das Geheimnis von dem Salz als dem edelsten Wesen der höchsten Wohlthat Gottes im Reich der Natur, beides in seinem Wesen und seinen Eigenschaften aus eigener Bemühung untersucht* (o. O, Steinkopf: 1770), dort bes. Dritter Theil, §§ 36–38, S. 64–71; Vierter Theil, § 72, S. 130–132.

gleichsam das Finale der hier untersuchten Motive.[101] Was hatten der Mensch und mit ihm der ganze Globus im Verlauf des Jüngsten Gerichtes zu erwarten? John Ray, neben William Derham der wichtigste Physiko-theologe, hatte im Feuer des Erdkerns, dessen Auswirkungen im Vulkanismus spürbar waren, das entscheidende Werkzeug der Apokalypse gesehen. Überdruck konnte die Erdoberfläche zum Bersten bringen und Magma den Planeten verheeren.[102] Es war jedoch schwer zu beantworten, woher der Druck, der das endzeitliche Feuer empordrücken sollte, rühren konnte. Johannes Heyn hatte in seinem 1742 entstandenen Trakat zur Physik der letzten Dinge einen besseren Vorschlag gemacht; ein gewaltiger Komet sollte mit der Erdkugel kollidieren, sie aus ihrer Bahn werfen und zur Gänze entflammen.[103] Auch ein Großgelehrter unter den Physiko-Theologen, Johannes Scheuchzer in seiner *Physica sacra*, hatte

101 Zur Gestalt des freigeistigen Astronomen Kindermann Mulsow M., *Freigeister im Gottsched-Kreis. Wolffianismus, studentische Aktivitäten und Religionskritik in Leipzig 1740–1745* (Göttingen: 2007) 100–103. Kurz zur Verklärung des Menschen aus der Asche auch Kindermann Eberhard Christian, *Die Geschwinde Reise auf dem Lufft-Schiff nach der obern Welt*, ed. H. Siebenpfeiffer (Hannover: 2010; zuerst 1744) 43–49.

102 Ray John, *Sonderbahres Klee-Blätlein, der Welt Anfang, Veränderung und Untergang; oder nebst allerhand mit eingeschalteten hochwichtig- und spitzfindig erörterten Fragen, hauptsächlich drei so wol aus Geist- und Weltlichen Schrifften gnugsam betrachtete als auch aus natürlichen Uhrsachen gründlich untersuchte neue gelehrte Handlungen* (Hamburg, Wiering: 1698), III. Handlung, cap. 5, III. Abtheilung, 355–362. Englisch z.B. als ders., *The physico-theological discourses, concerning the primitive chaos and creation of the world, the general deluge, and the dissolution of the world* (London, William Innys: 1713; zuerst 1692), Discourse III, chapt. 5, Section III, 375–380. Das Werk erhielt bis zum Ende des 18. Jahrhunderts in diversen Sprachen etliche Auflagen und darf als Klassiker der physikotheologischen Apokalypse gelten. Ein Nachfolgewerk, das allerdings unvollendet blieb, liefert z.B. Fabricius Johann Albert, *Pyrotheologie, oder Versuch durch nähere Betrachtung des Feuers den Menschen zur Betrachtung ihres gütigsten, weisesten, mächtigsten Schöpfers anzuflammen*, gedruckt als Anhang zu Derham Wilhelm, *Astrotheologie, oder Anweisung zur Erkenntniß Gottes aus der Betrachtung der himmlischen Körper* (Hamburg, Johann Carl Bohn: 1765), dort Buch VIII, cap. 14, S. 243.

103 Heyn Johann, *Versuch einer Betrachtung über die Cometen, die Sündflut und das Vorspiel des Jüngsten Gerichtes, nach astronomischen Gründen und der Heiligen Schrift angestellet* (Berlin, Ambrosius Haude: 1742), cap. 1, § 74, S. 109 f.; cap. 4, § 164, S. 246–251. Maßgeblich hatte diese Hypothese der Hamburger Geologe Detlev Clüver vorgebracht, dazu Clüver Detlev, *Geologia sive philosophemata de genesi ac structura globi terreni, oder: Natürliche Wissenschaft von Erschaffung und Bereitung der Erd-Kugel* (Hamburg, Liebezeit: 1700), cap. 14, S. 256–271. Eine ähnliche Auffassung, doch unter stärkerer Beteiligung des verdampfenden Wassers, vertritt z.B. auch Janssen Peter – J. Arnold Brans (resp.) *De novissima mundi deflagratione deque novo coelo terraque nova quae pios post deflagrationem istam ex promissione divina expectare ait Petrus Apostolus* (Duisburg, Straube: 1751), *passim*.

diese Auffassung favorisiert, wenn auch noch mit vorsichtiger Skepsis.[104] Für Kindermann, den frommen Astronomen, war hier die Lösung gefunden, um dem Vernichtung der Welt die angemessene Bühne zu geben. Vom inwendigen Schwefel entflammt und zugleich durch seinen Ausstoß raketengleich angetrieben, trieb ein Komet die Erde zunächst aus ihrer Rotation, um dann in einer Kettenreaktion Planet um Planet wie die Kerzen eines Geburtstagskuchens anzuzünden und in einem großen Feuerregen explodieren zu lassen.[105] Auf den ersten Blick musste dieses Flammenmeer nur Asche hinterlassen, doch war es diese Asche, aus deren Salzen der neue Mensch hervorgehen konnte, eine physikalische Wahrheit, die aus den Gesetzen der Natur resultierte. Alle Schlacken, Unreinheiten und Dunkelheiten waren, wie Kindermann betont, in den Ruinen dieses Sterntheaters zurückgeblieben; die alte Asche mußte in sich jedoch bereits das neue Licht enthalten, wie der Phosphor als das *Phlogiston*, das brennbare Prinzip, im Kampfer verborgen war.[106] Die Verwesung hatte den Menschen auf dieses Salz reduziert, Form und geistige Gewalt waren jedoch weiterhin in ihm verborgen. Dem göttlichen Chemiker oblag es vor dem Gericht, diese Lichtnatur wieder aus der Materie zutage zu fördern und ihn auf diese Weise mit einem neuen Körper, dem erneuerten alten Leib, zu versehen.

---

104 Scheuchzer Johann Jakob, *Kupfer-Bibel, in welcher Die Physica Sacra, Oder Geheiligte Natur-Wissenschaft deren in Heiliger Schrifft vorkommenden Natürlichen Sachen, Deutlich erklärt und bewährt*, 4 vols. (Augsburg, Wagner: 1731–1735), vol. IV, Tabula DCCCXLIV, 1404–1410, und in der lateinischen Ausgabe ders., *Physica sacra iconibus aeneis illustrata*, 4 vols. (Augsburg – Ulm, Wagner: 1731–1735), vol. IV, Tabula DCCXLIV, S. 1520–1524. Auch Laurentius Meijer, der Verfasser eines monumentalen Scheuchzer-Kommentars, hält zum Ende des 18. Jahrhunderts noch an der Kometentheorie fest, dazu *Bijbel der Natuur, ontworpen door de beroemde Geleerden J. J. Scheuchzer en M. S. G. Donat, met bijgevoegde Aanmerkingen, uit de beste Uitleggeren, nieuwste Historische Schriften en Reisbeschrijvingen*, 12 vols. (Amsterdam, Martinus de Bruyn: 1787–92), vol. XII, cap. 34, §§ 168–169, S. 280–282.

105 Kindermann Eberhard Christian, *Vollständige Astronomie, oder sonderbare Betrachtungen derer vornehmlichen an dem Firmament befindenden Planeten und Sternen* (Rudolstadt, Deer: 1744), cap. 23, S. 343–365.

106 Ebd., cap. 24, S. 374–383. Zum *phlogiston*, dem brennbaren Aggregat in der Materie, in dieser Zeit z.B. Wipacher David, *De phlogisto unionis rerum metallicarum medio disputatio physica* (Leipzig Georg Gottlob Kuchelbecker: 1752), §§ 9–12, S. 32–52; in Verbindung mit dem Salz z.B. Buchwald Balthasar – Gottfried Licht (resp.), *Meditationes de sale communi eiusque existentia in omnibus rebus mundanis* (Kopenhagen, E. H. Eberling: 1749), §§ 5–10, S. 3–5. Standardlehrbücher der Chemie weisen in der Zeit zudem darauf hin, daß auch Meteore ein schwefelhaltiges *phlogiston* enthalten mußten, um sich im Raum fortzubewegen, so z.B. Juncker Johann, *Conspectus chemiae theoretico-practica* (Halle, Orphanotropheus: 1738) 6 f.

Aus Asche und Salz mußte wieder leuchtender Phosphor werden. Gott würde dabei, wie Kindermann zum Ende deutlich macht, nicht anders vorgehen, als der Philosoph, der in der Lage war, die Rose aus ihrem aschernen Gefängnis zu befreien. Die Freunde Borels, denen die erweckten Seelen ein Jahrhundert zuvor scheinbar ihre Verzweiflung entgegengeschrien hatten, waren nur einem Fehler verfallen, sie hatten der eigentlichen Auferstehung vorgegriffen.[107]

Fazit

Vielleicht ist zum Ende deutlich geworden, wie sehr die spätmittelalterliche Philosophie durch ihren Willen zur Vollständigkeit und ihre Bereitschaft, auch alternative Konstruktionen auszubuchstabieren, die frühneuzeitliche Experimentierfreude hatte inspirieren können. Es war, so ist hier vielleicht gezeigt worden, nicht zuletzt die scotistische Lesart der Auferstehung gewesen und ihr Versuch, das Mirakel des wiederhergestellten Körpers zumindest theoretisch auf physikalische Gehalte zu reduzieren, die den Technokraten der Erlösung die entscheidene Grundlage zur Verfügung gestellt hatte. Die spekulative Naturphilosophie des Spätmittelalters und ihre frühneuzeitlichen scholastischen Rezipienten hatten die neuen Lehren des Paracelsismus in weiten Teilen antizipiert; auch ihre Aristoteleskritik mußte grundieren, was die scheinbar revolutionäre Praxis der Paracelsisten und Hermetiker der Frühen Neuzeit experimentell nachvollziehen konnte. Der entscheidende Unterschied lag in der Erfahrung: Im 17. Jahrhundert waren die Paracelsisten mit dem Evangelium ins Labor gegangen, um in die Tat umzusetzen, was für die Theologen des Mittelalters nur eine mögliche Alternative zu Aristoteles gewesen war. Ob Paracelsus selbst ein Zeuge der Modernisierung war oder seine Lehren, wie man behaupten könnte, in weiten Teilen doch eher eine romantische Reaktion auf die lange Dominanz der Aristoteliker und später der Cartesianer, soll hier nicht weiter diskutiert werden. Daß viele Beweisführungen der Physikotheologen des 18. Jahrhunderts sich darüber hinaus mit den Argumenten der heutigen Kreationisten decken, läßt auch ihre Gedanken in einem zwiespältigen Licht erscheinen. Von Bedeutung scheint es jedoch, festzuhalten, daß die Epochengrenzen, wenn man bereit ist, sich dem theologischen Pluralismus des Mittelalters zu öffnen, nicht mehr so scharf verlaufen, wie man vielleicht meinen könnte.

107 Kindermann, *Vollständige Astronomie*, cap. 25, S. 384–393.

Welches Ende läßt Borges seine Parabel nehmen, mit hier begonnen wurde? Man ahnt es vielleicht. Der junge Mann scheint zu erkennen, daß die Künste des Meisters ohne Wirkung waren. Er verspricht, zurückzukehren, wenn sein Glaube an das Mysterium stärker geworden ist, doch hat er jede Achtung vor Paracelsus verloren. Der Gelehrte war scheinbar doch ein Hochstapler gewesen. Mit dem Geld in der Hand verläßt Griesebach das Laboratorium. Paracelsus jedoch nimmt die Asche der Rose in die Hand, flüstert das entscheidende Wort und die Rose erhebt sich von neuem.[108]

---

108 Borges, „La rosa de Paracelso", 389 f., deutsch Borges, „Die Rose des Paracelsus", 108 f.

English Summary

In his famous parable of "The Rose of Paracelsus" the Argentinian writer Jorge Luis Borges recounts how the German theosophist brings a rose back to life from its ashes, despite the scepticism of the young Johannes Griesebach, who had tried to use money to pressurise him into doing so. Borges' story can only be understood correctly in the context of an early modern discussion, the debate on palingenesis, i.e., the idea of a natural resurrection. What was that? The present study reconstructs this long controversy and tries to uncover both its medieval roots and its early modern reception.

Aristotelian ontology, which saw every substance as a duality of form and matter, did not provide for the regeneration of a specific individual. To ensure the identity of a substance, both form and matter would need to be regenerated the same and united with each other, as Aristotle had insisted in his work *De generatione et corruptione*. If the connection between the two was broken, the substance, whether it be a plant or a person, was destroyed and could not be furnished with new life. The Aristotelian commentators and scholastic physics are for this reason unanimous in stressing that a natural resurrection is necessarily impossible. However, for Christian theology, both body and soul were indispensable, even after death, to ensure the unity of the person. In medieval and early modern theology, in which the debate about resurrection was dominated by Thomists, it was therefore God who must guarantee the regeneration of the body by a miracle. He united the soul-form with some characteristic particles of matter from the perished body and in this way ensured the continuity of the individual beyond death. However, not all medieval theologians accepted this consensus, as will be shown below. Especially the followers of Duns Scotus took the view that form and matter were characterised by a natural relation: whenever the soul-form encountered its body it unleashed its natural effect, for which no further miracle was needed. At least theoretically, Scotus had insisted, a natural resurrection, too, must be conceivable.

Already in the 17th century, scholars such as Caspar Posner had recognised that the Scotist hypothesis of a *resurrectio naturalis* could open a clear path to accepting palingenesis. If the connection of form and matter could operate by natural means at the moment of resurrection, then they were both perhaps continuously together even after death. Why should the form of the body not simply be re-activated as the life-principle in its remains after death? For Paracelsus, the form stayed in the remains of a creature and merely waited for its revivification. A whole generation of scholars at the end of the 16th and the early 17th century adopted Paracelsus' view and attempted to demonstrate its premises experimentally. Aristotelian ontology could be falsified by the resurrection of plants and animals by technical means.

The French doctor Joseph DuChesnes was the first to succeed in raising a rose to new life from its ashes. Other scientists subsequently managed to coax out poplars or violets from their remains. A whole wave of experiments followed. The soul of the plant, according to these scholars, had never left its body; it remained present in it as form or as spiritual atoms. As salt it waited in the ashes of the defunct remains of the old body. It only required a few ingredients, explained Johannes Tacke and Philipp Rosenberg, to prompt the soul to fulfil its old function once again. Scientists like Martin Kerger and Kenelm Digby even believed they had witnessed the revivification of a shellfish or a snake. Despite vehement criticism of resurrectionist practice from Aristotelians, Paracelsists like Pierre Borel and Robert Fludd went a step further and undertook the attempt to make human remains the starting point of at least a shadowy resurrection. The results of these laboratory efforts were disturbing, but they were greeted with scepticism and they definitively discredited the supporters of palingenesis in the eyes of their opponents.

In the 18th century, as is shown at the end of the paper, the physicotheologians would adopt the idea of a natural resurrection. Man, like the rest of creation, would be destroyed in the conflagration of the apocalypse. But from the salt, the life-giving particles of matter from the organic body that remained in the form of ashes, so figures like Johann von Loen and Eberhard Kindermann believed, the old, burnt body of the deceased could be brought to life again. It was thus a natural resurrection. In its nature, the physicotheologians were agreed, it corresponded to that regenerated rose that, by palingenesis, the alchemist called back to life.

Select Bibliography

Blümlein K., *Naturerfahrung und Welterkenntnis. Der Beitrag des Paracelsus zur Entwicklung des neuzeitlichen, naturwissenschaftlichen Denkens* (Frankfurt – Bern: 1992).

Borges Jorge Luis, „La rosa de Paracelso", in *Obras completas*, 4 vols. (Barcelona: 1996), vol. III, 387–389.

Bots J., *Tussen Descartes en Darwin. Geloof en natuurwetenschap in de achttiende eeuw in Nederland* (Assen: 1972).

George M.I., „Aquinas on Reincarnation", *The Thomist* 60 (1996) 33–52.

Heinzmann R., *Die Unsterblichkeit der Seele und die Auferstehung des Leibes. Eine problemgeschichtliche Untersuchung der frühscholastischen Sentenzen- und Summenliteratur von Anselm von Laon bis Wilhelm von Auxerre* (Münster: 1965).

Kämmerer E.W., *Das Leib-Seele-Problem bei Paracelsus und einigen Autoren des 17. Jahrhunderts* (Wiesbaden: 1971).

Lo Presti R. – Buchenau S. (eds.), *From ‚Animal' to ‚Man'? Human and Animal Perception in Early Modern Philosophical and Medical Anthropology (16th–18th century)* (Pittsburgh: 2015, im Druck).

Marx J., „Alchimie et Palingénésie", *Isis* 62 (1971) 274–289.

Michel P., *Physikotheologie: Ursprünge, Leistung und Niedergang einer Denkform* (Zürich: 2008).

Mulsow M., *Freigeister im Gottsched-Kreis. Wolffianismus, studentische Aktivitäten und Religionskritik in Leipzig 1740–1745* (Göttingen: 2007).

Pagel W., *Das medizinische Weltbild des Paracelsus, seine Zusammenhänge mit Neuplatonismus und Gnosis* (Wiesbaden: 1962).

Ruestow E.G., „Leeuwenhoeck and the Campaing against Spontaneous Generation", in *Journal of the History of Biology* 17 (1984) 225–248.

Ruestow E.G., *The Microscope in the Dutch Republic. The Shaping of Discovery* (Cambridge: 1996) 201–259.

Secret F., „Palingenesis, Alchemy and Metempsychosis in Renaissance Medicine", *Ambix* 26 (1979) 81–92.

Solère J.-L., "Was the eye in the tomb? On the metaphysical and historical interest of some strange quodlibetal questions", in Schabel C. (ed.), *Theological Quodlibeta in the Middle Ages. The Thirteenth Century* (Leiden/Boston: 2006) 507–558.

Stebbins S., *Maxima in minimis: Zum Empirie- und Autoritätsverständnis in der physikotheologischen Literatur der Frühaufklärung* (Frankfurt: 1980).

Telle J., „Chymische Pflanzen in der deutschen Literatur", *Medizinhistorisches Journal* 8 (1973) 1–34.

Walker Bynum C., *The Resurrection of the Body in Western Christianity, 200–1338* (New York: 1995).

Weber H.J., *Die Lehre von der Auferstehung der Toten in den Haupttraktaten der scholastischen Theologie von Alexander von Hales zu Duns Scotus* (Freiburg: 1973).

Weeks A., *Paracelsus. Speculative Theory and the Crisis of the Early Reformation* (New York: 1997).

Wilson C., *Early Modern Philosophy and the Invention of the Microscope* (Princeton: 1995).

Images of Genesis:
Intersections of the Visual Arts, Science, and Religion

CHAPTER 8

Rereading Dürer's Representations of the Fall of Man[1]

Paul J. Smith

Introduction

In representing the Fall, each early modern artist has to make a crucial choice: will he put, besides the almost obligatory snake, other animals in the Garden of Eden? The book of *Genesis* leaves him free to make this choice, because *Genesis* is not explicit about the presence of animals in the Garden. In *Genesis* 1, the Garden of Eden is not even mentioned, and there is no indication at all of where the creation of the animals took place. And according to *Genesis* 2, which gives a brief description of the Garden (mentioning briefly, among other things, the Tree of Life and the Tree of Knowledge of Good and Evil), it is not explicitly stated that the creation of the animals and the scene of Adam's naming of the animals took place inside or outside the Garden, or eventually on the Garden's threshold, as is shown in a series of illustrations in a play on the subject of *Genesis*, entitled *L'Adamo, sacra rapresentatione*, by the Italian playwright Giovanni Battista Andreini (1613) [Fig. 8.1].[2] Representing animals in Eden is therefore a deliberate choice, as is that of the different animal species to be represented.

This can be exemplified in the works by Albrecht Dürer and his immediate followers Lucas Cranach and Lucas van Leiden. Indeed, in his six works on the subject of the Fall, Lucas van Leiden never represents any animals (except for the compulsory serpent), whereas of Cranach's thirty-one works on the Fall, twenty-eight have animals, and only three do not. As for Dürer, three of his six

1 This article was written within the framework of the NWO programme *Cultural Representations of Living Nature: Dynamics of Intermedial Recording in Text and Image* (c. 1550–1670).

2 For these and other examples, see Prest J., *The Garden of Eden. The Botanic Garden and the Re-Creation of Paradise* (New Haven – London: 1981) 12–15. See also Anonymous (ed.), *Het Aards Paradijs. Dierenvoorstellingen in de Nederlanden van de 16de en 17de eeuw* (Antwerp: 1982).

 | DOI 10.1163/9789004279179_010

FIGURE 8.1 *Cesare Bassano,* Adam Naming the Animals. *Engraving in Gianbattista Andreini,* L'Adamo, sacra rapresentatione (*Milan, Geronimo Bordoni: 1613*).

works on the Fall have animals in Eden, and three do not (one of them does not even represent the serpent). In this article I will address the animals and their meaning in Dürer's representations of the Fall. Much has been written on Dürer's animals, mostly according to Erwin Panofsky's seminal and authoritative, but rather restrictive interpretations. In order to shed new light on this field, the present article will offer a reconsideration of Dürer's animals surrounding Adam and Eve.

Let us begin by recalling briefly, in chronological order, Dürer's six representations of the Fall. The first one is a woodcut in a devotional work by Marquart von Steyn, entitled *Ritter von Turn* (1493). In this woodcut no animals are represented, except for the serpent. In Dürer's second representation—a drawing, which was probably made after 1495 and is now in Paris [Fig. 8.2]—all animals, including the serpent, are absent: Adam and Eve are alone with their feelings, as well as the discovery of their mutual sexual attraction. As noticed by Christian Schoen, Eve's eye is focused not only on the apple, but also on Adam's genitals: 'Ambivalent wirkt ihr Blick, da er durchaus auch auf das ungewöhnlich auf-

FIGURE 8.2 *Albrecht Dürer,* The Fall of Man *(after 1495), drawing. Paris, Ecole des Beaux-Arts.*

fallend inszenierte männliche Geschlecht Adams gerichtet sein könnte'.[3] This eroticizing of the Fall will also be thematized in other works by Dürer, and by his followers Baldung and Cranach as well. Dürer's third representation of the Fall, a copper etching from 1504 [Fig. 8.3], is the most famous one: its eight animals—serpent, elk, parrot, cat, mouse, hare (or rabbit), cow (or ox), and he-goat (or ibex or chamois)—have been discussed by Panofsky,[4] who has set a trend in interpreting these animals. In 1507 Dürer painted two magnificent paired oil panels, now in the Prado in Madrid, without animals (except for the serpent). Dating from 1510 is a drawing in which three animals are present: the serpent, and a very traditional stag and lion [Fig. 8.4]. This drawing, now in the Albertina in Vienna, was probably meant to be a preparatory study for the opening woodcut of the series called the *Small Passion* (published in 1510). In this woodcut, however, it is not the serpent, lion, and stag that are represented, but four other animals—the serpent, badger, lion, and bison—two of which, the badger and the bison, are very untraditional in the iconography of the Fall [Fig. 8.5]. In order to appreciate Dürer's innovations in the field of animal symbolism, let us begin with the most traditional of Dürer's three animal representations: his drawing from 1510.

Serpent, Stag, and Lion in Dürer's 1510 Drawing

As *Genesis* 2 gives no detailed information about the serpent in Paradise, early modern artists were confronted with a fundamental question: What did the serpent of Paradise look like before becoming a real snake, that is, before receiving the punishment of God, who after the Fall said: 'on your *belly* you shall go' (*Genesis* 3:14)? Was the prelapsarian serpent a 'normal' snake, a seductive child or woman with a serpent's tail, or a hideous dragon, as can be seen in the famous engraving by Rembrandt?[5] Whereas in his other representations of the Fall Dürer chooses the first option—a normal snake, with only a little crown to distinguish him from other snakes—in this drawing he opts for the second possibility: a seductive female figure in miniature. Upon closer look the viewer is struck by the ambivalent eroticism of this creature: its bare

3 Schoen C., *Albrecht Dürer: Adam und Eva. Die Gemälde, ihre Geschichte und Rezeption bei Lucas Cranach d. Ä. und Hans Baldung Grien* (Berlin: 2001) 47.

4 Panofsky E., *The Life and Art of Albrecht Dürer* (Princeton: 1948 [1943]).

5 On Rembrandt's choice for the dragon, see Smith P.J., "Diersymboliek in Rembrandts *Zondeval* (1638) en in Vondels *Adam in ballingschap* (1664)", *De zeventiende eeuw* 26 (2010) 2–20.

FIGURE 8.3 *Albrecht Dürer,* The Fall of Man *(1504), copper etching.*

breasts denote female eroticism, but they also give to the serpent the form of an erect phallus. This phallic interpretation of the serpent is in accordance with 16th-century theological debates on the phallic nature of the serpent of Paradise. According to some theologians, the serpent symbolizes Adam's penis, and the significance of the Fall comes down to Man's discovery of sexuality—just as we saw in Dürer's 1498 drawing. One finds allusions to this theological

FIGURE 8.4 *Albrecht Dürer,* The Fall of Man *(1510), drawing. Vienna, Albertina.*

FIGURE 8.5 *Albrecht Dürer*, The Fall of Man (*1510*). *Woodcut taken from*: The Small Passion, *Chelidonius Benedictus*, Passio Christi ab Alberto Nurenbergensi effigiata cum varii generis carminibus (*Nuremberg, Albrecht Dürer: 1511*).

interpretation in the novels by François Rabelais, who also gives examples of the ambiguous gender, both male and female, of the serpent of Paradise. In his *Quart Livre* (1552), a liar's tale that was partly inspired by Lucianus' *True History*, Rabelais dedicates a whole episode to the Andouilles ('Chitterlings'), who are female creatures with a phallic form and whose queen is named 'Niphleseth', which is, as Rabelais explains, Hebrew for 'phallus'. In order to 'prove' this liar's tale, Rabelais (or rather his narrator) inserts a long pseudo-learned dissertation on female creatures with phallic forms:

> Encore maintient on en certaines Academies que ce tentateur [*i.e.* the Devil] estoit l'andouille nommée Ithyphalle, en laquelle feut jadis transformé le bon messer Priapus grand tentateur de femmes par les paradis en grec, ce sont Jardins en François.[6]

In this mock dissertation Rabelais mentions some other female phallic creatures: the Himantopodes, a monstrous race mentioned by Pliny; the fairy Melusine ('[elle] avoit corps foeminin jusques aux boursavitz [her genitals], et que le reste en bas estoit andouille serpentine, ou bien serpent andouillicque'); and the nymph Ora ('[elle] avoit pareillement le corps my party en femme et en Andouilles').

The authorities and examples quoted by Rabelais suggest that Dürer's viewers—at least the learned, humanist ones—could also have been acquainted with these kinds of ambiguous creatures, and would have understood the artist's theological allusions. This is also the case for the two other animals represented, the stag and the lion.

In medieval and early modern iconography, the stag is a christological symbol, for two main reasons. Firstly because of his antler, which the animal discards but which grows back again, and thus is the symbol for Christ's Death and Resurrection. As is visualized in Dürer's engraving *St. Eustace* (1501), the stag's antler and the Holy Cross are inextricably connected with each other. Secondly, because according to ancient zoology, the stag lives in antipathy with the snake.[7] The French architect Joseph Boillot cites the most evident ancient sources for this antipathy in his book on the use of animal antipathies in the decoration of columns (*Nouveaux pourtraitz et figures de termes pour user*

---

6 Rabelais François, *Œuvres complètes*, ed. M. Huchon (Paris: 1994) 628. In her glosses, Huchon notes: 'Pour Agrippa, *De originale peccato*, 1532, le serpent tentateur n'est autre que le pénis d'Adam' (p. 1552).

7 Regarding both reasons, see Bath M., *The Image of the Stag. Iconographic Themes in Western Art* (Baden-Baden: 1992).

en l'architecture (1592)): '[Le cerf] a forte guerre [avec les serpents], comme descript amplement Oppian liure second de la chasse, & en parlent Pline & Elian'.[8] One can also quote widely known medieval religious animal symbolism, like that found, for instance, in numerous medieval bestiaries, such as the 12th-century Latin bestiary known as the Aberdeen Bestiary (Aberdeen University Library MS 24), cited here in translation:

> Deer are the enemies of snakes; when they feel weighed down with weakness, they draw snakes from their holes with the breath of their noses and, overcoming the fatal nature of their venom, eat them and are restored.[9]
>
> They have another characteristic, that after eating a snake they run to a spring and, drinking from it, shed their long coats and all signs of old age.[10]

These 'features' automatically lead to a 'congruous and competent symbolism':

> after the incarnation of the Devil, that is, after committing a sin, they [the Christians] run, by their confession, to Christ, the true spring; drinking in his commandments, they are renewed, shedding their sin like old age.[11]

Through its features, the stag gives a typological meaning to the Fall, foreshadowing Christ's Death and Resurrection, and the Christians' Redemption as well.

This is also the case for the lion. These are the two features of the lion, relevant to our topic, in the words of the Aberdeen Bestiary:

> The second characteristic of the lion is that when it sleeps, it seems to have its eyes open. Thus our Lord, falling asleep in death, physically, on the cross, was buried, yet his divine nature remained awake; as it says in the Song of Songs: 'I sleep but my heart waketh' (5:2); and in the psalm: 'Behold, he that keepeth Israel shall neither slumber nor sleep' (121:4).
>
> The third characteristic of the lion is that when a lioness gives birth to her cubs, she produces them dead and watches over them for three days,

8 Boillot Joseph, *Nouveaux pourtraitz et figures de termes pour user en l'architecture, Langres 1592*, eds. P. Choné – G. Viard (Paris: 1995). In the absence of source texts prior to Dürer, I will quote a number of texts which, although they are posterior to Dürer, give an impression of how classical and medieval themes survive well into the 16th century.

9 See http://www.abdn.ac.uk/bestiary/translat/13r.hti (last consultation: 15 July 2013).

10 See http://www.abdn.ac.uk/bestiary/translat/13v.hti (last consultation: 15 July 2013).

11 Idem.

> until their father comes on the third day and breathes into their faces and restores them to life. Thus the Almighty Father awakened our Lord Jesus Christ from the dead on the third day; as Jacob says: 'He will fall asleep as a lion, and as a lion's whelp he will be revived' (see *Genesis* 49:9).[12]

The principle of antipathy, which we already noticed in the relationship between stag and serpent, is also at work between stag and lion. Both animals have antipathy toward each other, as can be read in ancient and early modern treatises on natural history. Boillot, for instance, says about the deer: 'cest la femelle du Cerf [...]. Ses contraires sont Lions, Serpens, Chiens, & aultres animaux'. The combination of two adversarial animals like the lion and the stag shows us the general peace reigning between the creatures in Eden—peace that is in stark contrast with the situation after the Fall, when the lion goes deer hunting, and all of the animals devour one another. This postlapsarian chaos is well illustrated in one of the woodcuts of the above-mentioned *L'Adamo, sacra rapresentatione* [Fig. 8.6], with the middle of the illustration showing a lion

FIGURE 8.6 *Cesare Bassano*, After the Fall. *Engraving in Gianbattista Andreini*, L'Adamo, sacra rapresentatione (*Milan, Geronimo Bordoni: 1613*).

12 See http://www.abdn.ac.uk/bestiary/translat/7v.hti.

killing a stag. Another interesting thing is that in Dürer's drawing, as in almost all of the representations of the Fall where lion and stag are combined, it is the lion that stands on the side of Eve, whereas the stag is on Adam's side. The symbolism is clear: Adam is intended to be the prey of Eve.

The Animals in the 1504 Print

There has been much comment upon the 1504 print. A lot of attention has been given to the corporal aspects of Adam and Eve; Dürer's preoccupation with the ideal human proportions and positions, inspired by statues of Antiquity (the Apollo Belvedere and Venus, for instance); and his characters' disposition and mutual interaction, which will be imitated by a number of other artists. This section will only focus on the representation, not of man, but of the animals and their meaning. Dürer has represented eight animals in this engraving, including the serpent, an elk passing behind Adam, a cat and a mouse at the feet of Adam and Eve, and a hare and an ox facing each other; in the Tree of Life there is a parrot, or rather a Ring-necked parakeet, to be precise; and on the top of the rock there is an ibex or chamois. The best-known interpretation of the animal aspects of this engraving is given by Erwin Panofsky:

> An educated observer of the sixteenth century [...] would have easily recognized the four species of animals in Dürer's engraving as representatives of the "four humors" and their moral connotations, the elk denoting melancholic gloom, the rabbit sanguine sensuality, the cat choleric cruelty, and the ox phlegmatic sluggishness.[13]

According to Panofsky, who based himself on medieval scholastic ideas on the four humours, these four humours, connected with the four elements, are in balance in the paradisiacal situation before the Fall. After the Fall, however, this equilibrium disappears rapidly.

Panofsky's interpretation became authoritative in the iconological interpretation of these animals. In 2001, two critical reactions were published independently from each other: the above-mentioned study by Christian Schoen and my own article, of which this present section is an updated synthesis.[14]

---

13 Panofsky, *Dürer* 85.

14 Smith P.J., "The melancholic Elk. Animal symbolism and linguistic ambiguity in Albrecht Dürer's *The Fall of Man* (1504) and Aegidius Sadeleer's *Theatrum Morum* (1609)", in Bogaards P. – Rooryck J.– Smith P.J. (eds.), *Quitte ou Double Sens. Articles sur l'ambiguïté offerts à Ronald Landheer* (Amsterdam – New York: 2001) 333–343.

My main criticism of Panofsky's interpretation is that it is limited to only four animals: the elk, the cat, the hare, and the ox. How should one interpret the three other animals represented, and the obvious mutual relationships between the cat and the mouse and the hare and the ox? Schoen seems to have the same criticism, albeit more implicit ('Maus, Papagei und Steinbock spielen bei dieser Interpretation kein Rolle'),[15] and comes to an alternative interpretation: 'Die Tiere wären somit verschlüsselte Hinweise auf spezifische Sünden, die auf die Ursünde zurückzuführen sind'.[16] This is how Schoen interprets the cat and mouse, whose confrontation is traditionally 'immer als Hinweis auf die Triebhaftigkeit des Menschen bzw. als Reflektion des antizipierten Geschlechterverhältnisses verstanden'. The parrot can be interpreted as a symbol of greed, and the ibex 'darf als vorausweisendes Zeichen des selbstverschuldeten Untergangs gedeutet werden'.[17] My own interpretation, proposed in 2001 and updated here, goes in another direction: I see Dürer's animal symbolism as a kind of puzzle or riddle to decipher, a puzzle that is based both on traditional animal symbolism and on learned linguistic remotivation of animal names and of proverbial material.

Let us start with the elk. The presence of this animal in the Earthly Paradise is surprising to say the least. One would expect to see a normal stag, which is, as we saw, a traditional christological symbol. As a kind of deer, the elk surely has the same typological symbolism, but why did Dürer choose precisely this kind of deer? The first interpretative clue lies in the animal's name in German: the *Elendt* (elk) is a foreshadowing of mankind's *Elendt* (sorrow) after the Fall.[18] Therefore, Panofsky's interpretation of the animal as a symbol of melancholy is valid, all the more because some early modern humanists mention the melancholic nature of this animal, illustrated by its name. Thus, the learned Swiss natural historian Conrad Gessner explains that this homonymy fits well with the elk's gloomy character: 'Germanicum nomen miseriam significat; & vere miserum est animal, si credendum est quod saepe audivimus' ('The German word means 'misery', and this animal really is sad, if it is true what we always have

15 Schoen, *Dürer* 113.

16 Schoen, *Dürer* 114.

17 Schoen, *Dürer* 114.

18 Dürer was possibly aware of the etymology of the German word *Elend*: the word is derived from the Old Saxon *elilendi*, which is composed of **alja* ('other') and *land* ('country'). The original meaning of the word is thus: living in another country, which is the cause of misery (cf. Kluge Friedrich, *Etymologisches Wörterbuch der deutschen Sprache*, 22nd ed. [Berlin – New York: 1989], s.v. Elend). In the context of the Fall, this means misery for Man caused by being chased from the Earthly Paradise (suggestion made by Merk Kingma).

heard').[19] But Dürer goes much further than his wordplay on the homonymy of the word *Elendt*. The Fall itself and Man's Redemption are suggested by the medical power bestowed on the animal. Indeed, the elk's hoof in powdered form presents a medication to treat epilepsy, the *falling* disease—just as Christ by his body cures mankind from the evil consequences of the Fall. In order to substantiate this hypothesis, we can once again quote Gessner, who in a later, much enlarged edition of his zoological encyclopedia, adds the authority of a Polish correspondent, a certain Johannes Bonerus: 'Alces [...] est animal valde melancholicum ac tetrum aspectu [...]. Ungula Alcis (ut scribit Bonerus) habetur apud Polonos gestata in digito vel brachio pro singulari remedio contra spasmum aut morbum caducum'.[20] ('The elk is a very melancholic animal of ugly appearance [...]. The hoof of the elk, as Bonerus wrote, put upon a finger or an arm, is held for an excellent remedy against spasm or the falling disease'). Dürer was very well aware of this widely known medical effectiveness of the elk's hoof, as can be read in his *Journal of the Voyage to the Netherlands*. During his travels in the Netherlands he bought an elk's hoof in powdered form,[21] and just before his death he ordered elk powder from Antwerp to cure his malaria.[22]

Dürer's playful ambiguity appears to be much more complicated if we take into consideration the drawing he made of an elk which he used as a model for his 1504 elk. At the bottom of this drawing, he wrote the name of the animal [Fig. 8.7], spelling it *heilant*. According to the standard work by Walter L. Strauss, this is 'archaic German for *elk*'.[23] But, of course, in German the word is homonymic with *Heilant*, meaning the Redeemer. *Nomen est omen*: by its ambiguous name, the elk is the most appropriate species of deer to symbolize Christ the Saviour. Moreover, *Heilant* (Redeemer) etymologically is the present participle of the verb *heilen* (to cure), so the connotation of curative power remains.

Thus it appears that German linguistic (re)motivation is of great importance for deciphering the typological message of Dürer's engraving. This is also the case for the two pairings of animals represented: the cat and mouse,

19 Gessner Conrad, *Historiae animalium* [...], 4 vols. (Zurich, Christoffel Froschauer: 1551–1558), vol. I *de quadrupedis viviparis* (1551) 3.

20 Gessner Conrad, *Historiae animalium* [...], vol. I, *de quadrupedis viviparis* (Frankfurt a.M., Johannes Wechel: 1620) 2, 4.

21 'Jch hab 20 stüber umb ein ellendsfus geben' (Antwerp, 10 October 1520). I quote from Dürer Albrecht, *Tagebuch der Reise in die Niederlande* [...], ed. F. Anzelewsky (Zurich: 1988) 23. There are two other mentions of elk's hoof in his Journal.

22 Eisler C., *Dürer's Animals* (Washington – London: 1991) 107.

23 Strauss W.L., *The Complete Drawings of Albrecht Dürer*, vol. 2 (New York: 1974) 568. Strauss dates this drawing to 1501.

FIGURE 8.7 *Albrecht Dürer,* Elk *(1504), drawing. London, British Museum.*

and the hare and ox. The mouse's tail touches Adam's foot, just as the cat's tail is twisted around Eve's ankle, thus suggesting a game of cat and mouse played between the two. Both animals literally visualize the proverbial material existing in the vernacular: *jouer au chat et à la souris* in French, and *mit jemandem Katz und Maus spielen* in German. Adam (the mouse) will be Eve's (the cat's) prey, which is the same message we saw in the traditional iconography of Dürer's 1510 drawing, which depicts the lion (Eve) preying upon the stag (Adam). One can, however, also more specifically think of the German proverb *Der Katze[n] Spiel ist der Mäuse Tod,* mentioned by Luther as an apparently widely known proverb, and used by him in the context of the Fall:

> Wir wissen fast wol, das des Teuffels scherz uns Christen ein ernst gilt, wie man spricht, der katzen spiel ist der meuse tod. [...] Und [der Teufel] möchte sich noch eben so wol inn seiner klugheit beschmeissen, als er sich im Paradies beschmeis, do er meinet, er hette nu gewonnen, Aber

> sich gar nicht versahe, das des weibes frucht solt so kurz hinder im her sein, und im das heupt zutretten.[24]

Luther applies the proverb to the game of cat and mouse the Devil played with Man in Paradise, but which will end in his own defeat.

The couple constituted by the hare and the ox, both sitting and strangely facing each other, also seems to point to some proverbial entity, just like the cat and the mouse: the Latin proverb *Bove leporem venari*, mentioned by Gessner, who is giving the German equivalent, *Man solt ehe ein hasen mit der trummen fangen*, in which the ox is replaced with a drum (literally: 'one could more easily catch a hare with a drum'). Erasmus, in the 1515 edition of his *Adagia*, gives the following explanation of the proverb: '*Bove leporem venari*, dicuntur qui rem absurdam, stultam ac praeposteram aggrediuntur' ('*To chase a hare with an ox* is said of those who undertake something absurd, foolish and wrong').[25] Erasmus quotes his main source as well: Plutarch's *De tranquillitate animi*. Plutarch's text is worth citing because it provides an interpretative clue to Dürer's use of the proverb:

> [...] in our expectations we aim at things too great; then, when we fail, we blame our destiny and our fortune instead of our own folly. For he is not unfortunate who wishes to shoot with his plough and *hunt the hare with his ox* [...]; it is through folly and stupidity that such men attempt the impossible. And self-love [*philautia*] is chiefly to blame, which makes men eager to be first and to be victorious in everything and insatiably desirous of engaging in everything.[26]

24 *Eine Wahrhafftige Historia geschehen zu Stasfart, am abend der geburt Christi, im MDXXXIIII jare. Mit einer schönen Vorrede* [*von*] *D. Mart. Luther* (Wittemberg [Nickel Schirlenz]: 1535) no pagination. See: http://digitale.bibliothek.uni-halle.de/vd16/content/pageview/4962894 (last consultation: 19 July 2013).

25 Erasmus, *Adagia* IV, iv, 44 (3344). I quote: http://sites.univ-lyon2.fr/lesmondeshumanistes/wp-content/uploads/2010/09/Adages-tome-4.pdf (last consultation: 19 July 2013).

26 *Moralia* 471d. I quote the Loeb edition: Plutarch, *Moralia* (ed. and trans. W.C. Helmbold), vol. 6 (London: 1939) 205. Without suggesting that Dürer himself had read Plutarch's treatise, we can plausibly admit that as *pictor doctus* he was acquainted with Plutarch, albeit indirectly, through his humanist contacts, for whom the discovery of Plutarch's works was one of the revelations of their time. Plutarch's treatise was well known in Dürer's time, although there seems to be no printed Latin translations before the one Guillaume Budé published and dedicated to Pope Julius II in 1505. However, it is quite possible that manuscript copies of this or other Latin translations of the treatise circulated

This quotation is very applicable to Adam's and Eve's *hubris*.

The Tree of Life with the parakeet at Adam's right side can be interpreted in the same onomastic way. Panofsky rightly identified this tree as a mountain ash (*Sorbus spec.*). The English name says nothing about an eventual relationship between the tree and the bird, but in German the name of the tree is *Vogelbeer*, which explains the presence of the bird in this particular tree, and not elsewhere in the illustration.

One of the reasons the Tree of Life is traditionally identified with a mountain ash is that, contrary to other trees, the mountain ash can simultaneously show the different stages of growing: buttons, young and full-grown leaves, flowers, and berries. This is exactly what Dürer is representing, thus turning the tree into a symbol of reflection on time—we will come back to this later.

The parrot is chosen for its imitative abilities, which made it a favourite symbol of the Renaissance artist and his ambitions to represent—to imitate—Nature. From a typological point of view, the bird (*avis—ave* in *ablativus* and *vocativus singularis*) functions as a *trait-d'union* between Eve and Mary through his ability to say *Ave* (the angel's greeting word to Mary in the Annunciation), which is the reverse of *Eva*.[27] Sixteenth-century zoologists like Ulisse Aldrovandi also stress the very peaceful character of the bird; it lives in sympathy with all of the other animals, even the wolf, one of the most antipathic animals in animal symbolism: 'Psittacus & Lupus simul pascuntur. Semper enim viridem hanc Avem amant lupi' ('The parrot and the wolf eat together. Wolves, indeed, always love this green bird').[28] Aldrovandi makes only one exception: the parrot is in antipathy with the serpent. This antipathy is perfectly applicable to Eve and her female offspring, who after the Fall are doomed to live in perpetual enmity with the serpent.[29]

---

widely among the humanists. For the early reception of Plutarch, see Aulotte R., *Amyot et Plutarque. La tradition des* Moralia *au XVIe siècle* (Geneva: 1965).

27 See Dittrich S. – Dittrich L., *Lexikon der Tiersymbole. Tiere als Sinnbilder in der Malerei des 14.–17. Jahrhunderts* (Petersberg: 2004) 322: 'Das Grußwort *Ave*, das Papageien nachsprechen lernten, die Umkehrung von *Eva*, stellte die Sinnbildbeziehung einerseits zu Eva her, mit der durch den Sündenfall der Tod in dem Welt gekommen war, und andererseits zu Maria als der neuen Eva, die Leben und Heil brachte'.

28 Aldrovandi Ulisse, *Ornithologiae, hoc est de avibus historia libri XII* (Bologna, Franciscus de Franciscis: 1599) 653.

29 One of the best-known examples, contemporary with Dürer, of the love symbolism of the parrot is *Les Epîtres de L'Amant Vert*, written c.1506 by the court poet Jean Lemaire de Belges for his patroness Marguerite d'Autriche. These poems on the "Green Lover" (i.e. Marguerite's favourite parrot) were printed in 1511.

Finally, there is the chamois, which seems to be about to throw itself down, and therefore is another symbol announcing the Fall of Man. To be convinced, one can once again read the Latin bestiary quoted above:

> There is an animal called the ibex, which has two horns of such strength that, if it were to fall from a high mountain to the lowest depths, its whole body would be supported by those two horns. The ibex represents those learned men[30] who are accustomed to manage whatever problems they encounter, with the harmony of the two Testaments as if with a sound constitution; and, supported as by two horns, they sustain the good they do with the testimony of readings from the Old and New Testament.[31]

The accompanying illumination is also very significant: it represents the falling chamois. Illustrations of other bestiaries represent the animal at the very moment it is about to jump—just as Dürer represents it. The chamois is therefore a perfect example not only of the visual rendering of the Fall, but also of the typological reading of it, by interpreting the Old Testament in light of the New.

Except for the active serpent and the moving elk, all animals seem to be at rest, waiting indolently for the inevitable: the chamois on the point of jumping into the depths, and the cat and mouse, and the hare and ox all on the point of chasing each other. Time seems to come to a standstill in a rather paradoxical way, as can be seen in the Tree of Life, itself a symbol, as we have seen, of plural temporality combined in one instant.

Both trees also merit a close inspection—closer than normal, because in most of Dürer's other engravings trees are not distinctively individualized, whereas here they seem to ask for a closer look. Indeed, the Tree of Life invites the viewer to seek a deeper meaning. At the height of Adam's leg it is partly stripped: this can, of course, be a purely ornamental coincidence, but it can also be in line with the well-known (since the time of the Church Fathers) image of the bark, whose rough exterior hides an inner message, a deeper meaning, an invitation to read 'sub cortice verborum hominum' ('under the bark of the words of men')—as is expressed in the oft-quoted sentence of Bonaventura: 'sub cortice litterae apertae occultatur mystica et profunda intelligentia' ('under the bark of the visible letter lies hidden a mystical and deep

30 http://www.abdn.ac.uk/bestiary/translat/11r.hti.

31 http://www.abdn.ac.uk/bestiary/translat/11v.hti.

understanding') (*Prologus in Breviloquium* § 4).[32] It is not impossible that there is also a reference being made to one of the numerous vernacular expressions on the bark and the tree, all expressions meaning a very difficult choice to make, a true dilemma: 'entre l'écorche et l'arbre', 'Put not your hand between the bark and the tree'. Modern German has an expression that is very applicable to the hesitant Adam: 'zwischen Baum und Borke wählen', i.e. to choose between two impossible alternatives, but I have not found any medieval or early modern attestation of this expression. There is, however, a Latin application of the expression in the much-read *Gesta Romanorum*, which can be applied to the whole scene of the Fall:

> Ponatur ergo mortis memoria inter arborem et corticem, i.e. inter animam et corpus. Corpus est quod cortex est, regens animam, et sic diabolus numquam prolem procreabit perverse operacionis.[33]
>
> Therefore the thought of death should be put between the tree and the bark, i.e. between the soul and the body. The body is the bark, which should be guided well by the soul, and therefore the devil should never have offspring by a malicious action.

There are several paradoxical aspects to the Tree of Life. We have already shown how this expresses and summarizes the cycle of life, from button to fruit. Time is, as it were, frozen and condensed into one single moment. This comes also to be expressed in a different way. The growing of the Tree of Life should, of course, be a slow vegetal process, yet it is at the same time a rapid movement, covering Adam in action. The branch of the Tree of Life makes an impossible curve from behind Adam to cover his genitals, depicted in such a way that the viewer sees the branch's leaves from below, and therefore his gaze coincides with the direction of the branch's growth—that is, to Adam's genitals. The relationship between Adam and the Tree of Life is very traditional—after all, mankind originates from Adam's genitals—and this can also be interpreted typologically: one can quite literally set up a genealogical tree that begins with Adam and ends with Christ, thus drawing a direct line from the Old to the New Testament. This is a well-known theme in iconology: from

32 Bonaventura, *Brevilogium* [...] (ed. and trans. J.C.M. van Winden – A.H. Smits), vol. 1 (Van Gorcum: 2000) 56.

33 *Gesta Romanorum*, ed. A. Keller, vol. 1 (Stuttgart – Tübingen: 1842) 71. See http://books.google.nl/books?id=xcp_HacrNsgC&printsec=frontcover&hl=nl&source=gbs_ge_summary_r&cad (last consultation: 15 July 2013).

the corpse of Adam, buried deep under the earth, springs a tree that turns into the wood of the Cross.[34]

The other tree, the Tree of Knowledge, is also paradoxical: the form of the leaves suggests that it is a fig tree, but the forbidden fruit looks more like an apple than a fig (which is mostly pear-shaped). The fig tree corresponds to the only botanical clarification that the book of *Genesis* gives: after the Fall Adam and Eve cover their nakedness with fig leaves (*Genesis* 3:7). The apple-shaped fruit is probably related to a traditional interpretation that identifies the forbidden fruit with an apple, according to the Latin homonymy between *melum* or *malum* ('apple tree') and *malum* ('evil', 'misfortune'). The roundness, according to the above-cited Aberdeen Bestiary, is most typical for an apple: 'The apple-tree, *malus*, was so called by the Greeks because its fruit was rounder than any other. From this comes the belief that real apples are those which are exceedingly well-rounded.'[35]

The position of the two trees relative to each other also seems to have a meaning. By its roots the Tree of Life seems to be linked to the Tree of Knowledge. This can, of course, be sheer coincidence or inspired by some aesthetic reason, but it is not impossible that here an echo of the Jewish Mystics or the Kabbalah resounds, in which the humanists took an interest at the time of Dürer, ever since the appearance of Reuchlin's *De verbo mirifico* (1494). On the symbolism of the two trees in Jewish mysticism Gershom Scholem writes the following:

> In the opinion of the Jewish mystics, both trees are in essence one. They grow out into two directions from a common trunk. *Genesis* tells us that the Tree of Life stood in the center of Paradise, but it does not indicate the exact position of the Tree of Knowledge. The Kabbalists took this to mean that it had no special place of its own but sprouted together with the Tree of Life out of the common matrix of the divine world. [...]
>
> The Tree of Life represents that aspect which has hitherto been unrealizable because, due to the sin of Adam, it remained virtually hidden and inaccessible, and we do not know the taste of its fruits. The law, which is concealed in the life of this tree, is that of a creative force manifesting itself in infinite harmonies, a force which knows no limitations or boundaries. The paradisiacal life under this law never came into being. The sin

34 See, for instance, the contemporary triptych on the *Crucifixion* (*c.*1510) by Cornelis Engebrechtsz. (*c.*1462–1527) (Museum De Lakenhal, Leiden).

35 See http://www.abdn.ac.uk/bestiary/translat/78v.hti. This etymology does not seem to have any linguistic grounds.

> of Adam was that he isolated the Tree of Life from the Tree of Knowledge, to which he directed his desire.[36]

With this quote by Scholem in mind, one can give an interpretation of the intermediate position of Dürer's Adam between both trees. Adam indeed seems to get loose from the Tree of Life and to focus on the Tree of Knowledge.

All of these paradoxes—past, present, and future solidified in one moment; the forbidden fruit, which is at the same time apple (the cause of evil) and fig (the cover of evil)—as well as the hermetic animal and tree symbolism, announce the hermetic density of Dürer's *Melencolia I* (1514), with its accumulation of enigmatic objects which, taken separately, require interpretation in order for the entire engraving to be understood. Through this Dürer presents himself as an exceptionally intellectualized artist. This also gives an indication of how we should interpret the wood engraving from 1510.

Badger and Bison in the 1510 Woodcut

As Christian Schoen rightly affirms: 'Seit Panofsky's Analyse von Dürer's Kupferstich von 1504 ist die Deutung von Tieren bei Adam und Eva-Darstellungen als Vertreter der Temperamente zu einem Automatismus geworden'.[37] Panofsky himself gave legitimacy to this interpretative automatism in interpreting Dürer's woodcut in the same manner as his 1504 etching:

> Only three beasts are depicted, and they symbolize the non-sanguine temperaments, the lion standing for 'choleric' wrath, the bison for 'melancholic' gloom and inertia, and the particularly conspicuous badger, notorious for his laziness, for 'phlegmatic' sloth. The 'sanguine' temperament is represented by Adam and Eve themselves; they illustrate its most characteristic feature, the capacity and inclination for love.[38]

This interpretation is as arbitrary as the previous one: I have not been able to find early modern reports of the melancholic gloom of the bison, and the interpretation of the badger as a 'lazy' beast appears anachronistic, as we shall see. And to put the fourth humour in Adam and Eve seems to me to be a distorted

36 Scholem G. "The trees of Eden in the Kabbalah", in Elon A. – Hyman N.M. – Waskow A. (eds.), *Trees, Earth, and Torah: A Tu B'Shvat Anthology* (Philadelphia: 1999) 115–120 (115).

37 Schoen, *Dürer* 160.

38 Panofsky, *Dürer* 143–144.

interpretation, imposed by Panofsky's immovable interpretative format, which is based on the scheme of the four humours.

In order to arrive at an interpretation of the woodcut, and of the animals depicted in it, we must see this woodcut not only as an isolated image, but as part of the larger whole for which it was intended. It is the second woodcut of a series known as *The Small Passion*. This series illustrates a small devotional book, entitled *Passio Jesu Christi*, with Latin poems written by Benedictus Chelidonius.[39] At the opening of this booklet the reader's eye is first struck by the image of the Man of Sorrows, and it then immediately passes to the second woodcut, which shows a huge badger, which, half pictured, looms up and goes from left to right, heading to the scene of the Fall. It is clear that this badger is not just a decorative element, but must have a symbolic meaning in relationship with both the scene from which he comes (the Man of Sorrows) and the scenes to which he goes: the Fall and the Expulsion from Paradise, represented in the third woodcut.

The movement of the badger, from left to right, reminds us of the movement of the elk in the 1504 etching. Does the badger have, by name and through its nature, the same positive iconological possibilities as the elk? Yes, indeed: in the Middle Ages and in the early modern period the badger did not have the dysphoric symbolism Panofsky believed it had. The animal was not 'notorious for his laziness', but, on the contrary, well known for his inventiveness and diligence in building his very complex subterranean den, or badger sett. This is mentioned by the great medieval encyclopaedists, such as Vincent of Beauvais, Thomas of Cantimpré, and Batholomeaus Anglicus, as we can read in a very useful article by Bohdana Librová.[40] Moreover, the badger is a very clean animal that cannot support any filth ('est mundissimum animal quod fetrum sustinere non potest'), according to Jacques de Vitry.[41] Therefore, the badger lives in antipathy with the fox. The fox sneaks into the badger's sett during the short absence of its rightful inhabitant, and he contaminates the den with urine and excrement, and therefore the badger, which cannot support the dirt, is forced to leave his home. This well-known story was given a religious interpretation in an anonymous 14th-century text:

39 Chelidonius Benedictus, *Passio Christi ab Alberto Nurenbergensi effigiata cum varii generis carminibus* (Nuremberg, Albrecht Dürer: 1511).

40 Librová B., "Le renard dans le *cubiculum taxi*: les avatars d'un *exemplum* et le symbolisme du blaireau", *Le Moyen Age* 109, 1 (2003) 79–111.

41 Cited by Librová, "Le renard".

> Taxus deus. Vulpes diabolus. Taxus [mordet graviter] et vulpes vult habere nidem sine labore. Nidus taxi est anima hominis, quam violat diabolus.[42]
>
> The badger is God. The fox is the devil. The badger [bites hard] and the fox wants to have his home without having to work for it. The badger's house is man's soul, which is violated by the devil.

The christological symbolism of the badger is explicitly mentioned in the following quotation:

> Taxus est Christus, caverna eius cor humanum, in qua [...] vulpes, i.e. diabolus, stercus divitiarum [...] imponit.[43]
>
> The badger is Christ. His hole is the human heart in which [...] the fox, i.e. the devil, deposits his excrement of richness.

That this story was also known in Dürer's time is evidenced by the fact that it is mentioned by Gessner, who quotes Isidorus and Albertus Magnus, although he does not mention the badger's moralizing religious meaning:

> Propriam sibi foveam vulpes non parat: sed melis foveae, absente ea, introitum excremento inquinat: rediens illa virus olentem deserit et vulpi habitandam relinquit. Isidorus et Albertus.[44]
>
> The fox does not make a hole for himself; but he fouls with his excrement the entry of the badger's hole while the badger is absent; when the badger returns, he runs away from the stinking dung and lets the fox live in his home. Isodorus and Albertus.

We can find the same story in Petrus Berchorius' *Dictionarii seu Repertorii moralis* (1574), now with a moral: 'Taxus naturaliter odit Vulpem. [...] Taxus, mundum animal, est Christus, Vulpes autem fraudulentus est Diabolus' (The badger hates the fox by nature. [...] The badger, a clean animal, is Christ; but the fox, a deceiver, is the Devil).[45]

---

42 Quoted by Librová, "Le renard".

43 Bersuire Pierre, *Reductorium morale* (*c.*1340), quoted by Librová, "Le renard".

44 Gessner, *Historiae animalium*, vol. I (1551) 780.

45 Berchorius Petrus, *Dictionarium seu Repertorium morale* [...] (Venice, Hieronymus Scotus's heirs: 1574), vol. I, 15.

How can we apply this symbolism to Dürer's Fall? Dürer represents the badger at the moment this animal shows up in full daylight; the animal seems to blink against the light. Because badgers are lucifugous and never show up in daylight, there should be a very good reason for this animal to do so. In light of the above-quoted medieval and early modern texts, this reason could be that the animal was chased by the fox out of his home, the fox being the traditional symbol of evil and the Devil ('Vulpes diabolus'). The fox itself does not show up. The Devil is, of course, present in the form of the serpent: he has penetrated into the earthly paradise, i.e. the human soul, from which God is absent. However, in the end the badger will be stronger than the fox: when it comes to a fight, the badger is always the winner, as is reported in works by Hildegard von Bingen and others.[46]

Where could Dürer have found the idea of using the symbolism of a badger? Maybe he was inspired by the frontispiece of the first edition of Chelidonius' *Passio Jesu Christi* [Fig. 8.8].[47] This frontispiece, made by Johann Wechtlin, was certainly a source of inspiration for Dürer. From this frontispiece he took two scenes—namely, the Fall and the Expulsion—and made two separate illustrations out of it. In the middle of the frontispiece there is a representation of the Garden of Eden [Fig. 8.9], traditionally surrounded by a wall, with the common source of the four rivers in it, as well as the two trees: the one on the left is the Tree of Life, and it is indicated by a *situla*, a drinking bucket, which is a liturgical object, holding the Holy Water. The other tree is, logically, the Tree of Knowledge of Good and Evil. At the foot of this tree lies the skin of a furred animal. The form of the skin indicates that this can only be the fur of either a wolf or a fox[48] (bear and badger are excluded, as is a feline or a marten). If this is indeed the skin of a fox, the woodcuts of Wechtlin and Dürer help to interpret each other mutually: the fox, a symbol for the devil, has penetrated into Paradise, but ultimately the fox will pay for this with his life.

It is likely that, as with the elk, Dürer also had an onomastic reason for depicting a badger. This animal has two names in Latin: *taxus* and *meles*. *Taxus* in Latin also means 'yew tree', a tree with a rich symbolism: through its old age (there are yew trees known to be 1500 years old) it symbolizes eternal life; through its toxicity (of berry and leaf) it is also the tree of death—this

46 Librová, "Le renard".

47 Chelidonius Benedictus, *Passio Jesu Christi salvatoris mundi, vario Carminum genere F. Benedicti Chelidonii Musophili doctissime descripta. Cum figuris artificiosissimis Ioannis Vuechtelin* ([Strasbourg], n.n.: [1506]).

48 See, for instance, http://upload.wikimedia.org/wikipedia/commons/e/ef/Red_fox_fur_skin,_white_spotted_3.JPG (date of consultation: 14 July 2013).

FIGURE 8.8 *Johann Wechtlin, frontispiece illustration to Benedictus Chelidonius,* Passio Jesu Christi *(Strasbourg, n.n.: 1506).*

dual symbolism is why the tree has been planted in graveyards everywhere in Europe. With regard to the Earthly Paradise, the yew tree unites both trees of Eden: the Tree of Life and the Tree of Knowledge of Good and Evil, which is also the Tree of Death—as such, some contemporaries of Dürer, such as Baldung Grien, depict the Tree of Knowledge of Good and Evil with a skeleton

FIGURE 8.9 *Johann Wechtlin, frontispiece illustration to Benedictus Chelidonius,* Passio Jesu Christi *(Strasbourg, n.n.: 1506), detail.*

in it, next to the serpent. By its other Latin name, *meles* or *melis*, the badger might have to be interpreted according to the same kind of wordplay. *Meles* is partly homophonic with the Latin and Greek words for *apple: melus* means 'apple', and *melum* or *malum* means 'apple tree'—which is, as we have noticed, homophonic with the Latin word *malum*, signifying 'evil' and 'misfortune'. One notices that, contrary to the 1504 etching, the woodcut presents no ambiguity in the tree species: the form of fruit and leaf indicates that the Tree of Knowledge of Good and Evil is indeed an apple tree. Thus, through one of its Latin names the badger is related to the apple, either by foreshadowing it or by warning Adam and Eve for it.

Let us take a quick look at the other animals. The lion, hidden between the trees in the background, will have the same typical symbolism as it does in Dürer's drawing. The third animal is a kind of cow, a bovine animal. It is hard to make out exactly what species it is. The horns and the long mane indicate that this animal is of a different species than the ox in the 1504 etching. On the basis of the mane, the animal could be a wisent, also known as European bison; the shape of the horns suggests that it is an aurochs. In Dürer's time both animals were already very rare: the bison lived only in the area where the species is still found today—in the primeval forests of Poland—and the aurochs was at that time confined to some hunting grounds, also in Poland, until 1627, when the last aurochs was shot. The species (and their names) were confused with each other in the early modern period.[49] The Latin name for both animals is *urus*, and one of the German names is *Ur*, which is homophonic with the prefix *ur-*, which means *primitive* or *original*. This relationship between the animal name and the prefix is also noted by Conrad Gessner, who was not only a zoologist but also a famous linguist:

> Dicimus [...] *urochs* urum bovem, [...] ut in nominibus *uralt, ur*[*h*]*ane, usprung, ursach, urhab*; et videri potest *u* in hoc sensu factum per syncopem a praepositione *vor*.[50]

> In German we call the 'urus bos' *urochs*, like in the words *uralt, ur*[*h*]*ane, usprung, ursach, urhab*, and in this sense the word *ur* can be seen as a syncope, a contraction of the German preposition *vor*.

One automatically thinks of the *original sin*, and the recently coined German word *Ur-Sünde* also comes to mind.

Conclusion

I hope to have shown that the clue to Dürer's animal symbolism is an erudite, sophisticated play on language, combined with traditional symbolism. Every animal should be interpreted that way; just like in Dürer's *Melencolia I*, every object asks for an individual interpretation that also, however, fits within the whole of the representation. However, it is important to not apply this method blindly to the work of Dürer's imitators. It is likely that Dürer's imitators had an inkling of the deeper meaning of Dürer's animals around Adam and

49 See Pyle C.M., "Some late sixteenth-century depictions of the aurochs (*Bos primigenius* Bojanus, extinct 1627): new evidence from Vatican MS Urb. Lat. 276", *Archives of Natural History* 21 (1994) 275–288.

50 Gessner, *Historiae animalium*, vol. I (1551) 780.

Eve, but they were not able to fathom this meaning in all of its richness and ambivalence.

What Dürer's animal symbolism has brought about is the fact that since Dürer a number of artists have had the tendency to give their own personal interpretation to traditional animal symbolism. This applies to Cranach and Baldung, as has been shown by Christian Schoen, and to Herri met de Bles, as has been recently demonstrated by Michel Weemans.[51] Elsewhere I have shown this to be true for later artists, such as Cornelis van Haarlem and Rembrandt,[52] for whom emblematics seem to play an important role, and for Jan Brueghel the Elder, for whom scientific topicality, especially in the field of natural history, appears to be important. All of these artists are inspired by Dürer but have their own interpretation of animal symbolism.

Bibliography

Aldrovandi Ulisse, *Ornithologiae, hoc est de avibus historia libri XII* (Bologna, Franciscus de Franciscis: 1599).

Anon. (ed.), *Het Aards Paradijs. Dierenvoorstellingen in de Nederlanden van de 16de en 17de eeuw* (Antwerp: 1982).

Aulotte R., *Amyot et Plutarque. La tradition des* Moralia *au XVIe siècle* (Geneva: 1965).

Bath M., *The Image of the Stag. Iconographic Themes in Western Art* (Baden-Baden: 1992).

Berchorius Petrus, *Dictionarium seu Repertorium morale* [...] (Venice, Hieronymus Scotus's heirs: 1574).

Boillot Joseph, *Nouveaux pourtraitz et figures de termes pour user en l'architecture, Langres 1592*, eds. P. Choné – G. Viard (Paris: 1995).

Chelidonius Benedictus, *Passio Jesu Christi salvatoris mundi, vario Carminum genere F. Benedicti Chelidonii Musophili doctissime descripta. Cum figuris artificiosissimis Ioannis Vuechtelin* ([Strasbourg], n.n.: [1506]).

Chelidonius Benedictus, *Passio Christi ab Alberto Nurenbergensi effigiata cum varii generis carminibus* (Nurnberg, Albrecht Dürer: 1511).

51 Weemans M., "The Earthly Paradise: Herri met de Bles's Visual Exegesis of *Genesis* 1–3", in Brusati C. – Enenkel K. – Melion W. (eds.), *The Authority of the Word: Reflecting on Image and Text on Northern Europe, 1400–1700* (Leiden – Boston: 2011) 263–312.

52 See my "Diersymboliek" and my "Fable and Emblem in *The Fall of Man* (1592) by Cornelis van Haarlem", in Manning J. – Porteman K. – Vaeck M. Van (eds.), *The Emblem Tradition in the Low Countries* (Turnhout: 1999) 281–302.

Dittrich S. – Dittrich L., *Lexikon der Tiersymbole. Tiere als Sinnbilder in der Malerei des 14.–17. Jahrhunderts* (Petersberg: 2004).

Dürer Albrecht, *Tagebuch der Reise in die Niederlande* [...], ed. F. Anzelewsky (Zurich: 1988).

Eisler C., *Dürer's Animals* (Washington – London: 1991).

Gessner Conrad, *Historiae animalium* [...], 4 vols. (Zurich, Christoffel Froschauer: 1551–1558), vol. I *de quadrupedis viviparis* (1551).

Gessner Conrad, *Historiae animalium* [...], vol. I *de quadrupedis viviparis* (Frankfurt a.M., Johannes Wechel: 1620).

Kluge Friedrich, *Etymologisches Wörterbuch der deutschen Sprache*, 22nd ed. (Berlin – New York: 1989).

Librová B., "Le renard dans le *cubiculum taxi*: les avatars d'un *exemplum* et le symbolisme du blaireau", *Le Moyen Age* 109, 1 (2003) 79–111.

Luther Martin, *Eine Wahrhafftige Historia geschehen zu Stasfart, am abend der geburt Christi, im MDXXXIIII jare. Mit einer schönen Vorrede [von] D. Mart. Luther* (Wittenberg, Nickel Schirlenz: 1535).

Panofsky E., *The Life and Art of Albrecht Dürer* (Princeton: 1948 [1943]).

Prest J., *The Garden of Eden. The Botanic Garden and the Re-Creation of Paradise* (New Haven – London: 1981) 12–15.

Pyle C.M., "Some late sixteenth-century depictions of the aurochs (*Bos primigenius* Bojanus, extinct 1627): new evidence from Vatican MS Urb. Lat. 276", *Archives of Natural History* 21 (1994) 275–288.

Rabelais François, *Œuvres complètes*, ed. M. Huchon (Paris: 1994).

Schoen C., *Albrecht Dürer: Adam und Eva. Die Gemälde, ihre Geschichte und Rezeption bei Lucas Cranach d. Ä. und Hans Baldung Grien* (Berlin: 2001).

Scholem G., "The trees of Eden in the Kabbalah", in Elon A. Hyman N.M. – Waskow A. (eds.), *Trees, Earth, and Torah: A Tu B'Shvat Anthology* (Philadelphia: 1999) 115–120.

Smith P.J., "Fable and Emblem in *The Fall of Man* (1592) by Cornelis van Haarlem", in Manning J. – Porteman K. – Vaeck M. Van (eds.), *The Emblem Tradition in the Low Countries* (Turnhout: 1999) 281–302.

Smith P.J., "The melancholic Elk. Animal symbolism and linguistic ambiguity in Albrecht Dürer's *The Fall of Man* (1504) and Aegidius Sadeleer's *Theatrum Morum* (1609)", in Bogaards P. – Rooryck J. – Smith P.J. (eds.), *Quitte ou Double Sens. Articles sur l'ambiguïté offerts à Ronald Landheer* (Amsterdam – New York: 2001) 333–343.

Smith P.J., "Diersymboliek in Rembrandts *Zondeval* (1638) en in Vondels *Adam in ballingschap* (1664)", *De zeventiende eeuw* 26 (2010) 2–20.

Strauss W.L., *The Complete Drawings of Albrecht Dürer*, 2 vols. (New York: 1974).

Weemans M., "The Earthly Paradise: Herri met de Bles's Visual Exegesis of *Genesis* 1–3", in Brusati C. – Enenkel K. – Melion W. (eds.), *The Authority of the Word: Reflecting on Image and Text on Northern Europe, 1400–1700* (Leiden – Boston: 2011) 263–312.

CHAPTER 9

Pioneers of the Printed Paradise: Maarten de Vos, Jan Sadeler I and Emblematic Natural History in the Late Sixteenth Century

Amanda K. Herrin

Introduction

In 1587 Jan Sadeler I (1550–1600) was appointed court engraver to Duke Wilhelm V in Bavaria (1548–1626).[1] Shortly thereafter he must have contacted his long-time collaborator, Maarten de Vos (1532–1603), one of the most sought-after print designers in Antwerp at the time.[2] Together they designed and executed an artistic project that would appeal to the aesthetic tastes and spiritual practices of Sadeler's new ducal patron. The fruit of their collaboration, a small, richly-detailed graphic series entitled *Imago Bonitatis* (*Image of Goodness*), traces God's creation of the world across Paradise landscapes diversely animated with living creatures.[3] Emblazoned on the series' title page is the ducal coat of arms with a dedication to Wilhelm V. The seven engravings in the series correspond to the six days of Creation as described in the Old Testament, with

* This article draws on my doctoral dissertation "Narratives of Origin in Netherlandish Art" at the Institute of Fine Arts, New York University, the research for which would not have been possible without the generous support of the Samuel H. Kress Foundation in the form of a two-year Kress Institutional Fellowship at Leiden University. I am indebted to Paul J. Smith in Leiden for offering me an affiliation with the NWO project *Cultural Representations of Living Nature: Dynamics of Intermedial Recording in Text and Image (ca. 1550–1670)*, from which the present volume comes to life. I especially thank Elizabeth A. Honig for her invaluable feedback and keen editorial eye on several drafts. Others whose comments have proven most useful include Erik Jorink and Marrigje Rikken.

1 The arrival of the Sadeler family in Munich in 1587 transformed the Bavarian court into a center of printmaking in Europe. With Jan came his brother Raphael I, and his nephews Aegidius II and Raphael II. See Maxwell S., *The Court Art of Friedrich Sustris: Patronage in Late Renaissance Bavaria* (Farnham: 2011) 200.

2 The three Hollstein volumes attribute the design of at least 1,598 engravings to Maarten de Vos. See Schuckman C. (comp.), *Maarten de Vos*, in Hoop Scheffer D. de (ed.), *Hollstein's Dutch & Flemish Etchings, Engravings and Woodcuts 1450–1700*, vols. 44–46 (Rotterdam: 1996).

3 Ibidem, vols. 44–45, nos. 11–18.

 | DOI 10.1163/9789004279179_011

the final day rendered in two separate prints (*Genesis* 1:1–2:3). Designed to impress, the *Imago Bonitatis* combines Maarten de Vos's ingenuity at rendering complex and lively compositions with Sadeler's own accomplished precision with the engraver's burin.

Several of the prints in the series accurately depict different species of birds, fishes, reptiles, and animals—an impressive accomplishment given that neither artist specialized in animal representation. In the print *The Sixth Day of Creation: The Creation of Adam, Eve and the Animals* [Fig. 9.1] familiar indigenous and domesticated species, such as horses, hares, dogs, and mice, appear alongside more recently-discovered exotic animals, such as the rhinoceros, camel, and llama. Within the series animals are grouped more or less according to their natural environments: ungulate quadrupeds graze in a meadow, fishes are displayed swimming in the ocean, and waterfowl stroll the seashore. In *The Fifth Day: The Creation of the Birds and Fishes* [Fig. 9.2] one can identify

FIGURE 9.1 *Jan Sadeler after Maarten de Vos, Detail from:* The Sixth Day: The Creation of Adam, Eve and the Animals (*ca. 1587*). *Engraving from*: Imago Bonitatis, *Amsterdam, Rijksmuseum, RP-P-OB-5395.*

FIGURE 9.2 *Jan Sadeler after Maarten de Vos*, The Fifth Day: The Creation of the Birds and Fishes (*ca.* 1587). *Engraving from*: Imago Bonitatis, *Amsterdam, Rijksmuseum, RP-P-OB-5394.*

various types of birds in the sky, including a heron (left margin, center), and an owl and a bird of paradise (upper margin, center). Even fantastical creatures, such as the griffin and unicorn, are depicted with striking specificity. The printmakers' articulation of every bristle on an ostrich's head in *The Fifth Day* and a chameleon's distinctive sagittal crest in *The Sixth Day* are all the more remarkable given the prints' scale (20.1 cm. × 25.6 cm.). The *Imago Bonitatis* is the earliest Creation series I know of in Netherlandish printmaking to display a kind of pictorial catalogue of birds, fishes and animals within a narrative biblical context.[4] Even as the display of animals serves the narrative setting, it becomes a subject unto itself. The description and arrangement of avian, marine and terrestrial species demonstrate that the artists—especially the designer, Maarten de Vos—were familiar with sixteenth-century naturalists' strategies for picturing, categorizing, and displaying *animalia*.

The attention to *animalia* in the landscapes of the *Imago Bonitatis* presents us with a case for examining the intersections between the visual arts and the rise of the natural sciences in early modern Europe. This essay considers the acculturation of natural history in religious printmaking, and accounts for the role that court collecting and display practices played in this process. The world of late sixteenth-century natural history was one in which animals were part of an intricate language of metaphors, symbols, and emblems. In his monumental *Historia Animalium* (1551–1558), for example, the Swiss naturalist Conrad Gessner (1516–1565) records not only firsthand observations of animals but also their forms, behaviors, habits, characteristics, etymological nomenclature, folkloric traditions, mythologies, and associative links with colors, astrological properties and other natural objects.[5] Gessner's sources include Aesopic fables, classical mythology, hieroglyphics, antique coins and medallions, adages, and emblems. The importance Gessner places on emblems in creating his web of associations for each particular animal has led one scholar to describe his style as 'emblematic natural history'.[6] The intermingling of known and imaginary animals in these natural histories reflects the associative thinking

4 Arianne Faber Kolb makes this claim for Jan Brueghel the Elder, crediting him with reinventing the Paradise landscape as a natural historical catalogue. However, Brueghel's earliest Paradise paintings date to the 1590s, several years after the production of the print series by Sadeler and De Vos discussed here. See Faber Kolb A., *Jan Brueghel the Elder: The Entry of the Animals into Noah's Ark* (Los Angeles: 2005).

5 Conrad Gessner, *Historia animalium Lib. I* [...] *de quadrupedibus viviparis* (Zurich, Christoffel Froschauer: 1551); *Lib. II* [...] *de quadrupedibus oviparis* (1554); *Lib. III* [...] *de avium natura* (1555); *Lib. IV qui est de piscium* (1558); and *Icones avium* (1560).

6 Ashworth W.B. Jr., "Emblematic natural history of the Renaissance", in Jardine N. – Secord J. – Spary E. (eds.), *Cultures of Natural History* (Cambridge: 1996) 17–37.

characteristic of Gessner and his contemporaries, for whom real, fantastical, and allegorical creatures were all considered an essential aspect of the study of the natural world, in large part because the existence of many wondrous beasts was still a matter of debate in the late sixteenth century. It was at this time that the representation of animals in art underwent an extraordinary flowering in the cities of Munich and Antwerp. The illustrated treatises of naturalists offered artists new pictorial formats for representing animals. Looking to recent publications of naturalists, artists began to devote increasing attention to the depiction of flora and fauna, both in the form of life study and within the narrative framework of biblical and genre imagery.

Jan Sadeler and Maarten de Vos's *Imago Bonitatis* points to the influence that Gessner's encyclopedic style of natural history had on the genre of the *Paradijslandschap* at the end of the sixteenth century. I shall argue here that Jan Sadeler, finding himself in a fertile environment for the collection, display and study of the natural world at the Bavarian court, conceived of a print series he believed would appeal directly to his new ducal patron's interest in the natural world, his collecting practices, and his spiritual devotion. But I contend that it was Maarten de Vos—working in the city Antwerp, where many of his fellow artists made animals the subject of their graphic studies and series—who brought natural history to the graphic *Paradijslandschap*. Together, I believe, Jan Sadeler and Maarten de Vos reinvented the Paradise print as a landscape of natural knowledge, adapting the predominant emblematic world view of sixteenth-century natural history to religious printmaking.

What can the *Imago Bonitatis* tell us about the nature, function and significance of the dialogue between the pictorial arts, natural history, and the culture of collecting at European courts during this period? How did the work of early naturalists directly and indirectly shape Paradise imagery at the end of the sixteenth century? After a brief consideration of Sadeler and De Vos's collaborative undertaking of the *Imago Bonitatis*, we will examine the animal imagery in the prints as it relates to the contemporaneous collection, illustration, and display of *naturalia*. The iconographic dimension of our study is served, in part, by looking at the various written and pictorial sources upon which Maarten de Vos relied in his design of the series. Specific specimens, such as a toucan and a blowfish, as well as imaginative details like a curious *tête-à-tête* between an ape and a fox, evidence that De Vos borrowed motifs from a number of different genres of animal representation. The visual repertoire from which he pulled includes recently-published editions of animal print series, as well as fable collections and emblem literature, all of which exhibit the strong influence of early modern natural history. While the physical production of the series will be tied to the rise of animal representation in Antwerp, the conceptual origins of the *Imago Bonitatis* will be addressed within the artistic and spiritual culture of the Bavarian court as a means to flesh out the significance of the animal imagery for its intended ducal audience.

An argument will be made for the series' conspicuous display of animals as visual exegesis of *Genesis* 1–2.[7] With the animal imagery the series' creators make a central exegetical claim: the Creation discloses the knowledge of all of the worldly expressions of God, and through knowledge of God's creations one might achieve one's ultimate spiritual goal: union with God Himself. The display of *animalia*, I believe, was intended not only to appeal to Wilhelm V's interest in the natural world, but also to encourage the pious beholder's reflection upon God through the variety and splendor of His creation. This meditative function of the image will be shown to be one sympathetic with the Counter-Reformation spiritual convictions of Duke Wilhelm V. Situating the *Imago Bonitatis* with respect to its production between the Munich court and the city of Antwerp—both flourishing centers for the scientific and artistic representation of animals—this essay unravels a complex picture of the uptake of early modern natural history in the visual arts late in the sixteenth century, and offers a glimpse at the possibilities for the artistic mobilization of natural history to follow in the seventeenth century.

Collaborative Origins

With Jan Sadeler installed at the Munich court and Maarten de Vos active in Antwerp the Flemish duo must have collaborated closely on the project's design and realization while working apart. The impetus for the project probably came from Sadeler, seeking to make a good first impression with his debut print series in his role as court engraver.[8] However, I believe we must consider De Vos to be an equal innovative partner in the production of the *Imago Bonitatis*, rather

7 For useful studies on visual exegesis see Berdini P., *The Religious Art of Jacopo Bassano: Painting as Visual Exegesis* (Cambridge, 1997) xii; and Weemans M., "The Earthly Paradise: Herri Met de Bles's Visual Exegesis of Genesis 1–3", in Brusati C. et al. (eds.), *The Authority of the Word: Reflecting on Image and Text in Northern Europe, 1400–1700*, Intersections 20 (Leiden – Boston: 2012) 281.

8 It was typical for the publisher, rather than the designer to take the initiative for a publication and to coordinate its execution. This was the case for many of Sadeler's contemporaries. For example, the Antwerp publisher Philip Galle (1537–1612) stipulated the manner in which designers working for him, including De Vos, were to produce preparatory drawings of subjects that he commissioned from them. See Sellink M. (comp.), *Philips Galle*, in Leesberg M. (ed.), *The New Hollstein Dutch & Flemish Etchings, Engravings and Woodcuts 1450–1700*, Vol. 10

than merely Sadeler's designer-amanuensis. De Vos executed the designs for the series in Antwerp and he then must have sent the drawings to the Bavarian court for Sadeler to engrave and publish.[9] Though the artists left the *Imago Bonitatis* undated, they must have produced it shortly after Sadeler's arrival in Munich in 1587. The title page prominently displays Sadeler's newly-acquired status as 'Celsitudinis suae chalcographus' (His Highness's Printer)—a title he held at the Bavarian court until 1595.[10] Sadeler's signature 'Joã[nnes] Sadeler aut[or] et scalpt[or] exc[cudit]' ('Johannes Sadeler, author and engraver, has made the engravings') on the final print in the series lays claim to his authoritative role in the process. Sadeler also inscribed a dedication to the duke on the title page ('D. D.', 'has given to'). It was no doubt Sadeler's hope that his new patron would subsequently purchase several impressions of the series to disperse within his circle of family and friends, both advertising the artists' skills and conferring a kind of official stamp of approval on the prints themselves.[11] A small *cartolino*, pictured at the lower left margin on the opening print in the series, repeats the ducal dedication alongside Sadeler's court title and De Vos's signature: 'M. de Vos figuravit' ('Maerten de Vos designed it').[12] Anticipating that impressions of the series might circulate both with and without the title page, the *cartolino* ensured that the collaborative and courtly origins of the *Imago Bonitatis* would not be lost on broader audiences.[13] As was typical for Sadeler, he engraved only one state of the prints, allowing for a large number of high-quality impressions to be pulled from each plate. The popularity of the series is evident from the fact that it was copied well into the seventeenth century.[14]

(Rotterdam: 1993) Pt. 1, xlii; and Stock J. van der, *Printing Images in Antwerp. The Introduction of Printmaking in a City: Fifteenth Century to 1585* (Rotterdam: 1998) 143.

9 From 1586 to 1587, prior to settling in Munich, Sadeler was primarily in Mainz and Frankfurt am Main, making it less likely that he carried De Vos's designs with him to Bavaria.

10 Sadeler began to use the title *Celsitudinis suae chalcographus* on prints he produced in the late 1580s, after arriving in Munich. The title appears to refer to his role at court in Bavaria, rather than to the royal publishing privilege granted him by Emperor Rudolf II in 1581.

11 Prints with dedications were typically bestowed as gifts upon the dedicatee, and it was often the case that the dedicatee would purchase copies to give as gifts to his own acquaintances; see Veldman I., *Crispijn de Passe and His Progeny (1564–1670): A Century of Print Production*, trans. M. Hoyle (Rotterdam: 2001) 68–81.

12 The term *figuravit* was used by printmakers during this period to identify the artist's specific role as the designer, rather than engraver or publisher of the print.

13 The full inscription on the *cartolino* reads: 'Sereniss[i]mo Bavariae Duci Chalcograph[us] I. Sadeler. Fec[it]: et excud[it]: M. de Vos figuravit'.

14 Anonymous copies after the series were published by the Amsterdam print publisher Claes Jansz. Visscher (1587–1652) in 1639, along with an undated copy by the French print publisher Pierre Firens. Another French print publisher, Thomas de Leu (1560–1612),

Sadeler's selection of the Creation theme for his first serial production at court, coupled with his long history of successful collaboration with De Vos, made the latter an obvious choice as the designer of the *Imago Bonitatis*. The pair had worked successfully with one another since the mid-1570s.[15] Sadeler was active in Cologne from 1579 to 1582 and thereafter in Antwerp between 1582 and 1586. In the year leading up to his court appointment, Sadeler was in Mainz and Frankfurt am Main. Throughout his peregrinations, he and De Vos continued to collaborate with one another. They dedicated several of their print series to important persons, revealing the sort of high-profile clientele that the market-savvy printmakers cultivated in these years.[16]

Neither the elite dedication, nor the subject matter of the *Imago Bonitatis* was foreign to the Flemish duo. Together Sadeler and De Vos had tackled the biblical subjects of the Creation and the Fall of Man a number of times. Their single-leaf print *The Fall of Man* (1579) [Fig. 9.3] shows the figures of Adam and Eve dominating a narrow landscape, in their company a few small animals including a lion, two hares, an elephant, and a deer.[17] Already in this early print one finds evidence of Maarten de Vos's familiarity with the animal imagery

---

produced a copy of the series in reverse, with two of those prints bearing the signatures of the engravers Elias van den Bosche and Edmundus Charpy. See Schuckman, *Maarten de Vos*, Vol. 44, 13. De Vos and Sadeler's series also served as the model for a series of paintings on large cooper panels by Willem van Herp the Elder (1613/14–1677) and Isaac van Oosten (1613–1661); see the catalogue of Johnny van Haeften, *Dutch and Flemish Old Master Paintings* (2008) no. 18. Jan Brueghel the Younger (1601–1678) based *The Creation*, a cycle of three copper paintings, on the print series; see the sale catalogue for Christie's, South Kensington, London, *Old Master Pictures and Drawings*, 10 December 2004, lot 3.

15 Their earliest prints together are dated 1576 and include: *The Christ-Child*; *Death of the Virgin*; and *Coronation of the Virgin*. See Schuckman, *Maarten de Vos*, vols. 44–46, nos. 692, 719, and 721. See Ramaix I. de, *Les Sadeler. Graveurs et éditeurs* (Brussels: 1992) 10; and Nagler G.K., *Neues allgemeines Künstler-Lexicon*, 22 vols. (Munich: 1835–1852) vol. 15, 548–558; and vol. 16, 1–23.

16 These series include: *Boni et Mali Scientia* (The Story of the First Men, 1583), with a dedication to Francesco Maria della Rovere, Duke of Urbino (1490–1538); *Seven Planets* (1585), with a dedication to Alessandro Farnese, Duke of Parma (1520–1589); and *Bonorum et Malorum Consensio* (The Story of the Family of Seth, 1586), with a dedication to Ferdinand II, Archduke of Austria and Count of Tyrol (1529–1595). Political motives and professional aspirations most likely lay behind their targeted audiences. For example, Sadeler dedicated the *Seven Planets* to Farnese in the year that the latter took control of Antwerp for the Spanish crown.

17 Schuckman, *Maarten de Vos*, vols. 44–45, no. 245.

FIGURE 9.3 *Jan Sadeler after Maarten de Vos,* The Fall of Man *(1579). Engraving. London, Victoria and Albert Museum, bequeathed by Eustace F. Bosanquet, E.93-1941. Image © Victoria and Albert Museum, London.*

of his Flemish colleagues. He directly copied a hare from an illustration by Marcus Gheeraerts the Elder (ca.1520–ca.1590). Gheeraerts's hare appeared in Eduard de Dene's (1505–1578) *De warachtighe fabulen der dieren* (1567), an illustrated Flemish edition of Aesop's fables first published in Bruges [Fig. 9.4].[18] De Vos and Sadeler returned to the Creation theme just a few years later in their print series *Boni et Mali Scientia* (1583), which illustrates the story of the first men, from Adam and Eve in Paradise (*Genesis* 2:8–9) to Tubalcain in his

FIGURE 9.4 *Marcus Gheeraerts,* The Hares and the Frogs. *Etched illustration to Eduard de Dene's* De warachtighe fabulen *(Bruges: 1567) fol. 64. The Hague, Royal Library.*

18 Dene Eduard de, *De warachtighe fabulen der dieren* (Bruges, Pieter de Clerck for Marcus Gheeraerts: 1567) 64. It is possible that De Vos copied the motif from the enlarged French edition of De Dene's *Fabulen*, printed in Antwerp in 1578 under the new title *Esbatement moral des animaux*; see *Esbatement moral des animaux* (Antwerp, Philip Galle: 1578). On the various editions of Gheeraerts's fable illustrations see Hodnett E., *Marcus Gheeraerts the Elder of Bruges, London, and Antwerp* (Utrecht: 1971); and Smith P.J., *Het schouwtoneel der dieren. Embleemfabels in de Nederlanden (1567–ca. 1670)* (Hilversum: 2006).

forge (*Genesis* 4:22).[19] The simultaneous scenes of the Creation depicted in the two initial prints of that series include a handful of animals. For instance, a fox sits at God's feet, a lion scratches its ear, and various other animals frame the water's edge in the background: a reclining stag, a long-eared goat, a hare, a giraffe and a camel.

Working independently from one another, Sadeler and De Vos had also executed a number of other Paradise print series. For example, Sadeler engraved a series of eight plates for the *Creation and Fall of Man* after designs by Crispin van den Broeck (1523–ca.1591). These prints were published in Gerard de Jode's (1509–1591) picture Bible, *Thesaurus Sacrarum Historiarum* (1579).[20] And just shortly before De Vos teamed up with Sadeler on the *Imago Bonitatis* he designed a Creation series for the engraver Crispijn de Passe (ca.1564–1637).[21] De Vos went on to portray the subject several more times.[22] These examples of Sadeler and De Vos's early Creation images demonstrate that, prior to their undertaking of the *Imago Bonitatis*, the artists used *animalia* primarily as staffage in their biblical scenes.

---

19 Schuckman, *Maarten de Vos*, vols. 44–45, nos. 245; and nos. 25–36.

20 Gerard de Jode, *Thesaurus Sacrarum historiarum veteris Testamenti* (Antwerp, Philip Galle: 1579). See Mielke H., "Antwerpener Graphik in der 2. Hälfte des 16. Jahrhunderts: Der Thesaurus veteris et novi Testamenti des Gerard de Jode (1585) und seine Künstler", *Zeitschrift für Kunstgeschichte* 38/1 (1975) 29–83; and Mielke U. (comp.), *Crispijn van den Broeck*, in Luijten G. (ed.), *The New Hollstein Dutch & Flemish Etchings, Engravings, and Woodcuts, 1450–1700* (Oudekerk aan den IJssel: 2011) vol. 21, nos. 1–8.

21 Schuckman, *Maarten de Vos*, vols. 44–45, nos. 19–24. De Passe's *Story of Adam and Eve* can be linked with De Vos and De Passe's earliest collaborative projects in 1585 for Philip Galle. The date '1585' is inscribed on four extant drawings for *The Story of Adam and Eve*, suggesting the likely year that De Vos worked on all three sets of design. It is, however, possible that De Vos designed *The Story of Adam and Eve* some years later. De Passe, responding to the religious situation in Antwerp, left that city for good in 1588. He moved his family to the Protestant city of Aachen, where he began designing his own prints, rather than merely engraving after the designs of others. He did, however, continue to engrave works after De Vos well into the 1590s. See Veldman, *Crispijn de Passe and his Progeny* 28 and Schuckman, *Maarten de Vos*, vols. 44–45, nos. 56–61, and 148–153.

22 A few years after producing the *Imago Bonitatis*, De Vos designed another Creation series for the young engraver Nicolaes de Bruyn (ca.1570–1656). Schuckman, *Maarten de Vos*, vols. 44–45, nos. 1–10. The series can be grouped according to a uniformity of style, technique, and format with a number of other series executed by the young engraver between 1594 and 1600. It was during this period that De Bruyn relied primarily on De Vos's designs and worked exclusively for the publisher Assuerus van Londerseel (ca.1572–1635). See Baines L., "Nicolaes de Bruyn (1571–1656) and the Art of the Professional Engraver" (Ph.D. Diss., University of Delaware: 2011) 28–29.

At the start of their *Imago Bontiatis* endeavor, however, Sadeler and De Vos must have discussed updating the pictorial content in their newest of Paradise print projects. Of their various Creation prints and printed series the *Imago Bonitatis* displays by far the artists' most descriptive landscapes. The proportion of pictorial space given over to fauna, and the detail with which the animal life is rendered, sets the series apart from their other works. Teeming with different species, three particular prints in the series—*The Fifth Day: The Creation of the Birds and Fishes* [Fig. 9.2]; *The Sixth Day: The Creation of Adam, Eve and the Animals* [Fig. 9.1] and *The Sixth Day: Adam, Eve and the Animals Blessed By God* [Fig. 9.5]—present a new artistic vision of Paradise. The zoological interest evidenced in these prints serves as an invitation to the viewer's eye to peruse the landscapes as a kind of pictorial catalogue of God's marvelous creatures, staging Paradise as the source for all knowledge of the natural world. Further, it reveals the moment of intersect between natural history and Paradise imagery in northern European printmaking.

FIGURE 9.5 *Jan Sadeler after Maarten de Vos*, The Sixth Day: Adam, Eve and the Animals Blessed by God (*ca. 1587*). *Engraving from*: Imago Bonitatis, *Amsterdam, Rijksmuseum, RP-P-OB-5396.*

The *Paradijslandschap* and Flemish Tapestry Production

The *Paradijslandschap* and the worldview it expressed changed over the course of the sixteenth century. One encounters an anthropocentric view of mankind's origins, with a pictorial focus on Adam and Eve in the moment of the Fall, in works by Albrecht Dürer (1471–1528) and Lucas van Leyden (1494–1533) at the beginning of the century. This vision of Paradise gives way in the *Imago Bonitatis* to an encyclopedic, nature-centric view of the Creation—a view that, despite the difference in the narrative moment portrayed, places greater visual emphasis on landscape and fauna than on the actions and bodies of the narrative's protagonists. Rather than dominating the landscape, the figures of God, Adam, and Eve become integrated into an image of nature's abundance. This vision of Paradise seems to have first found its expression in sixteenth-century Flemish tapestry. By mid-century Paradise had become a popular subject for tapestries produced in the Brussels workshops for elite audiences by artists who often also worked as print designers. This can be seen in *Adam Giving the Animals their Names*, one of a set of seven tapestries produced around 1550 showing the Creation and the Fall from the Brussels consortium of Jan de Kempeneer, Jan van Tieghem, and Frans Ghieteels. The design of the cartoons has been attributed to Jan Cornelisz. Vermeyen (ca. 1500–ca. 1559), a painter, draughtsman, etcher, and tapestry designer active in Flanders and Spain.[23] A strong zoological interest emerges Vermeyen's conception of *Adam Giving the Animals their Names*, which depicts several queues of animals, birds, reptiles and insects filing past Adam to receive their names. The groupings of animals suggest an attempt at species categorization: ostriches, eagles, owls, chickens, turkeys, butterflies, and other winged creatures take to the air; reptiles slither and creep alongside one another; and the smaller and larger mammals each adhere to separate lines. Vermeyen apparently made the cartoon for the tapestry before it was commonly known that ostriches do not fly. With its diverse arrangment of fauna, *Adam Giving the Animals their Names* reveals an awareness of early naturalists' encyclopedic efforts to categorize the natural world. This zoological interest frequently appears in Flemish tapestries, especially in tapestries depicting garden and hunting scenes.

It would take several decades for this vision of Paradise as a kind of catalogue of nature to find its complimentary expression in Flemish printmaking

23 For an illustration of the tapestry now in the Galleria dell'Accademia in Florence, see the exhibition catalogue *Het Aards Paradijs: Dierenvoorstellingen in de Nederlanden van de 16de en 17de eeuw* (Antwerp: 1982) 63, cat. no. 1.

and painting. Once it did, however, the zoological emphasis became a hallmark of Paradise paintings by such Antwerp artists as Jan Brueghel the Elder (1568–1625), and Roelandt Savery (1576–1639). The delay can be related in part to patronage and the rising status of printmaking in Northern Europe. In the golden age of Flemish tapestry production such sumptuous sets of tapestries were the delight of elite patrons. The Brussles set to which *Adam Giving the Animals their Names* belongs was commissioned by Cosimo I de'Medici in Florence and purchased in Antwerp in 1551.[24] This set of tapestries can be compared to a similar *Genesis* set in the royal castle of Wawel in Cracow, thought to have been produced for King Sigismund II Augustus of Poland (1520–1572) in the Brussels workshop of Pieter II van Aelst (before 1527–before 1559) after designs by Michiel Coxcie (1499–1592). Among the tapestries in Cracow, scenes for the *Story of Adam and Eve* and the *Story of Noah* exhibit a familiar zoological interest.[25] While the question remains open as to whether the king commissioned the great set of *Genesis* tapestries, or bought them on the Antwerp market, the intended audience of such grand sets as those in Florence and Cracow was without doubt a noble one. Few others could afford the precious gold and silver threads woven into such textiles.[26] It was with the rise of printmaking in northern Europe in the latter half of the sixteenth-century, and particularly the prolific production of serial prints during this period, that artsits began to produce print series for the same elite audiences.[27] Jan Sadeler and Maarten de Vos, two key figures in Antwerp's rapidly expanding print market, produced a print series for Duke Wilhelm V that picked up where the zoological interests of mid-century Flemish tapestry designers left off.

Maarten de Vos and the Antwerp Animalists

Before looking at the animal motifs in the *Imago Bonitatis* in greater detail, one question regarding Sadeler and De Vos's collaboration deserves further con-

24 Delmarcel G., *Flemish Tapestry* (London: 1999) 94, 121.

25 For example, see *Noah Embarking Animals in the Ark* as illustrated in Szalblowski J. (ed.), *The Flemish Tapestries at Wawel Castle in Cracow: Treasures of King Sigismund Augustus Jagiello* (Antwerp: 1972) 126–127; and in Delmarcel, *Flemish Tapestry* 118.

26 Delmarcel, *Flemish Tapestry* 134. For this discussion see also Duverger J., "Notes concernant les tapisseries du seizième siècle au château du Wawel", *Bulletin des Musées royaux d'art et d'histoire* 45 (1973) 65–76.

27 See Riggs T. – Silver L. (eds.), *Graven Images: The Rise of Professional Printmakers in Antwerp and Haarlem, 1540–1640* (Evanston: 1993).

sideration: To what degree is the naturalistic representation of animals in the series the designer's doing, or that of the engraver and publisher? Given that the whereabouts of the original drawings for the *Imago Bonitatis* are not known, a comparison of the detail in Sadeler's engravings with De Vos's designs eludes us. However, drawings for another series that De Vos designed around the same time offer a valid estimate of his share in the production process. The drawing *Asia* [Fig. 9.6] for the allegorical series *Four Continents* (1585), later engraved and published by his colleague Adriaen Collaert (ca.1560–1618) [Fig. 9.7], shows a female personification seated on the back of a camel.[28] The drawing captures the long tail hairs and cloven toes of a camel, details which also appear in a large

FIGURE 9.6 *Maarten de Vos,* Allegory of Asia *(1585). Drawing, Darmstadt, Landesmuseum, AE 440.*

28 Schuckman, *Maarten de Vos*, vols. 44 and 46, no. 1397. De Vos's original drawing for *Asia* is in Darmstadt; see Bergsträsser G., *Niederländische Zeichnungen 16. Jahrhundert im Hessischen Landesmuseum Darmstadt* (Darmstadt: 1979) 183, no. 137.

painting of a camel that De Vos executed in the 1570s [Fig. 9.7].[29] Yet Collaert's translation of De Vos's drawing onto the copperplate results in a camel with a tail and feet more closely resembling those of an elephant [Fig. 9.8]. This elision of details between De Vos's original drawing and Collaert's final product suggests that perhaps the designer was more familiar with the

FIGURE 9.7 *Adriaen Collaert after Maarten de Vos,* Allegory of Asia, *engraving. Leiden, Leiden University, Special Collections, PK-P-Th. 225.*

29 The painting is one of a series of large-scale 'animal portraits' executed for Johann Albrecht I of Meckelenburg. Six of these are known at present to be in Schwerin and Mainz. In Schwerin, Staatliches Museum: *Camel* (G 739), *Panther* (G 205), *Elephant* (G 740), and *Unicorn* (G 193). And in Mainz, Mittelrheinisches Landesmuseum: *Lion* (Inv. Nr. 666), and *Deer* (Inv. Nr. 667). On this series see Zweite A., *Marten de Vos als Maler: Ein Beitrag zur Geschichte der Antwerpener Malerei in der Zweiten Hälfte des 16. Jahrhunderts* (Berlin: 1980) 283–284, cat. nos. 48–53; and Erbentraut R., "Johann Albrecht I. zu Mecklenburg und die Grossen Tiere", in Blübaum D. – Erbentraut R. (eds.), *Die Erschaffung der Tiere: Tiere in der Niederländischen Kunst des Manierismus und Frühbarock und an den wandfesten Dekorationen des Schlosses zu Güstrow* (Schwerin: 2010) 21–45.

FIGURE 9.8 *Maarten de Vos,* Camel *(1572), Schwerin, Staatliches Museum, G 739/Art Resource, NY.*

dromedary's physiognomy than his engraver.[30] More importantly, De Vos's attention to anatomical detail in his draftsmanship indicates that he left little for his engravers to improve upon or improvise.

De Vos executed the drawings for the *Imago Bonitatis* in Antwerp, where he seems to have spent the entirety of his career, aside from a possible visit to

30 Although, as Daniel Margócsy has observed, the camel's head in the print is not that of a camel, but rather appears more horse-like, with its curved lips, elongated nose and rounded eyes; see Margócsy D., "The camel's head: representing unseen animals in sixteenth-century Europe", *Nederlands Kunsthistorisch Jaarboek / Netherlands Yearbook for History of Art* 61 (2011) 62–85, esp. 64.

Italy in the early 1550s.[31] The worlds of commerce, art, and natural history were closely linked in Antwerp, with ships arriving from all corners of the globe carrying precious goods, natural materials, and people from the New World and the East. Its bustling international port secured the city's status as a prominent commercial and artistic crossroads of Europe throughout much of the century, even after the siege of Antwerp and the blockade of the river Scheldt in 1585. Inundated with foreign objects and natural specimens, the port brought the people of Antwerp into contact with a great variety of flora and fauna, including common types native to Europe and exotic species from North and South America, Africa, and the East and West Indies. Many of these specimen entered private collections and royal menageries. This influx of wonders and curiosities never before seen in Europe led to a growing interest in understanding the natural world. The study of *naturalia* in the arts blossomed, with representations of plant and animal life reflecting a broader cultural impulse to systematically describe and categorize the known natural world.

A survey of the prints that Maarten de Vos designed over the course of his career reveals his attention to the natural world to have been largely limited to the portrayal of animal staffage as narrative enhancement in his religious imagery.[32] Two exceptions to this generalization stand out from his body of work: the *Imago Bonitatis*, and a series of large-scale animal paintings that De Vos executed for Johann Albrecht I, Duke of Mecklenburg (1525–1576) in the early 1570s.[33] The latter series situates him as a leading figure in the rise of ani-

31 Karel van Mander (1548–1606) mentions that De Vos visited Rome, Venice and other Italian cities; see Mander Karel van, *Het Schilder-Boeck* (Haarlem, Hans Passchiers van *Westbusch*: 1604; repr. Utrecht: 1969) fol. 265r. Two letters from the Roman scholar Scipio Fabio to his friend Abraham Ortelius (1527–1598) in Antwerp support this claim. In a letter dated 16 June 1561, Fabio asks Ortelius for news of Maarten de Vos (Martinus Vulpes) 'the most excellent painter, dear to him as a brother, and of Pieter Bruegel (Petrus Bruochl) equally dear to him'. In another letter of 14 April 1565 to Ortelius, Fabio sends greetings to De Vos and Bruegel, both of whom he had most likely met during their travels to Italy in 1553; see Hessels J.H., *Abrahami Ortelii (geographi Antverpiensis) et virorum eruditorum ad eundem et ad Jacobum Colium Ortelianum (Abrahami Ortelii sovoris filium) Epistulae* (Cambridge: 1887) 24–26, no. 11; and 32–33, no. 15. De Vos and Bruegel would have known one another through the guild, where Bruegel's name was inscribed in the *liggeren* in 1551; see Rombouts – Van Lerius, *De Liggeren en andere Historische Archieven der Antwerpsche Sint Lucasgilde* (Amsterdam: 1961) 175. See also Popham A.E., "Pieter Bruegel and Abraham Ortelius", *The Burlington Magazine* 59 (1931) 184–188; Kavaler E.M., *Pieter Bruegel: Parables of Order and Enterprise* (Cambridge: 1999) 37; and Zweite, *Marten de Vos als Maler* 21.

32 On De Vos's animal representation see Zweite A., "Exkurs I: Tierdarstellungen von Marten de Vos" in *Marten de Vos als Maler* 237–250.

33 See note 29 above.

mal representation in Antwerp. In last decades of the sixteenth century a number of Antwerp animalists, or artists highly skilled in depicting animals, began to produce series of drawing, watercolors and prints depicting birds, fishes, and animals, including animal fables and hunting scenes.[34] Maarten de Vos circulated both personally and professionally with many of these artists, among them Adriaen Collaert, Marcus Gheeraerts the Elder, Hans Bol (1534–1593), and Abraham de Bruyn (1538–ca.1587).[35] Yet De Vos never became a specialist in the depiction of animals in the vein of his Flemish colleagues. The absence of known life studies or preliminary sketches for both the Mecklenburg and *Imago Bonitatis* projects suggests that he perhaps relied on textual descriptions and artistic models for his animal representations.[36] Whether or not he also drew from living and preserved specimens, as did many of his fellow artists, remains unknown; if he did, it was an anomaly in his practice.[37] Influenced by his circle of friends and collaborators, De Vos's rendering of fauna in the *Imago Bonitatis* is in many respects on par with the works of his colleagues. In some cases he copied directly from their animal imagery. As we look more closely at the impressive array of animal motifs in the *Imago Bonitatis*, our attention turns to De Vos's direct and indirect exposure to sixteeenth-century natural historical information and imagery.

---

34 Hendrix coins this useful term without further explication; see Hendrix, "Joris Hoefnagel and the 'Four Elements': a study in sixteenth-century nature painting" (Ph.D. Diss. Princeton University: 1989), 13. The term as used here broadly encompasses artists who produced animal series and studies, without defining them as 'specialists'. On the development of animal imagery as an independent genre in Flemish art between 1550 and 1650, see Marrigje Rikken's forthcoming doctoral dissertation (Leiden University).

35 On Abraham Ortelius (1527–1598) as a central figure in this network of artists specializing in animal representation, see Rikken M., "Abraham Ortelius as Intermediary for the Antwerp Animal Trailblazers", *Jahrbuch für Europäische Wissenschaftskultur / Yearbook for European Culture of Science* 6 (2012) 95–128.

36 For example, De Vos's painting of a unicorn for the Meckelenburg series exhibits many of the same features described by Pliny in his popular account of the *Monoceros* in the encyclopedic *Historia naturalis*: 'The Orsaean Indians hunt an exceedingly wild beast, which has a stag's head, elephant's feet, and a boar's tail, the rest of its body being like that of a horse [...], and one black horn two cubits long projects from the middle of its forehead'. See Pliny the Elder, *Historia naturalis* VIII, 31.

37 Zweite refers to a possible drawing of a camel by Maarten de Vos which is recorded in the inventory of the painter and dealer Herman de Neyt for 15 and 21 October 1642: 'een stucxken van kemelshoofden van Maarten de Vos'; see Zweite, *Marten de Vos als Maler* 246; and Denucé J., *Inventare von Kunstsammlungen zu Antwerpen im 16. und 17. Jahrhundert* (Antwerp: 1932) 101.

Towards an Encyclopedic Naturalism

According to the Old Testament narrative of the Creation in the Book of *Genesis*, God created the fishes and birds on the fifth day:

> And God said, Let the waters bring forth abundantly the moving creature that has life, and fowl that may fly above the earth in the open firmament of heaven (*Genesis* 1:20)
>
> And God created great sea creatures, and every living thing that moves, which the waters brought forth abundantly, after their kind, and every winged fowl after its kind.... (*Genesis* 1:21)[38]

These scriptural passages form the subject of *The Fifth Day: The Creation of the Birds and Fishes*, the fifth print in the *Imago Bonitatis*. The four prints preceding it depict the creation of light and darkness, the creation of the firmament, the creation of the land and vegetation, and the creation of the sun, moon and celestial bodies according to *Genesis*. In *The Fifth Day*, God stands in an expansive coastal landscape amidst a *melé* of marine and avian creatures, swimming, jumping and flying in every direction [Fig. 9.2]. The animal life includes a comingling of the recognizable and the unfamiliar, including aquatic monstrosities and terrestrial fantasies. De Vos portrayed a number of European species mixed in with birds that were less well known at the time. Some of these seem to be designs born from the artist's imagination; others, though readily identifiable to the modern viewer, would have appeared as strange curiosities from distant lands—namely Africa and the New World—to the eyes of contemporary beholders. De Vos combines such elements of fiction and scientific study into a credible landscape, one that reads much like a sixteenth-century naturalist's catalogue of species.

The Fifth Day displays several types of birds that demonstrate that De Vos was aware of the very latest ornithological knowledge and that he was negotiating different scientific and artistic sources of information. For instance, his articulation of a crane [Fig. 9.9 = a detail of Fig. 9.2] includes traits characteristic of the grey crowned crane (*Balearica regulorum*) and its close relative, the black crowned crane (*Balearica pavonina*), both native to most of Africa south of the Sahara. These cranes typically exhibit a crown of crest feathers, a bare cheek patch, dark primary feathers, white wing coverts and a black tail.[39]

38 King James' version of the Bible.

39 Johnsgard P.A., *Cranes of the World: Crowned Crane* (Balearica pauonina); see http://digitalcommons.unl.edu/cgi/viewcontent.cgi?article=1012&context=bioscicranes; last accessed 5 January 2013.

FIGURE 9.9 *Jan Sadeler after Maarten de Vos, Crane and fantastic bird. Detail from idem,* The Fifth Day: The Creation of the Birds and Fishes *(ca. 1587). Engraving from:* Imago Bonitatis, *Amsterdam, Rijksmuseum, RP-P-OB-5394.*

De Vos's placement of the crowned crane at the water's edge reflects the bird's natural habitat of freshwater marshes, wet grasslands, and the peripheries of water bodies.[40] The specific type of crowned crane that De Vos had in mind is difficult to determine. If it was the black crowned crane with black-tipped crest feathers, then it could be one of two subspecies, *Balearica pavonina pavonina* or *Balearica pavonina ceciliae*, distinguished by the coloration of their cheek

40 Meine C.D. – Archibald G.W., *The cranes—status survey and conservation action plan.* International Union for Conservation of Nature and Natural Resources (Gland, Switzerland – Cambridge: 1996); see http://www.npwrc.usgs.gov/resource/birds/cranes/balepavo.htm; last accessed on 15 May 2013.

patches. If it was the grey crowned crane, then the bird would have a crown of stiff golden feathers and a red throat patch. The bichromatic nature of the print medium prevents the discerning eye from narrowing down the species identification any further.

From its crest down to the positioning of the legs, De Vos's bird closely relates to a depiction of a crowned crane [Fig. 9.10] in Adriaen Collaert's series of birds prints, the *Avium vivae icones* (ca.1580).[41] De Vos and Collaert began a fruitful working relationship in the early 1580s. It was at this time that Collaert designed, engraved and published the *Avium vivae icones*, along with three other print series on fishes, animals, and flowers.[42] Through Collaert's

FIGURE 9.10 *Adriaen Collaert,* Strutio ex China and Strutio *(ca. 1580). Engraving from:* Avium Icones, *Amsterdam, Rijksmuseum, RP-P-1892-A17420.*

---

41 Though the series is undated, Collaert probably published *Avium vivae icones* around 1580 in Antwerp. He followed it with a second edition, *Avium iconum editio secunda*, around 1600. For *Avium vivae icones*, see Diels A. – Leesberg M., *Collaert Dynasty*, Pt. 4, nos. 1404–1435. See also Jan Balis's introduction to the series in *Avium Vivae Icones in aes incisae et editae ab Adriano Collardo* (Brussels: 1967).

42 Along with the *Avium vivae icones*, Collaert produced the *Animalium quadrupedum*, a series of prints depicting four-legged animals, and the *Piscium vivae icones*, a series on aquatic species—both published after 1598. The floral series, *Florilegium*, is also undated, but perhaps coincides to a visit to Italy given the coat of arms of the De Medici family on the title page. See Diels – Leesberg, *Collaert Dynasty*, Pt 4, nos. 1462–1481; nos. 1436–1461; and nos. 1562–1585.

bird imagery De Vos was indirectly exposed to the dynamic and often circular progress of sixteenth-century natural history. Each of the *Icones* borrows heavily from the works of naturalists, including Gessner's *Historia animalium*, and the Flemish naturalist Rembert Dodoen's (1517–1585) *Cruydeboeck* (1554).[43] The *Avium vivae icones* also borrows several motifs from Hans Bol's albums of watercolor drawings.[44] Collaert's borrowings point to the acculturation of natural history in late sixteenth-century printmaking. Originally designed as model books, Collaert's series offered useful examples for artists like Maarten de Vos to follow. The *Avium vivae icones*, published in five separate editions, was widely copied in the north and in Italy. It proved so popular that its imagery eventually re-entered the circles of natural history when the Leiden botanist Carolus Clusius (1526–1609) consulted it, and Anselmus de Boodt (ca.1550–1632), a Flemish physician and Hapsburg collector in Prague, copied from it.[45]

Collaert's illustration of the crowned crane shows the bird standing in a marsh beside an ostrich (*Struthiocamelus*). The common ostrich is one or two species of large, flightless birds native to Africa. When the crowned crane and ostrich were first introduced to Europe early modern artists must have observed a strong physical relationship between the two. This is evident in Collaert's identification of the ostrich as '*Strutio*' and the crowned crane as '*Strutio ex China*' (ostrich from China). The origins of newly discovered species were many times unknown, because the trade routes from Asia, the Indies, and the Americas went through West Africa, where exotic specimens were transshipped en route to Europe.[46] We learn from the *Physiologus*—a small but influential classical treatise on animals, plants, and stones—that the ostrich's scientific name, *Struthio camelus*, derives from early zoologists' comparisons of the feet of the bird to those of a camel.[47] Collaert depicts the ostrich with a straight, elongated neck and short tail feathers. Following Collaert's example [Fig. 9.2], De Vos even captured the crisscrossed striations for the scaled tarsus on the upper leg, as well as his colleague's incorrect rendering of a toenail on the bird's bulbous, outer toe.

---

43 Margócsy, "The camel's head" 63.

44 Rikken, "Abraham Ortelius" 124.

45 Margócsy, "The camel's head" 63.

46 Smith P.J., "On Toucans and Hornbills: Readings in Early Modern Ornithology from Belon to Buffon", in Enenkel K.A.E. – Smith P.J. (eds.), *Early Modern Zoology: The Construction of Animals in Science, Literature and the Visual Arts* (Leiden – Boston: 2007) (75–119) 77.

47 I have used the English version provided by Curley M.J. (trans.), *Physiologus* (Austin – London: 1979) 55. For the influence of the *Physiologus* on medieval art, see Hicks C., *Animals in Early Medieval Art* (Edinburgh: 1993) 106–111.

De Vos borrowed selectively and creatively from Collaert's model book. Many of the detailed specimens in the *Avium vivae icones* make no reappearance in the paradisiacal compositions of the *Imago Bonitatis*. Others appear in fragments; for example, the fantastic crested specimen at the center foreground of *The Fifth Day* [Fig. 9.9] seems to be a combination of two different specimens in Collaert's series. De Vos accurately copied the body, legs, and tail feathers from Collaert's ostrich, but the unusual crown he derived from another specimen, which Collaert labeled among '*Rarae aves aquaticae*' (rare aquatic birds) [Fig. 9.11]. The latter is probably intended to represent a great crested grebe (*Podiceps cristatus*).[48] In the *Imago Bonitatis*, however, De Vos's crested ostrich appears to be an invented composite. This artistic strategy for representing less-familiar and imaginary creatures by juxtaposing the features of different known animals has been described by one author as a process of early modern metonymic composition.[49] The construction of hybrid, or composite animals harkens back to the ancient writings of Herodotus and was

FIGURE 9.11 *Adriaen Collaert,* Rarae Aves Aquaticae: Bustard and Great Grebe (*ca. 1580*). *Engraving from*: Avium Icones, *Amsterdam, Rijksmuseum, RP-P-1892-A17413*.

48 See Diels – Leesberg, *Collaert Dynasty*, Pt. 4, nos. 1416 and 1424.

49 Mason P., *Before Disenchantment: Images of Exotic Plants and Animals in the Early Modern World* (London: 2009) 87–123, esp. 112; and Margócsy, "The camel's head" 65–66.

perpetuated down through medieval and Renaissance bestiaries.[50] The composite avian motif in *The Fifth Day* suggests that De Vos sought a balance between representing those known species already illustrated by his more-practiced colleagues, and imagining possibilities for the many still-undiscovered species of God's Creation.

De Vos's portrayal of a toucan in the foreground of *The Fifth Day* [Fig. 9.12] also illustrates his engagement with different scientific and artistic sources of ornithological information. Characterized by an enormous, brightly-colored bill, the toucan is a tropical and subtropical bird that belongs to the genus *Ramphastos* and comes from the forests of Mexico, and Central and South America. Descriptions and illustrations of the toucan appeared in the writings of European naturalists and travelers during the sixteenth century, but the bird's anatomical disposition was not well understood until sometime later. One of the first descriptive accounts of the toucan comes from the Spanish naturalist Gonzalo Fernández de Oveido y Valdes (1478–1557) in his *Sumario de la natural y general historia de las indias* (1526):

FIGURE 9.12 *Jan Sadeler after Maarten de Vos, Toucan, blowfish and sawfish. Detail from:* The Fifth Day: The Creation of the Birds and Fishes (*ca. 1587*). *Engraving from:* Imago Bonitatis, *Amsterdam, Rijksmuseum, RP-P-OB-5394.*

50 See Karl Enenkel's contribution in this volume "The Species and Beyond".

> There is in Tierra Firme a bird that the Christians call *picudo* ['big bill', toucan] because it has a small body and large beak. The beak weighs more than the whole body.... It is a very strange bird, and different from any other that I have seen because of its feather tongue, its appearance, and its disproportionately large bill.[51]

The toucan entered the zoological canon through the influential *Histoire de la nature des oyseaux* (1555) of Pierre Belon (1517–1564).[52] Even though Belon had only the bird's bill at his disposal and the remainder of the toucan anatomy was unknown to him, the naturalist's description and illustration of the bird [Fig. 9.13] remained authoritative well into the eighteenth century. Belon hypothesized that the bill belonged to a web-footed river bird and he included the toucan among 'waterfowl', the second of his six ornithological groupings.[53] Belon's reason for this classification was probably the lengthy, saw-edged mandible, which the naturalist had also observed in web-footed fish-eaters, such as pelicans, mergansers, and ducks.[54] Though De Vos appears not to have copied the pictorial motif directly from Belon's or Oveido y Valdes's illustrations, he was at least familiar with their descriptions. He represented the toucan with webbed feet, positioning it in the water surrounded by sea creatures. This particular detail reveals that naturalists' groupings of species according to morphological and behavioral characteristics influenced De Vos's pictorial arrangement of specimens in the *Imago Bonitatis* as much as the biblical narrative.

Another significant detail is De Vos's accurate representation of the toucan's nostrils at the base of its bill. The question of whether or not the toucan had nostrils at all was one that puzzled naturalists in the sixteenth century. Belon remarks that he was unable to discover any nostrils: 'It is the only [bird] amongst all those that we have observed, in which we have not seen nostrils'.[55] He accounted for the absence of nostrils on the basis of the translucent material of the bill and its saw-like edges, which he judged made it impossible for

---

51 Fernández de Oveido y Valdéz G. *Natural History of the West Indies*, ed. and trans. by S.S. Stoudemire (Chapel Hill: 1959) 68. For the development of the toucan in natural history illustration see Paul Smith's study of early modern ornithology in Smith, "On Toucans and Hornbills" 77–88.

52 Belon, *Histoire de la nature des oyseaux* (Paris, Guillaume Cavellat: 1555) 184.

53 Belon groups birds into six categories: raptors, waterfowl, shore birds, terrestrial birds, large arboreal birds, and small brushwood birds.

54 Smith, "On Toucans and Hornbills" 80.

55 'Il est le seul [oiseau] entre touts ceux qu'avons observez, à qui n'ayons veu conduicts pour odorer'; see Belon, *Histoire de la nature des oyseaux* 184.

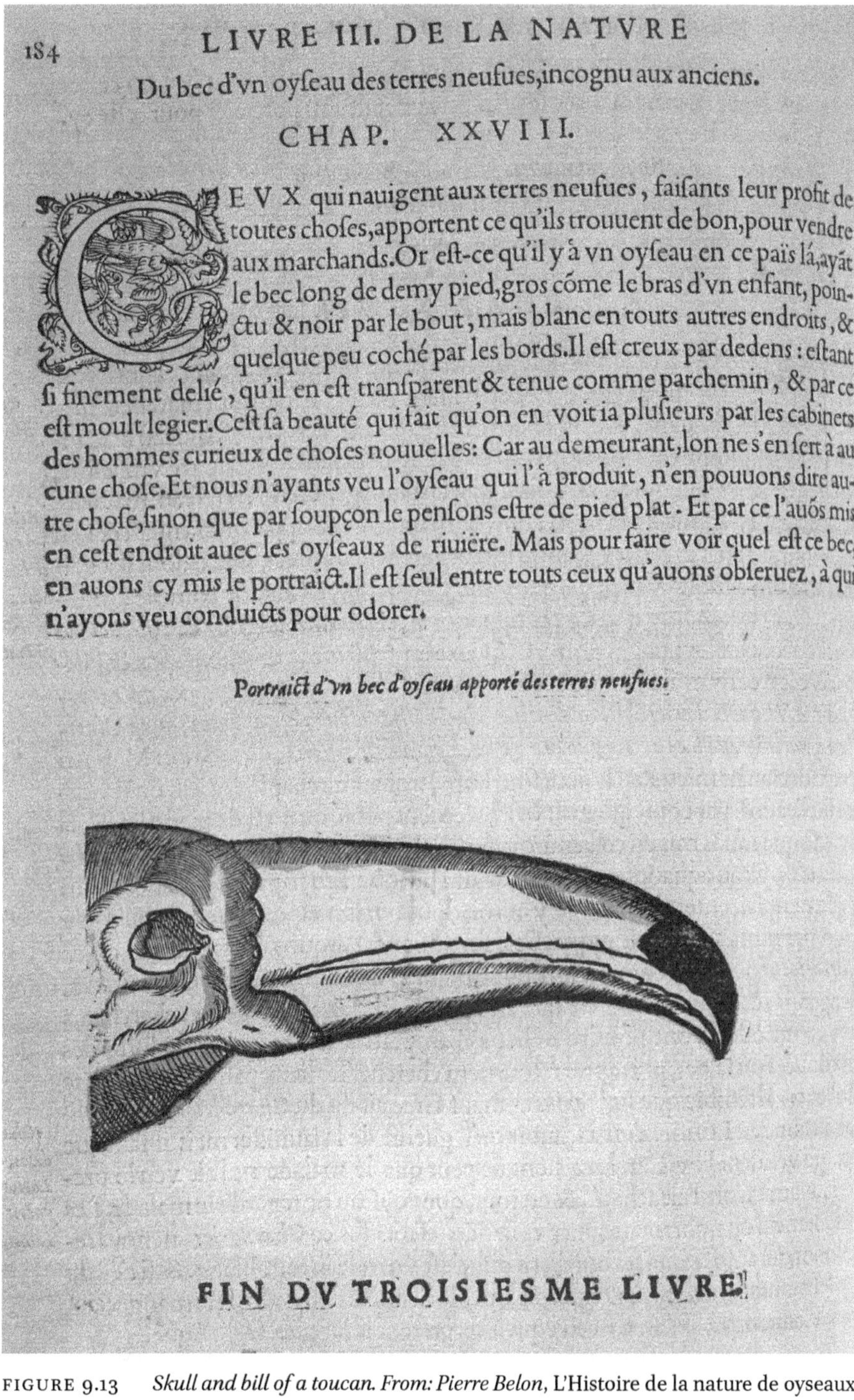

184 LIVRE III. DE LA NATVRE

Du bec d'vn oyſeau des terres neufues, incognu aux anciens.

CHAP. XXVIII.

CEVX qui nauigent aux terres neufues, faiſants leur profit de toutes choſes, apportent ce qu'ils trouuent de bon, pour vendre aux marchands. Or eſt-ce qu'il y à vn oyſeau en ce païs lá, ayãt le bec long de demy pied, gros cõme le bras d'vn enfant, poinctu & noir par le bout, mais blanc en touts autres endroits, & quelque peu coché par les bords. Il eſt creux par dedens : eſtant ſi finement delié, qu'il en eſt tranſparent & tenue comme parchemin, & par ce eſt moult legier. Ceſt ſa beauté qui fait qu'on en voit ia pluſieurs par les cabinets des hommes curieux de choſes nouuelles: Car au demeurant, lon ne s'en ſert à aucune choſe. Et nous n'ayants veu l'oyſeau qui l'à produit, n'en pouuons dire autre choſe, ſinon que par ſoupçon le penſons eſtre de pied plat. Et par ce l'auõs mis en ceſt endroit auec les oyſeaux de riuiere. Mais pour faire voir quel eſt ce bec, en auons cy mis le portraict. Il eſt ſeul entre touts ceux qu'auons obſeruez, à qui n'ayons veu conduicts pour odorer.

Portraict d'vn bec d'oyſeau apporté des terres neufues.

FIN DV TROISIESME LIVRE.

FIGURE 9.13 *Skull and bill of a toucan. From: Pierre Belon,* L'Histoire de la nature de oyseaux *(Paris: 1555) fol. 184r. Los Angeles, J. Paul Getty Museum, QL673.B45.*

the bill to close completely, and therefore rendered nostrils unnecessary.[56] The naturalist André Thevet (1516–1590) provided the first complete description and illustration of the toucan in his *Singularitez de la France Antarctique* (1557), where he incorrectly portrayed the bill with nostrils in the mid-section of the upper mandible [Fig. 9.14].[57] Probably facing criticism for his inclusion of nostrils, Thevet removed them altogether in his subsequent illustration of the bird in *Cosmographie universelle* (1575).[58]

De Vos's correct placement of nostrils on the toucan bill is therefore surprising. It is perhaps a reflection of the fact that, by the last decade of the sixteenth century, the toucan could be found more frequently in European *Kunst-* and *Wunderkammern* (art and curiosity collections). The motif is one of the few indications in the *Imago Bonitatis* that De Vos may have studied a preserved specimen in a local collection. That his depiction of the toucan's body is still wanting in its accuracy, however, suggests that the bill may have been all that he saw. Like naturalists of his day, De Vos too was not entirely certain of how the beak and head attached to the body. He adopted the naturalists' typical posture for toucan: a triangular configuration of the body with the bill tipped downwards. Recalling Oveido y Valdes's description of the toucan, the unnaturally long neck of De Vos's bird lends it a gangly, imbalanced appearance, as if its enormous, downward-curving bill outweighs its body.

Less hesitant is De Vos's portrayal of other species that were well known in Europe at the time. For example, a spoonbill (*Platalea*), seen in the sky at the far left, exhibits its characteristically large, flat, spatulate bill [Fig. 9.2]. To the right of God, De Vos also correctly represents a turkey (*Meleagris*) with a snood hanging from its beak and a fleshy wattle at the neck. The wild turkey, native to the forests of the Yucatán and North America, first arrived in Europe with Spanish explorers in the 1520s and, by the last quarter of the century, the bird was familiar to artists throughout Europe. In addition, De Vos's representation of a heron (*Ardea*) [Fig. 9.15] shows the crested bird in a natural manner: taking flight off the water in a natural manner. Belon observes in his description of the heron in *L'Histoire de la nature des oyseaux* that, on account of the heron's crest feathers, it was once confused with the crowned crane: 'there are certain feathers at the top of its head, and one finds that the ancients said that the *Grus Balearica* was marked

56 Smith, "On Toucans and Hornbills" 80.

57 'Sur la coste de la mer, la plus frequente merchandise est le plumage d'un oiseau, qu'ils appellent en leur langue *Toucan* [. . .] cest oyseau est de la grandeur d'un pigeon [. . .]'; see André Thevet, *Les Singularitez de la France Antarctique* (Paris: 1558) 90–91.

58 See André Thevet, *Cosmographie universelle* (Paris: 1575), vol. II, fol. 938v.

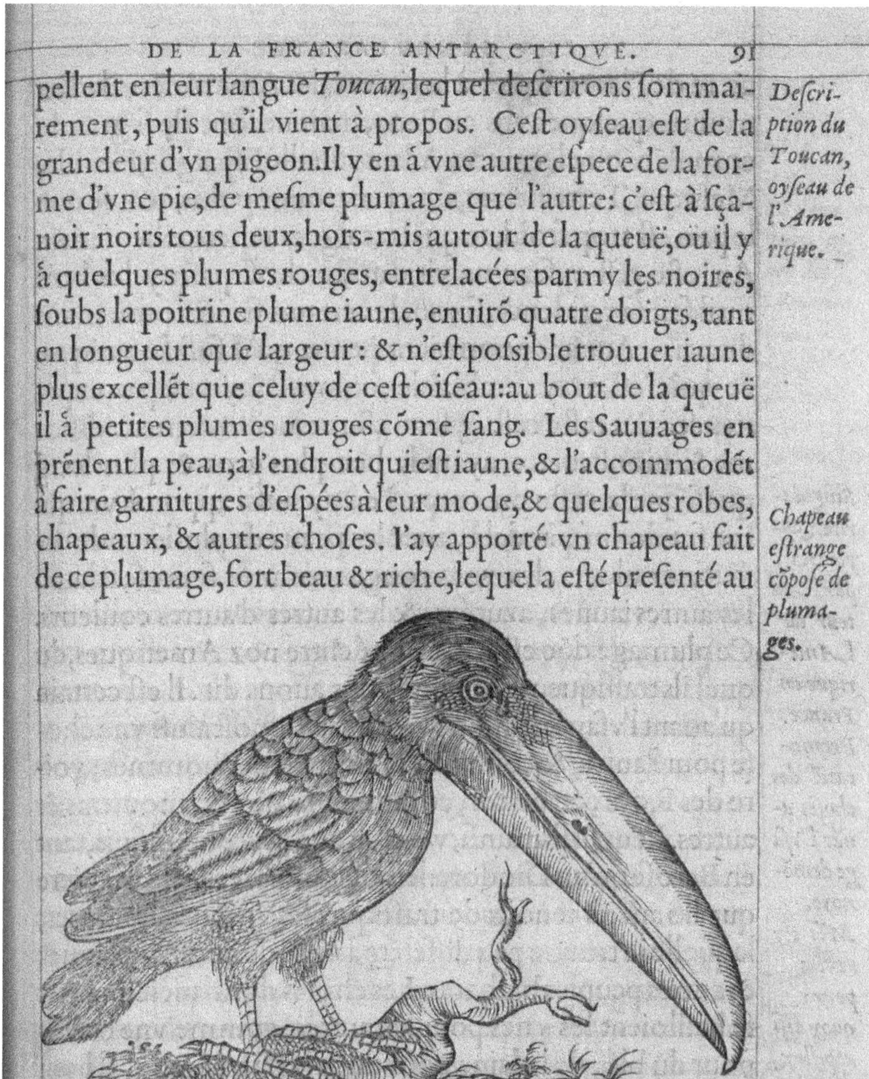

DE LA FRANCE ANTARCTIQVE. 91

pellent en leur langue *Toucan*, lequel deſcrirons ſommairement, puis qu'il vient à propos. Ceſt oyſeau eſt de la grandeur d'vn pigeon. Il y en á vne autre eſpece de la forme d'vne pie, de meſme plumage que l'autre: c'eſt à ſçauoir noirs tous deux, hors-mis autour de la queuë, ou il y á quelques plumes rouges, entrelacées parmy les noires, ſoubs la poitrine plume iaune, enuirõ quatre doigts, tant en longueur que largeur: & n'eſt poſsible trouuer iaune plus excellẽt que celuy de ceſt oiſeau: au bout de la queuë il á petites plumes rouges cõme ſang. Les Sauuages en prẽnent la peau, à l'endroit qui eſt iaune, & l'accommodẽt á faire garnitures d'eſpées à leur mode, & quelques robes, chapeaux, & autres choſes. I'ay apporté vn chapeau fait de ce plumage, fort beau & riche, lequel a eſté preſenté au Roy, comme choſe ſinguliere. Et de ces oyſeaux ne ſ'en trouue ſinon en noſtre Amerique, prenant depuis la ri-

Deſcription du Toucan, oyſeau de l'Amerique.

Chapeau eſtrange cõpoſé de plumages.

Z iij

FIGURE 9.14 *Toucan. From: André Thevet,* Les Singularitez de la France Antarctique *(Paris: 1558) fol. 91r. Courtesy of the Bancroft Library, University of California, Berkeley, E141.T39.*

FIGURE 9.15 *Jan Sadeler after Maarten de Vos, Heron, mayfly, and flying fish. Detail from:* The Fifth Day: The Creation of the Birds and Fishes *(ca. 1587). Engraving from:* Imago Bonitatis, *Amsterdam, Rijksmuseum, RP-P-OB-5394.*

by this sign'.[59] De Vos possibly meant to represent the grey heron (*Ardea cinerea*) native to wetlands and marshes in Asia and Africa, and eventually common in Europe. His familiarity with both of these birds is not surprising; the grey heron and the crane were valued in Europe from the fourteenth century up to the seventeenth century as delicacies for the table at feasts and other occasions, as well as for quarry for falconry.[60]

59 'Mais voyants qu'il y a certaines plumes sur le sommet de la tete, et trouvants que les anciens ont dit que Grus Balearica est marquee de tel signe [...]'; see Belon, *Histoire de la nature des oyseaux* 198.

60 Shrubb M., *Feasting, Fowling and Feathers: A History of the Exploitation of Wild Birds* (London: 2013) 51.

One bird-like creature less familiar to us now would have been easily recognizable to De Vos and Sadeler's contemporaries. To the right of the scene, De Vos depicts a griffin: a legendary beast with the head and wings of an eagle and the body of a lion [Fig. 9.2].[61] Already by the end of the sixteenth century naturalists understood the griffin to be a fictitious creature. Gessner observes that 'these birds are memorized by historians rather than by people versed in natural history'.[62] Another recognizable bird in the print is the phoenix, the actual existence of which was still questioned in the seventeenth century.[63] Belon had identified the phoenix with the *Rhyntaces*, described by Plutarch as a small Persian bird believed to subsist on air and dew; from his description of the latter it is clear that Belon confused the phoenix with the bird of paradise.[64]

In his representation of a bird of paradise [Fig. 9.16] in *The Fifth Day*, De Vos followed the consensus of ornithological knowledge at the time. The bird of paradise, believed to fly close to the heavens at all moments, appears without feet and positioned high in the sky at the upper margin of the print. Native to Papua New Guinea, the bird of paradise could also be found in the Moluccas and Australia. Sixteenth-century spice traders greatly prized the lesser bird of paradise (*Paradisaea minor*) and brought preserved specimens back with them to Europe on their voyages from the East. The birds became a sensation in Europe, because of their colorful plumage and curious lack of bones and legs. This anatomy was the result of the native Papuan technique for preparing the skins for sale; they removed the feet and sometimes also the wings, replacing the intestines with preservative spices.[65] The practice gave rise to the speculation that the bird of paradise spent its life in perpetual flight, living on dew, and that the female laid the egg in a cavity on the male's back. The Milanese physician Gerolamo Cardano (1501–1576) was one of the first to record this myth about the *Manucodiata* ('the bird of the gods'), and naturalists repeated

61 The griffin originates in ancient lore and persists in visual traditions reaching back to the Bronze Age and up through the Renaissance. According to Herodotus, the griffin lived in the mountains of India; see Gmelig-Nijboer C.A., *Conrad Gessner's 'Historia Animalium': An Inventory of Renaissance Zoology* (Meppel: 1977) 108.

62 Conrad Gessner, *Historia animalium, Liber III, De avium natura* (Zurich, Christoffel Froschauer: 1555) 521–523.

63 On the development of the myth of the phoenix, see Broek R. van den, *The Myth of the Phoenix, according to classical and early Christian Traditions* (Leiden: 1972).

64 Belon, *Histoire de la nature des oyseaux* 329–331; and Pierre Belon, *Les observations de plusieurs singularitéz et choses memorables* (Lyon, Gilles Corrozet: 1553) 421.

65 Gmelig-Nijboer, *Conrad Gessner's 'Historia Animalium'* 131–132.

FIGURE 9.16 *Jan Sadeler after Maarten de Vos, Bird of paradise. Detail from*: The Fifth Day: The Creation of the Birds and Fishes (*ca. 1587*). *Engraving from*: Imago Bonitatis, *Amsterdam, Rijksmuseum, RP-P-OB-5394.*

the story alongside their illustrations of the footless bird.[66] Gessner records that 'the people of the Mollucca islands bear witness that this very beautiful bird, which never sits upon the earth or any other thing, is born in Paradise'.[67]

Gessner modeled his own footless representation of the bird of paradise [Fig. 9.17] on a drawing sent to him by his correspondent Conrad Peutinger (1465–1547), who claimed firsthand knowledge of preserved specimens.[68] However, Gessner ends his chapter on the bird of paradise with contradictory information. He includes a description he received in a letter from Melchior Guillandinus in Padua mentioning an account written in 1521 by Antonio Pigafetta (1491–1534), Ferdinand Magellan's (1480–1521) voyaging companion; Pigafetta described the bird of paradise as having feet.[69] This incongruent information reveals that Gessner's contemporary correspondents were not all in agreement on the matter. The confusion over the foot issue lasted well

66 For example, in Ambroise Paré's *Des Monstres et Prodiges*; see the English translation with notes by Pallister J.L. (trans.), *Ambroise Paré On Monsters and Marvels* (Chicago: 1982) 140–141.

67 Gessner, *Historia animalium, lib. III*, 611–614.

68 Koreny F., *Albrecht Dürer and the Animal and Plant Studies* (Boston: 1988) 100.

69 Pigafetta Antonio, *Relazione del primo viaggio intorno al mondo* (Paris: 1524–1534).

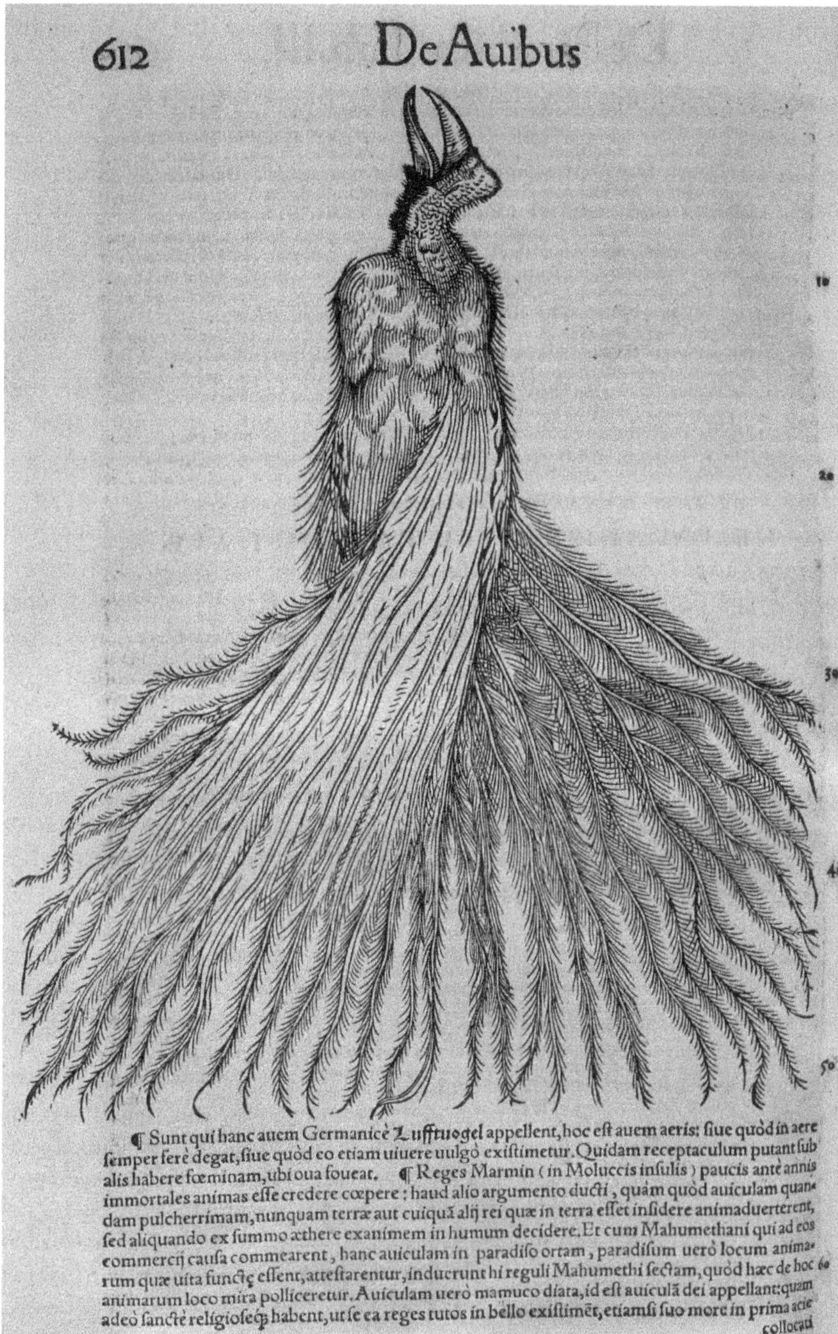

612 De Auibus

¶ Sunt qui hanc auem Germanicè Lufftuogel appellent, hoc eſt auem aeris: ſiue quòd in aere ſemper ferè degat, ſiue quòd eo etiam uiuere uulgò exiſtimetur. Quidam receptaculum putant ſub alis habere fœminam, ubi oua foueat. ¶ Reges Marmin (in Moluccis inſulis) paucis antè annis immortales animas eſſe credere cœpere; haud alio argumento ducti, quàm quòd auiculam quandam pulcherrimam, nunquam terræ aut cuiquã alij rei quæ in terra eſſet inſidere animaduerterent, ſed aliquando ex ſummo æthere exanimem in humum decidere. Et cum Mahumethani qui ad eos commercij cauſa commearent, hanc auiculam in paradiſo ortam, paradiſum uerò locum animarum quæ uita functę eſſent, atteſtarentur, induxerunt hi reguli Mahumethi ſectam, quòd hæc de hoc animarum loco mira polliceretur. Auiculam uerò mamuco diata, id eſt auiculã dei appellant: quam adeò ſanctè religioſeq; habent, ut ſe ea reges tutos in bello exiſtimẽt, etiamſi ſuo more in prima acie collocati

FIGURE 9.17 *Bird of Paradise. From: Conrad Gessner,* Historiae animalium, Liber III [...] de avium natura *(Zurich, Christoph Froschauer: 1555) 612. Los Angeles, J. Paul Getty Museum, QL41.G39.*

into the late seventeenth century. This was despite the fact that, by the middle of the sixteenth century, mounted birds of paradise could be found in many European curiosity cabinets, where they were recorded without feet or wings, without feet but with wings, with feet but without wings, and on at least one occasion with both feet and wings.[70] It was not until the beginning of the seventeenth century that another Fleming, Jan Brueghel the Elder, became the first artist to correctly depict the bird of paradise with feet in his 1611 painting of a Paradise landscape (see Marrigje Rikken's contribution in this volume).

The bird of paradise in *The Fifth Day* features an interesting detail: a long pair of curling, wire-like quills emerging from the tail feathers. The twin tail wires, indicative of the male greater bird of paradise (*Paradisaea apoda*) and lesser bird of paradise (*Paradisaea minor*), are not commonly emphasized in naturalists' depictions until the early seventeenth century, where they find their way into Ulisse Aldrovandi's (1522–1605) *Ornithologiae, hoc est de avibus historia* (1599) and Carolus Clusius's *Exoticorum libri decem* (1605).[71] Both of these works appeared in print well after Sadeler and De Vos executed the *Imago Bonitatis*. The quill-like feathers do appear, however, in depictions of birds of paradise in sixteenth-century emblem books, such as Joannes Sambucus's *Emblemata* (1564) and Juan de Borja's *Empresas Morales* (1581).[72] The tail wires

70 The 1607 inventory of the collection of Rudolf II in Prague refers to a large number of such varied specimens; see Schnapper A., *Le géant, la licorne, la tulipe. Collections françaises au XVIIe siècle* (Paris: 1988) 81. See also Mason P., "Americana in the Exoticorum", 134–136; and Swan C., "Birds of Paradise for the Sultan: Early Seventeenth-Century Dutch-Turkish Encounters and the Uses of Wonder", *De Zeventiende Eeuw* 29, 1 (2013) 49–63.

71 Carolus Clusius, *Exoticorum libri decem: quibus animalium, plantarum, aromatum, aliorumque peregrinorum fructuum historiae describuntur: Item Petri Bellonis observationibus* [...] (Leiden, Franciscus Raphelengius: 1605) 359–363: 'Manvcodiata', 'Paradisea Avis'; and 362 'Rex Avium Paradisearum majoris generis'. See also Ogilvie B., *The Science of Describing: Natural History in Renaissance Europe* (Chicago: 2006) 251.

72 Sambucus's *Emblemata* includes the bird of paradise in an emblem for "Vita Irrequieta" ("Restless Life"), where it is referred to as *Apus indica* (footless bird of India). The full text reads: 'India fert apodes, quas tantum sustinet aër/, Nusquam considunt, perpetuo error agit./ Parturiunt intra ventrem, si credere fas est,/ Ingens divini muneris istud opus./ Hic certent miras vates evolvere causas,/ Ne sileant, σοφίαν qui profitentur, habent./ Qui requiete caret, curis nec edacibus unquam/ Proficit, hunc avibus dic similem esse novis'—'The Indies bring (news of) the feetless bird, which only the air supports, they never sit down, a continuous wandering around drives it forth. They give birth inside the womb, if it is right to believe in this momentous work of divine quality. Let poets here contend to unfold the remarkable causes; those who lay claim to wisdom have this not in order to keep silent. Who lacks rest, and never makes progress because of consuming worries, call him similar to these new birds'. See Sambucus Joannes, *Emblemata* (Antwerp, Christoph Plantin: 1566) 132. The name Apus, derived from the Greek word 'apous', meaning 'without

suggest that De Vos formally drew as much from the pictorial models of naturalists as from contemporary emblem literature—a method akin to Gessner's own emblematic natural history.

Already we have encountered evidence that there was no fixed line between naturalists' so-called 'scientific' illustrations and the artistic representations of animals in printed series, emblem books and fables. Motifs circulated readily back and forth between these genres.[73] The fluidity of such imagery is coextensive of an 'emblematic world view', which William Ashworth defines as

> the single most important factor in determining late Renaissance attitudes toward the natural world [...]. The essence of this view is the belief that every kind of thing in the cosmos has myriad hidden meanings and that knowledge consists of an attempt to comprehend as many of these as possible.[74]

The bird of paradise in the central skyline of *The Fifth Day* exemplifies Ashworth's point. As a result of its name, celestial origins, and emblematic associations the bird of paradise became a standard motif in the image of Paradise, importing different layers of signification (mythic, scientific, and consumer commodity) into the biblical imagery.

But Ashworth's claim that such meanings are hidden is somewhat misleading, because for De Vos and Sadeler's audiences the religious significance and other associations of the bird of paradise would have been commonly known. As we already have observed, the bird of paradise held a distinctive place within the European culture of collecting, having become a hallmark of all excellent *Wunderkammern* of the period.[75] According to Belon in his *L'Histoire de la nature des oyseaux*, mounted birds of paradise had become a familiar

---

feet', was also given to one of the twelve constellations that the Flemish astronomer and cartographer Petrus Plancius (1552–1622) created in the late sixteenth century; he named it 'Paradysvogel Apis Indica' (footless bird of paradise of the Indies). On Plancius's constellations see Dekker E., "Early Explorations of the Southern Celestial Sky", *Annals of Science* 44, 5 (1987) 439–470. See also Juan de Borja, *Empresas Morales* (Prague, Jorge Nirgrin: 1581).

73 See Ashworth W.B., Jr., "Gheeraerts and the Aesopic connection in Seventeenth-Century Scientific Illustration", *Art Journal* 44, 2 (1984) 132–138.

74 Ashworth W.B., Jr., "Natural History and the emblematic world view", in Lindberg D.C. – Westman R.S. (eds.), *Reappraisals of the Scientific Revolution* (Cambridge: 1990) (303–332) 312.

75 See George W., "Alive or Dead: Zoological Collections in the Seventeenth Century", in Impey O. – MacGregor A. (eds.), *The Origins of Museums: The Cabinet of Curiosities in Sixteenth- and Seventeenth-Century Europe* (Oxford: 1985) 179–187.

sight in the cabinets of Europe and Turkey by the end of 1540s.[76] One of the earliest records of a bird of paradise in a European collection is the one recorded in Malines around 1523 in the collection of Margaret of Austria (1480–1530), who kept her precious specimen wrapped in taffeta in a wooden box in her library.[77] For many contemporary beholders, including the ducal dedicatee of the *Imago Bonitatis* series, the presence of the bird of paradise no doubt would have called to mind the bird's status as the most prized specimen in European collections of *naturalia*. De Vos's cognizance of the bird's value as a commodity, and his familiarity with sixteenth-century ornithological and emblematic descriptions and associations are packaged into one tiny, footless specimen at the print's margin. Not only does the detailed naturalism of De Vos's bird imagery serve an aesthetic purpose to impress the viewer, but it signals to the ways in which the animal motifs in the prints allow the beholder to import extra-narrative associations into his own reading of the Creation imagery.

A Paradise *Wunderkammer*

De Vos's compositions invite the viewer's eye to wander through paradisiacal terrains, discovering and identifying various individual specimens of God's manifold Creation in much the same way that one might peruse a sixteenth-century *Kunst*- or *Wunderkammer*.[78] Considering Sadeler and De Vos's intended ducal audience, this seems to be exactly what they proposed with the imagery in the *Imago Bonitatis*. The appearance of the bird of paradise and the toucan, for instance, link what is seen in the print with the manner of preservation and display of specimens in contemporary *naturalia* collections. With the exception of the toucan's correctly-positioned nostrils, there is little evidence in the *Imago Bonitatis* to indicate that De Vos based his anatomical descriptions on his own observations. Yet in several cases he describes specimens as one might have encountered taxidermic displays in a curiosity cabinet. In such displays birds were often mounted with feet splayed, one foot lifted, or wings fully extended. For example, in *the Fifth Day* De Vos's frontal

76 Belon, *Histoire de la nature des oyseaux* 329.

77 'Item, ung oyseau mort, appellé oyseau de paradis, envelopé de taffeta, mis en ung petit coffret de bois'; see Eichberger D., "*Naturalia* and *artefacta*: Dürer's nature drawings and early collecting", in Eichberger D. – Zika C. (eds.), *Dürer and his Culture* (Cambridge: 1998) (13–37), 26; Eichberger D., *Leben mit Kunst, Wirken durch Kunst. Sammelwesen und Hofkunst unter Margarete von Österreich, Regentin der Niederlande* (Turnhout: 2002) 185; as cited in "Americana in the Exoticorum libri decem", in Mason, *Before Disenchantment* 124–148, esp. 135.

78 On the distinction between the *Kunst*- and *Wunderkammer*, see Schlosser J. von, *Die Kunst- und Wunderkammer der Spätrenaissance* (Braunschweig: 1978).

portrayal of a bat with outstretched wings reveals the influence these display practices had on the visual arts. This type of taxidermic presentation can be seen in print series such as Collaert's *Icones*, as well as in Francesco Calzolari's (1522–1609) popular seventeenth-century cabinet print, which shows a similar bat suspended from the ceiling with wings fully extended.[79]

Various fishes in *The Fifth Day* are likewise indicative of early modern taxidermic displays [Fig.9.12].[80] De Vos rendered an inflated blowfish, shown with its mouth agape and floating atop the water in the lower left corner of the print. In the sixteenth century the blowfish, or porcupine fish of the order Tetraodontiformes, became a widely popular cabinet specimen. Native to the Pacific and Indian Oceans, the porcupine fish was especially prized for its exoticism. The stomach of two families of this fish (*Diodontidae*) can inflate to enormous sizes, hence the common name 'blowfish'. When the fish is frightened, it ingests water, causing inflation and the sharp spines to become erect. Inflation can also occur with air, as when the blowfish is removed from water, or upon death.[81] The picture of Calzolari's cabinet includes this type of inflated blowfish alongside a flying fish and a shark hanging from the ceiling. In *The Fifth Day* De Vos also depicts a sawfish, or carpenter shark (*Pristidae*). Records of *Wunderkammern* of the period typically refer only to partial remains of the specimen; its distinctive, saw-like rostrum lined with sharp, tooth-like denticles made the sawfish a particularly appealing object for collectors. An example appears hanging above the gallery visitors in the frequently-cited cabinet print of the Italian pharmacist Ferrante Imperato (1550–1625).[82] In *The Fifth Day* De Vos describes not only the rostrum of the sawfish, but correctly positions its eyes on the dorsal surface of its flattened head, suggesting that he possibly encountered a fresh-caught specimen of this edible ray at a local fish market in Antwerp.

The pictorial reference to the practice of collecting and preserving natural objects in the Paradise imagery of the *Imago Bonitatis* is significant. Naturalism here points to the extra-pictorial associations that such animal imagery might have had for contemporary audiences, particularly elite ones such as the duke.

---

79 Hieronymus Viscardus after Io. Bapt. Bertonus, "The Museum of Francesco Calzolari", published in Cerutus B. – Chiocco A., *Musaeum Francisci Calceolari* (Verona, Angelo Tamo: 1622).

80 On sixteenth-century taxidermy, see Schulze-Hagen K., et al., "Avian taxidermy in Europe from the Middle Ages to the Renaissance", *Journal für Ornithologie* 144 (2003) 459–478.

81 Nelson J.S., *Fishes of the World* (New York: 1984) 379–386; see also Claudia Swan's study on the blowfish in Swan C. "From Blowfish to Flower Still Life Paintings: Classification and Its Images, circa 1600", in Smith P.H. and Findlen P., *Merchants & Marvels: Commerce, Science and Art in Early Modern Europe* (New York: 2002) (109–136) 132, note 10.

82 The double-page, fold-out engraving was originally published in Ferrante Imperato's *Dell'Historia Naturale* (Naples, Constantino Vitale: 1599).

But was naturalism a means and an end in itself for De Vos and Sadeler in their representation of *animalia*? Or did De Vos's formal artistic borrowings and presentation of species according to naturalists' categories function in ways other than merely to imagine the encyclopedic diversity of the Divine Creation?

Interpreting the Natural World

A small insect in *The Fifth Day* catches the viewer's eye. Hovering over the coastal waves amidst various birds and sea creatures is a mayfly [Fig. 9.15]. Shown with its double wings and two long, filament-like tails, De Vos's mayfly is more anatomically accurate than his previous rendering of the same insect in *Air* (ca.1582), an allegorical print for the *Four Elements* series engraved by Adriaen Collaert.[83] De Vos's description of the mayfly appears to have improved over time, though the source of his intimate knowledge of the insect's anatomy is not known. It was not until several years after the publication of the *Imago Bonitatis* that a detailed illustration of an adult mayfly first appeared in Joris Hoefnagel (1542–1601) and Jacob Hoefnagel's (1575–1630) *Archetypa* (1592).[84]

Why did De Vos choose to include this particular insect in the Creation scene? Interest in the curious insect goes back to Aristotle's *ephemeron* and Pliny's *hemerobius* in classical antiquity.[85] Two early modern studies, dedicated solely to the mayfly, demonstrate the fascination that these creatures continued to hold for naturalists in the seventeenth century. The earliest work on the subject is *De hemerobio* (1634) by the Amsterdam physician and apothecary Augerius Clutius (1578–1636).[86] But the work of Johannes Swammerdam

83 Both the drawing for *Air* and Collaert's engraving after it lack the clearly-defined pair of filaments that De Vos depicts in *The Fifth Day*. See Diels – Leesberg, *Collaert Dynasty*, part 4, no. 1288. De Vos's original drawing for *Air* was sold in Paris by Hôtel Drouot (Adler, Picard and Tajan) on 22 November 1988; see *La Gazette de l'Hôtel Drouot*, no. 40 (11 Nov. 1988) 61. See also Le Bihan O., "Bestiaire et imaginaire de l'air, du XVIe au XVIIIe siècle. Le caméléon et l'oiseau de paradis", in *Les éléments et les métamorphoses de la nature: imaginaire et symbolique des arts dans la culture européenne du XVIe au XVIIIe siècle. Actes du colloque international de l'Opéra de Bordeaux, 17–21 septembre 1997* (Bordeaux: 2004) (139–152) 141.

84 Hoefnagel Joris – Hoefnagel Jacob, *Archetypa studiaque patris Georgii Hoefnagelii* (Frankfurt am Main: 1592) Pt. II, 10; see also Vignau-Wilberg T., *Archetypa studiaque patris Georgii Hoefnagelii, 1592: Natur, Dichtung und Wissenschaft in der Kunst um 1600 / Nature, poetry, and science in art around 1600*, trans. by M. Robertson (Munich: 1994) 69, 142–143.

85 Jorink E., *Reading the Book of Nature in the Dutch Golden Age, 1575–1715* (Leiden: 2010) 198–199.

86 Clutius Augerius, *Opuscula duo singularia. II. De hemerobio sive ephemero insecto, & majali verme* (Amsterdam, Jacob Charpentier: 1634). See also Francissen F.P.M. – Mol A.W.M., *Augerius Clutius and his 'De hemerobio'. An Early Work on Ephemeroptera* (Marburg: 1984).

(1637–1680), *Ephemeri vita* (1675), was the first to shed light on the mysterious generation of the mayfly.[87] An insect of the family Ephemeroptera, the mayfly is so named for its short life span, during which time it molts twice, mates, and, if a female, lays its eggs in freshwater. The larvae live for years at the bottom of river beds before swarming to the surface during a brief period in the warm summer months and dying within a few hours. The fleeting lifespan of the mayfly had been a metaphor for the tragic brevity of human existence ever since Cicero and Pliny.[88] Maarten de Vos seems to have included this lone insect in his representation of the fifth day of the Creation as a foreshadowing of this very fact of postlapsarian existence, brought on by Adam and Eve's actions in the Fall. The mayfly directs our attention from the spectacle of nature in the Paradise prints to its spiritual significance.

De Vos's selection of certain specimens, such as the mayfly, reveals the image to be visual exegesis, that is, a visual interpretation of the biblical narrative, rather than simply an illustration of it. More than just pictorial staffage, animals function in the *Imago Bonitatis* prints as interpretive glosses to the pictorial narrative of the Creation. The chameleon, pictured in the lower right corner of *The Sixth Day: The Creation of Adam, Eve, and the Animals* [Fig. 9.1], for example, was firmly embedded in the emblematic tradition of the sixteenth century, even before it entered the zoological studies of the early modern naturalists.[89] The various early editions of the Italian humanist Andrea Alciato's (1492–1550) *Emblemata* depict the chameleon as an emblem of duplicity because of its ability to change color, although there it appears as an opossum, or hedgehog-like creature [Fig. 9.18].[90] The earliest accurate representation of a chameleon showing its unusual foot structure—two outside toes and three inside toes on the front feet, and the complete reverse on

87 Swammerdam Jan, *Ephemeri vita of Afebeeldingh van 's menschen leven vertoont in de wonderbaarlijcke en nooyt gehoorde historie van het vliegent ende een-dagh-levent Haft of Oeveraas* (Amsterdam, Abraham Wolfgang: 1675). See also Francissen F.P.M., "Vroege Nederlandse bijdragen tot de kennis van ephemeroptera of eendagsvliegen", *Tijdschrift voor de geschiedenis van de geneeskunde, natuurwetenschap, wiskunde en techniek* 7 (1984) 113–128.

88 See Jorink, *Reading the Book of Nature* 199–200.

89 The evolution of the chameleon in natural history illustrations and emblem literature is treated at length in Smith P.J., "Een veranderlijk dier. De kameleon tussen natuurlijke historie en emblematiek." *De Boekenwereld*, 29, 1 (2012) 33–43.

90 Andrea Alciato, *Emblematum libellus* (Paris, Christian Wechel: 1534) 93; and idem, *Emblemata* (Lyon, Macé Bonhomme: 1551) 61. Paul Smith suggests that the chameleon in Alciato's emblem resembles a hedgehog; Ashworth compares it to a 'wet opossom'; see Smith, "Een veranderlijk dier" 33–43; and Ashworth, "Gheeraerts and the Aesopic connection" 134–135.

PERFIDIÆ. 61

In adulatores.

Semper hiat, semper tenuem qua vescitur auram,
Reciprocat Chamæleon,
Et mutat faciem, varios sumitq; colores,
Præter rubrum, vel candidum.
Sic & Adulator populari vescitur aura,
Hiansq; cuncta deuorat.
Et solùm mores imitatur principis atros,
Albi, & pudici nescius.

FIGURE 9.18 *Chameleon. From: Andrea Alciato,* Emblemata *(Lyon: 1551) fol. 61. Los Angeles, J. Paul Getty Museum, N7740.A34.*

the back feet—appeared in Belon's *De aquatilibus* (1553). Belon's woodcut was influential, becoming the prototype for Marcus Gheeraerts's illustration of a chameleon in *De warachtighe fabulen der dieren* (1567).[91] Gheeraerts's illustration, however, misconstrues the chameleon's toe structure.[92] De Vos appears to disregard the toe structure altogether; he was either unaware of the confusion on this matter, or simply unconcerned. Rather, he shows the chameleon walking on the ground, unlike most naturalists, who portrayed the creature on a branch. Following Gheeraerts, De Vos gives the reptile a distinct, semi-circular sagittal crest. Several years earlier, De Vos already had depicted virtually the same chameleon in the print *Air* (ca. 1582), where the reptile clings to the hand of the male personification for the element.[93] The allegorical meaning of the chameleon is explained by Pliny, who writes that the chameleon always keeps its mouth wide open and subsists entirely on air.[94] Although Pliny's account of the chameleon had been sufficiently dispelled by Belon, Gessner and others, its religious and allegorical significance persisted.[95] The latter certainly informed De Vos's placement of the reptile directly beneath God's creation of Adam in *The Sixth Day*. With the motif De Vos directs his audience to a relevant passage in the Bible, where the second act of the Creation in *Genesis* describes how God creates Adam by breathing air into him: 'And the Lord God formed man of the dust of the ground, and breathed into his nostrils the breath of life; and man became a living soul' (*Genesis* 2:7).

Several of the animals in the *Imago Bonitatis* prints indicate to De Vos's attention to typology in his visual interpretation of the Creation narrative. Because the saliva of the chameleon was thought to kill snakes, it was associated with Christ in his fight against the devil.[96] This helps to explain the appearance of a slithering snake next to the chameleon in *The Fifth Day* [Fig. 9.2]. The Italian scholar and physician Julius Caesar Scaliger (1484–1558) writes in his annotations on Pliny and Aristotle that the chameleon—the declared enemy of venomous serpents—represents Christ, who is the enemy of the Devil

91 Though there was no Aesop tale about the creature, Eduard de Dene invented a story to accompany Gheeraerts's illustration. See Dene, *De warachtighe fabulen der dieren* 72.

92 Ashworth, "Gheeraerts and the Aesopic connection" 135.

93 See note 83 above.

94 Pliny, *Naturalis historia*, VIII, 122.

95 Belon, *De Aquatilibus* 53–56; and Gessner, *Historia animalium*, Lib.I, p. 28.

96 On the iconography of sympathy and antipathy between animals in paradise imagery, see Smith P.J. "Sympathy in Eden: On *Paradise with the Fall of Man* by Rubens and Brueghel", in Göttler C. – Neuber W. (eds.), *Spirits Unseen: the representation of subtle bodies in early modern European culture* (Leiden: 2008) 211–244.

(the old serpent).[97] Together the chameleon and snake transform Adam into a typology for the New Adam, or Christ. De Vos reinforces this idea with his portrayal of a reclining unicorn on the other side of Adam. Medieval Christian authors described the unicorn as a symbol of Jesus incarnate.[98] In medieval descriptions of the Original Sin, the animal typically signifies the Redemption through Christ. In the thirteenth-century *Bestiaire* of Pierre de Beauvais, for example, the beast was identified with the Savior as the 'licorne céleste' (heavenly unicorn).[99] A unicorn is often seen drinking from the fountain of life in medieval imagery, emphasizing the idea of its original purity. This is probably a reference to Psalm 36: 7–10: 'man and beast you save, O Lord [. . .] and you give them drink from the river of your delights. For with you is the fountain of life'.[100] De Vos represents just such a unicorn in the final print in the series: *The Sixth Day: Adam, Eve and the Animals Blessed by God* [Fig. 9.5], where the animal drinks from a stream while at the same time dipping its horn into the water. The latter action occurs in an emblem for '*Nil inexplorata* (*Nothing unknown*)' in Joachim Camerarius the Elder's (1500–1574) emblem book *Symbolorum & emblematum* [Fig. 9.19].[101] There the unicorn's purity causes snakes, symbolic of evil, to flee the water.

The lions poised at the center foreground of De Vos's zoological configuration in *The Sixth Day: The Creation of Adam, Eve and the Animals* [Fig. 9.1), and again beside Adam and Eve in *The Sixth Day: Adam, Eve and the Animals Blessed by God* [Fig. 9.5], recall the primary role that medieval bestiaries grant the king of beasts. Such descriptions relate the lion to the death and resurrection of Christ, the two natures of Christ, the body of Christ, and the Redemption.[102] Through the books of the Prophets and the Apocalypse of John, the lion became the soteriological and eschatological symbol of Christ.[103] Folkloric claims that the lion slept with its eyes open also made it an apt symbol for

---

97 'Nous pouvons avec raison avancer que le caméléon, qui ne se nourrit que de la rosée du ciel, et qui est l'ennemi déclaré du serpent venimeux, nous représente le Fils de Dieu fait homme, l'ennemi implacable, du vieux serpent'; as cited in Bihan, "Bestiaire et imaginaire de l'air" 142.

98 Charbonneau-Lassay L., *Le Bestiaire du Christ: La mystérieuse emblématique de Jesus-Christ* (Bruges: 1940) 337–346.

99 Bianciotto G. (ed. and trans.), *Bestiaires du Moyen Âge. Pierre de Beauvais, Guillaume le Clerc, Richard de Fournival, Brunetto Latini et Jean Corbechon* (Paris: 1980).

100 Weemans, "The Earthly Paradise" 290.

101 Camerarius Joachim, *Symbolorum et emblematum ex aquatilibus et reptilibus desumptorum* (Nuremburg, Johann Hofmann – Hubert Caymox: 1590) fol. 24, pl. XII.

102 Charbonneau-Lassay, *Bestiaire du Christ* 36–53.

103 See Ferro X.R.M., *Symboles animaux: un dictionnaire des représentations et croyances en Occident* (Paris: 1996) 205.

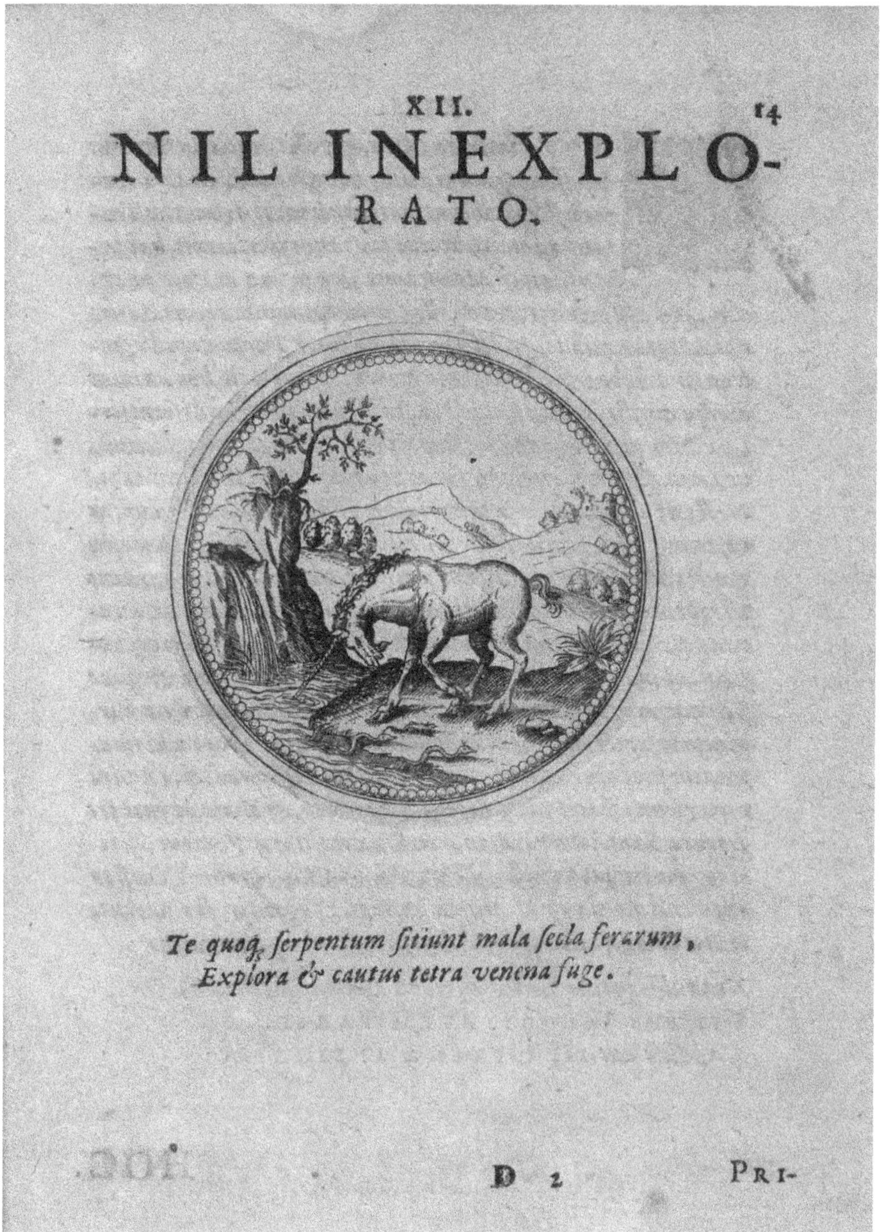

FIGURE 9.19 *Johann Siebmacher,* Unicorn. *Engraved illustration to Joachim Camerarius,* Symbola et emblemata ex aquatilibus et reptilibus desumpta [. . .] *(1604) p. 24, pl. XII. Courtesy of the Bancroft Library, University of California, Berkeley, QK41.C32.*

Christ's vigilance, while the strength of the lion's roar made it an appropriate sign for the power of the divine Word.[104] As one scholar has observed: 'Présent dans certaines représentations de la Création d'Adam, le lion possède certainement une double signification, à la fois figure symbolique du Christ rédempteur et membre de la Trinité, et signe du statut particulier de l'homme, vivifié par le souffle créateur comme le lionceau par sa mère'.[105] Indeed the lioness was believed to give birth to a dead cub, and three days later the father was said to bring it back to life with its breath. This soteriological significance explains the beast's appearance in De Vos's *The Sixth Day: The Creation of Adam, Eve, and the Animals*, where the lion is beside the figure of God, seen breathing life into Adam (*Genesis* 2:7).

De Vos's depiction of animals familiar to the Christian pictorial tradition was a calculated one. Many of the animals represented in the print series derive from the history of salvation: for example, the peacock (immortality and renewal), the unicorn (Incarnation and chastity), the fox (wisdom), the porcupine (greed), the chameleon (impermanence), and the heron (fidelity to Christ).[106] But nowhere are the specific animals mentioned in the *Genesis* passages. De Vos it seems expected his audiences to contemplate their varied symbolic associations in the context of the pictorial narrative. More than just iconographic embellishment, De Vos uses the *animalia* to interpret the biblical narrative of God's creation of the world. The animal life in the final three prints of the *Imago Bonitatis* thus points to the status of the image as visual exegesis, and defines the artist as exegete. De Vos's reading and portrayal of the narrative was deeply influenced by the way naturalists of his day categorized, described and portrayed animals, birds, fishes and reptiles; but it was equally influenced by the significance of animals within the Christian symbolic tradition.

Linking the symbolic world of *animalia*, which came before the rise of natural history and its acculturation in the visual arts, with the sixteenth century's growing interest in describing the natural world De Vos's compositions share an approach in common with early emblematic naturalists of his day. The belief that a contemplation of the things of the universe created in the Beginning could lead to a better understanding of God prompted Gessner to include a woodcut illustration of the Creation as a colophon to his first volume of the *Historia animalium*.[107] Like the work of Gessner, De Vos's animal imag-

104 See Weemans, "The Earthly Paradise" 301–302.

105 Ortega-Tillier V., *Le Jardin d'Éden: Iconographie et topographie dans la gravure* (*XV*[e]*–XVIII*[e] *siècles*) (Dijon: 2006) 191–192.

106 Erbentraut, "Als das Kamel Blätter vom Baum Erkenntnis Frass", 54–55, cat. no. 12.

107 As cited in Braun L., *Conrad Gesner* (Geneva: 1990) 15.

ery records the detail and diversity of the natural world, while also reminding the viewer that the purpose of such an endeavor is to achieve a union with God through one's knowledge of His creations. Having considered *what* some of the animals in De Vos's imagery mean in a religious context, it remains for us to determine *how* the motifs operate as signs which visually transform the *Imago Bonitatis* prints into visual exegesis of *Genesis* 1–2. That is, *how* does the animal imagery shift its status from simple pictorial detail, which fleshes out a narrative episode, to meaningful sign, which points to the artists' visual interpretation of that narrative? By what mechanism does Sadeler and De Vos's visualization of *animalia* mirror the way sixteenth-century biblical exegetes verbally interpret the *Genesis* narrative?

Naming and Meaning: Animals as Nature's Text

As we have already observed, De Vos borrowed a number of motifs from Gheeraerts's fable illustrations for his final two prints in the *Imago Bonitatis*: the chameleon, the hare, and the hedgehog.[108] Another quotation of Gheeraerts's work derives from the fable 'The Ape and the Fox' [Fig. 9.20].[109] This same vignette, showing a conversation between the two animals, reappears in revers at the lower right corner of *The Sixth Day: Adam, Eve and the Animals Blessed by God* [Fig. 9.21]. According to the fable, an envious ape implores a fox to share a piece of his bushy tail so that he might cover his bare bottom or, at the very least, to give it to his eldest son to do so. The fox refuses, saying that the ape had not labored, nor made his tail grow, and that he should be brought down to the ground to cover his shame. In the print a second ape, shown clinging to a tree trunk above the scene, verifies De Vos's familiarity with this fable. His borrowings from Gheeraerts's illustrations have a directness to them which suggests that De Vos may have intended them as a *hommage* to his fellow artist—an *hommage* that he anticipated might delight the *Imago Bonitatis*'s audiences. It is possible too that De Vos imagined his viewers would bring to their reading of the Paradise prints the moral associations of each of the fable characters; for example, he makes a visual connection between the ape's nakedness and shame and that of Adam and Eve after the Fall.

In the case of the ape and the fox, De Vos's imitation of Gheeraerts's motif is more than formal in nature, and more than an artistic *hommage*. He re-used

108 See De Dene, *Warachtighe Fabulen*, 64, 68, and 72. De Vos depicted the same hare in *The Fall of Man*; see note 18 above.

109 De Dene, *Warachtighe Fabulen der dieren* 74.

FIGURE 9.20 *Marcus Gheeraerts,* The Ape and the Fox. *Etched illustration to Eduard de Dene,* De Warachtighe Fabulen *(Bruges: 1567) fol. 74. The Hague, Royal Library.*

FIGURE 9.21 *Jan Sadeler after Maarten de Vos, Ape and fox. Detail from:* The Sixth Day: Adam, Eve and the Animals Blessed by God *(ca. 1587). Engraving from:* Imago Bonitatis, *Amsterdam, Rijksmuseum, RP-P-OB-5396.*

the vignette for another purpose as well. The ape and fox offer us the key to understanding *how* animal motifs function as signs in De Vos and Sadeler's print series. In copying Gheeraerts's vignette De Vos made a few minor changes: the ape no longer looks at the fox, its head turning instead to regard the viewer; and rather than gesturing to its own bare bottom, the ape points to the inscription on the ground between the two animal's: 'D figuravit'. The subtle changes are meaningful. Whether it was De Vos or Sadeler who added the inscription is uncertain, but it is hard to imagine that it was not De Vos himself who placed it there. In lieu of writing out 'Maarten de Vos figuravit', the designer transforms Gheeraerts's fable illustration into a pictogram for his own signature: the ape stands in for the artist's first name, 'Maarten'; the letter D stands in for 'de'; and the fox (literally *de vos* in Dutch) represents the designer's last name.[110] Dating from about 1460 to 1480, the Middle Dutch 'Martijn' was also the name for an ape in the fable *Reinaert de Vos*.[111] In 1574, the Flemish publisher Christoph Plantin (ca.1520–1589) published a Dutch-Latin dictionary, the *Dictionarium Teutonico-Latinum*, compiled by the poet Cornelis Kiliaan (ca.1529–1607), which relates the personal name 'Maarten' to the Latin term 'Simius' (ape or monkey): 'martin.holl.j.marte.Simius'.[112] The ape and fox in De Vos's print, therefore, become natural signs for spelling out the artist's signature.

The use of the ape as a pictogram for the artist's name relates to another sixteenth-century concept: *Ars simia naturae* (art is the ape of nature). Because the artist's skill was regarded as imitative it was easily associated with an animal renowned for its mimicry. The ape became a popular visual metaphor for the artist in general.[113] On one level, De Vos's pictogram points to the designer's self-awareness—an acknowledgement of his own imitative, artistic endeavor in copying Gheeraerts's imagery. On another level, however, De Vos directs our attention to the artist as an ape of God. Art imitates nature, while nature itself is God's work of art. Drawing a line of sight with the trunk of a tree to lead the beholder's eye from ape to God, De Vos seems to make the following point: artistic efforts to faithfully represent God's original creations have the

110 Zweite, *Marten de Vos als Maler* 239.

111 See Martin E. (ed.), *Reinaert: Willems Gedicht Van den Vos Reinaerde und die Umarbeitung und Fortsetzung Reinaerts Historie* (Paderborn: 1874) 107–341. In 1481 William Caxton translated *The Historie of Reynart the Foxe* into English.

112 See the third edition: Kiliaan Cornelis, *Etymologicum Teutonicae linguae: sive, Dictionarium Teutonico-Latinum: praecipuas Teutonicae linguae dictiones et phrases, Latine interpretatas, et cum aliis nonnullis linguis obiter collatas complectens* (Antwerp, Johannes Moretus: 1599).

113 On the development of *Ars simia naturae* in art and literature see Janson H.W., *Apes and Ape Lore in the Middle Ages and the Renaissance* (London: 1952) 287–325.

effect to define God as the Creator, while presenting the artist as the imitative 'Re-Creator'. De Vos's attention to animals in the print series can be characterized as something much more than staffage; it is both emblematic and metaphoric in nature.

By spelling out his own name with the objects of God's divine Creation, Maarten de Vos highlights the act of naming itself. Following God's creation of the animals in the Bible, Adam gave them each a name:

> Then the Lord God said, 'It is not good that the man should be alone; I will make him a helper fit for him'. So out of the ground the Lord formed every beast of the field and every bird of the air, and brought them to the man to see what he would call them; and whatever the man called every living creature, that was its name. (*Genesis* 2:18–19)

De Vos represents this episode in a subsidiary scene in the background of *The Sixth Day: The Creation of Adam, Eve and the Animals* [Fig. 9.5]. For Renaissance and early modern authors, the names given to the animals by Adam described their true essences, because it was the language of God that Adam spoke. His choice of words invested the animals' names with a divine quality and his language became synonymous with the vocabulary of nature.[114] Scholars believed that in naming the animals Adam had a true knowledge of God's creatures in Paradise, a knowledge that he subsequently corrupted in the Fall. The scene in *The Sixth Day* highlights the idea that in both a religious and natural historical context the act of naming, or identifying, God's creatures brings one closer to knowing the Holy Spirit through understanding nature itself.

While the subject of the Creation gave Sadeler and De Vos an occasion to showcase their ability to render an extraordinary abundance of creatures, it is their striking naturalism that beckons the viewer to lean in and look more closely, and to identify each individual specimen. In this process of looking and identifying, the beholder mimics Adam, performing his own act of naming the animals.[115] By identifying and cataloguing newly discovered species from the far corners of the earth, where they were displaced from Paradise after the Expulsion, early modern man hoped to regain Adam's prelapsarian knowledge.

---

114 Bono J., *The Word of God and the Language of Men: Interpreting Nature in Early Modern Science and Medicine* (Madison: 1995) vol. 1, 55, 73.

115 Erbentraut R., "Als das Kamel Blätter vom Baum der Erkenntnis Frass: Paradiesdarstellungen im niederländischen Manierismus", in Blübaum D. – Erbentraut R. (eds.), *Die Erschaffung der Tiere: Tiere in der Niederländischen Kunst des Manierismus und Frühbarock und an den wandfesten Dekorationen des Schlosses zu Güstrow* (Schwerin: 2010) 47–57, esp. 53–55, cat. no. 12.

The Creation story, understood to be the origin of all of Adam's encyclopedic knowledge, served as the perfect narrative framework for this endeavor. By studying nature alongside the Book of *Genesis* the individual might reconstitute Adam's once-close relationship with God.[116]

While the ape and the fox expose the animal-motif-as-sign, all of the animals represented in the *Imago Bonitatis* make up this text of nature. Together they function as something Ashworth described as 'living characters in the language of the Creator'.[117] Developing this notion further, Eric Jorink has observed that

> God's creatures are pages from the Book of Nature. Every creature, every 'member', 'page' or 'letter' is a wonder of God. The leaves of the trees, stars, and the phases of the moon: they all signify, they all mean something. Every facet leads to meditation on the *auctor intellectualis* of it all ... the Bible is not the only road to knowledge of God. There is a second divine text [...].[118]

De Vos's pictogram presents the ape and fox as words from this other text: the Book of Nature. Early modern naturalists, artists and theologians shared the view that nature itself was an encyclopedic book, whose pages were open for all. Nature was considered 'a book of God's wonders that lay waiting to be read, understood and pondered. A stroll through the Book of Nature could prompt the visitor to reflect on God's creation [...]'.[119] Nature was not just a reference to *the* Book—the Bible—but it was *itself* a book. And, like the Bible, it became an object of exegesis.

The description of the natural world, along with the collection and study of it, were at root theological activities in the late sixteenth century. To contemplate the physical manifestation of God's work in the Creation, whether in the form of objects or images, was to meditate on the Creator.[120] The pictogram of the ape and fox returns us to God's work at the world's origin: the *Dei verbum* (word of God). According to *Genesis*, each of the days of the Creation ensued from a divine utterance: 'And God said, Let there be light: and there was light'

116 For discussion of this idea, see Honig E., "Paradise Regained. Rubens, Jan Brueghel, and the Sociability of Visual Thought", *Nederlands Kunsthistorisch Jaarboek / Netherlands Yearbook for History of Art* 55 (2004) 271–301; and Faber Kolb, *Jan Brueghel the Elder* 48.

117 Ashworth, "Natural history and the emblematic world view" 308.

118 For an excellent study on the interpretation of the Book of Nature in the sixteenth and seventeenth centuries, see Jorink, *Reading the Book of Nature* 1–105.

119 Ibidem, 11–21.

120 Harrison P., *The Bible, Protestantism, and the Rise of Natural Science* (Cambridge: 1998) 199.

(*Genesis* 1:3). God's generative word was the foundation of Adam's prelapsarian knowledge of the natural world, and the language of Paradise.[121] Before the Fall, at a time when the Bible did not yet exist, these words and objects were identical; but after the Original Sin, the Fall, and the dissolution into confusion of Babel, the link between divine knowledge and human knowledge was severed.[122] Just as the text of the Bible revealed its hidden meanings to exegetes, biblical imagery had the power to reveal its meanings to its beholders. The representation of animals in the work of both naturalists and artists relates to this promise. Sadeler and De Vos's joint effort to capture the anatomical detail and natural behavior of God's creatures in their Paradise prints was as much an exegetical endeavor as it was a descriptive one.

In what follows, the relevance of the *Imago Bonitatis* for Duke Wilhelm V's spirituality and devotional practice comes into focus. We return to Munich as the site of the series' conception and publication, and to the arrival in Bavaria of a Flemish artist who quickly and seamlessly integrated himself into his new court surroundings. Jan Sadeler was an observant and strategic artist, taking great pains that his inaugural print series would appeal to his new ducal patron's artistic tastes, collecting interests, and spiritual attitude.

A Creation for the Duke: Courtly Origins & Christian Optimism

Under Wilhelm V (r. 1579–1597), the court in Munich became a thriving center for artists specializing in animal representation. The duke's patronage of the much-admired animal painters Joris Hoefnagel and Hans Hoffmann (ca.1550–1591) in the 1580s suggests that upon Sadeler's arrival at the Bavarian court the latter found himself in a fruitful artistic environment for the illustration of the natural world. It was possibly through Hoefnagel's miniature drawings for his *Four Elements* albums that Sadeler became acquainted with the duke's taste for natural historical knowledge and illustration. At the forefront of animal illustration in Bavaria, Hoefnagel's *Four Elements* albums portray hundreds of species of fauna in astonishing detail, interleaved with classical, biblical, and humanist texts describing the animals [Fig. 9.22]. Conceived as a pictorial compendium of the known animal world, the albums include a total of 277 miniatures in watercolor and gouache on vellum. Hoefnagel's watercolors borrow extensively from the work of Hans Bol, and reflect the advances of the major sixteenth-century naturalists, including Pierre Belon, Conrad

121 Bono, *Word of God*, 11–24, 45–63.

122 Jorink, *Reading the Book of Nature* 36.

FIGURE 9.22 *Joris Hoefnagel, "Elephant" (ca. 1575/1580), watercolor on vellum. From:* Four Elements [Animalia quadrupedia et reptilia (Terra)], *Plate I. Washington, D.C., National Gallery of Art, Gift of Mrs. Lessing J. Rosenwald, 1987.20.6.2.*

Gessner, and Guillaume Rondelet (1507–1566).[123] According to the artist and biographer Karel van Mander (1548–1606), Hoefnagel initially visited the ducal *Kunstkammer* in 1577 with the prominent Flemish cartographer, merchant, and scholar Abraham Ortelius (1527–1598).[124] At that time Hoefnagel sold 'a little piece with small animals and trees, illuminated on vellum' to Duke Albrecht V.[125] The small work appears to have been an illustration for the

123 Hendrix, "Joris Hoefnagel and the 'Four Elements'" 46–47.

124 Hoefnagel and Ortelius visited the *Kunstkammer* with a letter of introduction from the Fugger banking family—probably Hans Fugger (1531–1598), a financial and artistic advisor to Duke Albrecht V. See Hartig O., *Die Gründung der Münchener Hofbibliothek durch Albrecht V. und Johann Jakob Fugger* (Munich: 1917) 350–351, no. 2.

125 Karel van Mander writes: "Dus daer, ghereyst met eenen Recommandatie-brief van dese Heeren / heeftse de Hertogh alles laten sien / en ghebraeght aen Hoefnagel, of hy niet en hadde van zijn Const. Desen liet den Hertogh sien / zijn en zijns eerste Huysvrouwen Conterfeytsel / en noch een stucxké met beestgens en boomkens / van Verlichterije / op pergamijn. Doe sy in hun Herbergh waren / fondt den Hertogh door zijn Hofmeester / oft ander Heer vraghen / wat Hoefnagel begheerde voor dat Verlichterijken va de beestgens /

Four Elements, the earliest folios of which he seems to have carried with him from Antwerp on his travels with Ortelius.[126] Hoefnagel continued to add drawings to the *Four Elements* upon his return in 1578 as a court artist to Munich, where he remained in the service of Albrecht V and Wilhelm V until 1590/91.[127]

When Sadeler arrived in Munich almost a decade later there can be little doubt that he encountered Hoefnagel's striking flair for rendering *animalia*, including the latter's detailed naturalism, encyclopedic diversity, and catalogic arrangement of species in the *Four Elements*. Though they appear not to have collaborated artistically, it seems that Sadeler and Hoefnagel were personally acquainted at the Bavarian court.[128] Just before Hoefnagel left Munich to enter the service of the Holy Roman Emperor Rudolf II (1552–1612), Sadeler drew his friend's portrait and dated it 28 July 1591.[129] The following year he engraved and published the portrait with the inscription '*Ioann*[es] *Sadelerus Amicus Amico et Posteritati*' ('Jan Sadeler, gave this as a Friend to a Friend and to Posterity'),

dewijl hy de tronikens niet wauw quijt wesen. Hoefnagel, die hem noyt had uptghegheben voor een Schilder / noch laten voorsteaen pet te connen / was schromigh daer van pet te willen eysschen: Dan Ortelius hem moet ghebende, eyschte self voor hem een hondert gouden Croonen/ die hem den Hertogh stracx heeft gegeben / en begheert dat hy daer in zijnen dienst soude blijven [...]". See Van Mander, *Het Schilder-Boeck*, fols. 262v–263r. See also Hendrix, "Joris Hoefnagel and the 'Four Elements'" 92–93.

126 Rikken, "Abraham Ortelius" 106.

127 Hoefnagel received a token salary and the right to reside in town, to practice painting without paying taxes, and to receive payments for each separate commission from his private patrons. See Vignau-Wilberg T. "München—Stadt und Hof", in *In Europa zu Hause—Niederländer in München um 1600* (Munich: 2005) 26. Residing at the Bavarian court for over a decade, the artist divided his services between the Wittelsbach dukes and another eminent patron, Ferdinand II, Archduke of Austria. Hoefnagel possibly delivered other individual miniatures like those in the *Four Elements* to his other patrons during this period.

128 Sadeler appears never to have produced engravings after Hoefnagel's drawings. His nephew, the young Aegidius Sadeler (ca.1570–1629), collaborated with Hoefnagel and also Hans von Aachen (1552–1615) in Munich in the late 1580s. On Aegidius see Hoop Scheffer D. de (comp.), *Aegidius Sadeler to Raphael Sadeler II*, vol. 21, nos. 32 and 113. See also Fučikova E., "Prague Castle under Rudolf II, His Predecessors and Successors", in Fučikova E. et al. (eds.), *Rudolf II and Prague: the imperial court and residential city as the cultural and spiritual heart of Central Europe* (Prague – London: 1997) 39.

129 For the print, see Hoop Scheffer D. de (comp.), *Aegidius Sadeler to Raphael Sadeler II*, in Boon K.G. (ed.), *Hollstein's Dutch and Flemish Etchings, Engravings and Woodcuts ca. 1450–1700*, vols. 21–22 (Amsterdam: 1980), no. 599. The original drawing is in Bremen.

suggesting he intended it for the *album amicorum* (books of friends) of their mutual acquaintance Abraham Ortelius.

I believe that Sadeler, in an effort to conform to the trend in nature painting established by Hoefnagel and others at the Bavarian court, selected for his introductory series a theme that naturally called for the representation of animals: God's creation of the world. Indeed he appears to have thought through his thematic selection and presentation of the series very carefully. The full title of Jan Sadeler and Maarten de Vos's print series, '*Imago Bonitatis illius*', derives from a passage in the *Liber Sapientiae* (Book of Wisdom): 'candor est enim lucis aeternae, et speculum sine macula Dei majestatis, et *imago bonitatis illius*' ('for [Wisdom] is the brightness of the everlasting light, the unspotted mirror of the power of God, and the image of his goodness') (*Sap.* 7:26). The biblical passage defines the 'image of his goodness' as wisdom itself. This wisdom was gained through knowledge, in particular through knowledge of how God created the world and all the creatures in it. In the preceding lines of the seventh chapter, the *Liber Sapientiae* describes mankind's path to knowing God from man's birth—'I myself also am a mortal man, like to all, and the offspring of him that was first made of the earth' (7:1)—to his realization that in knowledge lies the power of God and the image of His goodness (wisdom):

> For he hath given me certain knowledge of the things that are, namely, to know how the world was made, and the operation of the elements: The beginning, ending, and midst of the times: the alterations of the turning of the sun, and the change of seasons: The circuits of years, and the positions of stars: The natures of living creatures, and the furies of wild beasts: the violence of winds, and the reasonings of men: the diversities of plants and the virtues of roots: And all such things as are either secret or manifest, them I know. For wisdom, which is the worker of all things, taught me: for in her is an understanding spirit holy [...] (Sap. 7:17–22)

While Martin Luther (1483–1546) and other Protestant reformers excluded the *Liber Sapientiae* from the biblical canon, relegating it to the Apocrypha, the Council of Trent (1546) dispelled any such doubt, including the Book of Wisdom in its list of sacred, deuterocanonical books. This reference to the *Liber Sapientiae* in the title of the *Imago Bonitatis* therefore was appropriate for the dedicatee of the print series; 'Wilhelm the Pious', as he came to be known, was a staunch champion of the Counter-Reformation renewal of the Catholic Church in Germany. Wilhelm V's appreciation for the works of his court animalists was linked to his theology, and his spiritual outlook provides the religious context within which Sadeler and De Vos's Paradise prints both were produced and understood at the Bavarian court. I suggest that Sadeler and De Vos's choice of series' title, relating the text of the *Liber Sapientiae* with animal-filled Paradise landscape prints, announces a carefully-conceived pictorial program sympathetic with Wilhelm V's spiritual devotion, patronage of the arts, collecting practices, and construction of gardens as landscapes for meditation and edification.

It was probably not a coincidence that Sadeler selected the *Imago Bonitatis* title from the *Liber Sapientiae*. The Wittelsbach *Kunstkammer*, which served as a gallery space for displaying *naturalia*, books, paintings, prints and other *objets d'art*, had been referred to as the '*Theatrum Sapientiae* (Theater of Wisdom)' since the reign of Albrecht V (r.1550–1579). By the late 1580s the Munich *Kunstkammer* had become famous among other princely collections throughout Europe, including those of Emperor Rudolf II in Prague, Ferdinand II, Archduke of Austria (1529–1595) in Tyrol, and Francesco de'Medici I (1541–1587) in Florence.[130] It was even said to possess a marvelous stuffed elephant, quite a spectacular feat of taxidermy for the period.[131] The size and value of the ducal collection can be gauged by Wilhelm's issue of an order to keep 3,407 precious items in the collection—over half of the collection's more than 6,000 objects—in perpetuity within the Wittelsbach family.[132] According to a passage devoted to the city of Munich in Braun-Hogenberg's great city atlas, *Civitas Orbis Terrarum* (1586), whosoever possessed curiosity could visit the extensive

130 Ibidem, 76–89. On the collection of Rudolf II in Prague, see Bukovinská B., "The *Kunstkammer* of Rudolf II", in Fučíková E. et al. (eds.), *Rudolf II and Prague: The Imperial Court and Residential City as the Cultural and Spiritual Heart of Central Europe* (London: 1997) 199–208. On the function of these collections in early modern court life, see DaCosta Kaufmann T., "Remarks on the Collections of Rudolf II: The Kunstkammer as a form of Representatio", *Art Journal* 38.1 (1978) 22–28; and DaCosta Kaufmann T. *Court, Cloister and City: The Art and Culture of Central Europe* (Chicago: 1995) 166–183.

131 Taylor F.H., *The Taste of Angels* (Boston: 1948) 49; as cited in Lach D.F., *Asia in the Making of Europe* (Chicago: 1993) 153.

132 For the inventory (Staatsbiliothek, Munich cgm 2133 and 2134), see Diemer P., *Johann Baptist Fickler: Das Inventar der Münchner herzoglichen Kunstkammer von 1598* (Munich: 2004) 48, no. 121a. See also Seelig L., "The Munich *Kunstkammer*, 1565–1807", in Impey O. – MacGregor A. (eds.), *The Origin of Museums: The Cabinet of Curiosities in Sixteenth- and Seventeenth-Century Europe* (Oxford: 1985) (76–89) 78; and Smith P.H., "Collecting Nature and Art: Artisans and Knowledge in the *Kunstkammer*", in Hanawalt B. – Kiser L.J. (eds.), *Engaging with Nature: Essays on the Natural World in Medieval and Early Modern Europe* (Notre Dame: 2008) (115–136) 117.

ducal collection.[133] As such, the Bavarian court attracted visitors from all over Europe, including princes, dukes, ambassadors, artists and scholars.[134]

One of the first things visitors encountered upon entry to the *Kunstkammer*'s exhibition gallery was the duke's print collection, kept in a chest in the north-western corner beside a book case holding the duke's most valuable volumes, many containing drawings and engraved illustrations.[135] When the collection's scholarly curator Johann Baptist Fickler (1533–1610) inventoried the *Kunstkammer* treasures at the time of Wilhelm's abdication in 1597 the inventory included a number of illustrated sixteenth-century natural history treatises. Among them were such works as the Italian physician Ippolito Salviani's *Aquatilium animalium historiae* (1554), a popular illustrated scientific work which includes eighty-one illustrations of fishes and cephalopods, engraved by Antoine Lafréry (1512–1577) and Nicolas Beatrizet (before 1520–after 1560), as well as Hans Bol's *Venationis, piscationis, et aucupii typi* (Hunting, Fishing and Fowling Scenes), published in Antwerp by Philips Galle (1537–1612).[136] Bound series of prints on spiritual themes and landscapes were also kept in the book-case, among them the various print series that Jan Sadeler dedicated to the duke, including the *Imago Bonitatis*.[137] Records show that Wilhelm's court artists had access to these valuable works; for example, the duke's artistic

133 Braun and Hogenberg, *Civitas Orbis Terrarum* Pt. 4, no. 43.

134 The numismatic specialist Adolf Occo visited the coin collection in 1579; see Occo A., *Imperatorum Romanorum Numismata a Pompeio Magno ad Heraclium* (Antwerp, Christoph Plantin: 1579) n.p. On Duke Augustus the Younger of Braunschweig-Lüneburg's visit in 1598, see Gobiet R. (ed.), *Der Briefwechsel zwischen Philipp Hainhofer und Herzog August d. J. von Braunschweig-Lüneburg* (Munich – Berlin: 1984); both as cited in Seelig, "Munich *Kunstkammer*" 77–78.

135 On the contents and organization of the print cabinet and its contents, see Diemer, "Verloren—verstreut—bewahrt. Graphik und Bücher der Kunstkammer", in Sauerländer W. (ed.), *Die Münchner Kunstkammer* (Munich: 2008) 225–252. On the arrangement of the *Kunstkammer* see Seelig, "Munich *Kunstkammer*" 79; and Kaliardos K.P., *The Munich Kunstkammer* (Tübingen: 2013) 11, 16.

136 Diemer, *Johann Baptist Fickler: Das Inventar*, 45, nos. 65 and 73. For Salviani's work, see Ippolito Salviani, *Aquatilium animalium historiae* (Rome, Ippolito Salviani: 1554). For Bol's *Venationis, piscationis, et aucupii typi*, see

137 See Diemer, "Verloren—verstreut—bewahrt", 225–252, esp. 234–235. Unlike several of Sadeler's other print series for the duke recorded in Fickler's inventory, the *Imago Bonitatis* does not appear by its title; rather he describes it as a series of Old Testament engravings bound in white copper: '(96) Figuren des Alten Testaments in kupffer gestochen, eingebunden in weiß Copert gesangsweiss'. See Diemer, *Johann Baptist Fickler: Das Inventar*, 47, no. 99.

advisor, Friedrich Sustris (1540–1599), borrowed Gessner's *Vogelbuch* (Book of Birds) from the library and used it as a model book for his design of avian wall paintings at the court.[138] Illustrated works like these were often displayed on tables and cabinet shelves beside natural artifacts, such as rhinoceros horns,[139] an old elephant's tooth,[140] ostrich feathers,[141] a giant tortoise shell,[142] a black gazelle horn,[143] claws and bones.[144]

According to a sixteenth-century manual for the ideal organization of a *Kunstkammer*—the *Inscriptiones vel Tituli Theatri Amplissimi* (1565), written by the Flemish advisor to Duke Albrecht V, Samuel Quiccheberg (1529–1567), graphic materials, works of art, and natural objects functioned together in a collection to further one's pursuit of a universal knowledge of the world.[145] Quiccheberg's treatise emphasizes the importance of engravings as tools for the acquistion of knowledge. In his conception of an ideal *Kunstkammer* images of sacred history take precendence. Given that the overall purpose of a collection was, according to his treatise, the production of knowledge, the importance placed on religious imagery suggests that all knowledge ultimately stemmed from divine origins. The prevalence of religious imagery throughout the collection supports Quiccheberg's understanding of the *Kunstkammer*'s central epistemological function, because at the time knowledge in general was inseparable from religious knowledge.[146] The title of Sadeler and De Vos's *Imago Bonitatis*, along with its religious subject matter and animal imagery, supports this very idea, proclaiming both the series' place and its function in the acquisition of knowledge within the context of the Bavarian *Kunstkammer*.

---

138 Gessner Conrad, *Vogelbuch* (Zurich, Christoph Froschauer: 1557). See Vignau-Wilberg, *In Europa zu Hause*, 137–138.

139 See Diemer, "Verloren—verstret—bewahrt", 55, no. 233.

140 Ibidem, 222, no. 3363.

141 Ibidem, 112, no. 1079; see also nos. 480, 513–515, 987–1002, tables 10, 15.

142 Ibidem, 208, no. 2974.

143 For mention of a black gazelle horn in Philipp Hainhofer's 1611 description of the *Kunstkammer*, see Doering O., "Das Augsburger Patriciers Philipp Hainhofer Beziehungen zum Herzog Philipp II von Pommern-Stettin, Correspondenzen aus den Jahren 1610–1619", *Quellenschriften für Kunstgeschichte und Kunsttechnik des Mittelalters und der Neuzeit*, n.s. 6 (1896) 140, 166.

144 See Seelig, "Munich *Kunstkammer*", 81.

145 Hajós E.M., "The Concept of an Engravings Collection in the Year 1565: Quiccheberg, *Inscriptiones Vel Tituli Theatri Amplissimi*", *The Art Bulletin* 40, 2 (June 1958) 151–156.

146 Kaliardos, *Munich Kunstkammer* 21.

The *Imago Bonitatis* was a series as aptly suited to the artistic and collecting culture at the Munich court as it was to the duke's fascination with creating paradisiacal gardens, which featured live exotic and domestic animals. From the 1570s onwards, Wilhelm began creating expansive botanical parklands at each of his ducal residences, amassing rare specimens of fauna to inhabit them.[147] For his gardens at Trausnitz, for example, he acquired giant turtles from Genoa, Babylonian fowl, parrots, and a monkey from Algeria.[148] He even had the medieval moat converted into a menagerie for bears, a lion, and a leopard, and he had enclosures built for peacocks, pheasants, foxes, and rabbits.[149] When asked to come up with a name for the Trausnitz gardens, Wilhelm's court humanist Anselm Stöckl suggested '*Paradisus* (Paradise)', an apt designation for a garden filled with the wondrous creatures of God's creation.[150]

The appeal of the natural world for Wilhelm continued throughout his reign. At each of his palaces he recreated edenic gardens in which to display the diversity of the Divine Creation. With the death of his father, Wilhelm moved to Munich where he repeated the Trausnitz parkland at the Bavarian court, later going on to build similar gardens at the Maxburg, Schleissheim Palace, and Neudecks Castle. Between 1582 and 1589, for example, Wilhelm had a large *Hofgarten* installed at his Munich residence, complete with running

147 In a letter to the duke on 24 October 1579, Friedrich Sustris writes that the 'Hermitorium', or 'Hermitage', had been constructed in the gardens at Trausnitz Castle, Wilhelm's residence in Landshut from 1568 and 1579. See Maxwell S., "The Pursuit of Art and Pleasure in the Secret Grotto of Wilhelm V of Bavaria", *Renaissance Quarterly* 61 (2008) (414–462) 426.

148 Baader B.Ph., *Der bayerische Renaissancehof Herzog Wilhelms V* (Leipzig: 1943) 297–300.

149 See Lietzmann H., *Der Landshuter Renaissancegarten Herzog Wilhelms V. von Bayern: Ein Beitrag zur Garten- und Kulturgeschichte der Frühen Neuzeit* (Munich: 2001) 31–40. It is likely that members of the Fugger banking family strongly influenced the formation of Wilhelm's menageries, particularly Hans Fugger (1531–1598), with whom Wilhelm corresponded on a daily basis. Around the same time that Wilhelm began expanding his parklands, Hans Fugger enlarged his own aviaries and animal enclosures so that he could exhibit live beasts from Africa, Central America, and India to the public. The Fuggers had long been involved with the trade and captivity of exotic animals; see Dembeck H., *Mit Tieren leben* (Düsseldorf – Vienna: 1961) vol. IV, 31; and Gorgas M., "Animal Trade Between India and Western Eurasia in the Sixteenth Century—The Role of the Fuggers in Animal Trading", in Matthew K.S. (ed.), *Indo-Portuguese Trade and the Fuggers of German, Sixteenth Century* (New Delhi: 1997) (199–225) 221. On the Fuggers and the Bavarian court, see Maxwell, *Court Art of Friedrich Sustris*; and Meadow M., "Merchants and Marvels: Hans Jacob Fugger and the Origins of the *Wunderkammer*", in Findlen P. – Smith P.H. (eds.), *Merchants and Marvels: Commerce and the Representation of Nature* (London: 2001).

150 See note 148 above.

creeks, fountains, large fishing ponds, and a growing collection of exotic plants and animals from as far away as Turkey and the Americas.[151] Parallel to the *Hofgarten* he constructed a small enclave known as the *Grottenhof*, an artificial grotto built of rocks, shells, coral, crystals and semiprecious gems.[152] The hunting scenes painted on the *Grottenhof* walls—designed by Friedrich Sustris after Gessner's *Vogelbuch*—displayed a variety of precisely-detailed bird species, ranging from owls to songbirds and even a turkey.[153] The *Grottenhof*, described by its earliest visitors as the duke's secret pleasure garden, stood in close proximity to the *Kunstkammer*, and together they harmoniously integrated the natural world with the man-made arts of painting, sculpture, and prints. In contrast to the grandeur and open nature of the *Hofgarten*, the *Grottenhof* at Trausnitz was a place of private retreat and esoteric pleasure.[154]

The various parklands that Wilhelm constructed at his Bavarian residences became landscapes of meditation and edification. They functioned as spaces of retreat, offering the duke privacy for solitary prayer, yet affording regular access to spiritual supervision. Meditative prayer had been an important aspect of Roman Catholic spirituality even before the Council of Trent, as made apparent in the writings of the Spanish founder of the Society of Jesus, Ignatius of Loyola (1491–1556). Ignatius's *Spiritual Exercises*, composed between 1528 and 1548, outline a program of meditation proposed to effect in the exercitant a life devoted to the greater service and praise of God.[155] In the introduction, Ignatius describes the most convenient locations for carrying out one's devotional exercises: a secluded house or room that would allow for privacy, absent of distractions.[156] Within the Christian monastic tradition at the time, this devotional practice mixing solitary existence with penitence and perpetual dedication to prayer was believed to lead to a purification of the heart and an ascent of the soul to its Creator.[157] Late in the sixteenth century the Jesuits began to set up retreats, locating them in the countryside or as

151 Diemer D., "Bronzeplastik um 1600 in München: Neue Quellen und Forchungen, II. Teil", *Jahrbuch des Zentralinstituts für Kunstgeschichte* 3 (1987) 109–168, esp. 118–119; as cited in Maxwell, *Court Art of Friedrich Sustris* 5.

152 On the Grottenhof, see Maxwell, "The Pursuit of Art and Pleasure"; and Vignau-Wilberg, *In Europa zu Hause*, 131–197.

153 See note 138 above.

154 Maxwell, *Court Art of Friedrich Sustris* 158.

155 Ignatius of Loyola, *Spiritual Exercises* (Westminster: 1951).

156 Witte A.A., *The Artful Hermitage: The Palazetto Farnese as a Counter-reformation diaeta* (Rome: 2008) 79.

157 Witte, *The Artful Hermitage* 126.

buildings detached from the main convents of the Society of Jesus, as was the case in Bavaria, where Wilhelm V championed the Jesuits, campaigning vigorously for the canonization of Ignatius of Loyola.[158]

Wilhelm's support of the Jesuits continued a deeply-rooted patrilineal legacy of Bavarian ducal patronage. It was his grandfather, Duke Wilhelm IV (1493–1550), who first brought the Jesuit order to Bavaria and in 1555 his father, Duke Albrecht V (1528–1579), introduced the important Jesuit Catholic priest Peter Canisius (1521–1597) to the university at Ingolstadt, transforming it over the next few decades into a Jesuit school.[159] Following this example, Wilhelm V kept a Jesuit confessor by his side from 1578 until his death in 1626, and was said to have catered to the poor, visited the sick in the hospital, and overseen the cleaning of their rooms.[160] The influence of Ignatius's *Spiritual Exercises* in Munich can be seen in the conceptual framework for the artistic program of Wilhelm V's most important architectural project: St. Michael's church, one of the earliest Jesuit churches north of the Alps.[161] Closely resembling the first Jesuit church of Il Gesú in Rome, St. Michael's and the Jesuit college were Wilhelm V's most visible manifestations of his support of the Society of Jesus in Munich. To St. Michael's he attached a large residential tract, the Wilhelminische Veste (later called the Maxburg), that became his ducal residence after his abdication in 1597.[162] He connected the Jesuit mission and ducal palace with an arch and elevated passageway called the Wilhelmsbogen, through which he passed daily, reportedly spending four hours a day in prayer and spiritual devotion.[163]

The duke's construction of gardens and his support of the Jesuits substantiate a reputed desire of the duke after his abdication in 1597 to step back from the world of politics in favor of dedicating his life to observing the anchorite

158 Chipps Smith J., *Sensuous Worship: Jesuits and the Art of the Early Catholic Reformation in Germany* (Princeton – Oxford: 2002) 58.

159 Albrecht's last will of 11 April 1578 asked that his heirs safeguard the two Jesuit colleges he had founded and increase their number. See Ziegler W., "Das Testament Herzog Albrechts V. von Bayern (1578)", in Greipl F.J. – Kraus A. (eds.), *Aus Bayerns Geschichte, Forschungen als Festgabe zum 70. Geburtstag von Andreas Kraus* (St. Ottilien: 1992) 268–270.

160 See Maxwell, *Court Art of Friedrich Sustris* 101.

161 Chipps Smith, *Sensuous Worship* 77–101.

162 A life-size clay portrait bust of the duke, originally placed in an exterior niche on the Wilhelmsbogen, publically signaled the state's support of the Society of Jesus. See reproduction of the bust by Hubert Gerhard as published in Chipps Smith, *Sensuous Worship*, 57, fig. 30.

163 See Chipps Smith, *Sensuous Worship*.

principles of retreat and penance.[164] His walled-in gardens sustained the concepts of isolation and contemplation that formed the basis of the anchortite ideal, which orders such as the Carthusians sought to revive within the limits of sixteenth-century monastic regulation. Philipp Hainhofer's (1578–1647) description of the Maxburg on the occasion of his visit in 1611 refers to a '*Grotte*' made up of a number of chapels and elaborately rusticated rooms with full-grown pines, a small stream and pond, altars made up of boulders, furniture crafted from straw and brush, and walls painted with scenes from the life of the hermit St. Francis.[165] The duke even installed two Carthusian monks in the Maxburg parklands, tranforming it into a true hermitage. After his abdication, Wilhelm retired to his country estate of Schleissheim, the gardens of which became his princely eremitic retreat, complete with nine chapels and attached hermitages.

Wilhelm V's personal tendency to retreat and his interest in garden culture, zoology, and the study of naturalia relates to the activities of a number of cardinals in Italy around 1600, likewise invested in the spiritual renewal of the Catholic Church.[166] The Jesuit Cardinal Roberto Bellarmino (1542–1621), for example, retreated annually to the Jesuit noviciate of Sant' Andrea al Quirinale for his spiritual exercises. The garden, situated on the slope of Quirinal Hill between the noviciate and the the church of San Vitale, is described in the introduction to Louis Richeome's (1544–1625) *La Peinture Spirituelle ou l'art d'admirer, aimer et louer Dieu en toutes ses oeuvres, et tirer de toutes profit salutere* (*The spiritual painting or the art of admiring and praising God in all his works and drawing healthy profit from all this*). With an account in the encyclopedic tradition of natural histories, the description links all of the flowers, plants, and trees to the Creation, and provides the reader with an allegorical interpretation of nature as a reflection of the Divine.[167] The belief that contemplation of the Creation and all things created could lead one to a union with God relates to an optimistic spiritual trend that one historian has described as *the* most characteristic strain of Counter-Reformation spirituality across

164 For the definition of 'anchorite', see Buchberger M. (ed.), *Lexikon für Theologie und Kirche* (Freiburg – Basel – Vienna: 1993–2001) vol. 1, col. 572.

165 Maxwell, *Court Art of Friedrich Sustris*, 199. For Hainhofer's original text see Andres H.M., *Die Rekonstruktion der Herzog-Maxburg in München* (Munich: 1987) 35–36.

166 See Witte, *Artful Hermitage*, esp. 89–123.

167 Richeome L., *La Peinture Spirituelle ou l'art d'admirer, aimer et louer Dieu en toutes ses oeuvres, et tirer de toutes profit salutere* (Lyon, Pierre Rigaud: 1611), 'Introduction'. As cited in Witte, *The Artful Hermitage* 104.

post-Tridentine Europe.[168] In his own writings, Cardinal Bellarmino considers the natural world to be a reflection of the Creator, suggesting that looking at nature could produce in the exercitant an understanding of the Divine. Bellarmino's book on the devotional interpretation of the visible world, *Scala di salire con la mente a Dio per mezo delle cose create* ('Ladder to ascend with the mind to God by means of the created things') (1615), cites both the Apostle Paul and Saint Augustine on this matter. In his letter to the Romans, Paul suggested that one of the ways that God revealed himself to man was through the Creation; Saint Augustine sustained this belief in his biblical exegesis, finding the Creation to be a condition for the Redemption of Man.[169] In the writings of individuals such as the Archbishop of Milan Federico Borromeo (1564–1631) and the priest Filippo Neri (1515–1595) one again finds the optimistic belief that all things are worthy of contemplation.[170] Endowing God's creatures with positive value, these men regarded the things of the earthly world as reminders of the Creator's generosity and as aids in constructing knowledge of Him through prayer as one sought the ultimate goal of union with God. Their writings express the conviction that original sin did not result in utter depravity, but rather in an opportunity to exercise free will in cooperation with God in order to obtain grace.

An important aspect of the optimistic spirituality of Bellarmino, Neri, and Borromeo was a love of solitude and dedication to meditation that involved the practice of retreat for the purpose of solitary prayer. In his description of Borromeo's spiritual life, biographer Francesco Rivola writes that Borromeo considered stepping back from his Episcopal obligations in order to retreat from public life altogether and enter a hermitage; in being prevented from doing so, Borromeo maintained a cell in the woods for regular retreat, so

168 Historians have suggested that the Counter-Reformation encompassed various spiritual trends which developed in phases, particularly in Italy: a period of stern and austere Counter-Reform Catholicism, prone to asceticism and contempt for earthly things, emerged from the Council of Trent in 1564 and continued through the 1570s; and a second period (1580–1600) emphasizes praise of God and the universe, including all mundane things. More recently historians of Roman Catholic reform have found a concurrence of the two trends. See the useful discussion in Jones P.M., *Federico Borromeo and the Ambrosiana: Art Patronage and Reform in Seventeenth-Century Milan* (Cambridge: 1993) 9–11.

169 Witte, *The Artful Hermitage* 100.

170 See Dupront A., "Autour de Saint Filippo Neri: De l'optimisme chrétien", *Mélanges d'archéologie et d'histoire* 49 (1932) 219–259.

that he could attend to his studies and the contemplation of Divine things.[171] In moments when spiritual retreat was not possible, Borromeo turned instead to natural imagery in the form of hermit landscapes and flower paintings. In 'Pro suis studiis', a manuscript collection of autobiographical notes, Borromeo alludes to the fact that such imagery made it possible for him to journey through nature and contemplate the natural world without having to go into the garden:

> I have had my room ornmaneted with paintings, and I have made sure that all of them are excellent; there is not one vulgar or cheap thing. And the pleasure I take in looking at these painted views has always seemed to me as beautiful as open and wide views [of nature].... Instead of them, when they are not had, paintings enclose in narrow places, the space of earth and the heavens, and we go wandering, and making long [spiritual] journeys standing still in our room.[172]

For this purpose Borromeo engaged Jan Brueghel the Elder and Paul Bril (1554–1626) to produce large hermit landscape paintings based on a series of hermit prints engraved by Jan Sadeler and his brother Raphael Sadeler (1560/61–1528 or 1530) after Maarten de Vos: *Solitudo sive vitae patrum eremicolarum* ('Solitude or the Life of the Hermit', 1585–1586), the first of four such hermit print series by the artists.[173] Faithfully inscribed on Bril's pendant paintings *Landscape with Anub* and *Landscape with Mutius* is the name of the hermit figure and the number of the relevant engraving from Borromeo's *Solitudo*.[174] Borromeo displayed the paintings publically in his Ambrosiana, suggesting that he viewed the works as pedantic guides for meditative prayer. In Borromeo's view, artistic representations enabled the worshipper to at least mentally feel the solitude of the natural space and the magnitude of God's creations. For example as a prelude to prayer Borromeo advocates that the exercitant first imagine all

171 Witte, *Artful Hermitage* 95.

172 Jones, *Federico Borromeo*, 64 citing Borromeo, *Pro suis studiis*, 1628, Bibl. Ambrosiana Ms. G.310inf., no. 8, fols. 252r–253r. See also Witte, *The Artful Hermitage* 96.

173 Schuckman, *Maarten de Vos*, vols. 44 and 46, nos. 964–993. See Brueghel's *Landscape with a Hermit Reading and Ruin* (1596) and *Mountain Landscape with a Hermit* (1597), both in Pinacoteca Ambrosiana, Milan; reproduced in Jones, *Federico Borromeo*, 78, figs. 33–34.

174 See Jones P.M., "Two Newly-Discovered Hermit Landscapes by Paul Bril", *Burlington Magazine* 130 (1988) 32–35; and Jones, *Federico Borromeo*, 132–133, figs. 68–69. See also Witte, *The Artful Hermitage* 132.

terrestrial and celestial creatures which contribute to God's glory.[175] Borromeo's attitude reveals that he understood nature paintings to operate in the same way as nature itself, that is, to stimulate contemplation. While Ignatius of Loyola never discussed the use of actual artworks in one's spiritual practice, he had insisted on the need to create mental images, thereby implying the utility of images in achieving union with God.[176] For Borromeo, hermit pictures along with paintings such as Brueghel's *Element of Earth*, in which the background figures of God, Adam, and Eve transform the picture into a Paradise landscape, served a purpose to inspire praise of God's marvelous ability to create an orderly and harmonious world.[177]

Whether or not Paradise imagery and hermit print series functioned for Duke Wilhelm V in the same devotional way as they did for Cardinal Federico Borromeo is undocumented. Certainly Wilhelm's appreciation of the natural world—from his creation of paradisiacal gardens filled with live animals and exotic plants, to his construction of chapels and hermitages in the parklands in each of his residences—can be traced in the print series that Jan Sadeler produced at the Munich court in collaboration with Maarten De Vos in Antwerp. The production of the *Imago Bonitatis* closely followed the duo's *Solitudo sive vitae patrum eremicolarum*, made in the year before Sadeler's court appointment, and it is likely that the broad success of this hermit print series is what brought the Flemish artists to the attention of the Bavarian duke in the first place—making their renewed joint venture with the *Imago Bonitatis* and its emphasis on the natural world a natural sequel. Years later Jan and Raphael collaborated with De Vos on the last of their four hermit print series, *Sylvae Sacrae Monumenta* ('Sacred Forest Monuments'), which Jan bestowed as his final gift to his ducal patron in 1594, at the end of his tenure as court artist.[178] The dedicatory poem for the *Sylvae Sacrae Monumenta*, penned by the erudite and eloquent Munich Jesuit Matthäus Rader (1561–1634), praises Wilhelm V as a keen admirer of such imagery. Indeed throughout his reign, Wilhelm's

175 Jones, "Two Newly-Discovered Hermit Landscapes by Paul Bril" 76; and Witte, *The Artful Hermitage* 96.

176 Jones, *Federico Borromeo* 68.

177 Jones, "Two Newly-Discovered Hermit Landscapes by Paul Bril" 79–80.

178 See Schuckman, *Maarten de Vos*, vols. 44 and 46, nos. 994–1023. Fickler's inventory records the print series as follows: 'Ein ander dergleichen buech intituliert, Sylvae Sacrae. Monumenta Sanctioris Philosophiae, quam severa Anachoretarum disciplina vitae et religio docuit. Joannis et Raphaelis Sadelerorum'; see Diemer, *Johann Baptist Fickler: Das Inventar*, 44, no. 53.

patronage of the arts, collecting practices, and building projects went hand in hand with his devotional activitites, so much so that when Philipp Hainhofer visited the Maxburg, with its hermitage and resident Carthusian monks, the latter remarked that 'everything in this grotto is done in the way seen in depictions of monks and hermits in paintings and engravings'.[179]

From our consideration of Counter-Reform spiritual attitudes alongside the, patronage, collecting, and building practices that characterize Wilhelm V's reign coalesces a vivid picture of the courtly world that Jan Sadaler navigated upon entry into his role as '*Celsitudinis suae chalcographus*'. As with the hermit series, the *Imago Bonitatis* speaks to Jan Sadeler's savvy as a court artist. Wilhelm's appreciation of the nature in connection with his devotional practice makes the title and theme of Sadeler and De Vos's *Imago Bonitatis* fitting in a work intended to showcase the court artist's talents, among them a keen sense of decorum. The passage in the *Liber Sapientiae* from which Sadeler extracted the series' title points to the spiritual framework for the duke's reception of the series, reminding audiences that the natural world and descriptions of it served an edifying purpose: to encourage one's contemplation of nature as a path to knowing God and achieving the goal of union with Him. In this light the *Imago Bonitatis* emerges as a work custom-tailored to the interests of Sadeler's ducal patron, while bringing to the fore the important role that collaboration and court patronage played in Maarten de Vos's activities as a print designer.

Conclusion

The idea that the image of animals in Paradise could edify its viewers was not one that Sadeler and De Vos imagined in a vacuum. It had its parallel already in sixteenth-century devotional practices and in natural historical treatises, as well as in contemporary literature. The celebrated hexaemeral epic by Guillaume de Saluste, Sieur du Bartas (1544–1590), *La Sepmaine, ou Creation du monde* (1578), retells the biblical story of the Creation in six days, evoking in

179 'und ist alle diese grotto zusamen gemacht, als wie mann inn den gemählen und küpferstucken die patres und Eremitas abconterfett sihet', as translated from Vignau-Wilberg T., "Eremiten und Einsiedeleien: Bilder zur Einkehr und Erbauung—Hermits and Hermitages: Images for Contemplation and Edification", in *In Europa zu Hause—Niederländer in München um 1600* (Munich: 2005) 373–374.

the reader a strong sense of the infinite array of created things.[180] Enumerating and describing many kinds of animals, birds and fishes in elaborate detail, *La Sepmaine* draws on information in the natural historical texts of Pliny, Rondelet, Belon, Gessner and others.[181] Du Bartas uses his descriptive lists of creatures to convey to his readers a sense of the greatness of their source: God the Creator. Du Bartas's epic was well-received and quickly appeared in many different languages throughout Europe. The affinity between Du Bartas's descriptive lists and De Vos and Sadeler's visual catalogue of *animalia* in the *Imago Bonitatis* was a connection observed by their contemporaries. The French publisher Simon Goulart (1543–1628) even had copies made after the final two prints in the *Imago Bonitatis* (*The Sixth Day of Creation: The Creation of Adam, Eve and the Animals* and *The Sixth Day: Adam, Eve and the Animals Blessed By God*) to use as illustrations to his 1611 and later editions of Du Bartas's *Les Oeuvres*.[182]

In devising the *Imago Bonitatis*, Jan Sadeler turned to his long-time collaborator, Maarten de Vos, to design landscapes filled with *animalia*—landscapes that could rival the exquisite renderings of the Bavarian court's celebrated animalists. Yet if it was Wilhelm V's appreciation for the natural world and the artistry of Hoefnagel and others that inspired Sadeler to create the *Imago Bonitatis* in the first place, then it was the artistic culture of animal representation in Antwerp to which De Vos turned in his execution of the *Imago Bonitatis*'s original designs. Working miles apart, Sadeler and De Vos's co-creation of the series synthesizes the artists' individual responses to their immediate environments in Munich and Antwerp. The subject of Paradise gave Sadeler, De Vos and later artists a narrative frame for the display of God's creatures in the service of spiritual edification.

At the turn of the seventeenth century, to know nature was to know God. Printmakers and painters alike began to identify an artistic knowledge of nature with the naturalist's view of the world. The collaborative work of De Vos and Sadeler on the *Imago Bonitatis* series redefined the *Paradijslandschap* as a synthesis of scientific inquiry and religious signification, two endeavors

---

180 Guillaume de Salluste, Sieur du Bartas, *La Sepmaine* (Paris, Jean Février: 1578).

181 Creore A.E., "The Scientific and Technical Vocabulary of Du Bartas", *Bibliothèque d'Humanisme et Renaissance* 21, 1 (1959) 131–160.

182 Goulart S. (ed.), *Les Oeuvres de G. de Saluste, Sr. Du. Bartas* (Paris, Toussainctz du Bray: 1611). See Schuckman, *Maarten de Vos*, vol. 44, nos. 11–18 (c). For illustrations of these copies see Williams A., *The Common Expositor: An Account of the Commentaries on Genesis 1527–1633* (Chapel Hill: 1948).

revealed to be sympathetic in the natural history of Conrad Gessner. More than just minutely-detailed images of Paradise, the *Imago Bonitatis* series reflects one of the ways that advances in scientific knowledge intersected with early modern religious printmaking. The new vision of Paradise presented in their series would find its most colorful expression in the work of Jan Brueghel the Elder in the early seventeenth century. Brueghel's numerous Paradise paintings depict vivid, pictorial catalogues of birds, fishes and animals with increasing scientific naturalism.[183] One scholar makes a useful argument for Brueghel's Paradise paintings, an argument that equally pertains to the *Imago Bonitatis* prints: the novelty of the Paradise landscapes lies not only in their impressive and detailed array of species, but also in the simultaneous presentation of these species both as figures of a religious narrative and as subjects of a scientific order. Maarten de Vos and Jan Sadeler's *Imago Bonitatis*—executed several years before Breughel's earliest Paradise painting, *The Creation of Adam* (1594), and decades before his first truly catalogic ones—anticipates this turn towards the spiritualized naturalism that Flemish animal imagery would take in the seventeenth century.

Bibliography

Andres H.M., *Die Rekonstruktion der Herzog-Maxburg in München* (Munich: 1987).

Ashworth W.B., Jr., "Gheeraerts and the Aesopic connection in Seventeenth-Century Scientific Illustration", *Art Journal* 44, 2 (1984) 132–138.

Ashworth W.B., Jr., "Natural History and the emblematic world view", in Lindberg D.C. – Westman R.S. (eds.), *Reappraisals of the Scientific Revolution* (Cambridge: 1990) 303–332.

Ashworth W.B., Jr., "Emblematic natural history of the Renaissance", in Jardine N. – Spary E. (eds.), *Cultures of Natural History* (Cambridge: 1996) 17–37.

Baader B. Ph., *Der bayerische Renaissancehof Herzog Wilhelms V* (Leipzig: 1943).

Baines L., "Nicolaes de Bruyn (1571–1656) and the Art of the Professional Engraver" (Ph.D. Diss. University of Delaware: 2011).

Bartas Guillaume De Saluste, Sieur Du, *La Sepmaine* (Paris, Jean Février: 1578).

Belon Pierre, *Les observations de plusieurs singularitéz et choses memorables* (Lyon, G. Corrozet: 1553).

183 See Rikken M. – Smith P.J., "Jan Brueghel's *Allegory of Air* (1621) from a natural historical perspective", *Nederlands Kunsthistorisch Jaarboek / Netherlands Yearbook for History of Art* 61 (2011) 87–115.

Belon Pierre, *L'Histoire de la nature des oyseaux* (Paris, B. Prevost – G. Cavellat: 1555).

Bianciotto G. (ed. and trans.), *Bestiaires du Moyen Âge. Pierre de Beauvais, Guillaume le Clerc, Richard de Fournival, Brunetto Latini et Jean Corbechon* (Paris: 1980).

Bihan O. Le, "Bestiaire et imaginaire de l'air, du XVIe au XVIIIe siècle. Le caméléon et l'oiseau de paradis", in *Les éléments et les métamorphoses de la nature: imaginaire et symbolique des arts dans la culture européenne du XVIe au XVIIIe siècle. Actes du colloque international de l'Opéra de Bordeaux, 17–21 septembre 1997* (Bordeaux: 2004) 139–152.

Blübaum D. – Erbentraut R. (eds.), *Die Erschaffung der Tiere: Tiere in der Niederländischen Kunst des Manierismus und Frühbarock und an den wandfesten Dekorationen des Schlosses zu Güstrow* (Schwerin: 2010).

Bono J., The Word of God and the Language of Men: Interpreting Nature in Early Modern Science and Medicine, vol. 1 (Madison: 1995).

Braun L., *Conrad Gesner* (Geneva: 1990).

Broek R. van den, *The Myth of the Phoenix, According to Classical and Early Christian Traditions* (Leiden: 1972).

Bukovinská B., "The *Kunstkammer* of Rudolf II", in Fučíková E. et al. (eds.), *Rudolf II and Prague: The Imperial Court and Residential City as the Cultural and Spiritual Heart of Central Europe* (London: 1997).

Camerarius Joachim, *Symbolorum et emblematum ex aquatilibus et reptilibus desumptorum* (Nuremburg, Johann Hofmann – Hubert Caymox: 1590).

Cerutus B. – Chiocco A., *Musaeum Francisci Calceolari* (Verona, Angelo Tamo: 1622).

Charbonneau-Lassay L., *Le Bestiaire du Christ: La mystérieuse emblématique de Jésus-Christ* (Milan: 1940).

Chipps Smith J., *Sensuous Worship: Jesuits and the Art of the Early Catholic Reformation in Germany* (Princeton – Oxford: 2002).

Clusius Carolus, *Exoticorum libri decem: quibus animalium, plantarum, aromatum, aliorumque peregrinorum fructuum historiae describuntur: Item Petri Bellonis observationibus* [...] (Leiden, Franciscus Raphelengius: 1605).

Clutius Augerius, *Opuscula duo singularia. II. De hemerobio sive ephemero insecto et majali verme* (Amsterdam, Jacob Charpentier: 1634).

Creore A.E., "The Scientific and Technical Vocabulary of Du Bartas", *Bibliothèque d'Humanisme et Renaissance* 21,1 (1959) 131–160.

Curley M.J. (trans.), *Physiologus* (Austin – London: 1979).

Dacosta Kaufmann T., "Remarks on the Collections of Rudolf II: The Kunstkammer as a form of Representatio", *Art Journal* 38, 1 (1978) 22–28.

Dacosta Kaufmann T., *Court, Cloister & City: The Art and Culture of Central Europe 1450–1800* (Chicago: 1995).

Dene Eduard de, *De warachtighe fabulen der dieren* (Bruges, Pieter de Clerck for Marcus Gheeraerts: 1567); *Esbatement moral des animaux* (Antwerp, Gerard Smits for Philip Galle: 1578).

Diels A. – Leesberg M. (comps.), *The Collaert Dynasty*, Leesberg M. – Balis A. (eds.), *The New Hollstein Dutch & Flemish Etchings, Engravings and Woodcuts* (Oudekerk aan den Ijssel: 2005).

Diemer P., *Johann Baptist Fickler: Das Inventar der Münchner herzoglichen Kunstkammer von 1598* (Munich: 2004).

Diemer P., "Verloren—verstreut—bewahrt. Graphik und Bücher der Kunstkammer", in Sauerländer W. (ed.), *Die Münchner Kunstkammer* (Munich: 2008) 225–252.

Duverger J., "Notes concernant les tapisseries du seizième siècle au château du Wawel", *Bulletin des Musées royaux d'art et d'histoire* 45 (1973) 65–76.

Eichberger D., "*Naturalia* and *artefacta*: Dürer's nature drawings and early collecting", in Eichberger D. – Zika C. (eds.), *Dürer and his Culture* (Cambridge: 1998) 13–37.

Eichberger D., *Leben mit Kunst, Wirken durch Kunst. Sammelwesen und Hofkunst unter Margarete von Österreich, Regentin der Niederlande* (Turnhout: 2002).

Erbentraut R., "Als das Kamel Blätter vom Baum der Erkenntnis Frass: Paradiesdarstellungen im niederländischen Manierismus", in Blübaum D. – Erbentraut R. (eds.), *Die Erschaffung der Tiere: Tiere in der Niederländischen Kunst des Manierismus und Frühbarock und an den wandfesten Dekorationen des Schlosses zu Güstrow* (Schwerin: 2010) 47–57.

Faber Kolb A., *Jan Brueghel the Elder: The Entry of the Animals into Noah's Ark* (Los Angeles: 2005).

Ferro X.R.M., *Symboles animaux: un dictionnaire des représentations et croyances en Occident* (Paris: 1996).

Fernández De Oveido Y Valdéz G., *Natural History of the West Indies*, ed. and trans. by S.S. Stoudemire (Chapel Hill: 1959).

Findlen P. – Smith P.H. (eds.), *Merchants & Marvels: Commerce, Science and Art in Early Modern Europe* (New York: 2002).

Francissen F.P.M. – Mol A.W.M., *Augerius Clutius and his 'De hemerobio'. An Early Work on Ephemeroptera* (Marburg: 1984).

George W., "Dead or Alive: Zoological Collections in the Seventeenth Century", in Impey O. – MacGregor A. (eds.), *The Origins of Museums: The Cabinet of Curiosities in Sixteenth- and Seventeenth-Century Europe* (Oxford: 1985) 179–187.

Gessner Conrad, *Historia animalium*, 4 vols. (Zurich, Christoph Froschauer: 1551–1558).

Gmelig-Nijboer C.-A., *Conrad Gessner's 'Historia Animalium': an inventory of Renaissance Zoology* (Meppel: 1977).

Gorgas M., "Animal Trade Between India and Western Eurasia in the Sixteenth Century—The Role of the Fuggers in Animal Trading", in Matthew K.S. (ed.), *Indo-Portuguese Trade and the Fuggers of Germany Sixteenth Century* (Delhi: 1997).

Gudger E.W., "The five great naturalists of the sixteenth century: Belon, Rondelet, Salviani, Gesner, Aldrovandi: a chapter in the history of ichthyology", *Isis* 22 (1934) 21–40.

Haeften J. van, *Dutch and Flemish Old Master Paintings* (London, Johnny van Haeften: 2008).

Hajós E.M., "The Concept of an Engravings Collection in the Year 1565: Quiccheberg, *Inscriptiones Vel Tituli Theatri Amplissimi*", *The Art Bulletin* 40, 2 (June 1958) 151–156.

Harrison P., *The Bible, Protestantism, and the Rise of Natural Science* (Cambridge: 1998).

Hendrix M.L., *Joris Hoefnagel and the 'Four Elements': a study in sixteenth-century nature painting* (Ph.D. Diss., Princeton University: 1989).

Het Aards Paradijs: Dierenvoorstellingen in de Nederlanden van de 16de en 17de eeuw (Antwerp: 1982).

Hicks C., *Animals in Early Medieval Art* (Edinburgh: 1993).

Hodnett E., *Marcus Gheeraerts the Elder of Bruges, London, and Antwerp* (Utrecht: 1971).

Honig E., "Paradise Regained. Rubens, Jan Brueghel, and the Sociability of Visual Thought", *Nederlands Kunsthistorisch Jaarboek / Netherlands Yearbook for History of Art* 55 (2004) 271–301.

Hoop Scheffer D. de (comp.), *Aegidius Sadeler to Raphael Sadeler II*, in Boon K.G. (ed.), *Hollstein's Dutch and Flemish Engravings, Etchings and Woodcuts ca. 1450–1700*, vols. 21–22 (Amsterdam: 1980).

Imperato Ferrante, *Dell'Historia Naturale* (Naples, Constantino Vitale: 1599).

Jorink E., *Reading the Book of Nature in the Dutch Golden Age, 1575–1715* (Leiden: 2010).

Kaliardos K.P., *The Munich Kunstkammer* (Tübingen: 2013).

Kavaler E.M., *Pieter Bruegel: Parables of Order and Enterprise* (Cambridge: 1999).

Koreny F., *Albrecht Dürer and the Animal and Plant Studies* (Boston: 1988).

Lach D.F., *Asia in the Making of Europe* (Chicago: 1993).

Lietzmann H., *Der Landshuter Renaissancegarten Herzog Wilhelms V. von Bayern:Ein Beitrag zür Garten- und Kulturgeschichte der Frühen Neuzeit* (Munich: 2001).

Margócsy D., "The camel's head: representing unseen animals in sixteenth-century Europe", *Nederlands Kunsthistorisch Jaarboek / Netherlands Yearbook for History of Art* 61 (2011) 62–85.

Mason P., *Before Disenchantment: Images of Exotic Plants and Animals in the Early Modern World* (London: 2009).

Maxwell S., "The Pursuit of Art and Pleasure in the Secret Grotto of Wilhelm V of Bavaria", *Renaissance Quarterly* 61 (2008) 414–462.

Maxwell S., *The Court Art of Friedrich Sustris: Patronage in Late Renaissance Bavaria* (Surrey – Burlington: 2011).

Meadow M., "Merchants and Marvels: Hans Jacob Fugger and the Origins of the *Wunderkammer*", in Findlen P. and Smith P. (eds.), *Merchants and Marvels: Commerce and the Representation of Nature* (London: 2001).

Mielke H., "Antwerpener Graphik in der 2. Hälfte des 16. Jahrhunderts: Der Thesaurus veteris et novi Testamenti des Gerard de Jode (1585) und seine Künstler", *Zeitschrift für Kunstgeschichte* 38/1 (1975) 29–83.

Mielke U. (comp.). *Crispijn van den Broeck*, in Luijten G. (ed.), *The New Hollstein Dutch & Flemish Etchings, Engravings, and Woodcuts, 1450–1700*, vol. 21 (Oudekerk aan den IJssel, Sound & Vision: 2011).

Nelson J.S., *Fishes of the World* (New York: 1984) 379–386.

Ogilvie B., *The Science of Describing: Natural History in Renaissance Europe* (Chicago: 2006).

Ortega-tillier V., *Le Jardin d'Éden: Iconographie et topographie dans la gravure (XVe–XVIIIe siècles)* (Dijon: 2006).

Paré Amboise, *On Monsters and Marvels*, transl. by P.L. Pallister (Chicago: 1982).

Pigafetta Antonio, *Relazione del primo viaggio intorno al mondo* (Paris, Simon de Colines: 1524–1534).

Ramaix I. de, *Les Sadeler. Graveurs et éditeurs* (Brussels: 1992).

Richeome Louis, *La Peinture Spirituelle ou l'art d'admirer, aimer et louer Dieu en toutes ses oeuvres, et tirer de toutes profit salutere* (Lyon, Pierre Rigaud: 1611).

Riggs T. – Silver L. (eds.), *Graven Images: The Rise of Professional Printmakers in Antwerp and Haarlem, 1540–1640* (Evanston: 1993).

Rikken M., "Abraham Ortelius as Intermediary for the Antwerp Animal Trailblazers", *Jahrbuch für Europäische Wissenschaftskultur / Yearbook for European Culture of Science* 6 (2012) 95–128.

Rikken M. – Smith P.J., "Jan Brueghel's *Allegory of Air* (1621) from a natural historical perspective", *Nederlands Kunsthistorisch Jaarboek / Netherlands Yearbook for History of Art* 61 (2011) 87–115.

Salviani Ippolito, *Aquatilium animalium historiae* (Rome, Ippolito Salviani: 1554).

Schlosser J. Von, *Die Kunst- und Wunderkammer der Spätrenaissance* (Braunschweig: 1978).

Schnapper A., *Le géant, la licorne, la tulipe. Collections françaises au XVIIe siècle* (Paris: 1988).

Schuckman C. *Maarten de Vos*, D. de Hoop Scheffer (ed.), *Hollstein's Dutch & Flemish Etchings, Engravings and Woodcuts 1450–1700*, vols. 44–46 (Rotterdam: 1996).

Schulze-Hagen K., et al., "Avian taxidermy in Europe from the Middle Ages to the Renaissance", *Journal für Ornithologie* 144 (2003) 459–478.

Shrubb M., *Feasting, Fowling and Feathers: A History of the Exploitation of Wild Birds* (London: 2013).

Seelig L. "The Munich *Kunstkammer*, 1565–1807", in Impey, O. – MacGregor A., (eds.), *The Origin of Museums: The Cabinet of Curiosities in Sixteenth- and Seventeenth-Century Europe* (Oxford: 1985) 76–89.

Sellink M. (comp.). *Philips Galle*, in Leesberg M. (ed.), *The New Hollstein Dutch & Flemish Etchings, Engravings and Woodcuts 1450–1700*, vol. 10 (Rotterdam: 1993).

Smith P.H., "Collecting Nature and Art: Artisans and Knowledge in the *Kunstkammer*", in Hanawalt B. – Kiser L.J. (eds.), *Engaging with Nature: Essays on the Natural World in Medieval and Early Modern Europe* (Notre Dame: 2008) 115–136.

Smith P.J., *Het schouwtoneel der dieren. Embleemfabels in de Nederlanden (1567–ca. 1670)* (Hilversum: 2006).

Smith P.J., "On Toucans and Hornbills: Readings in Early Modern Ornithology from Belon to Buffon", in Enenkel K. – Smith P.J. (eds.), *Early Modern Zoology: The Construction of Animals in Science, Literature and the Visual Arts*. Intersections: Yearbook for Early Modern Studies 7 (Leiden – Boston: 2007) 75–119.

Smith P.J., "Sympathy in Eden: On Paradise with the Fall of Man by Rubens and Brueghel", in Göttler C. – Neuber W. (eds.), *Spirits unseen: the representation of subtle bodies in early modern European culture*. Intersections: Yearbook for Early Modern Studies (Leiden – Boston: 2008) 211–244.

Smith P.J., "Een veranderlijk dier. De kameleon tussen natuurlijke historie en emblematiek", *De Boekenwereld*, 29, 1 (2012) 33–43.

Stock J. van der, *Printing Images in Antwerp. The Introduction of Printmaking in a City: Fifteenth Century to 1585* (Rotterdam: 1998).

Swan C., "From Blowfish to Flower Still Life Paintings: Classification and Its Images, circa 1600", in Smith P.H. – Findlen P. (eds.), *Merchants & Marvels: Commerce, Science and Art in Early Modern Europe* (New York: 2002) 109–136.

Swan C., "Birds of Paradise for the Sultan: Early Seventeenth-Century Dutch-Turkish Encounters and the Uses of Wonder", *De Zeventiende Eeuw* 29, 1 (2013) 49–63.

Szalblowski J. (ed.), *The Flemish Tapestries at Wawel Castle in Cracow: Treasures of King Sigismund Augustus Jagiello* (Antwerp: 1972).

Thevet André, *Singularitez de la France Antarctique* (Paris, Maurice de la Porte: 1558).

Thevet André, *Cosmographie universelle* (Paris, Pierre L'Huilier: 1575).

Veldman I. *Crispijn de Passe and His Progeny (1564–1670): A Century of Print Production*, trans. M. Hoyle (Rotterdam: 2001).

Vignau-Wilberg T., *Archetypa studiaque patris Georgii Hoefnagelii, 1592: Natur, Dichtung und Wissenschaft in der Kunst um 1600 / Nature, poetry, and science in art around 1600*, trans. by M. Robertson (Munich: 1994).

Weemans M., "The Earthly Paradise: Herri Met de Bles's Visual Exegesis of Genesis 1–3", in Brusati C. – Enenkel K. – Melion W. (eds.), *The Authority of the Word: Reflecting on Image and Text in Northern Europe, 1400–1700, Intersections: Yearbook for Early Modern Studies* 20 (2011) 263–312.

Williams A., *The Common Expositor: An Account of the Commentaries on Genesis 1527–1633* (Chapel Hill: 1948).

Witte A.A., *The Artful Hermitage: The Palazetto Farnese as a Counter-reformation Diaeta* (Rome: 2008).

Zweite A., *Marten de Vos als Maler: Ein Beitrag zur Geschichte der Antwerpener Malerei in der Zweiten Hälfte des 16. Jahrhunderts* (Berlin: 1980).

CHAPTER 10

Exotic Animal Painting by Jan Brueghel the Elder and Roelant Savery

Marrigje Rikken

During the second half of the sixteenth century, knowledge of European flora and fauna grew swiftly and natural history changed fundamentally. Moreover, the many expeditions to the New World led to the discovery of numerous species. Not only did natural historians display a great interest in these new and exotic species, but artists did as well. Some artists even specialized in animal imagery.[1] From the 1550s onwards, artists from the Southern Netherlands started depicting the animal kingdom in drawings, which were usually assembled in albums. Two decennia later, from the 1570s on, the animal kingdom was recorded in animal print series, often divided into different series of prints grouped according to genus; quadrupeds, birds and fish. Just before the turn of the seventeenth century exotic animal paintings emerged as a new genre in painting.[2] In these paintings artists depicted numerous animal species, displaying nature in all its variety in a single work of art. The two main innovators of this genre were Jan Brueghel the Elder (1568–1625) and Roelant Savery (1576–1639).[3]

1 This article was written within the framework of the NWO program *Cultural Representations of Living Nature: Dynamics of Intermedial Recording in Text and Image (ca. 1550–1670)*, also known as *Rereading the Book of Nature*. In my forthcoming dissertation I analyze the development of Southern-Netherlandish animal imagery between 1550–1650 as an independent genre in art in relation to developments in natural history, networks between artists and natural historians and elite collecting practices. I would like to thank Eric Jan Sluijter and Paul J. Smith for reading an earlier draft of this article and Erin Downey for her helpful feedback.

2 Animal painting is still somewhat underexposed in Dutch and Flemish art history. Numerous studies have appeared on Dutch and Flemish portraits, landscapes, history paintings and still lives, yet hardly any surveys of animal painting have been published.

3 In the revised oeuvre catalogue of Brueghel's work by Klaus Ertz most exotic animal paintings have been listed, but they have not been analyzed as a subgenre. See Ertz K. – Nitze-Ertz C., *Jan Brueghel der Ältere (1568–1625): kritischer Katalog der Gemälde*, 4 vols. (Lingen: 2008). In 2010 an exhibition of Savery's work was organized by Isabelle de Jaegere. See De Jaegere I., *Roelandt Savery 1576–1639* (Kortrijk: 2010). Although several catalogue entries were devoted to exotic animal paintings, they were not discussed as an independent genre.

 | DOI 10.1163/9789004279179_012

They were the first artists who specialized in exotic animal paintings and as such shaped the genre.[4]

In this article, I will discuss the early development of the new genre, by means of the oeuvres of Brueghel and Savery. The genre is strongly rooted in a political and religious context. I will argue that exotic animal painting developed within the Habsburg courts—the political and cultural centers of Europe around the turn of the seventeenth century—yet it derived from a religious (and to a lesser extent mythological) pictorial tradition. Both Brueghel and Savery were associated with the Habsburg courts. Savery worked at the court of Rudolf II (1552–1612) in Prague from 1603 or 1604 until 1613. Brueghel served unofficially as court painter to the Archduke Albrecht of Austria (1559–1621) and the Infanta Isabella of Spain (1566–1633) in Brussels.[5] Yet before Brueghel started working for the Archduke, he had visited the court of Rudolf II in Prague in 1604. Especially the court of Rudolf II in Prague—but the court of the Archduke in Brussels as well—played a key role in the early development of the genre of exotic animal paintings, since it stimulated the creation of new types of painting.[6] The first exotic animal paintings arose from religious and mythological subjects, such as the Earthly Paradise, in which Adam and Eve are surrounded by animals, and Orpheus charming the animals with his music. The theme of Orpheus was particularly popular among princely patrons; Rudolf II appears to have had a special predilection for the theme, which reflected his political aspirations, and he thus likely encouraged representations of this subject in painting at his court. Albrecht of Austria, on the other hand, favored paintings with a religious theme.

Furthermore, the Hapsburg court environment enabled Brueghel and Savery to study exotic animals that were kept in the menageries for their paintings. It is also highly probable that the court environment allowed the artists to come into contact with new developments in natural history. Both artists deliberately demonstrate their natural historical knowledge through their exotic animal paintings, which is an important characteristic of the new genre.

---

4 Roelant Savery received his training from his older brother Jacob Savery (ca. 1565–1603), who also produced a few animal paintings although Jacob did not specialize in it. Both Brueghel and Savery had many followers who subsequently also specialized in animal paintings. Yet Brueghel and Savery were the only two artists in the first two decades of the seventeenth century producing exotic animal paintings on a large scale.

5 Maeyer M. de, *Albrecht en Isabella en de Schilderkunst. Bijdrage tot de Geschiedenis van de XVIIe-Eeuwse Schilderkunst in de Zuidelijke Nederlanden* (Brussel: 1955) 147–149.

6 See for the role that Prague played in creating new types of painting and in innovating of older types of painting, DaCosta Kaufmann T., *The Eloquent Artist: Essays on Art, Art Theory and Architecture, Sixteenth to Nineteenth Century* (London: 2004) 71–89.

A New Genre in Painting

Jan Brueghel the Elder and Roelant Savery shared much in common. In addition to having worked at prestigious courts, they had similar artistic backgrounds and were very prolific painters.[7] As such, it is interesting to analyze how the two artists related to one another in consideration of their significance for the development of exotic animal paintings. Yet, the exotic animal paintings did not constitute the majority of either artist's oeuvre. Aside from exotic animal paintings, they painted various types of landscapes, including mountainous landscapes, riversides and village views. They were also important specialists of flower still lifes, another new genre of painting.[8] Nevertheless, their exotic animal paintings are a very significant part of their artistic output, as they both shaped that genre. Brueghel painted at least 25 original exotic animal paintings, Savery produced more than double that amount, with at least 55 paintings that are known to us, which results in a corpus of a total of 80 paintings (see Table).[9]

The thematic formats for paintings with exotic animals in oeuvres of these two artists are rather limited, comprising of only four different categories, that can be broken down into several types or subjects. The religious paintings consist of the Earthly Paradise—as mentioned above—and the Ark of Noah.[10]

---

7 Both artists were born in the Southern Netherlands into an artistic family. They both received an artistic training form early on. For information of the Brueghel dynasty, see Brink P. van den, *De firma Brueghel* (Amsterdam: 2001). The Savery dynasty has not yet received much scholarly attention.

8 Much more has been written on Jan Brueghel the Elder and Roelant Savery and their role in the development of flower painting than on their role in exotic animal painting. See for their flower painting for example, Taylor P., *Dutch Flower Painting, 1600–1720* (New Haven: 1995) 115–193. For that reason, I will not elaborate on their flower still lifes in the present article.

9 The corpus consists of all known exotic animal paintings by the two artists in which a large variety of animals—including numerous exotic species—are the main subject. This means that several paintings by Savery with indigenous animals such as cows, deer and horses, have been excluded. Brueghel often painted the same composition more than once. Only the paintings that are considered to be the principle version have been included. Some of the paintings in the corpus were a cooperation between either Brueghel or Savery and another painter. The animals however are always by Brueghel or Savery. Of course the paintings in the corpus do not make up for their entire artistic output in this genre. Some paintings will have been lost over time, or new paintings may still be discovered.

10 The Earthly Paradise can be divided into subcategories such as the Creation, the Name giving of the Animals, and the Fall. However, since it is not always clear which moment is depicted, I did not make such a subdivision. In order to establish a painting as an Earthly Paradise, Adam (and Eve) must be present. Paintings of the Ark of Noah can either portray the animals entering the Ark or the animals leaving it; or they could represent the Deluge itself. However, the Ark always has to be present.

TABLE

No.	Artist	Title	Date	Medium	Seize	Place	Collection
1	Brueghel	Orpheus and the Animals		panel	71.5 × 125	Rodez	Musée des Beaux-Arts
2	Brueghel	Orpheus and the Animals	1600	copper	25.5 × 35.5	Paris	Ader Picard Tajan, 12-12-1989, 82
3	Brueghel	Orpheus playing for the animals	1605–1609	panel	34.5 × 47	Paris	Ader Picard Tajan, 29-6-1989, 47
4	Brueghel	Orpheus and the Thracian Maenads	1605–1609	panel	51.5 × 67		Private Collection
5	Brueghel	Abduction of Ganymede	c. 1606	copper	10.5 × 10.5	London	Rafael Valls
6	Brueghel	Earthly Paradise		copper	28.8 × 36.7	Frankfurt	Städel
7	Brueghel	Earthly Paradise with Creation of mankind	1594	copper	26.5 × 35	Rome	Doria Pamphilj Gallery
8	Brueghel	Earthly Paradise with Fall	c. 1596	copper	27.4 d.	Neuburg	Staatsgalerie
9	Brueghel	Allegory of Earth or Earthly Paradise	1607–1608	copper	45 × 65	Paris	Louvre
10	Brueghel	Earthly Paradise with Fall	before 1612	copper	40 × 50	Madrid	Prado
11	Brueghel	Earthly Paradise (with Fall in the background)	1612	copper	50.3 × 80.1	Rome	Doria Pamphilj Gallery
12	Brueghel	Garden of Eden	1613	canvas	23.7 × 36.8	Brentford	Duke of Northumberland
13	Brueghel	Adam and Eve in the Garden of Eden	1615	copper	48.6 × 65.6	London	Royal Collection
14	Brueghel	Creation of Eve	c. 1615	panel	50 × 87	Pommersfelden	Schloss Weissenstein
15	Brueghel	Earthly Paradise with Fall	c. 1616	panel	64 × 114	The Hague	Mauritshuis
16	Brueghel	The temptation in the Garden of Eden	c. 1617	panel	53 × 84	London	Victoria and Albert Museum

TABLE

No.	Artist	Title	Date	Medium	Seize	Place	Collection
17	Brueghel	The Fall of Man	1616	panel	51 × 93	Budapest	Szepmuveszeti Museum
18	Brueghel	Entry of Noah's Ark	1596	copper	27 × 35.5	Italy	Private Collection
19	Brueghel	Entry of Noah's Ark	1613	panel	54.5 × 87	Malibu	Getty Museum
20	Brueghel	Entering the Ark	1615	copper	25.7 × 37	London	Wellington Museum
21	Brueghel	Allegory of air	c. 1608	panel	53 × 95	Rome	Doria Pamphilj Gallery
22	Brueghel	Allegory of air	1611	panel	46 × 83	Lyon	Musée des Beaux-Arts
23	Brueghel	Allegory of air	1621	copper	45 × 65	Paris	Louvre
24	Brueghel	Bird Concert	1616?	copper	13 × 18	Switzerland	Private Collection
25	Brueghel	Earthly Paradise without human figures	1615	copper	19 × 15	Canada	Private Collection
26	Savery	Orpheus and the Animals		panel	65 × 110	Pommersfelden	Schloss Weissenstein
27	Savery	Orpheus Charming the Animals		canvas	60 × 114	Montauban	Musée Ingres
28	Savery	Landscape with Orpheus		panel	56.5 × 100.5	London	Christie's 6-7-2000, 42
29	Savery	Orpheus and the Animals		canvas	172 × 281.5	Kassel	Gemäldegalerie
30	Savery	Orpheus and the Animals		panel	17 × 20	Göttingen	Gemäldesammlung
31	Savery	Orpheus' death				London	Philips 2-12-1997, 89
32	Savery	Orpheus and the Animals	1610	panel	51 × 66	Frankfurt	Städel
33	Savery	Orpheus and the Animals	1611	panel	39 × 53	Berlin	Gemäldegalerie

TABLE (*cont.*)

No.	Artist	Title	Date	Medium	Seize	Place	Collection
34	Savery	Orpheus Charming the Animals	1617	panel	62 × 131.5	The Hague	Mauritshuis
35	Savery	Orpheus and the Animals	1617	canvas	101.5 × 207	Rotterdam	Collection of Mrs. Kronenberg
36	Savery	Orpheus and the Animals	c. 1617	canvas	84 × 139	Antwerp	KMSK
37	Savery	Poultry with Orpheus being killed by the Thracian Maenads in the background	1618	panel	35 × 47	Antwerp	KMSK
38	Savery	Orpheus and the Thracian Maenads	1618–1620	panel	27.9 × 41.8	Maastricht	Bonnefantenmuseum
39	Savery	Orpheus and the Animals	c. 1620	canvas	114 × 192	St. Petersburg	Hermitage
40	Savery	Orpheus and the Animals	1621	panel	50.8 × 91	London	Christie's 22-4-1988, 85
41	Savery	Orpheus with Animals and Birds	1622	copper	22.6 × 26.4	Cambridge	Fitzwilliam Museum
42	Savery	Orpheus and the Animals	1625–1628	copper	28 × 36	Vienna	Kunsthistorisches Museum
43	Savery	Orpheus Charming the Animals	1625	canvas	55 × 85	Prague	Narodni Gallery
44	Savery	Orpheus Charming the Animals	1626	panel	32 × 42	Paris	Louvre
45	Savery	Orpheus and the Animals	1628	panel	53 × 81.5	London	National Gallery
46	Savery	Landscape with Orpheus (being killed by the Thracian Maenads)	c. 1628	panel	35 × 49	Vienna	Kunsthistorisches Museum
47	Savery	Orpheus and the Animals	c. 1630	panel	92 × 185	Utrecht	Centraal Museum
48	Savery	Earthly Paradise		panel	54 × 87	London	Philips 5-12-1995, 69
49	Savery	Earthly Paradise with Fall	1614?			Pommersfelden	Schloss Weissenstein

TABLE

No.	Artist	Title	Date	Medium	Seize	Place	Collection
50	Savery	Adam naming the animals	1618	panel	55 × 107	Prague	Narodni Gallery
51	Savery	Earthly Paradise with Fall	1618	panel	81 × 138	Vienna	Liechtenstein Collection
52	Savery	Landscape with Animals	1620	panel	55 × 90	Maastricht	Bonnefantenmuseum
53	Savery	Garden of Eden with Fall	c. 1620	panel	83.3 × 139.2		Private Collection
54	Savery	Earthly Paradise	c. 1622	panel	54 × 90	Antwerp	KMSK
55	Savery	Earthly Paradise	1625	canvas	86 × 142	Munich	Sale, 2-7-1931
56	Savery	Earthly Paradise	c. 1625	canvas	244 × 375	Munich	Bayerische Staatsgemäldesammlungen
57	Savery	Earthly Paradise	1625	panel	59.3 × 122.4	Washington	Private Collection
58	Savery	The temptation in the Garden of Eden	c. 1625	panel	78.7 × 133.4	Buscot	Faringdon Collection
59	Savery	Earthly Paradise with the Four Elements	c. 1625	panel	47 d.	Amsterdam	Waterman Gallery
60	Savery	Earthly Paradise	1626	panel	78 × 135	Berlin	Gemäldegalerie
61	Savery	Earthly Paradise	1628	copper	42 × 57	Vienna	Kunsthistorisches Museum
62	Savery	Noah's Ark	1610?	panel	56 × 101	Paris	Drouot, 3-12-1990
63	Savery	Before the Deluge	1620	panel	82 × 137	Dresden	Gemäldegalerie
64	Savery	Noah's Ark	c. 1625	panel	52 × 97	Reims	Musée des Beaux-Arts

TABLE (*cont.*)

No.	Artist	Title	Date	Medium	Seize	Place	Collection
65	Savery	Noah's Ark	c. 1625	canvas	175 × 212	Warschau	Muzeum Narodowe
66	Savery	Deluge	c. 1625	canvas	137 × 190	Nuneaton	Museum and Art Gallery
67	Savery	Bird Paradise	c. 1616	panel	39 × 50	Antwerp	KMSK
68	Savery	Bird Paradise	1618	panel	30 × 42	Berlin	Grunewald
69	Savery	Bird Paradise	1618	panel	29.5 × 42	Dresden	Gemäldegalerie
70	Savery	Landscape with Animals	c. 1618	panel	56 × 80.5	Brussels	KMSK
71	Savery	Creation of birds	1619	panel	20.6 × 25.4	Cambridge	Fitzwilliam Museum
72	Savery	Landscape with Animals	c. 1620	panel	53 × 91	Courtrai	Broelmuseum
73	Savery	Water birds	1622	panel	28 × 42	Brussels	KMSK
74	Savery	Birds	1622	panel	54 × 108	Prague	Narodni Gallery
75	Savery	Birds and Exotic Animals	1623	panel	32 × 41.5	Verviers	Musée des Beaux-Arts et de la Céramique
76	Savery	Birds and Exotic Animals	1623	panel	53.8 × 107	Munich	Pinakothek
77	Savery	Water birds	1624	panel	56 × 88	Munich	Private Collection
78	Savery	Water birds	c. 1626	canvas		New York	Christies, 1-11-1991, 17
79	Savery	Landscape with Birds	1628	copper	42 × 57	Vienna	Kunsthistorisches Museum
80	Savery	Birds	c. 1629	panel	40 × 64.5	London	Zoological Society

In the mythological works Orpheus—again already stated—and Ganymede are portrayed among animals.[11] One category consists of allegories, mostly Allegories of Air in which numerous bird species—exotic as well as indigenous—are depicted.[12] The last category involves non-narrative paintings, in which no human, mythological or other figure is present; only animals are represented. This category can be subdivided further into paintings with an array of animals (both mammals and birds) and paintings with only birds.[13]

In comparison to Brueghel, Savery depicted the Earthly Paradise less frequently in his exotic animal paintings, even though Savery had twice the output of Brueghel (see the histogram of Fig. 10.1). Orpheus, on the other hand, was portrayed much more often by Savery than by Brueghel, relatively speaking. While Savery never created an allegorical exotic animal painting, Brueghel rarely painted exotic animals without a narrative subject.[14] The different sub-

11 In most paintings of Orpheus, the Thracian singer is charming the animals. However, Savery also depicted Orpheus being chased by the Thracian Maenads or even of him being killed, yet even in these there are still many exotic animals present as witnesses. There is only one painting in the corpus of Ganymede. This painting is by Brueghel. The depiction of Ganymede with exotic animals is very atypical; all other paintings by Ganymede never feature more than a handful of animals (the eagle and a dog most often) to my knowledge. For more information about the subject of Ganymede in paintings see, Kempter G., *Ganymed: Studien zur Typologie, Ikonographie und Ikonologie* (Cologne: 1980). Ertz lists the Ganymede painting as an allegory, since he describes it as an Allegory of Air, see Ertz – Nitze-Ertz, *Jan Brueghel der Ältere (1568–1625)* 1058. This is also the only known painting of Ganymede by Brueghel.

12 Brueghel painted three different Allegories of Air. Although these paintings belong to series of the Four Elements, Brueghel included far fewer animals in the Allegories of Earth from the same series, making them not so much animal paintings, since the animals figure merely as "bijwerk" [secondary motifs] as it was called at the time. The Allegories of Water feature numerous fish and other aquatic animals, but since it is much more difficult to establish which fish were exotic, I chose to leave them out for the present study.

13 Especially Savery seems to have made a deliberate choice to produce paintings with only birds.

14 There is only one painting by Brueghel that could be characterized as a bird painting (without any figures) and one painting that could be considered an animal painting, with both exotic animals and birds. However, it is possible that the latter painting with mammals and birds originally contained a scene with Adam and/or Eve, since it has an atypical standing format and composition, suggesting it could have been cut off at some point. This painting is dated 1615 and is on copper and measures 19 × 15 cm. It is in a private collection in Canada. Ertz describes it as a 'Paradieslandschaft ohne Menschen', see Ertz. – Nitze-Ertz, *Jan Brueghel der Ältere (1568–1625)* 430.

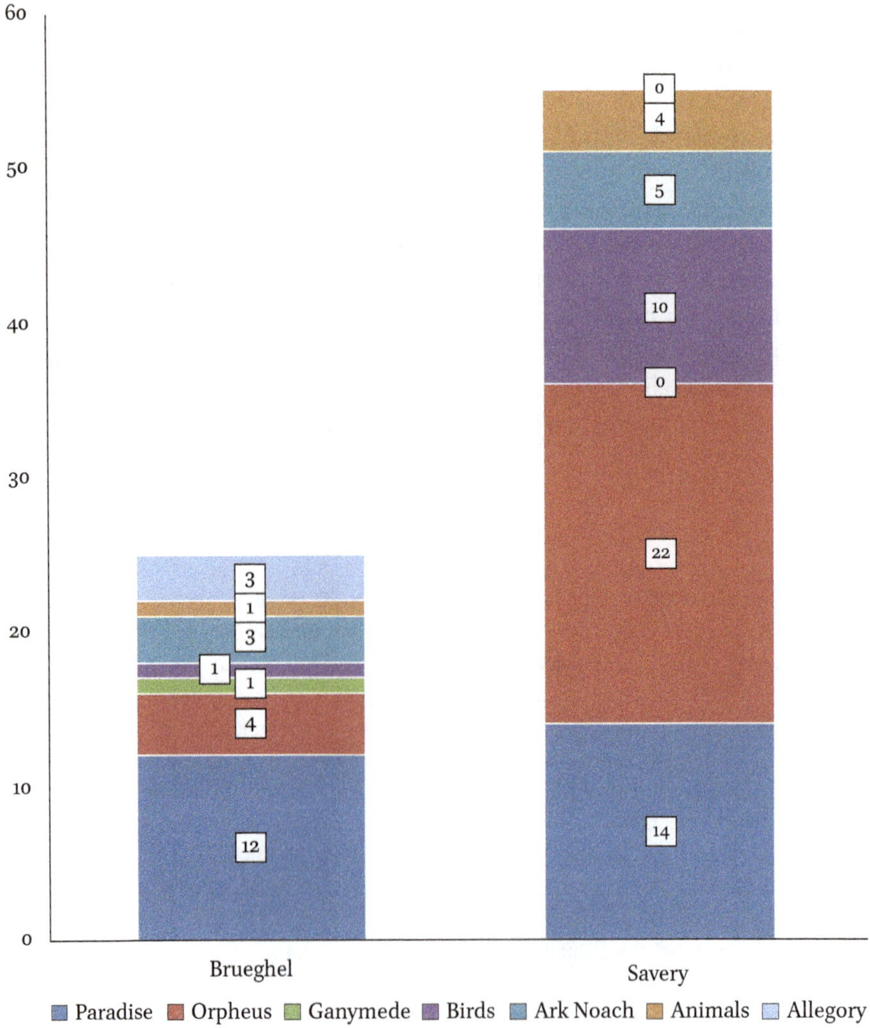

FIGURE 10.1 *Histogram of exotic animal paintings by Brueghel and Savery per theme.*

ject matter both painters chose most frequently for their exotic animal paintings has to do with the development of the genre.

Brueghel led the way in creating this new genre, by depicting exotic animals as part of existing and well-known religious subjects. His earliest dated painting of the genre is an Earthly Paradise from 1594, which shows the creation of Eve from Adam's rib in the background (see Table no. 7).[15] In the

15 The painting measures 26.5 × 35 cm and is painted on copper. It is in the collection of the Galleria Doria Pamphilj in Rome, inv. no. 278.

foreground, numerous fish are depicted on the left side and mammals are portrayed on the right side. Many of the represented animals are indigenous, such as cats, a horse and deer, but several exotic species have also been included, with lions and an ostrich most notably among them. Savery, on the other hand, seems to have taken up the theme of the Earthly Paradise only much later in his career. His earliest dated paintings of the subject date from 1618, in which various indigenous and exotic mammals and birds are mixed together (see Table nos. 50, 51).[16] Noah's Ark appears for the first time in a painting dated 1596 by Brueghel (see Table no. 18).[17] Again, this theme features only much later in paintings by Savery (see Table nos. 62, 63).[18]

While these two religious subjects have a long pictorial tradition—in paintings and in prints—Brueghel transforms the subject in such a way that the viewer's attention is no longer focused on the figures, but instead on the abundant and naturalistically rendered animals. Both subjects were well-suited for depicting exotic animals. However, until the late sixteenth century, only a few exotic species were depicted in these subjects, and if animals were included at all it was usually as mere decoration.[19] The first art works in which numerous exotic animals are combined, can be found in prints and paintings that originated in Antwerp during the 1570s by artists such as Maerten de Vos (1532–1603) and Hans Bol (1534–1593).[20] Thus, Brueghel, also from Antwerp, seems to have deliberately placed himself in the Antwerp tradition of depicting exotic

16 Two dated paintings from 1618 by Savery are known with this theme. Table no. 50 measures 55 × 107 cm and is painted on panel. It is in the collection of the Narodni Gallery in Prague, inv. no. 4245. The other painting, Table no. 51, measures 81 × 138 cm and is also on panel. It is in the Liechtenstein collection in Vienna. Cornelis Cornelisz. van Haarlem painted the figures of Adam and Eve in the latter painting.

17 The painting is on copper and measures 27 × 35.5 cm. It is in a private collection in Italy.

18 The Ark of Noah is depicted in a painting on panel that measures 56 × 101 cm. It has been auctioned by Drouot, Paris, 3 December 1990. The date of 1610 is uncertain. Another painting with the Ark by Savery has been dated to around 1616. The only certain dated painting with the Ark is from 1620 and on panel, measuring 82 × 137 cm. It is in the collection of the Gemäldegalerie in Dresden, inv. no. 932.

19 Dürer only included two exotic species in his famous print of 1504, yet the animals he depicted are more than decoration and have specific meanings in relation to the theme, see Panofsky E., *Albrecht Dürer* (Princeton: 1948) and Smith P.J., "Diersymboliek in Rembrandts *Zondeval* (1638) en in Vondels *Adam in ballingschap* (1664)", *De Zeventiende eeuw* 26, 1 (2010) 9–11. In Cranach's *Fall* of 1509, a lion is the only exotic species.

20 The first part of my forthcoming dissertation is devoted to the Antwerp artists in the second half of the sixteenth century who specialized in animal imagery. These artists frequently copied motifs of exotic animals from one another, see Rikken M., "Abraham Ortelius as Intermediary for the Antwerp Animal Trailblazers", *Jahrbuch für Europäische Wissenschaftskultur* 6, 2011 (2012) 95–128.

animals. Savery subsequently followed suit. However, it is important to note that even though Brueghel acknowledged his Antwerp predecessors, neither he nor Savery copied motifs from other Antwerp artists.[21] This is even more remarkable considering the fact that Rudolf II in Prague had several of the animal drawings and series by the sixteenth-century Antwerp artists in his collection, which Savery, and perhaps Brueghel as well, could have seen.

The court of Rudolf II in Prague played an important part in the development of this new genre of exotic animal paintings, in particular in representations of Orpheus. Again, Antwerp artists from the second half of the sixteenth century, such as Adriaen Collaert (ca. 1560–1618) and Nicolaes de Bruyn (1571–1656) first portrayed this theme with numerous exotic animals. For example, the print engraved and published by Collaert, after a design by Adam van Noort (1562–1641) depicts Orpheus playing his lyre, surrounded by a lion, leopard, elephant, dromedary, monkey and turkey, among others.[22] The first dated painting of Orpheus with animals by Brueghel is from 1600 and renders an even much greater variety of exotic species (see Table no. 2).[23] In contrast to most other subjects, Savery soon followed Brueghel and began portraying Orpheus in landscapes with exotic animals. Dated paintings by Savery of this subject survive from 1610, 1611, 1617 and 1618 (see Table nos. 32–35, 37).[24]

Savery may have been encouraged to paint Orpheus quite early in his career by Rudolf II, for whom the theme would have particularly resonated.[25] On the one hand, the subject often was considered as a metaphor for the unification of mankind with the animal kingdom or for the domination of mankind over nature—an important theme in the Kunstkammer and collections of Rudolf II.[26] The musical connotations would have also appealed to Rudolf II, as well as many other European rulers. Rudolf II, moreover, owned several paint-

21 For my dissertation I did extensive research on the drawings and prints by the Antwerp artists.

22 A copy of this print is in the collection of the Rijksprentenkabinet in the Rijksmuseum, Amsterdam, inv. no. RP-P-BI-6083.

23 The painting is on copper, measuring 25.5 × 35.3 cm. It was auctioned at Ader Picard Tajan, Paris on 12 December 1898, lot 82.

24 The 1610 painting is on panel, measuring 51 × 66 cm. It is in the collection of the Städel Museum in Frankfurt, inv. no. 977.

25 This was also noted by De Jaegere, see De Jaegere, *Roelandt Savery 1576–1639* 142.

26 See for more information on the pictorial tradition of Orpheus and the connotations of the subject: Semmelrath H., *Der Orpheus-Mythos in der Kunst der italienischen Renaissance: eine Studie zur Interpretationsgeschichte und zur Ikonologie* (Cologne: 1994) and Preiss P., *Vaìclav Vavrinec Reiner (1689–1743): Landscapes with Animals: Motifs of Orpheus and Animals and Birds Flocking around an Owl in the Painting of Central Europe and the Szech Lands in the Baroqu* (Prague: n.d.).

ings of Orpheus. The 1621 inventory of the Kunstkammer in Prague indicates that Rudolf II owned an Orpheus painting by Perino del Vaga (1501–1547).[27] More interestingly for the present article, the inventory also lists several paintings by Savery, including one painting of Orpheus, described as 'Orpheus mit den wilden thieren in einer Landschaft'.[28]

It is probable that Savery knew of a painting with Orpheus by Brueghel and that this inspired him to start painting the theme early on in his career.[29] The artists could only have met once, in the summer of 1604, when Brueghel visited Prague while Savery was working there at the court. It seems plausible that Brueghel either sold or gifted a painting to Rudolf II, or that he received a commission from the Emperor for a painting. Even though the 1621 inventory does not mention any paintings by Brueghel, Rudolf II most likely owned a painting by Brueghel portraying Orpheus among animals. An undated painting of this theme by Brueghel depicts the naked Orpheus playing his lute surrounded by a peacock, ostrich, horse, deer, badger, swan, turkey, heron, dog, lion, fox, leopard, scarlet macaw, ox, goat, porcupine, and guinea pig, among other animals [Table no. 3, Fig. 10.2].[30] In the background we can further discern an elephant and a dromedary. The painting has been dated to around 1605.[31] However, I would like to suggest that it was painted in 1604, in—or just before Brueghel left for—Prague.

---

27 See Zimmermann H., "Das Inventar der Prager Schatz- und Kunstkammer vom 6. Dezember 1621: (nach Akten des k. und k. Reichsfinanzarchivs in Wien)", *Jahrbuch der Kunsthistorischen Sammlungen des Allerhöchsten Kaiserhauses* 25, 5 (1905) no. 848.

28 The 1621 inventory lists 17 paintings by Savery, inv. nos. 876, 909, 930, 932, 997, 1005, 1007, 1071, 1078, 1082, 1086, 1146, 1150, 1182, 1185, 1323, 1324. The Orpheus painting is 1082. Müllenmeister states that 16 paintings by Savery are listed in this inventory, which must be considered to be a mistake, see Müllenmeister K.J., *Roelant Savery: Kortrijk 1576–1639 Utrecht: Hofmaler Kaiser Rudolf II. in Prag: Die Gemälde mit Kritischem Œuvrekatalog* (Freren: 1988) 344. The description of the paintings is usually too imprecise to identify the exact subject. Five of the paintings portrayed landscapes, and it is possible these landscapes also featured exotic animals.

29 Although Savery followed in Brueghel's footsteps in creating exotic animal paintings, he seems to have done so quite independently from his colleague. He never copied motifs from Brueghel and Savery probably could not see many paintings by Brueghel.

30 The guinea pig seems to have been Brueghel's "signature animal". He was the first to depict the South American species and he was quite known for his portrayal of the small rodent. The painting is on panel, measuring 34.5 × 47 cm. It was auctioned at Ader Picard Tajan in Paris on 29 June 1989, lot 47.

31 Ertz – Nitze-Ertz, *Jan Brueghel der Ältere (1568–1625)* 767. The resemblance to two Earthly paradises of ca. 1605–1610 and of before 1612 (Madrid, Prado) leads Ertz to date this Orpheus painting to around 1605.

FIGURE 10.2 *Jan Brueghel the Elder*, Orpheus Playing Music for the Animals (*c.* 1604). *Oil on panel, 34.5 × 47 cm. Paris, Ader Picard Tajan 29 June 1989, lot 47.*

FIGURE 10.3 *Jacob Hoefnagel,* Orpheus Charming the Animals *(1613). Point of brush and watercolor and bodycolor on vellum, 16.6 × 21 cm. New York, The Morgan Library and Museum.*

Jacob Hoefnagel (1575–ca. 1630) copied this composition in a painting dated 1613 [Fig. 10.3]. Hoefnagel's miniature (painted on a piece of vellum half the size of Brueghel's panel) shows the exact same species in a nearly identical composition. He only changed the background landscape and moved the elephant further to the back of the scene, to the other side of a lake, thereby rendering the animal much smaller, as if he was doubting whether it was portrayed accurately by Brueghel.[32] Hoefnagel worked at the court in Prague from 1602 to 1617, first under Rudolf II and after his death in 1612, under the Emperor Matthias of Austria (1557–1619). If Brueghel had painted his Orpheus in or just before 1604, the painting presumably would have become part of the collection of Rudolf II, enabling Hoefnagel to copy it almost ten years later. If this indeed was the case, Savery also could have studied Brueghel's painting in Prague.

It seems that Savery subsequently concentrated on painting exotic animals with Orpheus, since this subject occurs most often in his body of exotic animal paintings. Unlike Savery, Brueghel probably abandoned the theme of Orpheus quite early in his career, as no dated paintings of Orpheus by Brueghel are known after 1600. This could be due to the fact that Brueghel worked as a court painter—albeit not officially—from 1606 to 1621 for Albrecht of Austria and Isabella of Spain, who had a particular predilection for religious subjects.[33] Although no inventories from the collection of Albrecht and Isabella survive, it is known from documents that Brueghel received several large commissions from the Archduke.[34] In 1617–1618 Brueghel painted a series of the Five Senses for Albrecht and Isabella and it has been argued that the paintings show the collection they owned.[35] In the *Allegory of Hearing*, which Brueghel painted in collaboration with Rubens, a large painting of Orpheus is depicted at the top right, in which Orpheus is surrounded by a lion, elephant, ostrich and other

32 Further on in this article I will discuss the degree of accuracy of the exotic paintings by both Brueghel and Savery.

33 Smolar-Meynart A., *Het Paleis van Brussel: Acht eeuwen kunst en geschiedenis* (Brussel: 1991) 227.

34 Ibid., 222, 238. Brueghel received a commission for 11 large scale paintings in 1609–1610 and a commission for 38 miniature paintings in 1619. In 1615 the city council of Antwerp donated four paintings by Brueghel to the Archdukes and in 1618 they donated two more paintings of the Five Senses to Albrecht and Isabella. According to Faber Kolb it is highly likely that Brueghel's *Ark of Noah* of 1613, now in the collection of the J. Paul Getty Museum (inv. no. 92.PB.82, Table no. 19), was painted for the Archduke, see Faber Kolb A.L., *Jan Brueghel the Elder: The Entry of the Animals into Noah's Ark* (Los Angeles: 2005) 4, 71–74.

35 Smolar-Meynart, *Het Paleis van Brussel* 226. However, it has not yet been possible to identify art objects or paintings in the *Senses* with objects they owned for certain.

exotic animals.[36] The Orpheus painting betrays certain characteristics found in other paintings by Brueghel, but the large scale is very unusual for Brueghel. Brueghel probably depicted the Orpheus painting not so much because Albrecht and Isabella owned it, but rather because it referred to music and thus to the sense of hearing.

Savery continued to depict Orpheus among exotic animals after he left Prague for Amsterdam and later Utrecht. Yet during the later stages of his career he apparently found it less important to situate his exotic animal depictions within a narrative. The earliest dated paintings with only birds are from 1618 (see Table nos. 68, 69) and his paintings with various kinds of animals (mostly mammals) without a figure have been dated to the 1620s (see Table nos. 72, 75). By this time, it seemed no longer necessary to include human figures, because the genre had developed to such an extent that it derived its appeal solely from the exotic animals portrayed. The development of exotic animal paintings without a narrative was probably also stimulated by the new situation of the art market in the Northern Netherlands.[37] Instead of painting on commission, painters more often produced works for the open market. Such works featured religious or mythological subject matter less frequently than works that were commissioned by the church or by political institutions or corporations.

Exotic Animals at Court

Even though the appointment as a court painter was not necessary for either Brueghel or Savery to begin or to continue painting exotic animals—Brueghel painted his earliest exotic animal paintings before he went to Prague or worked for the court in Brussels and Savery kept practicing the genre while working in the Northern Netherlands—the appointment certainly contributed a great deal to the development of the genre. It would have been primarily at court that the artists came into contact with the exotic animals they portrayed. Brueghel certainly studied animals at the court in Brussels, since he wrote in a letter of 1621 to Federico Borromeo, one of his most important patrons, that he painted animals from life in a garland of flowers on the basis of animals he had seen in the menagerie.[38] Furthermore, the court was the environment

36 The painting is on panel, measuring 64 × 109.5 cm. It is in the collection of the Museo del Prado, Madrid, inv. no. 1395.

37 Numerous studies have appeared on this topic, see for example: North M., *Art and commerce in the Dutch Golden Age* (New Haven: 1997) 82–130.

38 Rikken M. – Smith P.J., "Jan Brueghel's 'Allegory of Air' (1621) from a Natural Historical Perspective", *Nederlands Kunsthistorisch Jaarboek* 61 (2011) 108.

par excellence for these artists to become acquainted with natural historical knowledge. Natural historians frequented courts and their publications would have been available there as well. There were, of course, variances between the courts of Prague and Brussels, which could explain particular differences between the exotic animal paintings by Brueghel and Savery. Yet, they also shared important similarities.

Some species occur frequently in the paintings of both Brueghel and Savery. Brueghel depicted lions most often: in 18 of the 27 paintings. Savery, on the other hand, portrayed ostriches most frequently: in 42 of the 55 paintings. Both species also occur in numerous paintings by the other artist, indicating that these were favorite species for both artists. Other 'favorites' were the elephant, dromedary, leopard, cassowary, pelican, turkey and blue-and-yellow macaw. This leads to the question of whether Brueghel and Savery were actually able to study these species at court. By studying the paintings that are dated during the years in which the artists had easy access to the menageries and the collections of the Emperor and the Archduke, it is possible to draw some preliminary conclusions as to which animals they saw in the menageries. Of course one has to be cautious about using the paintings as primary sources, since both Brueghel and Savery could have seen the animals somewhere else, as stuffed specimen, or they could have used pictorial sources or written descriptions to depict certain species. Yet by combining other sources with the evidence from the paintings themselves the working method and preferences of Brueghel and Savery can be revealed.

Unfortunately, no inventories for the menageries in Prague or Brussels have survived from the years that Brueghel and Savery were in contact with the courts. Various documents, however, do specify which exotic species were kept at specific times in the different menageries. In the menageries in Brussels, for example, there were tanks in the pond that held tortoises.[39] Brueghel depicted tortoises in ten paintings, far more often than Savery, who included tortoises only in three paintings even though he had almost twice the artistic output as Brueghel. Thus, it is possible that Brueghel studied live tortoises, while Savery had to rely on a stuffed specimen or a pictorial source. Brueghel probably also studied exotic bird species at the court in Brussels. In 1615 the Archduke acquired a toucan. Brueghel depicted the toucan only in paintings from the end of his career. Savery never painted a toucan. Since Brueghel could only have seen the menagerie in Prague briefly, and there is no indication that Savery visited the menagerie in Brussels, Brueghel probably saw the toucan that the Archduke purchased in 1615 and then started to depict the species in his paintings. Albrecht of Austria also purchased a scarlet macaw in the

39 Smolar-Meynart, *Het Paleis van Brussel* 110.

same year, but this parrot species already features in paintings by Brueghel that are dated earlier, indicating he must have seen one at an earlier time somewhere else.[40]

Drawings of animals by either Brueghel or Savery can also help establish whether the artists saw living specimens of particular animals. Savery depicted the ostrich—the exotic animal he painted most often—in numerous poses, which suggests he was able to study a live animal for a longer period of time. Brueghel on the other hand, always portrayed the bird in the same pose, although sometimes in reverse. This exact same pose appears in a drawing of an ostrich that is attributed to Brueghel [Fig. 10.4]. It is probable that Brueghel could not study the flightless bird for very long, and as a consequence he had to rely on this one drawing he made, perhaps in the menagerie in Prague. The elephant, as mentioned, also features quite often in the exotic animal paintings by Brueghel and Savery. Whereas Brueghel always rendered the mammal in the background or on the outer edge of the composition, Savery depicted the animal in the foreground. We do not know of any drawings of an elephant by Brueghel, but Savery drew the animal at least twice [Fig. 10.5].[41] Since the pose of the elephant is different in both drawings and in Savery's paintings, Savery probably was able to study the animal alive and frequently over a long period of time. Brueghel, on the other hand, appears not to have had an opportunity to study the quadruped alive, leaving him perhaps somewhat reluctant to represent the elephant in the foreground of his paintings. Indeed, no elephant is known to have been present in Brussels in the years that Brueghel worked in that city.

Neither painter seems to have studied a live giraffe. The ungulate is depicted in only two paintings by Brueghel, which are dated 1596 and 1612, and in only e painting by Savery, dated 1628.[42] Thus it seems that Brueghel depicted the large animal in the beginning of his career, whereas Savery did so mostly at the end of his. Moreover, both artists painted very small examples of giraffes and only in the background. In fact, no giraffe was recorded on European soil between 1486, when Lorenzo d'Medici (il Magnifico, 1449–1492) received one

---

40 Faber Kolb gives a good list of animals purchased by the archdukes. See Faber Kolb, *Jan Brueghel the Elder* 14–15.

41 This drawing is in black and red chalk on paper, measuring 43.3 × 55.7 cm. It is in the collection of the Albertina, Vienna, inv. no. 15092. The other known drawing of an elephant also shows a monkey on the back of the elephant. It is in black chalk with brown wash, measuring 14 × 21 cm. It is in the collection of the Crocker Art Museum in Sacramento, inv. no. 1871.101. See for more information on drawings of animals by Savery: Spicer-Durham J.A., *The Drawings of Roelandt Savery* (Connecticut: 1979).

42 Cuttler remarks that Savery never depicted a giraffe in his oeuvre, which is thus a mistake. See Cuttler C.D., "Exotics in Post-Medieval European Art: Giraffes and Centaurs", *Artibus Et Historiae* 12, 23 (1991) 170.

FIGURE 10.4 *Jan Brueghel the Elder*, Two Studies of an Ostrich. *Pen and watercolor on paper, 20.4 × 14.1 cm. Boston, Harvard Art Museum/ Fogg Museum.*

FIGURE 10.5 *Roelant Savery,* An Elephant Rubbing against a Tree *(1608–1612). Black and Red chalk on paper, 43.3 × 55.7 cm. Vienna, Albertina.*

in Florence, and 1827, when Charles X (1757–1836) of France was presented with the rare animal.[43] Brueghel and Savery probably relied on a pictorial source for their depictions of a giraffe.[44] However, they only seem to have opted for a pictorial source when necessary. This may have been the case as well for the rendering of the pelican by Brueghel, who depicted it in seven paintings. Savery, on the other hand, painted the pelican often, in at least 33 paintings and in many different poses. Brueghel's pelican is always identical in his paintings, and it shows striking similarities with a drawing from an album compiled by Anselmus de Boodt (1550–1632), the private physician of Rudolf II, who worked

43 Ringmar E., "Audience for a Giraffe: European Expansionism and the Quest for the Exotic", *Journal of World History* 17, 4 (2006) 377.

44 Gessner had already depicted a giraffe in his first volume of the *Historiae animalium* of 1551. See Gessner Conrad, *Historiae animalium Lib. I. de Quadrupedibus Viviparis* [...] (Zurich, Christoph Froschauer: 1551) 160. It is, however, not possible to identify the exact pictorial source that Brueghel and Savery used for their rendering of giraffes, since the animal is painted too small and shows no characteristic features on which to base an identification.

FIGURE 10.6 *Anonymous,* Pelican. *Part of Anselmus de Boodt's albums with animal drawings, album III, fol. 40. Black chalk and watercolor on paper. Private collection.*

in Prague between 1583 and 1612.[45] In particular, the strangely and quite inaccurate curved beak is characteristic for both the pelican found in the album and those in Brueghel's paintings [Fig. 10.6]. It is not known when De Boodt

45 See for more information on the albums of De Boodt: Maselis M.-C. – Balis A. – Marijnissen R., *De Albums van Anselmus de Boodt (1550–1632): Geschilderde Natuurobservatie aan het Hof van Rudolf II te Praag* (Tielt: 1989). The albums have been reassembled in the nineteenth century in 11 volumes and are now in a private collection.

FIGURE 10.7 *Roelant Savery,* Ark of Noah *(1620). Oil on panel, 82 × 137 cm. Dresden, Gemäldegalerie.*

assembled the drawings for his albums, but Brueghel very well may have seen the drawing of the pelican when he visited Prague in 1604. However, there is a problem with identifying this as the pictorial source for Brueghel's pelicans. Brueghel already represented his characteristic pelican in a painting that is dated 1600, before he had gone to Prague.[46] Possibly, the drawing in De Boodt's albums and the pelican depicted by Brueghel are both copied from another pictorial source that is no longer known.

Some species have only been depicted by either Brueghel or Savery, but not by both. For example, in certain works by Savery a dodo, rhinoceros, civet, nasua, armadillo and seal are portrayed, while these species do not occur in the extant exotic animal paintings by Brueghel. Three of these rarely depicted species can be found in Savery's painting of Noah's Ark of 1620 [Table no. 63, Fig. 10.7]. Savery included a bear, tortoise and porcupine in this painting too, and these species are also quite rare in his oeuvre. This painting in particular, in which he represented the greatest number of different animal species, seems to have been a key work for Savery, probably made on commission. He appears to have experimented with depicting species he was not that familiar with, possibly because he had only seen them once or twice, or perhaps he had only seen them as stuffed specimen or in illustrations. Savery probably did not

46 See Table no. 2. The painting depicts Orpheus and measures 25.5 × 35.5 cm. It was auctioned at Ader Picard Tajan in Paris on 12-12-1989, lot 82.

FIGURE 10.8 *Jan Brueghel the Elder,* The Temptation in the Garden of Eden *(c. 1617). Oil on panel, 53 × 84 cm. London, Victoria and Albert Museum.*

make sketches of these animals, otherwise he would have been able to depict them more frequently.

Brueghel also rendered species that Savery did not paint at all, which is the case for the tiger, the yellow crested cockatoo, lovebirds, the toucan and the penguin. Most of these species only appear in Brueghel's later paintings, beginning around 1615. This makes it possible to correct the date of a specific painting that shows several exotic species that belong to Brueghel's later career, the *Temptation in the Garden of Eden* which has been dated incorrectly to 1600 [Table no. 16, Fig. 10.8].[47] In the tree on the left side of the painting two toucans can be discerned, which Brueghel probably first studied after Albrecht acquired one in 1615. The penguin and green winged macaw additionally feature only in late works. A bird of paradise is perched next to the toucan on the upper branch, which furthermore helps to establish an accurate date for the painting. The bird of paradise features in at least 17 exotic animal paintings in Brueghel's oeuvre. In his early works, the bird of paradise is always floating in the air, without visible feet or wings. The idea of a floating bird was the result of the skins—which were prepared by the native Papuans without feet or wings—that sailors brought back with them from

47 This date is found in the online catalogue of the Victoria and Albert museum: http://collections.vam.ac.uk/item/O17809/the-temptation-in-the-garden-oil-painting-brueghel-jan-the/.

their voyages in the east. According to the myth the bird of paradise always floated in heaven, lived on dew and could not built a nest because it had no feet.[48] The natural historian Carolus Clusius (1526–1609) was the first to note in his description in the *Exoticorum libri decem* of 1605 that he had heard from a reliable witness that the bird of paradise did have feet in the wild.[49] Brueghel, however, was the first to actually depict the bird with clearly discernible feet and wings, and always rendered the bird perched or standing on the ground in his later works from 1615 onwards, as if to emphasize the demystification of the myth. Savery, on the other hand, always depicted the bird of paradise floating in the air. Thus, Brueghel must have been apprised by someone of the most recent developments in natural history. The *Temptation in the Garden of Eden* thus should not be dated to around 1600, but to around 1617.

Exotic Animals and Natural History

The case of the bird of paradise in the oeuvres of Brueghel and Savery indicates their respective knowledge of natural history. The fact that some species only appear in Brueghel's late works, suggests that he strove to expand his repertoire of exotic animals over the course of his career. Brueghel thus remained innovative and current in his exotic animal paintings, while Savery displayed less development over the course of his career. Savery depicted the same animals, such as ostriches and lions, over and over again. Furthermore, since Brueghel showed an increasing interest in painting rare and exotic species, prioritizing them over more common and domestic species that appear less often in his later paintings, he also demonstrated a growing interest in natural history. Yet the question whether Savery similarly had any awareness of the state of natural history has not been dealt with properly.

Parrot species are especially revealing with regard to not only Brueghel's, but also Savery's knowledge of natural history. More and more parrot species were discovered at the beginning of the seventeenth century and natural historians tried to outdo one another when describing and illustrating new species. Conrad Gessner (1516–1565) and Pierre Belon (1517–1564) only illustrated

48 Rikken – Smith, "Jan Brueghel's 'Allegory of Air' (1621) from a Natural Historical Perspective" 95.

49 Clusius Carolus, *Exoticorum libri decem: quibus animalium, plantarum, aromatum, peregrinorum fructuum historiae describuntur. Item Petri Bellonii Observationes, eodem Carolo Clusio interprete* [...] ([Antwerp], Franciscus Raphelengius: 1605) 361. Clusius' informant is Johannes de Weely, a merchant from Amsterdam. He even gives the date of the merchant's statement: late June 1605.

two species in their bird volumes of 1555.[50] Ulisse Aldrovandi (1522–1605) described fourteen parrot species, thirteen of which were illustrated.[51] Carolus Clusius yet again added descriptions of new species in 1605.[52] Brueghel and Savery both depicted the blue-and-yellow macaw (*Ara ararauna*). Brueghel only painted the bird from 1615 onwards, in his later career. Savery also seems to have started to render the bird around that time, but for him this was still in the beginning of his career, since he commenced around 1610 with his exotic animal paintings. The blue-and-yellow macaw was still quite rare at the time in Europe and by representing it, Brueghel and Savery each emphasized that they were very much aware of topical knowledge and recent discoveries. Both painters also depicted the scarlet macaw (*Ara macao*), but Brueghel shifted to the rarer green-winged macaw (*Ara chloropterus*) around 1613, which seems to have almost replaced the scarlet macaw in his oeuvre from that point on and again can be regarded as a sign of his striving to depict the latest developments in natural history. Savery never portrayed the green-winged macaw.

Cockatoos were similarly still very rare in Europe in the first quarter of the seventeenth century. Aldrovandi was the first of the authoritative natural historians to describe—and illustrate—a cockatoo, which he called the 'psittacus albus cristatus' [Fig. 10.9]. Brueghel depicted the yellow-crested cockatoo (*Cacatua sulphurea*), and the earliest dated painting in which it appears is from 1615.[53] Savery's cockatoo has a big, bright red and white striped crest, which is probably a salmon-crested cockatoo (*Cacatua moluccensis*). Whether one of the two cockatoo species found in Brueghel's and Savery's paintings is the same species as described and illustrated by Aldrovandi remains unclear. The original drawing in Aldrovandi's collection on which the woodcut was based, shows a cockatoo with a grey crest, which I have not been able to identity.[54]

50 Gessner Conrad, *Historiae animalium liber III qui est de avium natura* [...] (Zurich, Christoph Froschauer: 1555). Gessner however did point out that many more species existed. Belon Pierre, *L'histoire de la nature des oyseaux, avec leurs descriptions, et naïfs portraicts retirez du naturel* [...] (Paris, Guillaume Cavellat: 1555).

51 Aldrovandi Ulisse, *Ornithologia, hoc est de avibus historiae libri XII* (Bologna, Franciscus de Franciscis: 1599).

52 Clusius, *Exoticorum libri decem*.

53 The painting is on copper, measuring 19 × 15 cm. It is in a private collection in Canada. Paul Smith and I argued that the yellow-crested cockatoo is represented for the first time in Brueghel's Allegory of Air of 1621, see Rikken – Smith, "Jan Brueghel's 'Allegory of Air' (1621) from a Natural Historical Perspective" 101. At the time of writing that article, we were not yet familiar with the 1615 painting.

54 The inventory number of the drawing is Tavola 006-2 054. The drawings on which the woodcuts in Aldrovandi's publications were based, are available online: www.filosofia.unibo.it/aldrovandi/pinakesweb/main.asp?language=it. Possibly the drawing was made after a badly stuffed specimen, of which the crest was already discolored.

FIGURE 10.9 *Anonymous, woodcut illustration to Aldrovandi's* Ornithologia, hoc est de avibus historiae libri XII *(Bologna, Franciscus de Franciscis: 1599), book XI, p. 667.*

Thus it cannot be established with certainty if Brueghel and Savery depicted a new species that was not yet described by natural historians. However, their rendering of cockatoos does underline their knowledge, or perhaps the knowledge of their patron and the importance of his collection and menagerie.

Scholars have already pointed out that Brueghel displayed his natural historical knowledge in his paintings.[55] For Savery this has not yet been established, but he certainly showed himself to be aware of the state of natural history too. Savery depicted the cassowary, always a southern cassowary (*Casuarius casuarius*), proportionately much more often in his paintings than Brueghel did. Clusius was the first natural historian to describe and illustrate a specimen, which had been brought from New Guinea in 1597, in his *Exoticorum libri decem*. The flightless bird that Clusius described was first in possession of the landgrave of Solms in The Hague, who bequeathed the specimen to the Elector Ernst of Cologne, who then handed it over to Rudolf II, in whose menagerie Savery had ample chance to study it. Savery also depicted the crowned crane—either a black crowned crane (*Balearica pavonina*) or a grey crowned crane (*Balearica regulorum*), it is not possible to identify which subspecies he depicted—much more often, portraying it in at least 34 paintings, whereas Brueghel depicted crowned cranes on at least 5 paintings. Moreover, Savery captured the bird in many different poses while Brueghel always depicted it in the same static position. The crowned crane originated from Africa and was first described by Aldrovandi in his *Ornithologia* of 1603 [Fig. 10.10].[56]

However, the species apparently remained unknown for some time thereafter, for Clusius only had an illustration of the head of the bird at his disposal and wrongly called it a 'sea peacock' [Fig. 10.11]. Savery's rendering of the bird in multiple poses suggests that he was able to study a live crane, probably in the menagerie in Prague, although evidence of its existence there is lacking. Brueghel's static portrayal of the bird, in a single pose, indicates that he was unable to see the bird alive. Savery seems to have made a deliberate choice to portray some species in different poses, even within one painting, not only to emphasize that he could and did study live specimens, which of course reflected positively on the patron who could afford to keep these animals, but also to show that he could do more than the natural historians, who always showed an animal in a single pose in an illustration.

55 See for example Faber Kolb, *Jan Brueghel the Elder* 26–27.

56 The crowned crane had already been depicted by the Flemish artist Joris Hoefnagel in his *Aier* album (leaf 10), which is now in the collection of the National Gallery of Art, inv. no. 1987.20.8.

FIGURE 10.10 *Anonymous, woodcut illustration to Aldrovandi's* Ornithologia, hoc est de avibus historiae libri XIX–XX *(Bologna, Giovan Battista Bellagamba: 1603), book XX, p. 362.*

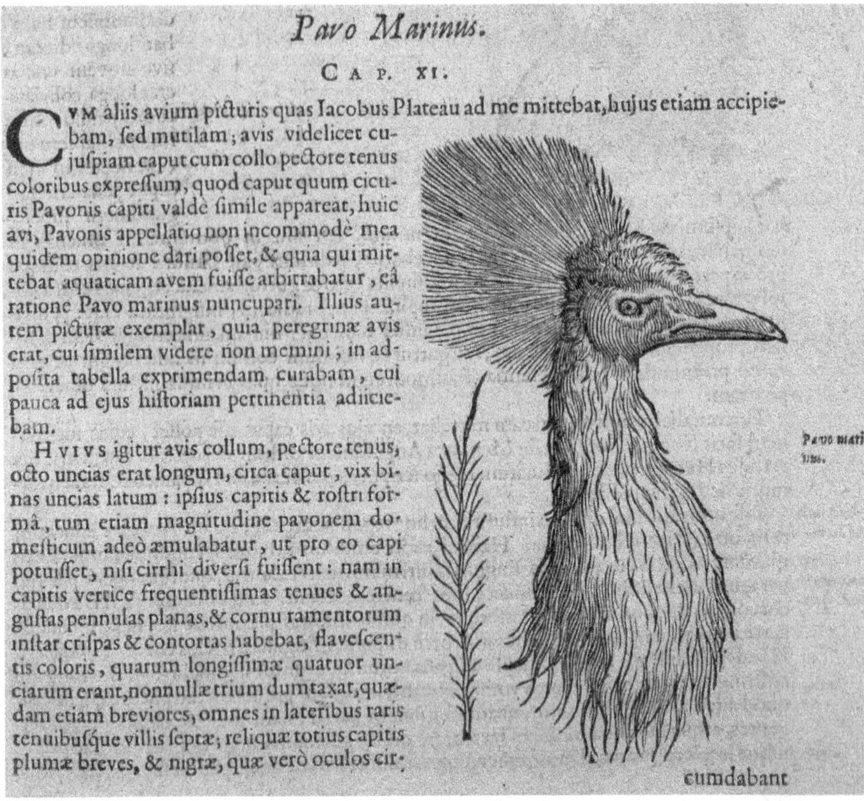

Pavo Marinus.

CAP. XI.

CVM aliis avium picturis quas Iacobus Plateau ad me mittebat, hujus etiam accipiebam, ſed mutilam; avis videlicet cujuſpiam caput cum collo pectore tenus coloribus expreſſum, quod caput quum cicuris Pavonis capiti valdè ſimile appareat, huic avi, Pavonis appellatio non incommodè mea quidem opinione dari poſſet, & quia qui mittebat aquaticam avem fuiſſe arbitrabatur, eâ ratione Pavo marinus nuncupari. Illius autem picturæ exemplar, quia peregrinæ avis erat, cui ſimilem videre non memini; in adpoſita tabella exprimendam curabam, cui pauca ad ejus hiſtoriam pertinentia adiiciebam.

HVIVS igitur avis collum, pectore tenus, octo uncias erat longum, circa caput, vix binas uncias latum: ipſius capitis & roſtri formâ, tum etiam magnitudine pavonem domeſticum adeò æmulabatur, ut pro eo capi potuiſſet, niſi cirrhi diverſi fuiſſent: nam in capitis vertice frequentiſſimas tenues & anguſtas pennulas planas, & cornu ramentorum inſtar criſpas & contortas habebat, flaveſcentis coloris, quarum longiſſimæ quatuor unciarum erant, nonnullæ trium dumtaxat, quædam etiam breviores, omnes in lateribus raris tenuibuſque villis ſeptæ; reliquæ totius capitis plumæ breves, & nigræ, quæ verò oculos circumdabant

Pavo marinus.

FIGURE 10.11 *Anonymous, woodcut illustration to Clusius'* Exoticorum libri decem *(Antwerp, Christopher Plantin: 1605), book V, p. 105.*

Conclusion

Brueghel and Savery both shaped the new genre of exotic animal painting and each played their own part in its development. Brueghel created the genre by elaborating on religious and mythological themes such as the Earthly Paradise and Orpheus, in which he gave the exotic animals a much more prominent position. He showed himself to be a very innovative painter, demonstrating a continued interest in not only depicting new themes, but in rendering new species as well. He constantly strove to include current knowledge of natural history. Savery at first elaborated on the themes that Brueghel had depicted before him, with Orpheus most notable among them, but later on Savery focused his attention on the actual animals themselves by omitting figures and a narrative. Thereby, Savery initiated a new course with the genre. His interest lay not so much in upgrading his repertoire of exotic animals over the course of his career, but in highlighting the different poses of animals and thus emphasizing their liveliness.

Consequently, very important for the development of the genre was access to the menageries of the courts in Prague and Brussels, which the painters must have enjoyed to study the exotic animals. The court in Prague seems to have played a key role; perhaps it was even on the instigation of Rudolf II that the new genre came into vogue, since the theme of Orpheus seems to have been of special interest to him and reflected his political aspirations. However, Brueghel was only in Prague for a very short time, and Savery kept producing exotic animal paintings long after leaving Prague in 1613. Brueghel developed the genre further in the Southern Netherlands, where he painted most exotic animal paintings on commission. Savery settled in the Northern Netherlands, where he would have mostly worked for the open market. Because of the different circumstances for both painters after that time, the genre of exotic animal painting developed into different directions in the second quarter of the seventeenth century.[57] However, if these two animal painters had not crossed paths roads in Prague, the new subject matter may not have evolved into an independent genre that many later painters also took up, and as such, it deserves much more attention than it has received thus far.

Selective Bibliography

Aldrovandi Ulisse, *Ornithologia, hoc est de avibus historiae libri XII* [...] (Bologna, Franciscus de Franciscis: 1599).

Belon Pierre, *L'histoire de la nature des oyseaux, avec leurs descriptions, et naifs portraicts retirez du naturel* [...] (Paris, Guillaume Cavellat: 1555).

Brink P. van den – Allart D. – Glas M. van der – Wegen D.H. van, *De firma Brueghel* (Amsterdam: 2001).

Clusius Carolus, *Exoticorum libri decem* [...] ([Antwerp], Franciscus Raphelengius: 1605).

Cuttler C.D., "Exotics in Post-Medieval European Art: Giraffes and Centaurs", *Artibus et Historiae* 12, 23 (1991) 161–179.

57 In my forthcoming dissertation I will elaborate on the different directions into which exotic animal painting developed after the first quarter of the seventeenth century.

DaCosta Kaufmann T., *The Eloquent Artist: Essays on Art, Art Theory and Architecture, Sixteenth to Nineteenth Century* (London: 2004).

De Jaegere I., *Roelandt Savery 1576–1639* (Kortrijk: 2010).

Ertz K. – Nitze-Ertz C., *Jan Brueghel der Ältere (1568–1625): kritischer Katalog der Gemälde*, 4 vols (Lingen: 2008).

Faber Kolb A.L., *Jan Brueghel the Elder: The Entry of the Animals into Noah's Ark* (Los Angeles: 2005).

Gessner Conrad, *Historiae animalium Lib. I. de quadrupedibus viviparis* [...] (Zurich, Christoph Froschauer: 1551).

Gessner Conrad, *Historiae animalium liber III qui est de avium natura* [...] (Zurich, Christoph Froschauer: 1555).

Kempter G., *Ganymed: Studien zur Typologie, Ikonographie und Ikonologie* (Cologne: 1980).

Maeyer M. de, *Albrecht En Isabella En de Schilderkunst Bijdrage Tot de Geschiedenis van de XVIIe-Eeuwse Schilderkunst in de Zuidelijke Nederlanden* (Brussels: 1955).

Maselis M.-C., Balis A., and Marijnissen R., *De Albums van Anselmus de Boodt (1550–1632): Geschilderde Natuurobservatie aan het Hof van Rudolf II te Praag* (Tielt: 1989).

Müllenmeister K.J., *Roelant Savery: Kortrijk 1576–1639 Utrecht: Hofmaler Kaiser Rudolf II. in Prag: Die Gemälde mit Kritischem Œuvrekatalog* (Freren: 1988).

North M., *Art and commerce in the Dutch Golden Age* (New Haven: 1997).

Panofsky E., *Albrecht Dürer* (Princeton: 1948).

Preiss P., *Vaìclav Vavrinec Reiner (1689–1743): Landscapes with Animals: Motifs of Orpheus and Animals and Birds Flocking around an Owl in the Painting of Central Europe and the Szech Lands in the Baroqu* (Prague: n.d.).

Rikken M., "Abraham Ortelius as Intermediary for the Antwerp Animal Trailblazers", *Jahrbuch Für Europäische Wissenschaftskultur* 6, 2011 (2012) 95–128.

Rikken M. – Smith P.J., "Jan Brueghel's 'Allegory of Air' (1621) from a Natural Historical Perspective", *Nederlands Kunsthistorisch Jaarboek* 61 (2011) 87–115.

Ringmar E., "Audience for a Giraffe: European Expansionism and the Quest for the Exotic", *Journal of World History* 17, 4 (2006) 375–397.

Semmelrath H., *Der Orpheus-Mythos in der Kunst der italienischen Renaissance: eine Studie zur Interpretationsgeschichte und zur Ikonologie* (Cologne: 1994).

Smith P.J., "Diersymboliek in Rembrandts *Zondeval* (1638) en in Vondels *Adam in ballingschap* (1664)", *De Zeventiende eeuw* 26, 1 (2010) 2–20.

Smolar-Meynart A., *Het Paleis van Brussel: Acht eeuwen kunst en geschiedenis* (Brussels: 1991).

Spicer-Durham J.A., *The Drawings of Roelandt Savery* (Connecticut: 1979).

Taylor P., *Dutch Flower Painting, 1600–1720* (New Haven: 1995).

Zimmermann H., "Das Inventar der Prager Schatz- und Kunstkammer vom 6. Dezember 1621: (nach Akten des k. und k. Reichsfinanzarchivs in Wien)", *Jahrbuch der Kunsthistorischen Sammlungen des Allerhöchsten Kaiserhauses* 25, 5 (1905) XIV–LXXXVIII.

Symbolic Use of Animals and Political Education

CHAPTER 11

Are Cranes Republicans? A Short Chapter in Political Ornithology

Sabine Kalff

In his *Historia Animalium*, Aristotle discussed the social behavior of animals with regard to their characteristics and life-style, not to forget their habits of communication. With regard to the animal's social skills, Aristotle noticed:

> Here are some further differences with respect to animals' manner of life and activities. Some are gregarious, some solitary: this applies to footed animals, winged ones, and swimmers alike; others are dualizers [show both characteristics]. Some of the gregarious animals are social, whereas others are more dispersed. Examples of gregarious animals are: birds—the pigeon class, the crane, the swan (N.B. no crook-taloned bird is gregarious); swimmers—many groups of fishes, e.g., those called migrants, the tunnies, the pelamys, and the bonito. And man dualizes.[1]

Investigating further into the species' political predilections, Aristotle concluded:

> The social animals are those which have some one common activity; and this is not true of all the gregarious animals. Examples of social animals are man, bees, wasps, ants, cranes. Some of them live under a ruler, some have no ruler; examples: cranes and bees live under a ruler, ants and innumerable others do not.[2]

While many of the Aristotelian political animals such as the bee and the ant have been studied thoroughly, others have been neglected. This holds particularly true for those animals which displayed flocking behavior, but lacked

1 Aristotle, *History of Animals*, transl. A.L. Peck. 3 vols. (London-Cambridge, Mass: 1965), vol. I, 1, 488a33 ff. etc.

2 Ibidem.

 | DOI 10.1163/9789004279179_013

swarm behavior and therefore were not suited for the popular studies of swarm theory, unlike many fish and birds.[3]

Hence, the fate of the crane as a political animal remained overshadowed by the far more popular bee, which became an icon of social and political rule. But in the seventeenth century, the crane figured prominently as a symbol of democracy. In his treatise on reason of state, the *Considerationi Politiche, e Morali sopra Cento Oracoli d'Illustri Personaggi antichi* (1621),[4] Ludovico Zuccolo pointed out:

> Those who hold that there is only one God and not more to govern the world, and that there is only one King of the bees to govern his colony must be either kidding or do not understand much of logic, since they cannot make out the difference between a naturally superior being governing inferiors and one who rules over one's equals.[5]

Here, Zuccolo attacked a hypothesis put forward by theorists of sovereignty, such as Jean Bodin, who in his *Six Livres de la République* (1576) famously maintained:

> All the laws of nature point towards monarchy, whether we regard the microcosm of the body, all of whose members are subject to a single head on which depend will, motion, and feeling, or whether we regard the macrocosm of the world, subject to the one Almighty God. If we look at the heavens we see only one sun. We see that even gregarious animals never submit to many leaders, however good they may be.[6]

According to Zuccolo, how God ruled the world was politically irrelevant, since God and mankind were of a different species. And no one (besides some absolutists with deficient logical skills) doubted that rulers generally were of the

---

3 Cf. for instance Heiden von der Heiden A. – Vogl J. (eds.), *Politische Zoologie* (Zurich – Berlin: 2007); Horn E. – Gisi L.M. (eds.), *Schwärme—Kollektive ohne Zentrum. Eine Wissensgeschichte zwischen Leben und Information* (Bielefeld: 2009).

4 Reference is made to the year of publication, unless otherwise stated.

5 'Quelli, che dicon poi, che un solo Iddio, e non più governa il Mondo, et un solo Rè delle api il suo sciame, ò burlano, ò non s'intendono troppo bene di Logica, non conoscendo la differenza, la quale è tra il governare il maggiore di natura gli inferiori, od uno eguale il suoi pari'. Zuccolo Ludovico, *Considerationi Politiche, e Morali sopra Cento Oracoli d'Illustri Personaggi antichi* (Venice, Marco Ginami: 1621) 6.

6 Bodin Jean, *The Six Bookes of a Common-Weale*, transl. E. Knolles (London, Adam Islip: 1606) VI, 4, 735.

same species as their subjects. Their earthly rule necessarily differed from the godly one. Zuccolo thus concluded that the most exemplary animal to imitate were cranes and storks:

> And if we absolutely have to take an animal as a model, it is more convenient to imitate the government of the ants, the cranes or the storks, who are naturally equal [among themselves] and of equal value, and who also do not live in a monarchy.[7]

Notably, Zuccolo turned to the crane as an example of the rule of a *primus inter pares* according to the model of the Republic of Venice, which was what early modern political philosophers usually intended for a democratic form of government. While Zuccolo based his hypothesis of the cranes' republican inclination on Aristotle's famous remarks on gregarious animals, he also reversed the Aristotelian argument entirely. Between the fifth century BC and the early seventeenth century AD, the cranes apparently switched sides. The once monarchic animal had been transformed into a republican species. But how could that happen? Most of the other animals evoked by Aristotle remained stable with regard to their presupposed political inclination. For instance, Aristotle took the ants as an example of non-monarchic government, and they remained faithful to this judgment until the late eighteenth century. In 1769, Voltaire still claimed that 'Les fourmis passent pour une excellente démocratie.'[8] This paper will thus investigate the factors that enabled the cranes to switch sides in political zoology. In which way did the crane figure in the different discourses in the seventeenth century? What was its status in natural philosophy, and what was its fate in emblematics, symbology and literature?

Zuccolo's Political Zoology

Ludovico Zuccolo (1568–1630) was obviously no advocate of monarchy. He was the son of a nobleman of Faenza, who died during a five-year sentence to the galleys after being convicted of heresy. Zuccolo studied philosophy, probably also natural philosophy, in Bologna, where he also lectured, as well as at the

7 'Però se habbiamo à pigliare esempio da gli animali, converrà più tosto imitare il governo delle formiche, ò delle grù, ò delle cicogne, le qual per essere di natura, et di valore eguali, vivono anche non à Monarchia'. Zuccolo, *Considerationi* 6.

8 Voltaire, *Dictionnaire philosophique. La raison par alphabet*, 2 vols. (Geneva: Gabriel Cramer: 1769) vol. II, "Des Loix" 14.

University of Padua.[9] Zuccolo served as a secretary for several high clergymen and at the court of Francesco Maria II della Rovere, Duke of Urbino.[10] He was not quite satisfied with court life and remarked drastically: 'Having finished my studies, I went into slavery.'[11] He wrote on the poetics of the Italian language and wrote poems himself. He was most renowned for his political writings, consisting in three utopian texts and the influential treatise on reason of state. In the *Considerationi politiche*, Zuccolo advocated that political theory should be considered a science, a concept which exerted influence even on geographically distant political philosophers, such as Henning Arnisaeus and Hermann Conring. These followed Zuccolo's suggestion by establishing political theory as a discipline at the University of Helmstedt. Zuccolo very likely also came to know contemporary concepts of natural philosophy and medicine during his studies, since his treatises demonstrate a solid knowledge of these topics.

Zuccolo's hypothesis of the democratic inclination of the cranes seems to rely heavily on natural philosophy, since he put forward: 'it is not the case that a stork reigns over the other storks, or that a lion rules the other lions, or even that a dolphin commands the other dolphins'.[12] The lion and the dolphin were usually considered examples of monarchic animals, too. But Zuccolo did not refer to their emblematic status, thus not addressing the lion as the king of the animals or the dolphin as king of the sea. He pointed instead to their interspecies or social behavior. This, as Zuccolo noted, was based rather on cooperation than on subordinance.

In political ornithology, as well as in the genre of fables, cranes and storks seem to have been interchangeable.[13] They were not referred to frequently in political treatises, but Thomas Browne mentioned in his *Pseudodoxia epidemica* the widely held opinion that storks insisted so strongly in republicanism that they refused to dwell in monarchic states: 'Storks are to be found, and will only live in Republikes or Free States'. Browne denied this argument on the basis of contrary empirical evidence and held: '[it] is a pretty conceit to advance the opinion of popular policies, and from Antipathies in nature, to

9 Cf. Pissavino P.C., *Lodovico Zuccolo: Dall'audizione a corte alla politica* (Florence: 1984) 32.

10 Ibidem 11, 22.

11 'Mi posi, compito lo studio, alla servitù'. Zuccolo, *Considerationi* 290.

12 'Però non fà, che una cicogna sola governi l'altre cicogne, ne un leone tenga l'imperio sopra gli altri leoni, ò pure un delfino à gli altri delfini commandi'. Zuccolo, *Considerationi* 7.

13 For the interchangeability between crane and stork see for instance the Aesop fable of the fox and the stork which Plutarch referred to as the fox and the crane.

disparage Monarchical Government'.[14] He put forward that storks lived and were still living in monarchic states throughout history, from ancient Egypt to contemporary France. Notably, Browne based his biological reflection on an authoritative ornithological source—Pierre Belon's *L'histoire de la nature des oyseaux* (Paris, 1555). Browne's argumentation proves that the idea of the freedom-loving storks was equivalent to the one of the cranes' republicanism and that it was well diffused.

Unlike the hypothesis reported by Browne, Zuccolo's assertion of the free-spirited cranes was not based on simple analogy (that cranes lived only in states which conform with their democratic inclination) but on his knowledge on the social or 'political' behavior of cranes. They constituted a fitting example of the rule of the few,[15] which according to Zuccolo was generally recommendable: 'And therefore, in order to be ruled in a good way, a people also needs several rulers, since one hardly suffices for such an exhaustive enterprise'.[16] Cranes were exemplary for their cooperation and sharing of demanding tasks. While Zuccolo's political ideas concerning cranes differed from the political theory of Aristotle, they agreed with his natural philosophy as put forward in the *Historia animalium*:

> Many instances of intelligence seem to occur among the cranes too. For they migrate a long way, and fly to a great height in order to survey the distance, and if they see clouds and bad weather they fly down and stay quiet. Further, the fact that they have both a leader and signallers that whistle among the end birds so that their call is heard. And when they settle, while the others sleep with their head under the wing, standing on one foot and then on the other, the leader with his head uncovered keeps a look-out, and whenever he perceives something he signals with a cry.[17]

Most of the information given here by Aristotle has been repeated and enriched with details over and over again in zoological, agricultural, historical

14 Browne Thomas, *Pseudodoxia epidemica: Or, Enquiries Into Very many Received Tenents, And commonly Presumed Truths* (London, Edward Dod: 1658) 218.

15 What Zuccolo terms a republic would be an aristocracy in modern terms. Since Zuccolo does not speak of the rule of the best, I use here the Aristotelian term 'rule of the few' in a neutral sense. Cf. Aristotle, *Politics*, transl. H. Rackham (London – Cambridge, Mass.: 1967) III,7, 1279a34–37, 207.

16 'Et così anco un popolo per essere ben retto […] richiede più governatori, per bastar malamente un solo à tanta impresa'. Zuccolo, *Considerationi* 8.

17 Aristotle, *History of Animals*, transl. D.M. Balme, 3 vols. (London – Cambridge, Mass.: 1991), vol. III, 614b18, 267.

and emblematic works from antiquity onwards. The crane watching over his flock by night was already equipped by Pliny with a stone in his foot in order that its dropping would prevent the crane from falling asleep:

> At night time they have sentries who hold a stone in their claws, which if drowsiness makes them drop it falls and convicts them of slackness, while the rest sleep with their head turned under their wing, standing on either foot by turns; but the leader keeps a lookout with neck erect and gives warning.[18]

The crane taking turns to keep vigil became the symbol of vigilance and prudence in early modern books of emblems and symbols. As prudence was a genuine political virtue, the vigilant crane on raised foot with a stone made its appearance in Albrecht Dürer's woodcut print of the triumphal arch for Maximilian I (1515) among other significant political animals such as the lion and the eagle. Dürer was strongly influenced by the animal symbolism as put forward by Horapollo's *Hieroglyphica* (1505), a Greek book of symbolic images from the fourth century.[19]

The crane also figured in an influential early modern iconographic treatise, Cesare Ripa's *Iconologia* (1593) as a symbol of vigilance (*gru simbolo della vigilanza*). Notably, Ripa underpinned the crane's symbolic value with a sentence from the Song of Solomon (5,2), to which he referred in its Latin version 'Ego dormio, et cor meum vigilat'.[20] In the Late Middle Ages, this sentence assumed juristic and political importance. For instance, the jurist Baldo degli Ubaldi contextualized the biblical sentence with the concept of the ruler, whose subjects could trust his watching over their sleep, himself remaining perpetually insomnious as a personified *lex animata*.[21] Thus, the frequent early modern references to the crane as a symbol of vigilance may be understood as a new means of expressing a conventional juristic notion. In other words, political ideas which in the Late Middle Ages were preferably expressed through juristic constructions could also be sustained by natural philosophy in the early modern period. The fact that one of the first and most influential authors of

---

18 Pliny, *Natural History*, transl. T.E. Page et al., 10 vols. (London – Cambridge, Mass.: 1967), vol. III, X, 58–61, 329–330.

19 Cf. Boas G., "Introduction", in Horapollo, *The Hieroglyphics*, transl. George Boas (New York: 1950) 13–54, 28.

20 Cf. Ripa Cesare, *Iconologia* (Venice, Cristoforo Tomasini: 1645) 668.

21 Kantorowicz E.H., *Die zwei Körper des Königs. Eine Studie zur politischen Theologie des Mittelalters* (Munich: 1990) 148.

emblem books, Andrea Alciato, was actually a jurist, points to the same conclusion. Emblem books which spread animal symbology as well as knowledge on zoology,[22] may be regarded as lingering on the threshold between the spheres of law and natural philosophy.

The Cranes in Aldrovandi's *Ornithologiae*

The watchful crane as a model of vigilance was not necessarily a symbol of republicanism. In Ripa's *Iconologia* it symbolized even the omnipresence of the monarchic ruler. Thus, Zuccolo must have been inspired by other sources. The most impressive contemporary ornithology was Ulisse Aldrovandi's encyclopedic *Ornithologiae* which appeared in three volumes between 1599 and 1603. Although nothing is known about Zuccolo's readings in natural philosophy, it is impossible that he was unaware of Aldrovandi, who was a widely known professor at the University of Bologna while Zuccolo studied there. Aldrovandi's *Ornithologiae* was part of an encyclopedia covering all branches of zoology. Aldrovandi devoted 41 pages to the species of the crane, including five illustrations,[23] in which he addressed the crane in terms of 31 different aspects. Only some of the categories were strictly zoological, among them the crane's different subspecies (*genera differentiae*), its outer (*forma descriptio*) and inner (*anatome*) appearance, its biological development from birth to old age (*coitus/partus, exclusio/ educatio; vox/aetas*) and of course its flight habits (*volatus*). Besides biological categories, Aldrovandi addressed numerous linguistic aspects of the crane, some of them placed prominently at the beginning of the article. He reflected at length on the nomenclature of the crane in a variety of languages (*aequivoca*) and its philological meaning, in synonymous expressions (*synonyma*) and in the names given by other zoologists (*denominata*). Furthermore, the crane's appearance in history (*historica*) was highlighted, focusing on the story of the cranes of Ibycus. Aldrovandi then gave an overview on what the crane signified meteorologically (*praesagia*), morally (*moralia*) and iconographically (*hieroglyphica, emblemata*), as well as what the crane was used for in literature—for proverbs (*proverbia*) and fables (*fabulosa*). But he also discussed its strictly material profit, such as culinary benefits

22 Cf. Harms W., "On Natural History and Emblematics in the 16th Century", in Ellenius A. (ed.), *The Natural Sciences and the Arts. Aspects of Interaction from the Renaissance to the 20th Century* (Uppsala: 1985) (67–83) 69.

23 Aldrovandi Ulisse, *Ornithologiae*, 3 vols. (Bologna, Giovan Battista Bellagamba: 1603), vol. III, 324–365.

(*usus in cibis*) and pharmaceutic use (*usus in medicina*), applied internally or externally (*usus in externis*). As medical and culinary processing first required a crane to be caught, hunting strategies were also discussed (*capiendi ratio*).

Under the heading of *moralia*, Aldrovandi discussed the cranes' social behavior and pointed out that cranes were eager to take turns keeping night watches for the common good. Since they did so without coercion, this seemed indicative of their military virtues as well as of their responsibility for the community as a whole. Aldrovandi's judgment on the cranes' admirable martial discipline was based on a rather unlikely authority—the *Hexaemeron* of Saint Ambrose. In his *Hexaemeron*, cranes, like other birds, emerged on the fifth day of the process of creation. Ambrose held: 'In their [the cranes'] state there is a natural military service, which in ours is coerced and slave-like'.[24] Ambrose spoke in high terms of the crane's division of labor with regard to their night watches and flight formation and considered them exemplary with regard to the republican virtues which he saw embodied in the Roman Republic.

Although it is well known that contemporary emblem books, fables, treatises of mythology and symbology exerted a strong influence on most of the early modern zoologists,[25] it is somewhat surprising that even Ambrose's early medieval didactic account of creation, which was not particularly inspired by empirical observation, was still considered a valid zoological source by a highly learned seventeenth century natural philosopher like Aldrovandi. With regard to the republican virtues of the cranes, Aldrovandi also referred to other remote sources such as Aelian's late antique *De natura animalium*. Aldrovandi extensively cited another early medieval source, Cassiodorus's *Variae*, a collection of letters and documents from the time of Theoderic the Great. Cassiodorus held: 'Live then in justice and moderation. Follow the example of the cranes, who change the order of their flight, making foremost hindmost, and hindmost foremost, without difficulty, each willingly obeying its fellow—a commonwealth of birds'.[26] Cassiodorus considered the practice of rotating the flight formation exemplary with regard to the political rotation principle which had been constitutive for the Roman Republic with its limited terms of

24 'In illis (gruibus) politica quaedam, [...] est, et militia naturalis, in nobis coacta et servilis'. Aldrovandi, *Ornithologiae* 351; cf. Ambrose, *Exameron* V, 50–51.

25 Cf. Harms, "Natural History" 67, and Ashworth W.B., "Natural History and the Emblematic World View", in Ellenius A. (ed.), The Natural Sciences and the Arts. Aspects of Interaction from the Renaissance to the 20th Century (Uppsala: 1985) (303–332) 313.

26 Cassiodor, "Edict of King Athalaric", in Cassiodor, *The Letters of Cassiodor* (*Variae Epistolae*) *Being a Condensed Translation of the Variae Epistolae of Magnus Aurelius Cassiodorus Senator*, trans. T. Hodgkin (London: 1886) IX, 2, 386.

political office. Not only Cassiodorus, but also Aldrovandi stressed their cooperative and non-hierarchical practices.

Particularly their flight formation seems to have inspired Cassiodorus and the early medieval authors to conclude that the cranes lived in a community of equals in which each individual contributed to the common good, being free from constraint and obligation to obey, since they lived not under a permanent ruler.[27] Thus, Cassiodorus seems to have been the crucial authority for the early modern authors' concept of the republican government of cranes. Combined with martial discipline, their species displayed almost all virtues of ancient republicanism, which was embraced by many of the early modern political writers, for instance by Zuccolo.

Without quoting authorities, Aldrovandi went on to explain that cranes fled superbia, embodied in the species of the eagle.[28] This was a moral elaboration of a reflection put forward by Aelian, who held, according to Aldrovandi, that there was an essential enmity between cranes and eagles. This idea might be based on some remarks by Aristotle concerning basic enmities between species. Aristotle mentioned only the enmity between a certain owl named hybris and the eagle, and reported that the birds were so absorbed while fighting that they could easily be picked up by shepherds.[29] He went on to write that cranes could be caught in the same way, as they were equally desperate fighters among themselves—a finding that contrasted notably with the idea of the cranes' being thoroughly cooperative. Probably, the assumption of the enmity between eagle and crane results from the sheer textual proximity of these species in Aristotle. Aldrovandi concluded that cranes were hostile to eagles, because eagles morally symbolized superbia. This observation was based on political symbolism—the eagle was not only considered arrogant, but also tyrannical, and superbia was a classical vice of tyrants. In the Renaissance, eagles as well as other 'monarchic animals' such as lions, were sometimes depicted as attributes of superbia. Equating eagles with tyrants seems to have been popular, since it occurs also in Erasmus's *Institutio Principis Christiani*

27 'His etiam volatus vicaria equalitate disponitur, ultima fit prima, et qui primatum tenuit, esse posterior non recusat; sic quadam communione societatae sibi sine regibus obsequuntur, sine dominatu parent, sine terrore famulantur. voluntariae serviendo liberae sunt et invicem se diligendo muniuntur'. Cf. Cassiodor, *Variarum libri XII*, ed. Å.J. Fridh – J.W. Halporn (Turnholt: 1973) IX, 2, 347–348. Cited almost word for word in Aldrovandi, *Ornithologiae* 351.

28 'statim clamemus, aquilam, id est superbiam et tyrannidem, ante omnia fugiamus'. Aldrovandi, *Ornithologiae* 351–352.

29 Cf. Aristotle, *History of Animals*, vol. III, 615b10–19, 273–275.

(1516), where he put forward: 'If you want a comparison for the tyrant, take the lion, bear, wolf, or eagle, all of which live on their mangled prey'.[30]

Hence, the assumption that storks fled monarchic states reported by Browne was based on a variety of sources and authorities. The political identification of storks and cranes seemingly led to the conclusion that storks also abhorred eagles and everything they stood for, such as tyranny and superbia. To call monarchies tyrannical was obviously not uncommon for advocates of republicanism. Aldrovandi's compilation of an enormously wide range of information on cranes might also have provided the basis for Zuccolo's assertion that cranes were genuinely devoted to republicanism. Thus it was, to no small extent, remote and half-forgotten sources, and not at all new empirical findings, that led to the rather new assumption that cranes were a vivid example of democracy. It is interesting to see that even Saint Ambrose has contributed to early modern republicanism, since he has hardly ever been considered in this context.

The Crane as Symbol of Democracy

Cranes were frequently supposed to maintain among themselves a social peace, an observation usually based on etymological evidence, which is that the name of the species *grue* brought about the Latin verb *congruere*, to agree or to concord. This statement can be found almost anywhere, in Aldrovandi[31] as well as in Piero Valeriano (1477–1558), author of one of the most influential early modern dictionary of symbols, the *Hieroglyphica* (Basel 1556), which appeared in many editions and was translated into several modern languages, such as French (1576 and 1615) and Italian (1602). Valeriano explained: 'And in fact, for the Romans, the consensus and union among the different parties concerning one counsel seems to have been inspired by the crane itself, so that we may say "to congregate" when we mean "to agree"'.[32] In his *Hieroglyphica*, Valeriano also provided a very charming illustration of the convening cranes [Fig. 11.1]. Moreover, besides Cassiodorus, Valeriano seems to be the author who most explicitly held that cranes lived under democratic rule:

30 Erasmus Desiderius, *The Education of a Christian Prince*, transl. L.K. Born (New York: 1936) 166.

31 Cf. under the heading "denominate", Aldrovandi, *Ornithologiae* 349.

32 'Sane quidem apud nostros, consensus unioque, partium in eandem sententiam aut rem, Latino vocabulo formam ab ipsis gruibus videtur accepisse, ut inde "congruere" pro "convenire" dicamus'. Valeriano Giovanni Pierio, *Hieroglyphica Sive De Sacris Aegyptiorum aliarumque gentium literis* [...] (Basel, Leo Curio: 1575) fol. 128v.

FIGURE 11.1 *Woodcut illustration to Valeriano's* Hieroglyphica sive De sacris Aegyptiorum aliarumque gentium literis [...] *(Basel, Leo Curio: 1575) fol. 128v.*

> Due to the veritable assembly that is undertaken by the cranes, some maintain that the crane hieroglyphically represents democracy, which is to say, the popular state or government by the people.[33]

Here, the crane figured outright as a symbol of democracy, besides representing other political virtues, such as vigilance, prudence, order, industriousness,

33 I cite the French translation which gives the most elaborate account of the democratic order among the cranes: 'Quelques-uns tiennent que l'assemblee qu'on void tenir aux Grues represente hieroglyphiquement la Democratie, c'est à dire l'Estat populaire, ou gouverné par le peuple'. Valeriano Giovanni Pierio, *Les hiéroglyphes*, transl. J. de Montlyard (Lyon, Paul Frellon: 1615) 219.

and perseverance.[34] Valeriano, like his contemporaries, understood symbols not as artificial tokens, but rather as the essence of things. He thus held: 'to speak hieroglyphically is nothing other than to discern the true nature of things divine and human'.[35] Also Aldrovandi considered Valeriano a valuable source concerning zoological knowledge and repeated his judgment that cranes were an image of democracy, word for word.[36]

The Cranes and Navigation

Due to their admirable sense of orientation, cranes came to figure both as a symbol and as a means of orientation. Cranes did not only migrate in a highly organized and egalitarian manner, they also knew their way. Cranes communicate a lot among themselves during flight, which in the antique myth of deluge was much to the benefit of Deucalion, who could reach land after hearing voices of cranes above him, as Andrea Alciato reported in his emblem book.[37] In the Deucalion myth, the crane replaced the raven of the Judeo-Christian deluge as a signaler of land. Cranes thus became linked to navigation. Initially, they signaled acoustically, not visually. In the myth, Deucalion was saved by climbing up a hill named Mount Gerania, which means crane mountain. This etymological association between the crane and the geranium plant is also the reason why cranes are frequently depicted together with geraniums, for instance in Aldrovandi's *Ornithologiae* [Fig. 11.2].[38]

More frequently, cranes figured as symbols of orientation. Already Aristotle reported that cranes were among the birds who migrated farthest. Although he refuted the opinion that cranes followed their course infallibly with the aid of a stone they swallowed in order to ballast themselves,[39] this idea became repeated over and over again from antiquity onwards. Pliny had it that cranes not only took a stone in their claws before crossing the straits between Europe

34 Cf. Valeriano, *Les hiéroglyphes* 212.

35 'Quippe cùm hieroglyphice loqui nihil aliud sit, quam divinarum humanarumque rerum naturam aperire'. Valeriano, *Hieroglyphica* Title page.

36 'Ex eo in primis collegio, quod grues celebrant inter se, nonnulli dicunt democratiam [...] hieroglyphice significari'. Aldrovandi, *Ornithologiae* 353. For Valeriano cf. *Hieroglyphica* fol. 128v.

37 Alciato Andrea, *Emblemata* (Antwerp, Christoph Plantin: 1577) 113.

38 Aldrovandi, *Ornithologiae* 329.

39 'Now among the birds the cranes, as was said earlier, migrate from the farthest points to the farthest. They fly into the wind. The report about the stone is false: the report is that they contain a stone as ballast, which when it falls out becomes useful for testing gold'. Aristotle, *History of Animals*, vol. III, 597a30–32, 135.

FIGURE 11.2 *Woodcut illustration to Aldrovandi's* Ornithologia, *3 vols. (Bologna, Giovan Battista Bellagamba: 1603), vol. III, 329.*

and Africa, but that they additionally swallowed sand.[40] But most authors contented themselves with the report that cranes took pebbles in their claws and dropped them after having crossed the open sea. It was always implied that this strange strategy was adopted in order not to be led astray by the wind or by impeded sight while crossing the sea. Seemingly, cranes were imagined as a sort of flying loadstone. Early modern writers were probably not completely

40 'It is certain that when they are going to fly across the Black Sea they first of all make the straits between the two promontories of Ramsbrow and Carambis, and proceed to ballast themselves with sand; and that when they have crossed the middle of the sea they throw away the pebbles out of their claws and, when they have reached the mainland, the sand out of their throats as well'. Pliny, *Natural History*, vol. III, X, 58–61, 329 f.

mistaken, since it is very probable that magnetism plays a role in the orientation of migrating birds, albeit modern authors would rather claim that birds had a magnetic sense than that they were literally loadstones. Although it was by no means clear (and is still not clear) how cranes navigate so well, it was obvious that they did so. This also led to their transformation into a symbol of navigation. From early on, the crane's flight formation was described as a Y or V formation. Aldrovandi claimed that the flight formation looked most similar to a Y,[41] although he also discussed other alphabetical letters, such as alpha and lambda, referring to a variety of authorities.

In his encyclopedic book of symbols, the *Mondo simbolico* (1653), Filippo Picinelli unhesitatingly described the flight formation as a Y, which he called the Pythagorean letter.[42] Since antiquity, the *littera Pythagorae* was commonly considered a symbol that indicated a crossroad, demanding a moral choice between virtue and vice. Early modern symbology, as well as navigation and astronomy, obliterated the moral aspect and referred to the crane's flight formation simply as a representation of the right way, in contrast to disorientation.

Not surprisingly, the crane symbolized by a Y was actually transposed to the sky, in the form of a new star constellation in the southern hemisphere. Based on the observations made by two Dutch explorers, Pieter Dirkszoon Keyser and Frederick de Houtman, the astronomer Petrus Plancius introduced 12 new constellations of the southern hemisphere in his celestial globe of 1598. Plancius's interest in filling the stellar void of the southern hemisphere was primarily nautical, being himself one of the founders of the Dutch East India Company.[43] Several of the newly created constellations assumed names of exotic, only recently discovered birds, such as the toucan and the bird of paradise. The transfer of exotic birds to stellar constellations is also indicative of the link between the trading exploration voyages through which these birds became known in Europe, usually in the form of stuffed specimens, and the nautical skills and practices via which these explorations were accomplished.

The first celestial atlas which introduced the 12 new southern constellations was Johann Bayer's *Uranometria* (1603) [Fig. 11.3]. Bayer wrote almost nothing about the new constellations and gave only a plate of all new figures together.[44]

41 Aldrovandi, *Ornithologiae* 334.

42 'Le schiere delle grue, volando sogliono formare la lettera di Pitagora Y.' Picinelli Filippo, *Mondo simbolico* (Milan, Francesco Mognaga: 1653) 123.

43 For the history of constellations in the southern hemisphere see also http://members.westnet.com.au/Gary%2DDavid%2DThompson/page11-32.html. Consulted on April 7, 2013.

44 Bayer Johann, *Uranometria* (Augsburg, Christoph Mang: 1603) Plate 49v.

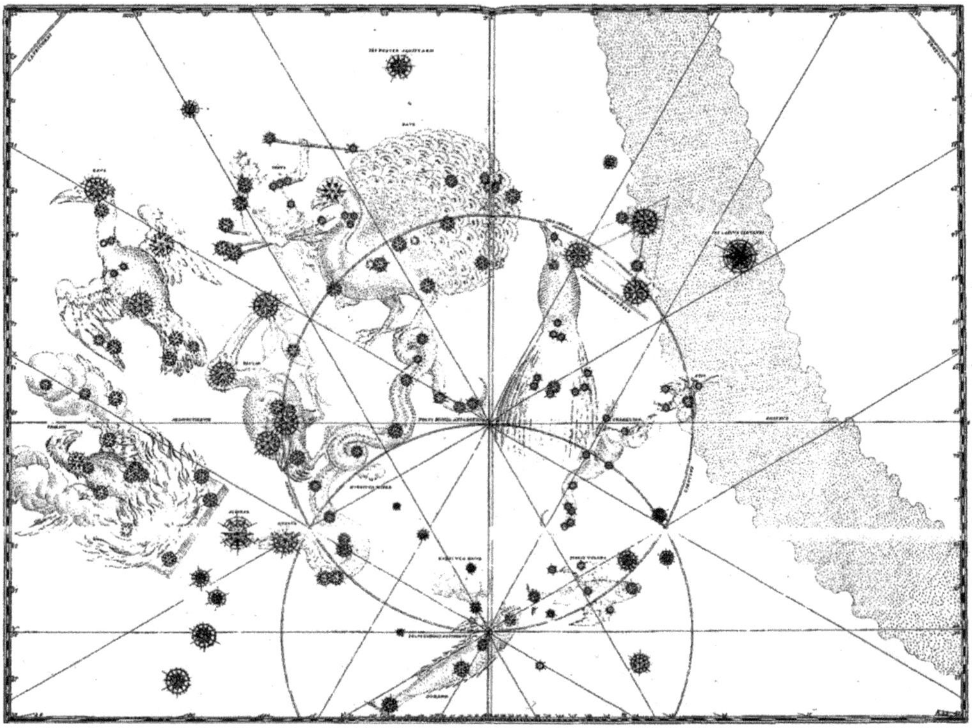

FIGURE 11.3 *Alexander Mair, engraving to Bayer's* Uranometria *(Augsburg, Christoph Mang: 1603; Reprint Dortmund: 1981) 199–200.*

Some of the birds are hardly recognizable in these illustrations, among them the crane. But the atlas surely helped to canonize the new constellations.[45] A star atlas which was very accurate, not only with regard to stellar positions, but also to constellation figures, was the *Firmamentum Sobiescianum sive Uranographia* (Gdansk, 1690) by Johann Hevelius, the astronomer of the Gdansk observatory. The reason why Hevelius's stellar positions were so accurate, was that he had used the most recent data available: Edmund Halley's 1679 star catalog for the southern hemisphere.[46] In Hevelius's atlas, the southern polar map was well executed and the quality of the crane's illustration [Fig. 11.4],[47] like that of the other stellar bird figures, had improved

45 Cf. Allen R.H., *Star Names and Their Meanings* (New York: 1936²) 237 f.

46 Halley Edmund, *Catalogus stellarum Australium sive Supplementum Catalogi Tychonici* (London, Thomas James – R. Harford: 1679).

47 Hevelius Johannes, *Firmamentum Sobiescianum sive Uranographia* (Gdansk, Johann Zacharias Stolle: 1690) pl. 55.

FIGURE 11.4 *Engraving to Hevelius's* Firmamentum Sobiescianum sive Uranographia *(Gdansk, Johann Zacharias Stolle: 1690), pl. 55.*

considerably, to rival that of contemporary ornithological works. Hence, if one wants to look for early modern ornithological illustrations, one should also take into account celestial atlases.[48]

48 This is particularly evident in the case of the bird of paradise which is depicted as footless by Bayer and Hevelius, as well as in the contemporary ornithological treatises. The species was supposed to be footless, since the stuffed birds arrived in Europe in a footless state, caused by preservation problems. Repercussions of this hypothesis also affected, as we have seen, early modern astronomical literature. Cf. for instance Smith P.J., "On Toucans and Hornbills: Reading in Early Modern Ornithology from Belon to Buffon", in Enenkel K. – Smith P.J. (eds.), *Early Modern Zoology. The Construction of Animals in Science, Literature and the Visual Arts*, 2 vols., Intersections 7,1 (Leiden et al.: 2007) vol. I, 75–117.

Through the transformation into a star constellation, the crane finally became an actual device by which ships could navigate.

Hybrids: The Crane-Men

In the early modern period, humans were not only exhorted to follow the example of cranes, but were also supposed to be transformable into cranes. This could be accomplished optically, by means of a distorting mirror, as Athanasius Kircher explained in his *Ars magna lucis et umbrae* (1646). He instructed the reader on how to achieve a crane's neck:

> Take a mirror and shape it cylindrically close to the bottom [...]. If you look straight at yourself, you will see your face change into the head of a crane and your neck will become very long. If you look at it obliquely there will be a stream of water with a crag or a rhinoceros horn that protrudes out of your head.[49]

But this metamorphosis was not restricted to catoptrics, since already Aelian had it that cranes actually transformed into humans at the end of their life cycle.[50] This process was not necessarily one-directional, as Athanasius Kircher asserted, that man could be transmuted into any possible form.[51] Complete transformation was just as possible as a half-way process that led to the emergence of hybrids. In his *Monstrorum historia* (Bologna 1642), Aldrovandi discussed and illustrated a variety of animal human hybrids, such as the fish-man,[52] the ape-man,[53] the worm-man,[54] the bear-man[55] and the satyr.[56] Naturally, also a crane-man made its appearance among these

49 'Hoc enim speculum si recta aspexeris, faciem tuam in Gruis caput, et collo longissimo mutatam videbis; si obliquè id inspexeris, flumen cum crepidine, vel etiam facies cornu Rhinocerotis ex fronte excrescente, deturpabitur.' Kircher Athansius, *Ars magna lucis et umbrae* (Amsterdam: J. Jansson van Waesberge and Elizeus Weyerstraeten, Widow: 1671) 786.

50 Lanata G., "Credenze e saperi sugli animali", *I quaderni del ramo d'oro*, 3 (2000) (7–38) 20.

51 'Uno verbo, nullum mostrum tam turpe est, sub cujus formâ te in speculo simili industriâ adornato non respicias'. Kircher, *Ars magna* 786.

52 Aldrovandi Ulisse, *Monstrorum historia* (Bologna, Nicolò Tebaldini and Marco Antonio Bernia: 1642) 27.

53 Ibidem 17.

54 Ibidem 219.

55 Ibidem 581.

56 Ibidem 23.

remarkable monstrosities [Fig. 11.5].[57] Aldrovandi considered them primarily in terms of medical aspects. Hence, his illustration of the crane-man might be understood as depicting a bodily abnormality, the goiter. This is also implied in another (symbolic) reference to the crane made by Ripa, who laconically remarked: 'Gru, simbolo della gola.'[58] While Ripa's *Iconologia* gave no illustration, this could be found in several vernacular translations of Alciato's *Liber emblematum*. Emblem 91 showed under the motto 'Gola' a man with a crane's neck. The epigram said: 'He has a swollen belly and the neck of a crane; and in his hands he carries two greedy birds—this man who eats and devours all his fortune, and shortens his life. Thus do gluttons, who are never satisfied

FIGURE 11.5 *Woodcut illustration to Aldrovandi's* Monstrorum historia *(Bologna, Nicolò Tebaldini and Marcantonio Bernia: 1642) 14.*

57 Ibidem 14.

58 Ripa, *Iconologia* 564.

and never have enough, behave.'[59] But it was not Aldrovandi's sole interest to report on medically astonishing deformations like 'monstrous' ape-men and women, such as the famous Gonzales family, who were actually afflicted by a rare disease (*hypertrichosis*).[60] With his long list of animal human hybrids, Aldrovandi also partook in a utopian tradition. Many of his hybrids appeared as different peoples in the early modern utopia most obsessed with hybrids: Margaret Cavendish's *The Description of a New World, Called the Blazing World* (1666). Almost all hybrid creatures mentioned and depicted in Aldrovandi's book of monstrosities emerged here as different strata of the utopian society:

> The rest of the inhabitants of that world, were men of several different sorts, shapes, figures, dispositions, and humours, as I have already made mention heretofore; some were bear-men, some worm-men, some fish- or mear-men [mermen], otherwise called syrens, some bird-men, some fly-men, some fox-men, some ape-men, some jackdaw-men, some magpie-men, some parrot-men, some satyrs, some giants, and many more, which I cannot all remember [...].[61]

Cavendish's utopian society was professionally highly specialized, as the narrator somewhat humorously emphasized:

> The bear-men were to be her [the Empress's] experimental philosophers, the bird-men her astronomers, the fly-, worm- and fish-men her natural philosophers [...], the satyrs her Galenic physicians, the fox-men her politicians, the spider- and lice-men her mathematicians, the jackdaw-, magpie- and parrot-men her orators and logicians [...].[62]

---

59 "Ha il ventre gonfio, & ha il collo di Grue,
E ne le man duo ingordi uccelli porta
Questo, che tutte le sustanze sue
Mangia e divora, e la sua vita accorta.
Cosi fanno i Golosi, che giamai
Satij non sono, e lor non basta assai". Reference is made to the 1551 Lyon edition. Cf. Alciato Andrea, *Index emblematicus. Emblems in translations*, ed. P.M. Daly et al. (Toronto: 1985) n. pag., emblem 91.

60 For the fate of Pedro Gonzales cf. Zapperi R., *Der wilde Mann von Teneriffa. Die wundersame Geschichte des Pedro Gonzales und seiner Kinder* (Munich: 2004).

61 Cavendish Margaret, *The Description of a New World, Called The Blazing World*, in Cavendish Margaret, *Political Writings*, ed. S. James (Cambridge et al.: 2003) 17.

62 Ibidem 17.

Although the bird-men described by Cavendish did not particularly resemble cranes, but rather geese, they shared qualities which were usually attributed to them in the symbolic, emblematic and zoological literature. Thus, the bird-men were not only endowed with astronomical knowledge, but were also excellent navigators, who had the gift of prevision:

> [...] very good navigators they were, and though they had no knowledge of the loadstone, or needle, or pendulous watches, yet [...] they had subtle observations, and great practice; [...] besides, they were excellent augurers, which skill they counted more necessary and beneficial than the use of compasses, cards, watches, and the like.[63]

The idea that cranes represented astronomical skills was based on information given already by Horapollo, who reported that the crane was a symbol of star-observers in Egypt.[64]

Crane-men were not solely introduced to utopian writing by Cavendish, but appeared much earlier in travel writing. Already in the thirteenth century, on his oriental travels, the protagonist of the epic romance *Herzog Ernst*[65] encountered a whole population of crane-men who resided in a place called Grippia. Through the reception of *Herzog Ernst*, crane-men became popular and were depicted frequently in the Late Middle Ages. Illustrations of crane-men were also spread through influential prints such as Hartmann Schedel's *World Chronicle* (1493).[66] In the early modern period, crane-men, together with other animal human hybrids, became an issue in books of monstrosities like Aldrovandi's *Monstrorum historia*. Aldrovandi considered these bodily abnormalities mainly from a medical and scientific standpoint. Cavendish shared Aldrovandi's interest in natural philosophy, and the highly hybrid society of her *New World* seems directly inspired by Aldrovandi's book of monstrosities. This points to the typical early modern admixture of symbolic and 'real' animal characteristics. Cavendish's utopian bird-men on the one hand symbolically stood for excellent nautical skills, while real birds such as cranes actually

63 Ibidem 10–11.

64 Cf. Allen, *Star Names* 237.

65 Reference is made to version B which is written around 1210. Cf. Brunner H., "Der König der Kranichschnäbler. Literarische Quellen und Parallelen zu einer Episode des Herzog Ernst", in Brunner H., *Annäherungen* (Berlin: 2008) (21–37) 23.

66 Cf. Brunner, "Kranichschnäbler" 31.

displayed these skills. Equally, according to Zuccolo and Valeriano, cranes represented democratic rule, while on a less symbolic level, during migration, cranes actually behaved highly cooperatively with regard to communication and flight formation. Ultimately, the crane remained an ambivalent creature, both at the concrete and the symbolic level. Some of its characteristics such as vigilance or prudence could be used to justify monarchy. Focusing on the crane's cooperative skills, one could come to the opposite conclusion—that the crane represented democracy. Valeriano, Zuccolo, Aldrovandi and the unknown source of Browne advocated the latter view, not because the crane was doubtlessly a 'democratic' animal, but because these authors politically sided with the democratic aspects of the crane. Since symbols of democracy were rare birds in the early modern period, it is noteworthy that the crane figured prominently as a democratic political animal in the seventeenth century. It is even more fascinating that this interpretation was based on a variety of sources, among them Cassiodorus and Saint Ambrose, who are rarely associated with early modern republican thought.

Bibliography

Alciato Andrea, *Index emblematicus. Emblems in translations*, ed. P.M. Daly et al. (Toronto: 1985).

Aldrovandi Ulisse, *Ornithologiae*, 3 vols. (Bologna, Giovan Battista Bellagamba: 1603), vol. III.

Aldrovandi Ulisse, *Monstrorum historia* (Bologna, Nicolò Tebaldini and Marco Antonio Bernia: 1642).

Allen R.H., *Star Names and Their Meanings* (New York: 1936^2).

Aristotle, *History of Animals*, transl. A.L. Peck. 3 vols. (London – Cambridge, Mass: 1965).

Aristotle, *History of Animals*, transl. D.M. Balme, 3 vols. (London – Cambridge, Mass.:1991),

Aristotle, *Politics*, transl. H. Rackham (London – Cambridge, Mass.: 1967).

Ashworth W.B., "Natural History and the Emblematic World View", in Ellenius A. (ed.), *The Natural Sciences and the Arts. Aspects of Interaction from the Renaissance to the 20th Century* (Uppsala: 1985) 303–332.

Bayer Johann, *Uranometria* (Augsburg, Christoph Mang: 1603).

Boas G., "Introduction", in Horapollo, *The Hieroglyphics*, transl. George Boas (New York: 1950) 13–54.

Bodin Jean, *The Six Bookes of a Common-Weale*, transl. E. Knolles (London, Adam Islip: 1606).

Browne Thomas, *Pseudodoxia epidemica: Or, Enquiries Into Very many Received Tenents, And commonly Presumed Truths* (London, Edward Dod: 1658).

Brunner H., "Der König der Kranichschnäbler. Literarische Quellen und Parallelen zu einer Episode des Herzog Ernst", in Brunner H., *Annäherungen* (Berlin: 2008) 21–37.

Cassiodorus, *The Letters of Cassiodor (Variae Epistolae) Being a Condensed Translation of the Variae Epistolae of Magnus Aurelius Cassiodorus Senator*, trans. Hodgkin T. (London: 1886).

Cassiodorus, *Variarum libri XII*, ed. Å.J. Fridh – J.W. Halporn (Turnholt: 1973).

Cavendish Margaret, *Political Writings*, ed. S. James (Cambridge et al.: 2003).

Erasmus Desiderius, *The Education of a Christian Prince*, transl. L.K. Born (New York: 1936).

Halley Edmund, *Catalogus stellarum Australium sive Supplementum Catalogi Tychonici* (London, Thomas James – R. Harford: 1679).

Harms W., "On Natural History and Emblematics in the 16th Century", in Ellenius A. (ed.), *The Natural Sciences and the Arts. Aspects of Interaction from the Renaissance to the 20th Century* (Uppsala: 1985) 67–83.

Hevelius Johannes, *Firmamentum Sobiescianum sive Uranographia* (Gdansk, Johann Zacharias Stolle: 1690).

Horn E. – Gisi L.M. (eds.), *Schwärme—Kollektive ohne Zentrum. Eine Wissensgeschichte zwischen Leben und Information* (Bielefeld: 2009).

Kantorowicz E.H., *Die zwei Körper des Königs. Eine Studie zur politischen Theologie des Mittelalters* (Munich: 1990).

Kircher Athansius, *Ars magna lucis et umbrae* (Amsterdam: J. Jansson van Waesberge & Elizeus Weyerstraeten, Widow: 1671).

Lanata G., "Credenze e saperi sugli animali", *I quaderni del ramo d'oro* 3 (2000) 7–38.

Picinelli Filippo, *Mondo simbolico* (Milan, Francesco Mognaga: 1653).

Pissavino P.C., *Lodovico Zuccolo: Dall'audizione a corte alla politica* (Florence: 1984).

Pliny, *Natural History*, transl. T.E. Page et al., 10 vols. (London – Cambridge, Mass.: 1967).

Ripa Cesare, *Iconologia* (Venice, Cristoforo Tomasini: 1645).

Smith P.J., "On Toucans and Hornbills: Reading in Early Modern Ornithology from Belon to Buffon", in Enenkel K. – Smith P.J. (eds.), *Early Modern Zoology. The Construction of Animals in Science, Literature and the Visual Arts*, 2 vols., Intersections 7 (Leiden – Boston: 2007) vol. I, 75–117.

Valeriano Giovanni Pierio, *Hieroglyphica sive De sacris Aegyptiorum aliarumq[ue] gentium literis* [...] (Basel, Leo Curio: 1575).

Valeriano Giovanni Pierio, *Les hiéroglyphes*, transl. J. de Montlyard (Lyon, Paul Frellon: 1615).

Voltaire, *Dictionnaire philosophique. La raison par alphabet*, 2 vols. (Geneva: Gabriel Cramer: 1769).

Von der Heiden A. – Vogl J. (eds.), *Politische Zoologie* (Zurich – Berlin: 2007).

Zapperi R., *Der wilde Mann von Teneriffa. Die wundersame Geschichte des Pedro Gonzales und seiner Kinder* (Munich: 2004).

Zuccolo Ludovico, *Considerationi Politiche, e Morali sopra Cento Oracoli d'Illustri Personaggi antichi* (Venice, Marco Ginami: 1621).

CHAPTER 12

Tierallegorie als ein Mittel der Fürstenerziehung. Die *Theriobulia* des böhmischen Humanisten Johannes Dubravius *With an English Summary*

Alexander Loose

Nach Bruno Singer ist es für die Fürstenspiegel aus der Zeit des Humanismus kennzeichnend, dass gegenüber den mittelalterlichen Fürstenspiegeln staatstheoretische Betrachtungen zurücktreten, die Herrschertheologie ersatzlos ausfällt, stattdessen die Tugendlehre beträchtlich ausgeweitet und die Geschichte als eine den Wechselfällen des Lebens am besten gerecht werdende *Magistra vitae* hochgeschätzt wird.[1] Anders als bei den Fürstenspiegeln des Hochmittelalters, die in streng systematischer Form eine Fülle staatstheoretischer Themen wie z.B. die Hierarchien- und Ständelehre, das Naturrecht, das Lehnsrecht, das Widerstandsrecht, den Tyrannenmord, Verfassungsfragen u. dgl. abhandelten,[2] sind die Fürstenspiegel der Humanisten, beginnend mit Petrarcas Brief an Francesco Carrara, sehr oft als eine lockere Folge von moralischen Ermahnungen sowie Klugheits- und Anstandsregeln konzipiert, durch die die jeweiligen *viri litterati* auf ihren Regenten Einfluss zu nehmen sich berufen fühlten. An die Stelle einer distanziert nüchternen und sorgfältig gegliederten theoretischen Abhandlung, wie sie für die Fürstenspiegel des Hochmittelalters typisch ist, tritt bei Ihnen die eindringlich mahnende und eher assoziativ strukturierte persönliche Ansprache an den Adressaten. Sie griffen dabei auf eine von Isokrates durch sein Lehrschreiben an den zypriotischen Königssohn Nikokles geprägte Form des Fürstenspiegels als einer persönlichen Mahnrede zurück und wählten dabei sehr oft den Brief als die ihrem Anliegen angemessenste literarische Gattung.[3]

1 Vgl. Singer B., *Der Fürstenspiegel in Deutschland im Zeitalter des Humanismus und der Reformation* (München: 1981) 28 und 34.

2 Vgl. Singer B., Art. „Fürstenspiegel“, in: *Theologische Realenzyklopädie*, Bd. 11 (Berlin: 1993) 708.

3 Vgl. z.B. Petrarca, Seniles XIV,1 (1373); Piccolomini, *De liberorum educatione ad Ladislaum regem* (1450); Pontano, *De principe* (1468); Bohuslaw Hassenstein, *De re publica* 3 (1499).

 | DOI 10.1163/9789004279179_014

Mit der *Theriobulia*, einem polymetrischen Gedicht, das der böhmische Humanist Johannes Dubravius[4] im Jahre 1518 für Ludwig II., den damals gerade erst zwölfjährigen König von Böhmen und Ungarn, verfasst hat, liegt uns ein literarisches Werk vor, das in mancherlei Hinsicht einen durchaus typischen, in einer besonderen Hinsicht aber einen einzigartigen humanistischen Fürstenspiegel darstellt. So wird man wie in anderen humanistischen Fürstenspiegeln auch in der *Theriobulia* ausführlichere staatstheoretische Betrachtungen oder eine Herrschertheologie nach der Art mittelalterlicher Fürstenspiegel vergeblich suchen. Überhaupt werden in ihr Fragen der Religion zugunsten einer Beschränkung auf das Diesseits nur selten und dann sehr knapp thematisiert. Stattdessen dominiert in ihr überall die moralische Unterweisung, die Mahnung zur Tugend und die Warnung vor dem Laster. Dass Dubravius nach Humanistenart zudem der Geschichte eine wichtige Funktion für die Bildung des Fürsten zumaß, wird sowohl durch die überaus häufige Verwendung historischer Exempel als auch eine ausdrückliche Forderung deutlich, mit der er den König zum Studium der Geschichte als einer *disciplina regia* ermuntert.[5] Neben diesen Charakteristika, in denen die *Theriobulia* anderen Fürstenspiegeln ihrer Zeit sehr ähnlich ist, besteht ihre Besonderheit, ja Einzigartigkeit in ihrer literarischen Form: Die mahnend eindringliche Ansprache an den Regenten wird im Fürstenspiegel des Dubravius in die literarische Fiktion eines Tierparlamentes gekleidet, in dem verschiedene Tiere nacheinander und scheinbar frei assozierend ihrem König dem

4 Zur Biographie dieses vornehmlich wegen seiner *Historia Bohemica* und einer Abhandlung über Fischteiche bekannten Autors vgl. Skutil J., Jan Dubravius. Biskup, státník, ekonom a literát (Kroměříž: 1992).

5 Vgl. *Theriobulia* 1,95–108: ‚Obibis autem doctrinae id genus / easque disces litteras, / quae principum regumque et gentium omnium / cum moribus facta adnotant. / Miraberis Bruti legens constantiam / Fabiique felicem moram, / duri Catonis durius sectae genus / et fata vinci nescia. / Occurret et bello Camillus utilis / animaeque Paulus prodigus, / quid ponte Cocles vel quid igne Mucius / fecerit ob ipsam patriam. / Ea perleges ac disces multa eiusmodi / exempla vitae congrua‘ („Du sollst aber diejenige Art der Unterweisung in Angriff nehmen und diejenigen Wissenschaften lernen, die die Sitten und Taten aller Fürsten, Könige und Geschlechter schriftlich vermerken. Bewundern wirst du beim Lesen die Standhaftigkeit des Brutus, das glückliche Zaudern des Fabius, die harte Lebensweise des harten Cato und das Schicksal, das nicht besiegt werden kann. Begegnen wird auch der zum Krieg taugliche Camillus, der sein Leben aufopfernde Paulus, was mit der Brücke Cocles oder was mit dem Feuer Mutius fürs Vaterland getan hat. Durchlesen und kennenlernen sollst du diese und viele ähnliche mit dem Leben übereinstimmende Beispielgeschichten“).

Löwen Ratschläge unterbreiten, durch die dieser in der rechten Lebens- und Regierungsführung unterwiesen werden soll.

Im Proöm seines Fürstenspiegels erklärt Dubravius, dass er sich bei seinem Dichten auf ein in seiner Muttersprache verfasstes Werk gestützt habe. Es handelt sich bei diesem von Dubravius nur unbestimmt erwähnten *libellus patrio sermone scriptus* um die *Nová Rada*, ein Tierparlament, das der böhmische Edelmann Smil Flaschka von Pardubitz gegen Ende des 14. Jahrhunderts in Anlehnung an die spätmittelalterliche Gattung der Vogelparlamente[6] in alttschechischer Sprache gedichtet hat. Vogelparlamente sind lehrhafte, in Paarreimen verfasste Gedichte, in denen ein Vogelkönig (oft ein Zaunkönig) den Rat der vor ihm versammelten Vögel erbittet und daraufhin von diesen darüber belehrt wird, wie er sein Leben zu führen und seine Herrschaft auszuüben habe. Ratschläge tugend- und lasterhafter Vögel pflegen dabei einander so abzuwechseln, dass es unter den Ratgebern nicht selten zum direkten Widerstreit zwischen vorbildlichen und verwerflichen Vogelcharakteren kommt. In Übereinstimmung mit diesen Gattungsmerkmalen ist auch die Handlung der *Nová Rada* äußerst knapp und daher schnell erzählt: Ein junger, eben erst auf den Thron gelangter König Löwe schickt Boten aus mit der Order, dass sich ein jeder seiner Untertanen, er sei Vogel oder Vierfüßer, an den Königshof zu verfügen habe. Der eifrig herbeigeeilten Schar eröffnet er dort den Zweck ihres Kommens. Da er noch jung und unerfahren sei, möge ihn ein jedes Tier nach seiner jeweiligen Veranlagung und Natur in der rechten Lebens- und Regierungsführung unterweisen. Nach dieser kurzen Setzung der Szene halten, jeweils einander abwechselnd, die Vögel und Vierfüßer ihre lehrhaften Mahnreden.

Die Grundkonzeption seines Tierparlaments verdankt Dubravius zwar seinem Landsmann Smil Flaschka von Pardubitz. Er nimmt jedoch gegenüber seiner Vorlage zahlreiche Veränderungen vor, die sein Gedicht zu weit mehr machen als einer bloßen Übersetzung. Einige der bei Smil ratenden Tiere ließ er weg, andere fügte er hinzu, wieder andere behielt er zwar bei, ließ sie aber einen vollständig anderen Rat vorbringen. Die Folge der Ratschläge unterteilt er nach den vier Kardinaltugenden in vier Hauptteile und gibt ihr einen knappen erzählerischen Rahmen. Als Humanist und *vir litteratus* überarbeitete er seine volkstümliche Vorlage unter Verwendung vieler antiker griechischer und latei-

6 Eine Zusammenstellung des Überlieferungsmaterials findet man bei Busch P., *Die Vogelparlamente und Vogelsprachen in der deutschen Literatur des späten Mittelalters und in der frühen Neuzeit. Zusammenstellung und Beschreibung des Textmaterials—Rekonstruktion der Überlieferungsgeschichte—Untersuchungen zu Genese, Ästhetik und Gebrauchsfunktion der Gattung*, Beihefte zu Poetica 24 (München: 2001) 57–75.

nischer, aber auch neulateinischer Autoren und gab ihr dadurch das Gepräge einer humanistischen Gelehrtendichtung. Durch den genauen Vergleich mit seinen vielfältigen Quellen kann der Schaffensprozess, an dessen Ende die *Theriobulia* als fertiges Gedicht steht, wie folgt rekonstruiert werden:[7] Zunächst ließ sich Dubravius von einigen wenigen Leittexten das zu behandelnde Thema, d.h. die zu behandelnde Tugend bzw. das zu behandelnde Laster, vorgeben; er folgte hierbei einerseits seiner muttersprachlichen Vorlage, andererseits aber auch dem vierten Buch der Nikomachischen Ethik des Aristoteles sowie einem lateinischen Fürstenspiegel seines Landsmannes Bohuslaw Hassenstein von Lobkowitz und dem Fürstenspiegel *De principe* des italienischen Humanisten Giovanni Pontano. Anschließend suchte er zu jedem vorgegebenen Thema ein Tier, zu dessen Charakter die jeweilige Mahnrede in eine bedeutungsvolle Beziehung gesetzt werden konnte. Sodann zog er zur Behandlung des vorgegebenen Themas die Schriften weiterer Autoren heran und griff hierbei vor allem auf Werke Ciceros, Senecas und des Erasmus zurück. Daraufhin fügte er zur Veranschaulichung und zur Steigerung der Überzeugungskraft Exempel hinzu, die er vor allem aus Suetons Kaiserbiographien, der *Historia Augusta* und den Schriften Plutarchs entlehnt hat. Schließlich brachte er die übernommenen Textpassagen in die metrische Form verschiedener antiker Versmaße und versetzte sie mit lateinischen Dichterzitaten, wobei er eine besondere Vorliebe für Plautus, Horaz und Seneca tragicus erkennen lässt.

In diesem Aufsatz soll nur jener Arbeitsschritt untersucht werden, durch den Dubravius einem ihm vorliegenden Thema ein passendes Tier zugeordnet hat. Indem Dubravius jeden einzelnen Ratschlag stets in eine deutlich erkennbare Beziehung zu einer bestimmten, dem jeweils ratschlagenden Tier unterstellten Eigenschaft setzt, erhebt er jedes der auftretenden Tiere zu einem Inbegriff der Tugend, zu der es mahnt, bzw. des Lasters, vor dem es warnt. Er stützt sich dabei in der Regel auf eine der folgenden drei Quellen. Erstens übernimmt er einige Tiercharaktere seiner muttersprachlichen Vorlage, in der bereits verschiedene Überlieferungen wie Fabeltraditionen, der *Physiologus* und Vogelparlamente aufgegriffen und bearbeitet sind. Zweitens macht er umfangreichen Gebrauch von den *Adagia* des Erasmus, indem er seine Tiere wiederholt Sprichwörter, die sie jeweils selbst betreffen, in Beziehung auf das von ihnen behandelte Thema auslegen lässt. Drittens stützt er sich auf die *Historia naturalis* des älteren Plinius, indem er die dort überlieferte Tierbeschreibung zum Ausgangspunkt und gewissermaßen zur emblematischen Zentrierung der Rede einiger seiner Tiere macht.

---

7 Eine detaillierte Quellenanalyse findet sich bei Loose A. (Hrsg.), *Johannes Dubravius, Theriobulia*, Bd. 1, (Hildesheim: 2011) XIX–LVI.

Die Charakterisierung der Tiere nach der *Nová Rada*

Dubravius übernimmt die allegorische Deutung einiger seiner Tiere aus Smils Dichtung. So rät bei ihm, ebenso wie bei Smil, der Adler zur Frömmigkeit, die Taube gegen den Zorn, das Lamm zur Demut, der Hahn als Weckrufer zur Wachsamkeit, der Hund als Jagdhund zur Jagd und als Wachhund zur Sicherheit des Königs, die keusche Turteltaube zur Keuschheit, der Pfau zur rechten Kleidung und der Packesel zum geduldigen Ertragen von Lasten und Mühen. Wenngleich in der *Theriobulia* die Ratschläge dieser Tiere anders akzentuiert werden, so ist doch Dubravius, was die Tierallegorie betrifft, in den meisten dieser Fälle lediglich ein Nachahmer seiner Vorlage. Im Folgenden sei nur auf diejenigen Fälle hingewiesen, in denen Dubravius als ein eigenständiger Dichter in Erscheinung tritt, indem er gegenüber seiner Vorlage den Bezug zwischen Tiercharakter und vorgebrachten Ratschlag verändert.

Diejenigen Tiere, denen man nach alter Fabeltradition einen moralisch verwerflichen Charakter zuzuschreiben pflegt, bringen in der *Nová Rada* verwerfliche und in ihrem Aufforderungscharakter ironisch zu verstehende Ratschläge vor.[8] Gemäß der für einen Fürstenspiegel typischen direkten Unterweisung tilgt Dubravius die Ironie seiner Vorlage und verkehrt den Ratschlag eines jeden dieser Tiere in sein Gegenteil. So wird in der *Nová Rada* etwa der Wolf in altbekannter Fabelmanier als ein gieriges Raubtier vorgestellt: Er rät dem Löwen zu gewaltsamer Selbstbereicherung und Völlerei und bietet ihm zu diesem Zweck seine Dienste an.[9] In Übereinstimmung hierzu schildert auch der Wolf des Dubravius sich selbst ausführlich als einen Wolfscharakter, Räuber und Gewalttäter:

> Nostrum est nova invenire vectigalia,
> tributa non audita, nec tolerabiles
> collationes pauperum gementium.
> Nostrum impias admoliri sacris manus,

8 Zur Ironie im Werk des Smil Flaschka von Pardubitz vgl. Čapek J.B., „Die Ironie des Smil Flaška", *Slavische Rundschau* 10 (1938) 68–79.

9 Smil Flaška z Pardubic, *Nová Rada* 708–715: „i co tobě, králi, škodí, / na svú moc se rozpomena, / že mimo tě nenie pána, / muóžeš, což chceš, komu vzéti. / Pakli mně ráčíš popřieti / krávy i ovce na dvořě, / vydruť, což jest i v komořě. / Naleznuť já cestu k tomu" („Was kann dir, o König, schaden? Wenn du dich an deine Macht erinnerst und daran, dass es außer dir keinen Herrn gibt, dann kannst du, was du willst, wem du willst, wegnehmen. Wenn du mir dann auf einem Gehöft Kühe und Schafe zu gönnen geruhst, werde ich auch herausschinden, was in der Kammer ist. Ich werde dazu einen Weg finden"). Aus der *Nová Rada* wird hier und im Folgenden zitiert nach Daňhelka J. (Hrsg.), *Smil Flaška z Pardubic, Nová Rada* (Prag: 1950).

simulacra facta auro conflare donaque
auferre in illis et thesauros maximos.
Nostrum potestates, honores vendere,
damno esse gentibus, regnis, provinciis,
nostrum subinde facta infecta reddere.[10]

Unsere Aufgabe ist es, neue Staatseinkünfte zu erfinden, unerhörte Steuern und ächzender Armeleute unerträgliche Geldentrichtungen. Unsere Aufgabe ist es, frevlerische Hände nach den Heiligtümern auszustrecken, in ihnen goldene Bildnisse einzuschmelzen, Weihgaben und riesige Schätze fortzuschleppen. Unsere Aufgabe ist es, Machtpositionen und Ehrenämter zu verkaufen und den Völkern, Königreichen und Provinzen zum Schaden zu gereichen. Unsere Aufgabe ist es, Getanes sogleich wieder ungetan zu machen.

Nachdem der Wolf des Dubravius die verderbenbringenden Werke eines Wolfes aufgezählt und sich seiner Taten gerühmt hat, mahnt er in plötzlichem Gesinnungswandel und gegen seine Natur, der Löwe solle nicht ebenso in verruchter Wolfsmanier ruhmreich sein wollen:

Quorsum ista? Ne Leonis nomen regium
sanctumque moribus Lupi inquinaveris.[11]

Warum erzähle ich das? Besudele den königlichen und heiligen Namen des Löwen nicht durch die Sitten eines Wolfes.

Desgleichen tritt in Vogelparlamenten oft auch der Geier als ein gieriges Tier auf, das zur Selbstbereicherung rät.[12] Ist nach dieser Tradition auch bei Smil von Pardubitz der Geier gierig und lüstern nach jedem Kadaver und rät er dementsprechend, dass der König Erbschaften an sich reißen, Witwen und Waisen der ererbten Güter berauben solle,[13] so stellt sich bei Dubravius der

---

10 *Theriobulia* 1,611–619.

11 *Theriobulia* 1,620–621.

12 Vgl. z.B. das Vogelparlament der Stuttgarter Handschrift Württemb. LB, *cod. med. et phys.* 30, 203r–v (15. Jhd.): „Der Gire: Herre, du sollt nemen waz man dir git, / gloube (*l.* Gelobe) wol, gip nieman nüt, / ahte nüt waz man von dir klage, / durch daz man dir daz guot zuo trage, / da mit sich din schatz gemeret, / als mich mein vernunft leret“. Zitiert nach Pfeiffer F., „Das Märchen vom Zaunkönig“, in: *Germania. Vierteljahrsschrift für deutsche Alterthumskunde* 6 (1861) 80–106, hier 84.

13 Vgl. Smil Flaška z Pardubic, *Nová Rada* 747–774.

Geier zwar ebenfalls als ein Aasfresser vor, um dann aber doch in tugendhafter Inkonsequenz dem Löwen ans Herz zu legen, er solle anders als die *Nerones* der Geschichte immer darauf bedacht sein, dass Erben ihrer Erbschaften gerade *nicht* beraubt würden:

> Tantum autem abest, ut regum facta eiusmodi
> ego approbem, quamquam meum
> inter rapaces ponam nomen alites,
> ut censeam dignissimos,
> qui dicerentur praedones, non principes.
> Tu rex futurus cogita,
> dum solus ipse hereditate indebita
> libenter abstines, tui
> idem facere condiscant.[14]

> Wenngleich ich mich selbst zu den Raubvögeln zähle, liegt es mir dennoch so fern, derartige Taten der Könige zu billigen, dass ich meine, sie seien überaus würdig, ‚Räuber', nicht ‚Fürsten' genannt zu werden. Solange nur du selbst dich einer dir nicht zustehenden Erbschaft gern enthältst, sinne als künftiger König auch darauf, dass die Deinen dasselbe zu tun lernen.

Nach der vor allem aus dem Tierepos bekannten Charakterisierung des Fuchses tritt in der *Nová Rada* ferner der Fuchs als ein verschlagener Schmeichler auf, der den Löwen intrigant von seinen königlichen Ratgebern entfremden möchte.[15] Die bei Smil ironisch zu verstehende Mahnung, dass sich der Löwe dem Einfluss seiner Ratgeber eigenwillig entziehen solle, hat Dubravius unter Tilgung aller Ironie zu der direkten und ernsten Ermahnung geändert, dass der

---

14 Vgl. *Theriobulia* 1,565–573.

15 Smil Flaška z Pardubic, *Nová Rada* 1388–1401: „Protož cožť jest třěba rady, / by jim sě tak chtěl poddati / a jich rady poslúchati? / Neb kniežata, mocní, páni, / ježtoť jsú k radě vydáni, / cožť nepodobného zvědie, / i v uočiť pravdu povědie, / uchopieť tě po své vóli, / čehožť se uradie koli, / chtiec, aby poslúchal rady. / A tys člověk bujný, mladý, / neroď sě tak zavázati, / daj sě jim, což chtie, tázati, / a ty buď živ k svéj libosti" („Was bedarf es also einer Ratsversammlung? Willst du dich ihnen etwa unterwerfen und ihrem Rat gehorchen? Denn die Fürsten und mächtigen Herren, die zum Ratschlagen erschienen sind und die Wahrheit offen ausplaudern, wenn sie etwas Unschönes erfahren haben, krallen dich nach ihrem Willen und wollen, dass du, was immer sie auch beschlossen haben, ihrem Ratschlag gehorchst. Doch du bist ein junger, ungestümer Mensch. Lass dich nicht so einschnüren, lass sie fragen, was sie wollen, und lebe zu deinem Vergnügen").

Löwe verständigen und rechtsgelehrten Ratgebern ein offenes Ohr leihen solle. Er lässt diese Mahnung allerdings nicht seinen Fuchs, sondern einen eigens dafür eingeführten Fasan vorbringen, dessen Rede er der Rede seines Fuchses unmittelbar voranstellt.[16] Den Fuchs stellt Dubravius nach seiner muttersprachlichen Vorlage zwar ebenfalls als einen verschlagenen Gauner vor, der andere Tiere betrügerisch hinters Licht führt. Der Ratschlag, den er seinem Fuchs in den Mund legt, ist jedoch nicht der eines Betrügers, sondern verrät in jeder Hinsicht die ernste Sorge um das Wohl des Königs. Während Smils Fuchs dem Löwen tyrannischen Eigensinn nahelegt, fordert der Fuchs des Dubravius den Löwen zu selbstbestimmter und entschiedener Herrschaftsausübung auf. Smils Satire, in der angeprangert wird, dass sich der König aufgrund der Einflüsterungen intriganter Emporkömmlinge dem heilsamen Einfluss seiner Ratgeber entziehe, wird so zu der Paränese umgeformt, dass der König unter Ausspielung seiner Machtbefugnisse den unheilvollen Ränken eigenmächtiger Untertanen ein Ende machen solle:

> Sed nec Leoni pectore simplici
> hic consulens: „Cum tanta potentia
> et tantae opes rebusque agendis
> ingenium validum tibi adsit,
> o rex“, ait, „quo, quo imperium tibi,
> semper tuorum consilio regi
> si mavoles passus quod illa,
> flumine quae fluitant: Feruntur,
> quippe haud eunt. Tu te tibi vindica
> regnique partem suscipias tui
> per te regendam, scilicet sunt
> propria multa tibi“.[17]

Indem es aber hier nicht einmal dem Löwen mit arglosem Sinn riet, sprach es: „Wozu, o König, wozu ist dir die Herrschaft gegeben, wenn du, obwohl dir so große Macht, so große Mittel und für Taten eine starke Natur zur Verfügung stehen, stets lieber vom Rat der Deinen wirst regiert werden wollen und dabei das erleidest, was im Fluss das Treibgut

16 Vgl. *Theriobulia* 1,849–880; zur Charakterisierung des Fasans s.u.

17 *Theriobulia* 1,890–900. Dubravius verwendet in diesen Versen Passagen, die er Senecas Briefen an Licinius entnommen hat, vgl. Seneca, *Epistulae ad Lucilium* 23,8: „Pauci sunt qui consilio se suaque disponant: ceteri, eorum more quae fluminibus innatant, non eunt sed feruntur“; ebd. 1,1: „Ita fac, mi Lucili: vindica te tibi“.

> erleidet: Es wird fortgetragen, denn es geht ja nicht (selbst). Erhebe du für dich deinen Anspruch auf dich selbst und nimm auf dich den von dir auszuübenden Teil deiner Herrschaft, es gibt nämlich vieles, was ausschließlich dir eigentümlich ist".

Wie man an den angeführten Beispielen erkennen kann, hatte Dubravius für die in einer Tierdichtung typischerweise begegnende Satire und Ironie keine Verwendung. Aufgrund der Erfordernisse, die ihm durch die Gattung des Fürstenspiegels vorgegeben waren, musste er alle ironisch und satirisch gemeinten Ratschläge seiner Vorlage in das direkte, ohne Hintersinn zu verstehende Gegenteil verkehren. Smil von Pardubitz schrieb als ein Gegner Wenzels IV. nicht an diesen König, sondern über ihn.[18] In den direkt lehrhaften und nichtironischen Ratschlägen seines Gedichtes entwirft er das Bild eines idealen böhmischen Königs; in den ironischen Ratschlägen spielt er satirisch auf den König seiner Zeit an. Sein Werk ist daher durch den Kontrast von Ideal und Wirklichkeit, durch das Gegeneinander guter und böser Kräfte gekennzeichnet, und die Wahl der literarischen Form eines Tierparlamentes war seinen Aussageabsichten durchaus angemessen. Die Situation, in der Dubravius schreibt, ist eine ganz andere. Dubravius wendet sich als ein loyaler Untertan an seinen König mit der Absicht, ihn durch seine Schrift in der Lebens- und Regierungskunst zu unterweisen. Nicht nur die Rücksicht auf den von der Etikette verlangten Ton, auch die Rücksicht auf das zarte Alter und die vermutlich naive Auffassungsart seines Adressaten, eines Kindes von zwölf Jahren, verboten ihm jede Zweideutigkeit und Ironie. Indem er für seinen Fürstenspiegel die auch ein Kind ansprechende Einkleidung in die Form eines Tierparlamentes wählte, musste er allen Tiercharakteren, auch den nach verbreiteter Auffassung zwielichtigen und moralisch verwerflichen, einen Rat zuordnen, der ohne Hintersinn vor dem Laster warnt und zur Tugend mahnt. Die ironische Königs- und Gesellschaftskritik, die in Smils Tierparlament eine wichtige Rolle spielt, tritt daher bei Dubravius vollständig hinter die unmissverständliche moralische Unterweisung zurück. Bei Dubravius raten somit auch solche Räuber wie der Wolf, der Geier und der Fuchs gegen ihren Charakter zur Unterlassung und Unterbindung räuberischer Übergriffe. In den Reden dieser Tiere stellt der Übergang vom beschriebenen oder als bekannt vorausgesetzten Tiercharakter zur Paränese geradezu einen Bruch dar, an dem deutlich wird, dass das Motiv eines vom Löwen einberufenen Hoftags, in dem alle, auch die zwielichtigen Tiere geladen sind, mit der für Fürstenspiegel gemeinhin üblichen direkten Unterweisung und Paränese unvereinbar ist.

18 Zur Begründung dieser These vgl. Loose (Anm. 7) XXII–XXVII.

Zwar keine Verkehrung, so aber doch eine bemerkenswerte Verkürzung und Einschränkung der traditionellen allegorischen Deutung eines Tiercharakters lässt sich bei Dubravius in der Rede des Adlers beobachten. Im *Physiologus* wird die Ansicht vertreten, dass ein Adler, wenn er altert, hoch emporfliege, um dort im Feuer der Sonne, das für den Heiligen Geist steht, verjüngt zu werden.[19] Als dieses vom Heiligen Geist durchglühte Tier gedenkt bei Smil der Adler in seinen langen Ausführungen über die Gottesliebe auch des Heiligen Laurentius, in dem das Feuer der Gottesliebe so sehr gebrannt habe, dass er das Feuer, mit dem er gemartert wurde, nicht mehr spürte.[20] Anders als bei Smil von Pardubitz spielt jedoch bei Dubravius die auf den *Physiologus* zurückgehende Tradition, nach der der Adler das Sinnbild des durch Taufe und Heiligen Geist wiedergeborenen Christen sei, keine Rolle. Von den 238 Versen, über die sich bei Smil die Rede des Adlers erstreckt, hat Dubravius gerade einmal ein gutes Dutzend für sein Dichten herangezogen. Während Smil von Pardubitz weitschweifig der Sündigkeit des Menschen, des Kreuzes Jesu, der Erlösung durch Gottes Liebe, des Schreckens ewiger Verdammnis und der Hoffnung auf ewige Seligkeit gedenkt, ist Gott für den Adler des Dubravius vor allem eine Kategorie des irdischen Machterhalts: Er ist der allmächtige Sieger schlechthin, dem der König als dem obersten Machthaber untertänige Ehrerbietung entgegenzubringen habe, wenn er von ihm seinerseits protegiert und in seiner irdischen Machtposition erhalten werden will.[21] Dieser Unterschied zur muttersprachlichen Vorlage lässt sich nicht nur in der Rede des Adlers, sondern in der ganzen *Theriobulia* feststellen: Während Smil in seinem ganzen Gedicht immer wieder auf das Drama von Sünde und Erlösung, von Himmel und Hölle, von Tod und ewigem Leben zu sprechen kommt, beschränkt Dubravius leidenschaftslos und wortkarg alle religiösen Betrachtungen auf ein Minimum, durch das lediglich nominell die Christlichkeit des Verfassers festgestellt werden soll. An mehreren Stellen der *Theriobulia* verrät Dubravius stattdessen eine gewisse Neigung zur heidnischen Antike. Der Adler ist nach heidnisch-antiker Tradition als die *ales magni Iovis*,[22] als der Kriegsknecht eines mächtigen Kriegsgottes charakterisiert. Der Ibis spricht in heidnisch-polytheistischer

---

19 Vgl. *Physiologus latinus*, cap. 8: „De Aquila".

20 Vgl. *Nová Rada* 269–270: „Milost Vavřinci svatému / ta osladila rošt jemu." („Die Liebe versüßte dem Heiligen Laurentius den Rost").

21 Eine detaillierte Analyse der Rede des Adlers findet man bei Loose (Anm. 7) Bd. 2, S. 20–29.

22 Er wird vom Löwen mit folgenden Worten zum Reden aufgefordert: *Tuque ante cunctos, ales o magni Iovis, / in verba rostrum tempestiva solvito!* (*Theriobulia* 1,53–54: „Und öffne zuerst du, o Vogel des großen Jupiter, deinen Schnabel für eine passende Rede!").

Weise vom „Gesetz des Schicksals und den Göttern".[23] Das Kreuz wird nur, und dies gleich zweimal, als Marterinstrument für Verbrecher erwähnt.[24]

Die Charakterisierung der Tiere nach den *Adagia* des Erasmus

Neben den genannten Beispielen, in denen Dubravius die Tierallegorie seiner muttersprachlichen Vorlage lediglich seiner Aussageabsicht angleicht, begegnet in der *Theriobulia* auch eine Reihe von Ratschlägen, in denen er nach eigenen Assoziationen durchaus originell, mitunter jedoch sehr versteckt und konstruiert, Ratschlag und ratschlagendes Tier miteinander verbindet. Wie erwähnt, stützte er sich dabei oft auf Sprichwörter, die er ausnahmslos den *Adagia* des Erasmus entnommen hat. Der folgende Überblick soll einen Eindruck davon vermitteln, wie Dubravius bestimmte Tiere aufgrund von Sprichwörtern zu den Inbegriffen bestimmter Tugenden erhoben hat.

Nachdem in der *Theriobulia* der Büffel zu Milde und Nachsicht in der Behandlung straffälliger Untertanen gemahnt und vor wütender Grausamkeit im Betragen des Königs gewarnt hat, fasst er mit folgenden Worten seine Rede zusammen: „Cave, ut ad facta omnia ne naribus ducaris tamquam Bubalus".[25] Nach einer Auskunft des Erasmus ist die Metapher „naribus trahi" von den Büffeln entlehnt, die man zur Bändigung ihrer Wildheit an einem Nasenring zu führen pflege.[26] Der Löwe soll somit nach dem Ratschlag des Büffels nicht erst mit Gewalt, durch den Widerstand seiner Untertanen, gebändigt und zur Milde gezwungen werden müssen.[27]—Ferner heißt es im Sprichwort vom

23 Vgl. *Theriobulia* 1,137: „lege fatorum divisque urgentibus" („weil das Gesetz des Schicksals und die Götter dazu drängen").

24 Vgl. *Theriobulia* 1,778; 2,602.

25 *Theriobulia* 1,367–368: „Pass auf, dass du zu allen Taten nicht an der Nase gezogen wirst wie ein Büffel".

26 Vgl. Erasmus, *Adagia, chil.* 2,1,19.

27 Vgl. folgende Passage aus der Rede des Büffels (*Theriobulia* 1,324–330): „Porro cum in omnes facilis esse debeat / regis animus, tum fugere praecipue decet, / ne quid temere in amicos consulas tuos, / ne in genere claros neve in factis nobiles. / Praeterquam enim quod inter illustres viros / pulchrum eminere sit, non est tutum quoque / in capita desaevire praepotentium" („Während ferner die Gesinnung des Königs allen gegenüber freundlich sein muss, so schickt es sich doch vor allem, dies zu vermeiden, dass du gegen deine Freunde, die durch Herkunft Berühmten oder die durch Taten Ausgezeichneten leichtfertig etwas beschließt. Abgesehen nämlich davon, dass es schön ist, zwischen erlauchten Männern hervorzuragen, ist es auch nicht sicher, gegen die Häupter der Mächtigen zu wüten").

Ochsen, er schreite, wenn er müde werde, nur desto festeren Schrittes einher: „Bos lassus fortius figit pedem“, und wo jemand beharrlich sich mühe, sage man mit der Wendung „Bos sub iugum“, er arbeite wie ein Ochse unterm Joch.[28] In Übereinstimmung hierzu fordert der Ochse des Dubravius, dass der Löwe all seine Unternehmungen und Amtsgeschäfte mit Beharrlichkeit und Beständigkeit verfolgen solle.[29]—Das Schaf warnt sodann den Löwen vor den sprichwörtlichen „ovium mores“,[30] die man dummen Menschen nachsagt: Er solle genau erwägen, was er sagt, und nur sagen, wovon er etwas versteht, andernfalls werde man ihn als ein in Purpur gekleidetes Schaf, einen prachtvollen Dummkopf, verlachen können.[31]—Unter den *Adagia* des Erasmus findet sich auch das Sprichwort „Formica camelus“ („Ein Kamel—eine Ameise“).[32] Es wird auf die ungleiche Verteilung von Gütern und Lasten angewendet, so dass einer „ein Kamel“, d.h. sehr viel, ein anderer hingegen nur „eine Ameise“, d.h. sehr wenig, zugeteilt bekommt. Dementsprechend spricht

---

28 Vgl. Erasmus, *Adagia, chil.* 1,1,47; 2,6,74.

29 Vgl. folgende Passage aus der Rede des Ochsen (*Theriobulia* 1,724.227–733): „Si me audire voles Bovem / [...] / o rex, non modo vestigia regia / et lento et placido pede / incessuque gravi ponere gesties, / sed constare tibi quoque / ac consistere firmo incipies gradu, / perdurans rigidus tua / mente in proposito, quod semel approbas“ („Wenn du auf mich Ochsen wirst hören wollen, [...] o König, so wirst du nicht nur verlangen, die königlichen Schritte mit langsamem, ruhigem Fuß in würdevoller Gangart zu setzen, sondern auch anheben, konsequent zu sein und festen Schrittes Fuss zu fassen, indem du unerschütterlich in deinem Geist bei dem Vorsatz bleibst, den du ein für allemal gutheißt“).

30 Vgl. Erasmus, *Adagia, chil.* 3,1,95.

31 Vgl. folgende Passage aus der Rede des Schafes (*Theriobulia* 1,817–831): „[...] tacere praestiterit, nisi / meliora possis proloqui / taciturnitate. Regi Alexandro fuit / hic mos, ut iret saepius / in officinam Apellis comitateque / et arte nobilissimi / pictoris oblectatus. Accidit semel, ut disserentem plurima, / sed imperite, commoneret talibus / pictor: „Loqui, oro, desine, / rex magne, nam quem suspexere servuli / nuper tacentem purpuram / aurumque demirati, iidem non temperant / nunc a cachinno, quod super / iis, quae minus novisti, pergis dicere“ („wenn man nichts sagen kann, was besser wäre als Schweigsamkeit, wird es besser sein zu schweigen. König Alexander hatte die Angewohnheit, des öfteren in die Werkstatt des Appelles zu gehen, denn er erfreute sich an der Geselligkeit und der Kunst des vornehmen Malers. Einmal geschah es, dass er über mehrere Dinge, doch ohne Verstand, Erörterungen anstellte und ihn der Maler deswegen mit folgenden Worten ermahnte: ‚Höre bitte zu sprechen auf, großer König, denn dieselben Sklaven, die zu dir aufblickten, als du neulich schwiegst, voll Bewunderung für den Purpur und das Gold, können sich jetzt das Gekicher nicht verkneifen, weil du nicht aufhörst, über Dinge zu reden, die du nicht verstehst‘“). Zum erzählten Exempel vgl. Plinius, *Naturalis historia* XXXV,85–86; Plutarchus, *Moralia* 58d.

32 Vgl. Erasmus, *Adagia, chil.* 1,5,47.

auch das Kamel des Dubravius von der ungleichen Verteilung von Lasten, und zwar den Lasten der Landesverteidigung gegen die Türken und fordert dazu auf, alle Untertanen gleichermaßen zur Landesverteidigung heranzuziehen.[33]—Überaus arme Menschen, Bettler und Landstreicher wurden nach einer bäuerlichen Redensart Κίγκλοι (*cincli*, „Wasserstare") genannt, denn der Wasserstar habe kein eigenes Nest und müsse seine Eier in fremde Nester legen.[34] Der Wasserstar des Dubravius fordert dementsprechend den Löwen auf, im Krieg Milde walten zu lassen und niemanden durch unmäßige Gewalt und Zerstörung zum Heimatlosen, zum Wasserstar in diesem sprichwörtli-

---

33 Vgl. folgende Passage aus der Rede des Kamels (*Theriobulia* 2,527–529): „Volo ego cura, industria / te pervidere, ait, rex o noster Leo, tempore duelli commune ut periculum / communiter agitetur opera et sumptibus" („Ich will dass du mit Sorge, mit Fleiß darauf achtest, o Löwe, unser König, dass zur Zeit des Krieges der gemeinsamen Gefahr gemeinsam durch Mühe und Kosten widerstanden wird"). Nachdem das Kamel die ungleiche Verteilung der Lasten der Landesverteidigung angeprangert hat, klagt es über die ebenso ungleiche und von Gewalt geprägte Güterverteilung im Inneren des Landes, vgl. *Theriobulia* 2,534–540: „Nunc vero ad hunc cum semper accidat modum, / non modo nihil, qui sint dites, ut conferant / in Martios usus, sed rideant quoque, / qui symbola dederint,—quid dico rideant, / immo auferant, harpagent, raptent, distrahant, / atque, ut vetus verbum decantat, antea / quam sortiantur, devorent communia" („Jetzt aber, da es immer so zugeht, dass die Reichen nicht nur nichts zur Ausübung des Krieges beitragen, sondern diejenigen, die ihren Beitrag geleistet haben, auch noch verlachen—was sage ich ‚verlachen', vielmehr nehmen, rauben, raffen, entreißen sie und, wie ein alter Spruch sagt, bevor sie das gemeinsame Gut teilen, verschlingen sie es"). Dubravius zitiert in diesen Versen ein anderes, ebenfalls in den *Adagia* verzeichnetes Sprichwort, vgl. Erasmus, *Adagia*, 3,6,1. Bemerkenswert ist zudem auch die Ähnlichkeit zu Petrus Alfonsi, vgl. *Disciplina clericalis, exemplum* 19: „Qui totum voluit, totum perdidit. Haec autem natura est canis, cui faverunt illi: quorum unus alii cibum auferre cupit. Sed si naturam cameli sequerentur, mitiorem naturam imitarentur. Nam talis est natura cameli: quando insimul datur praebenda multis, quod nullus eorum comedet, donec omnes insimul edant" („Wer alles wollte, hat alles verloren. Dies aber ist die Natur eines Hundes, der jene sich zugeneigt haben: Bei ihnen trachtet einer dem anderen die Speise wegzunehmen. Doch wenn sie der Natur eines Kamels folgten, würden sie eine milderes Wesen nachahmen. Denn dies ist die Natur eines Kamels: Wenn vielen Nahrung zugleich gegeben wird, wird keines von ihnen essen, bis nicht alle zugleich essen"). Das Kamel des Dubravius beklagt, dass sich der Adel in Böhmen und Ungarn wie eine Hundemeute streitet und unterdessen seine Heimat an die Türken verliert, vgl. *Theriobulia* 2,541–543: „Quid deinde miramur, si factionibus, / contumeliis et opprobrio gravi / certemus invicem, praeda interim hostibus?" („Was wundern wir uns also, wenn wir im Parteienzwist, unter Schmähungen und schlimmer Beschimpfung uns gegenseitig bekämpfen und unterdessen den Feinden eine Beute sind?").

34 Vgl. Erasmus, *Adagia, chil.* 2,2,78.

chen Sinn, zu machen.[35]—Nach dem Zeugnis des Erasmus nannte man in der Antike einen Sklaven, der von seinem Herrn grün und blau geprügelt worden war, einen *attagen*, ein Haselhuhn, denn dieser Sumpfvogel zeichnete sich durch ein in ähnlicher Weise buntgefiedertes Federkleid aus.[36] Das Haselhuhn des Dubravius bittet dementsprechend den Löwen, dass er gütig und jedenfalls nicht hart und rau mit seinen Höflingen umgehe.[37]—Der Phönix entwirft das Ideal eines guten Ratgebers und macht sich selbst den Einwand: „Si dixeris: ‚Fingit citius sibi hunc quispiam, quam comperit', haud ego hic pugno".[38] Indirekt ist dabei auf das Sprichwort „Phoenice rarior"[39] angespielt. Die Seltenheit ist es demnach, die der Phönix mit dem guten Ratgeber gemeinsam hat; ihretwegen erteilt er Ratschläge bezüglich der Auswahl der königlichen Ratgeber.—Nach der Redensart „Surdior turdo" ist es die Drossel, die den Löwen auffordert, gegenüber übler Nachrede unempfindlich zu sein; dringen beleidigende Worte an seine Ohren, soll er in königlicher Langmut „surdior turdo", tauber und unempfindlicher sein als eine Drossel. Erasmus erläutert in seinen *Adagia* die Redensart „Surdior turdo" wie folgt:

> Adscribit [sc. Zenobius] surditatem huic avi peculiarem, quum sit loquacissima, vel proverbio teste. Unde concinne dicetur in eos, qui perpetuo blaterantes ipsi non auscultant, quid vicissim ab aliis dicatur.[40]

---

35 Vgl. folgende Passage aus der Rede des Wasserstars (*Theriobulia* 2,553–561): „Quoties exitio immines / victor nobilis urbis, / id prudens cave, ne quid temere audeas / neu crudeliter, ipsas / caedes accumulans caedibus, ignibus / ignes stupraque stupris. / Hoc namque officium est magnanimi viri / multis parcere, solos condemnare nocentes" („Sooft du als Sieger einer namhaften Stadt mit Untergang drohst, sei klug auf der Hut, dass du nicht irgendetwas unbesonnen oder grausam zu tun dich erdreistest, Gemetzel auf Gemetzel, Feuersbrunst auf Feuersbrunst, Vergewaltigung auf Vergewaltigung häufend. Dies ist nämlich die Pflicht eines großmütigen Mannes: viele zu schonen, allein die Schadenden zu verurteilen").

36 Vgl. Erasmus, *Adagia, chil.* 4,1,5; 4,3,73.

37 Vgl. folgende Passage aus der Rede des Haselhuhns (*Theriobulia* 2,673–676): „Non ergo huic generi tristis et asper eris, / non saevus aut superbus. Oderunt enim / et vix non gravius, quam faciant domini, / quaecumque regum servuli fastidia" („Du sollst also zu diesem Geschlecht nicht unerfreulich und rauh, nicht hart oder hochmütig sein. Es hassen nämlich die Diener—und schwerlich nicht mehr, als es die Herren tun—jedweden verdrießlichen Dünkel der Könige").

38 *Theriobulia* 1,207–209: „Wenn du vielleicht sagst: ‚Den erdichtet man sich schneller, als man ihn findet', so streite ich hier nicht".

39 Vgl. Erasmus, *Adagia, chil.* 2,1,57.

40 Erasmus, *Adagia, Chil.* 2,9,9. Vgl. Zenobius 4,66.

> Er [sc. Zenobius] schreibt diesem Vogel, der doch überaus geschwätzig ist, eine einzigartige Taubheit zu, wie ja auch das Sprichwort bezeugt. Daher wird es (sc. das Sprichwort) passenderweise auf diejenigen bezogen werden, die, indem sie selbst fortwährend schwatzen, nicht zuhören, was umgekehrt von anderen gesagt wird.

Der Drossel des Dubravius geht es freilich nicht um die hier getadelte plappernde Taubheit, sondern um eine *surditas* eigener Art: Der König wird ermahnt, *surdior turdo* zu sein, und zwar nicht etwa aufgrund eigener Geschwätzigkeit *tauber* als eine Drossel für das, was andere sagen, sondern aufgrund seiner Geduld und Langmut *unempfindlicher* als eine Drossel für das, was an den Worten anderer beleidigend ist.[41]

Die Charakterisierung der Tiere nach der *Naturalis historia* des Plinius

Schließlich tragen die Tiere ihren Rat oft auch nach Eigenschaften vor, die ihnen vom älteren Plinius nachgesagt werden. So rät der Elefant, dem Plinius ein Gerechtigkeitsempfinden attestiert, zur Gerechtigkeit.[42]—Der Fasan, von dem Plinius sagt, dass er zwei Ohren aus seinem Gefieder hervorstrecke, ermahnt den Löwen, immer aufmerksam auf seine bewanderten Rechtsgelehrten zu hören.[43]—Bei Plinius ist zu lesen, dass die Vögel und vor allem die Papageien „humanas voces reddunt“; Dubravius sagt dementsprechend vom Papageien, er sei „solertissimus humanas voces reddere“, und lässt ihn daraufhin den Löwen ermuntern, bei geselligen Anlässen sich fein, witzig und gebildet zu betragen; Dubravius spielt dabei mit der doppelten Bedeutung des Wortes „humanus“, besagt dieses Wort doch nicht nur so viel wie „menschlich“, sondern auch so viel wie „gebildet, kultiviert, fein“.[44]—Nach Plinius ist bei der Vogelschau allein der Rabe verständiger Künder von Künftigem: Indem

41 Vgl. folgende Passage aus der Rede des Drossel (*Theriobulia* 1,703–709): „Instrumentum ego iudico tuendo / regno ingens patientiam modestam, / illam scilicet, ebrii procacem / linguam, simplicium oscitationes / suorumque iocos licentiores, / nonnumquam vel amariora dicta / quae non suppliciis statim coercet“ („Nach meinem Urteil ist bescheidene Geduld ein außerordentlich wirksames Mittel zum Herrschaftserhalt, jene nämlich, die die freche Zunge des Trunkenen, das fade Gerede der Einfältigen und die losen Scherze oder oftmals verletzenden Sprüche der Ihren nicht sofort mit Strafen zügelt“).

42 Vgl. Plinius, *Naturalis historia* VIII, 1; *Theriobulia* 1,243–279.

43 Vgl. Plinius, *Naturalis historia* X, 132; *Theriobulia* 1,849–880.

44 Vgl. Plinius, *Naturalis* historia X, 117; *Theriobulia* 2,30–71.

er selbst versteht, was er vorhersagt, wisse er auch selbst, was kommen werde; nach dem Ratschlag des Dubraviusschen Raben soll der Löwe ebenfalls wissen, was kommen wird, und sich darauf gefasst machen: Der Rabe des Dubravius rät somit zur tapferen Gefasstheit auf ein Unheil, das man kommen sieht.[45]—Vom Hirsch sagt u.a. Plinius, er sei das friedliebendste und sanfteste aller Tiere; dementsprechend ist es bei Dubravius der Hirsch, der den Frieden preist und das Unheil des Krieges beklagt.[46]—Plinius bezeugt, dass es den Staren eigentümlich ist, scharenweise und in geordneter Formation zu fliegen; der Star des Dubravius rät dementsprechend, dass der Löwe sich um den militärischen Drill seiner Untertanen, die strenge Ordnung und Formation seines Heeres kümmern solle.[47]—Nach dem Zeugnis des Plinius ist von allen Tieren das Schwein das roheste; das Leben sei ihm anstelle des Salzes allein zur Konservierung seines Schweinefleisches gegeben. Bei Dubravius ergreift das Schwein eben mit dieser Selbstempfehlung das Wort, um den Löwen vor schweinischer Rohheit, Schmutzigkeit und Wollust zu warnen.[48]

Bemerkenswert ist zudem die Rede der Haubenlerche, die ihren Ratschlag vor dem Hintergrund einer historischen Begebenheit erteilt:

Alauda
Alauda: „Quanto," ait, „rem mecum saepius
hanc cogito, quam nunc tibi narravero,
o rex Leo, animo tanto maeror auctior
inest ad illum prorsus impium modum
tot agmina novae sectae regnum Boemiae
fovere, quae si memorem ego, et quae cuilibet
insit seorsum prava relligio, diem
nox adimat, ipse ego servatum excessero
modum. Hoc tamen caput rei haecque summa sit,
fides tot esse, quot voluntates, item
tot dogmata, quot in regno mores perfidi.
Blasphemiarum deinde tot causae vigent,

45 Vgl. Plinius, *Naturalis historia* X, 33; *Theriobulia* 2,379–414.

46 Vgl. Plinius, *Naturalis historia* VIII, 112; *Theriobulia* 2,415–434.

47 Vgl. Plinius, *Naturalis historia* X, 73; *Theriobulia* 2,489–526.

48 Vgl. Plinius, *Naturalis historia* X, 207; *Theriobulia* 2,250–254: „Non est, ex Sue temperans aut castum, Leo, ut audias, / tam brutum siquidem genus / est nostrum, ut data pro sale / sit nobis anima" („Es ist nicht der Fall, Löwe, dass du vom Schwein etwas Maßvolles oder Züchtiges zu hören bekommst. Denn unsere Gattung ist bekanntlich so roh, dass uns die Seele anstelle des Salzes gegeben ist").

quot vitia, scilicet dum pro libidine
vel sic fides scribuntur, ut nobis placet,
aut ut volumus, ita hoc sensu stultae palam
enuntiantur plebi. Nec sic constamus tamen
nobis fidesque certa non spatio annuo
sibi congruit mutantibus sententiam
animis vagis exaestuante saepius
Euripo. Ad idque postremo ventum est, nova
dum quaeritur semper fides, ut nulla sit.
Memini rogationes factas plurimas
ut ne coirent, rite qui facti parum
sint Christiani, neve perfidiam malam
male perfidis monstrarent, cum socordia
illi interim leges rogatas rumperent,
hodieque rumpunt foribus effractis simul
pudoris. Ergo quantum olim muscarum adest
aestate, quo die caletur maxime,
tantis profecto praesens haec licentia
genus auxit infidelis turbae copiis.
Ideoque moritur nunc quodam morbo novo
regnum Bohemiae atque tali scilicet,
ut, cum illa, quae fiunt, facta omnes improbent,
querantur, animo iniquo perferant, gemant
clare, loquantur clarius, nec sit gravi
in re ulla varietas, opem medicam tamen
qui ferre curet huic dolori, nullus est.
Proinde propere ni succurris idque agis
(donec veniat aetas maior), mores novi
sed impii ut rogationi obtemperent,
vereor, nimis ne divisum in semet male
propere pereat ipsum illud regnum Boemiae".

Die Haubenlerche sprach:
„Je häufiger ich die Sache bei mir bedenke, die ich dir jetzt erzählen werde, [605] o König Löwe, um so betrübter ist mein Gemüt darüber, dass das Königreich Böhmen auf geradezu gottlose Weise so viele Scharen einer neuen Sekte hegt und begünstigt. Wenn ich ihrer und der verkehrten Religion gedächte, die jeder beliebigen von ihnen jeweils für sich genommen innewohnt, so würde den Tag [610] die Nacht wegnehmen und würde ich selbst das eingeräumte Maß überschreiten. Doch ist dies der Hauptpunkt und der Kern der Sache, dass es so viele Glaubensrichtungen

> gibt wie Willensneigungen und so viele Dogmen wie im Reich treulosen Lebenswandel. Zudem sind ebenso viele Veranlassungen zur Blasphemie im Schwange wie Laster, solange die Glaubensinhalte nach Gutdünken so ausgearbeitet werden, wie es uns gefällt, oder sie dem dummen Volk öffentlich so, in dem Sinn, verkündigt werden, wie wir es wollen. So bleiben wir uns jedoch nicht gleich und auch der unzweifelhaft gewisse Glaube stimmt nicht für den Zeitraum eines Jahres mit sich überein, da doch unstete Geister ihre Meinung häufiger ändern als der aufbrandende Euripus. Schließlich ist es dahin gekommen, dass es, während stets ein neuer Glaube gesucht wird, gar keinen mehr gibt. Ich erinnere mich, dass etliche Gesetze eingebracht wurden, die gebieten, dass diejenigen, die in nur unzureichend richtiger Weise Christen geworden sind, sich nicht versammeln und Ungetreuen ihre ruchlose Untreue nicht ruchlos zeigen dürfen, während unterdessen ebenjene Ungetreuen mit Sorglosigkeit die eingebrachten Gesetze brachen,—und auch heute brechen sie sie, da einmal die Tore der Scham aufgebrochen sind. Wieviel Fliegen es gewöhnlich am wärmsten Sommertag gibt, um wahrlich so große Scharen hat diese gegenwärtige Freizügigkeit den Stand des ungläubigen Haufens vermehrt. Deshalb stirbt jetzt das Königreich Böhmen an einer neuen Krankheit, die solcherart ist, dass zwar alle jene begangenen Taten missbilligen, beklagen, unwillig ertragen, deutlich beseufzen, deutlicher besprechen und es in den wichtigen Gesichtspunkten keine Meinungsverschiedenheiten gibt, sich aber dennoch keiner findet, der dafür sorgt, dass diesem Leiden ein Heilmittel verabreicht wird. Wenn du demnach nicht schnell (bevor das höhere Alter [sc. der Krankheit] kommt) zu Hilfe eilst und dich darum kümmerst, dass der neue, aber gottlose Lebenswandel den eingebrachten Gesetzen gehorcht, dann fürchte ich, dass, in sich selbst schlimm zerteilt, ebenjenes Königreich Böhmen allzu rasch zugrunde geht".

Nach diesem Rat der Haubenlerche soll der Löwe die „agmina novae sectae", die Scharen einer neuen Sekte, unterdrücken bzw. zum traditionellen Glauben zurückführen. Andernfalls werde es zur religiösen und damit auch zur politischen Spaltung des Königreichs kommen. Wie die im Text erwähnten gesetzlichen Verbote unmissverständlich deutlich machen, handelt es sich bei der bekämpften religiösen Gemeinschaft um die Böhmische Brüdergemeinde.[49]

---

49 Dubravius erwähnt in den Versen 624–627 die 1508 auf dem sog. St.-Jakobslandtag gefassten Beschlüsse gegen die Brüdergemeinde, die sog. „Pikarden". Den Mitgliedern der Brüdergemeinde wurde durch ein Landtagsmandat untersagt, ihren Gottesdienst zu

Die Frage, weswegen Dubravius ausgerechnet eine Haubenlerche diesen Rat hat vorbringen lassen, kann beantwortet werden, wenn man bedenkt, dass es, wie u.a. auch Plinius bezeugt, eine römische Legion gegeben hat, die den Namen dieses Vogels trug.[50] Die Geschichte dieser Legion weist bedeutsame Parallelen zur Rede der Haubenlerche auf und sei daher hier kurz erinnert. Als Julius Cäsar zur Verstärkung der ihm vom Staat übergebenen Streitmacht auf eigene Kosten Legionen warb, bildete er eine dieser Legionen aus transalpinischen Galliern und gab ihr, vermutlich wegen des Helmschmucks ihrer Soldaten, den gallischen Namen *Alauda*. Im Bürgerkrieg nach Cäsars Ermordung stand die *Legio Alaudae* unter dem Befehl des Marcus Antonius; sie bildete den Kern seines Heeres und hielt ihm bis zur endgültigen Niederlage bei Actium die Treue. Bei der Wiederherstellung eines einheitlichen Reichsheeres nach dem Jahr 30 übernahm Augustus die alte Legion Cäsars und gliederte sie wieder in sein stehendes Heer ein.[51] Die Analogie zur Rede der Haubenlerche ist nun folgende: Wie einst Augustus die abtrünnige *Legio Alaudae* in das römische Heer wieder eingliederte, so soll Ludwig II. die *agmina novae sectae* in das Heer der *Christiani rite facti* wieder eingliedern. Wie dort der Bürgerkrieg durch die Rückführung abtrünniger Soldaten beendet wurde, so soll er hier durch die Zurückführung abtrünniger Glaubensgenossen vermieden werden.

Dass Dubravius im Gegensatz zu seiner muttersprachlichen Vorlage Fragen der Religion zugunsten einer Beschränkung auf das Diesseits nur sehr selten, nur sehr knapp und unter fast vollständigem Verzicht auf eine pronon-

---

feiern (vgl. *Theriobulia* 2,625: „ut ne coirent") oder überhaupt öffentlich in Erscheinung zu treten (vgl. *Theriobulia* 2,626–627: „neve perfidiam malam male perfidis monstrarent"). Dubravius gibt diese Beschlüsse auch dreißig Jahre später in seiner *Historia Bohemica* wieder, vgl. Dubravius J., *Historia Bohemica ab origine gentis, per diversas temporum et familiarum vices, usque ad Ferdinandi Imperatoris et Regis auspicia, deducta*, in: *Rerum Bohemicarum antiqui scriptores aliquot insignes, partim hactenus incogniti*, Bd. 2 (Hannover, Claudius Marnius und Johannes Aubry: ^{2}1602) 262,25–28: „Ab iisdem legatis mandato regis peculiari hoc quoque de Piccardis actum fuit, ut ne in coetus publice aut privatim coirent, synagogas suas omnes occluderent, a contionibus scriptisque de sua secta plane abstinerent, et intra certum tempus ut ad alterutram partem, Romanam aut Calixtinam, discederent" („Von ebendiesen Gesandten wurde auf Sonderbefehl des Königs über die Pikarden auch dies erlassen, dass sie sich weder öffentlich noch privat zu ihren Versammlungen zusammenfinden dürften, dass sie all ihre Versammlungsorte schließen, sich gänzlich aller Predigten und Schriften über ihre Lebensweise enthalten und innerhalb einer bestimmten Zeit zu einer von beiden Glaubensparteien, der römischen oder kelchlerischen, auseinandergehen sollten").

50 Vgl. Plinius, *Naturalis historia* XI, 121, ferner Suetonius, *De vita Caesarum*, „Divus Iulius" 24.

51 Vgl. *RE* 12,1565,18–19.29–39, s.v. „Legio" (V Alaudae).

ciert christliche Dogmatik thematisiert, ist oben bereits erwähnt worden. In der Rede der Haubenlerche gedenkt Dubravius der Religion allein unter dem Gesichtspunkt ihrer politischen Bedeutung für die Integration und den Zusammenhalt des Staates. Mit Blick auf die noch nicht lange zurückliegenden Hussitenkriege und die ohnehin noch nicht überwundene religiöse Spaltung in Böhmen fürchtet Dubravius aufgrund der Erstarkung einer neuen religiösen Sonderbewegung um den Frieden im Land. Daher ist ihm wichtig, dass die Religion nur eine sei und also im Staat neben dem traditionell überkommenen Katholizismus (und dem als unvermeidlichem Übel verschwiegenen Hussitentum) keine Häretiker geduldet werden dürften. So verbindet sich bei ihm religiöse Leidenschaftslosigkeit und Vorliebe für die heidnische Antike mit einem strengen religiösen Konservatismus, der religiöse Dissidenten rechtzeitig und wenn nötig auch gewaltsam zu unterdrücken für geboten hält.[52]

Dieser Rigorismus mutete bereits einem der älteren Herausgeber der *Theriobulia* befremdlich und untragbar an und ließ ihn stillschweigend tiefgreifende Veränderungen am Text der Rede der Haubenlerche vornehmen. Zum besseren Verständnis des Folgendem sei hier ein kurzer Blick auf die Druckgeschichte der *Theriobulia* geworfen: Nachdem die *Theriobulia* in den Jahren 1520 und 1521 zwei Auflagen von geringer Stückzahl erfahren hatte und daraufhin bald schon in Vergessenheit geraten war, war es im Jahr 1614 der Breslauer Arzt und Dichter Caspar Cunradus,[53] der eine Neuauflage dieses Gedichts besorgte. Caspar Cunradus war als ein Mitglied des weltoffenen Breslauer Bürgertums um eine Verständigung zwischen den Konfessionen bemüht. Er strich daher in der Rede der Haubenlerche alle eifernden Ausfälle gegen die Brüdergemeinde und spielt in seiner Überarbeitung des Textes auf diese überhaupt nicht mehr an. So tilgt er die Verse 2,624–633, in denen die Durchsetzung der gegen die Brüdergemeinde eingebrachten Gesetze gefordert wird, gänzlich und beklagt im Übrigen nicht mehr die vermeintliche Verkehrtheit bestimmter religiöser Überzeugungen, sondern die Tatsache

52 Eine detaillierte Analyse der Rede der Haubenlerche findet man bei Loose (Anm. 7) Bd. 2, S. 239–246.

53 Zu Caspar Cunradus vgl. Pfotenhauer P., „Cunradus Vratislaviensis", in: *Zeitschrift des Vereins für Geschichte und Altertum Schlesiens* 26 (1892) 328; Martínek J., Art. „Cunradus, Caspar", in: Truhlář J. – Hrdina K. – Hejnic J. – Martínek J. (Hrsg.), *Rukověť humanistického básnictví v Čechách a na Moravě / Enchiridion renatae poesis latinae in Bohemia et Moravia*, Bd. 1 (Prag: 1966–1982) 516–518; Garber K., Art. „Cunrad, Caspar", in: Killy W. (Hrsg.), *Autorenlexikon. Autoren und Werke deutscher Sprache*, Bd. 2 (Gütersloh: 1989) 486–487; Folaron S., „Polen und polnische Kultur im Werk des Breslauer Gelehrten und Philosophen Caspar Cunradus (1571–1633)", in: Engel W. – Honsza N. (Hrsg.), *Kulturraum Schlesien. Ein europäisches Phänomen* (Breslau: 2001) 221–225.

religiöser Streitigkeiten überhaupt: Nach seiner Bearbeitung ist die Haubenlerche nicht darüber betrübt, dass „das Königreich Böhmen so viele Heerscharen einer neuen Sekte hegt", sondern darüber, „dass man von so vielen Kämpfen in der Glaubensgemeinschaft hört": statt der Worte des Dubravius *tot agmina novae sectae regnum Boemiae fovere* (*Theriobulia* 2,607–608) schreibt er: *tot in fidei consortio certamina audirier*. Da überdies die Unterscheidung zwischen „wahrer" und „entarteter" Religion der irenischen Gesinnung des Cunradus widerstreben musste, ersetzte er in Vers 2,609 das von Dubravius verwendete Wort *prava* durch das Wort *mixta* und sagt somit nur noch, dass die synkretistische Vielfalt unter den Glaubensgemeinschaften und religiösen Splittergruppen unüberschaubar groß sei. Indem Cunradus das Thema der Rede der Haubenlerche dahingehend ändert, dass bei ihm nicht mehr gegen eine besondere religiöse Gruppierung geeifert, sondern allgemein die religiöse Zerstrittenheit beklagt wird, erweitert er auch den Horizont seiner Klage: Es geht ihm nicht mehr nur um die in Böhmen, sondern um die in der ganzen Welt (Christenheit) herrschende Zerstrittenheit: in Vers 2,613 ersetzt er das von Dubravius verwendete Wort „regno" durch „mundo" und in Vers 2,635 spricht er nicht mehr wie Dubravius vom „regnum Bohemiae", sondern von der „orbita isthaec" (von „diesem Weltlauf"). Die von Dubravius vorgebrachte Polemik gegen eine bestimmte religiöse Gruppierung in Böhmen wird bei ihm somit zu einer Klage über den Zustand der Christenheit in Europa am Vorabend des Dreißigjährigen Krieges.

English Summary

According to Bruno Singer, it is characteristic of the mirrors of princes written by Renaissance Humanists that in comparison to medieval mirrors of princes much less attention is paid to political theory and the theological justification of the power of the regent. Instead of composing comprehensive volumes, which deal in a strict systematic way with such topics as the doctrine of hierarchies, natural law, the feudal system, the right to resist or even to kill a tyrant, Renaissance Humanists, beginning with Petrarca's letter to Francesco Carrara, wrote loosely composed exhortatory letters, by means of which they endeavored to advise their regent on prudent government as well as a life of decency and virtuosity befitting of a king.

The *Theriobulia*, a polymetric poem written in 1518 by the Bohemian Humanist Johannes Dubravius, is in many respects a typical humanistic mirror of princes. However, its literary form makes it unique among others of its genre. In *Theriobulia*, the instruction and admonition is transposed into literary fiction in the form of an animal parliament convened by the Lion, the king of all animals. Very much in accordance with the taste of the time, Dubravius characterized each animal distinctly as an emblematic personification of the virtue it exhorts or the vice it cautions against. In many cases, he based the animal allegory on the *Nová Rada* by the Czech nobleman Smil Flashka of Pardubic, which combines many traditions as fables, animal epics, bird parliaments and the *Physiologus*. In other cases, Dubravius established a relationship between the advice given by an animal and a proverb in which the animal is mentioned. Without exception, these proverbs are extracted from the *Adagia* by Erasmus of Rotterdam. Moreover, Dubravius drew on Pliny's *Historia Naturalis* making Pliny's description the emblematic foil against which some of his animals give their advices.

In respect to religion, the author of the *Theriobulia* is a reserved and passionless man, who finds contentment by declaring to be a Christian, while at the same time showing some affinity for ancient paganism. This disposition, however, does not prevent him from cherishing a strict religious conservatism, which leads him to attack the Bohemian Brotherhood with scathing polemic. In the *Theriobulia*, it is the crested lark, who admonishes the king to suppress and to destroy this religious movement, which in the eyes of Dubravius posed a severe threat to the unity and very existence of the Kingdom of Bohemia. In this case, Dubravius contrived the animal allegory with reference to Roman history, comparing the Bohemian Brotherhood with a Roman Legion, which due to the crests on the helmets of its soldiers was called "the legion of the crested lark".

About a century after the first publication of the *Theriobulia*, the Silesian humanist Caspar Cunradus re-edited Dubravius's poem, which at this time had fallen into total oblivion. Due to his irenic stance in questions concerning the Christian denominations, he converted Dubravius's polemic against the Brotherhood into a lament over the enmity and the antagonism, which was prevalent in Christianity on the eve of the Thirty Years' War.

Bibliographie

Busch P., *Die Vogelparlamente und Vogelsprachen in der deutschen Literatur des späten Mittelalters und in der frühen Neuzeit. Zusammenstellung und Beschreibung des Textmaterials—Rekonstruktion der Überlieferungsgeschichte—Untersuchungen zu Genese, Ästhetik und Gebrauchsfunktion der Gattung*, Beihefte zu Poetica 24 (München: 2001) 57–75.

Čapek J.B., „Die Ironie des Smil Flaška“, *Slavische Rundschau* 10 (1938) 68–79.

Daňhelka J. (Hrsg.), *Smil Flaška z Pardubic, Nová Rada* (Prag: 1950).

Dubravius Johannes, *Historia Bohemica ab origine gentis, per diversas temporum et familiarum vices, usque ad Ferdinandi Imperatoris et Regis auspicia, deducta*, in: *Rerum Bohemicarum antiqui scriptores aliquot insignes, partim hactenus incogniti*, Bd. 2 (2. Aufl., Hannover, Claudius Marnius und Johannes Aubry: 1602).

Loose A. (Hrsg.), Johannes Dubravius, *Theriobulia* (Hildesheim: 2011).

Pfeiffer F., „Das Märchen vom Zaunkönig“, *Germania. Vierteljahrsschrift für deutsche Alterthumskunde* 6 (1861) 80–106.

Singer B., *Der Fürstenspiegel in Deutschland im Zeitalter des Humanismus und der Reformation* (München: 1981).

Singer B., Art. „Fürstenspiegel“, in: *Theologische Realenzyklopädie* 11 (Berlin: 1993) 707–711.

Skutil J., *Jan Dubravius. Biskup, státnik, ekonom a literát* (Kroměříž: 1992).

Physiology and Political Ideology

CHAPTER 13

From Physiology to Political Ideology: The Images of Man in Early Modern Scotland

Tamás Demeter

Introduction

Much of Enlightenment Scottish philosophy, natural and moral equally, have developed in a social and political context dominated by problems surrounding Jacobitism. Jacobitism emerged in Britain supporting the Stuart cause in response to the Glorious Revolution, i.e. to deposing the Catholic James II (of England and Ireland) and VII (of Scotland) and offering the crown to the Protestant William of Orange and his wife Mary in 1688. A reaction to these events, Jacobitism arose as a distinctive stance on the questions of monarchical succession, the nature of political legitimacy, obligation, and religious tolerance. Although these were important issues throughout Britain in the period, they had a distinctive flavour in Scotland, the primary seat of Jacobitism, where it proved to be a constant source of political instability and religious turbulence well after it had ceased to be a real threat in England.

On the political map of England, a Jacobite could be placed as a radical Tory committed to the idea of passive obedience to the monarchical will and to the monarch's indefeasible and hereditary right. As they took it, the king ruled by divine right, and hereditary succession was taken to be divinely sanctioned. Therefore the subjects had no right actively to resist monarchical will—just as they had no right to defy God's commands. At most, passive resistance was acceptable only when the king had manifestly breached the laws of God. In the Whig alternative, absolute monarchical power was rejected in principle in favour of a rule properly constrained by law. Succession was conceived to depend not on indefeasible right, but on some sort of contract with the political nation and thus God played no direct part in the legitimacy of any particular monarch.

Jacobites were an important faction in English politics that gave a radical answer to the Tory dilemma that arose after the succession of 1688: according to the Tory principles of political legitimacy, William and Mary simply were not legitimate monarchs. However, the Stuart alternative was to them dangerous, threatening with the establishment of a Catholic succession and risking

 | DOI 10.1163/9789004279179_015

civil war. The Tories' resulting bad conscience was eased by acknowledging allegiance to William and Mary as *de facto* rulers, without compromising on the Stuarts' divinely ordained rights to throne. English Jacobites were Tories who did not need this kind of unstable manoeuvring, and they did not bent their political sensitivity with the more favourable wind.

In Scotland where, as David Hume observes, no true Tory party existed,[1] Jacobites, mostly Episcopalian and Catholic, were the main ideological opponents of mostly Presbyterian Whigs. Scottish Jacobitism was different from its English counterpart mainly in being a much more overarching view of the world not restricted to matters of politics. As Frank McLynn puts the difference, while '[t]he ideology of English Jacobites was a 'partial' view of the world [...], [t]he ideology of Scottish Jacobitism came nearer to being a *Weltanschauung*'.[2] While agreeing on monarchical issues with English Jacobites, this *Weltanschauung* also served as an ideology of nationalism, and especially after the Union of 1707.[3] Jacobitism merged with questions of national independence, ancient traditions, and the glorious past, and against the Union. This led to the 'Highlandization of Scottish national identity, so typical of Jacobite propaganda', a tendency so prevalent in e.g. Allan Ramsey's poetry.[4] In its various manifestations it meant a plea for ancient virtues to be restored in the Lowlanders and a worldview imbued with sympathies for the mythical, spiritual, and supernatural. This *Weltanschauung* had soon to be faced with strong opposition by those who are now widely considered the main figures of Scottish Enlightenment natural and moral philosophy. They were typically Lowlanders, supporters of the Union, and opponents of the Jacobite cause, albeit not necessarily Whig.

In this paper I am going to argue that in the late seventeenth- to the mid-eighteenth-century Scottish context the theoretical work of those studying various aspects of human phenomena reflected their respective social imagery, i.e. the way they more or less explicitly envisaged social order and governance, and the proper condition of human beings within these structures. My main point is that those developing theories about human functioning, from physiology through the 'science of man' to political contexts, had not only

1 Hume David, "Of the Parties of Great Britain", in *Essays: Moral, Political and Literary* (Indianapolis: 1985) 64–72, the quoted passage appears in the appendix of this edition on page 615, and in editions prior to 1768.

2 McLynn F., "Jacobitism and David Hume", *Hume Studies* (1983) 171–199, 172.

3 For a comprehensive introduction see McLynn F., *The Jacobites* (London: 1985).

4 Pittock M.G.H., *Poetry and Jacobite Politics in Eighteenth-Century Britain and Ireland* (Cambridge: 1994) 158.

purely cognitive agendas but also a political one broadly understood: theories about various facets of human nature also served political purposes, and exhibited affinities to the social-political situation in which they originated, and gestured toward specific stances on questions of political and religious significance. So texts addressing problems of purely theoretical issues frequently had other layers of significance in the social-political-religious sphere, mainly through an alleged congruence of epistemic and political values.[5]

All these theories had been developed under the label of Newtonianism whose various forms were put to apologetic use in extra-theoretical contexts.[6] Newton's writings proved to be a Rorschach test not only for those working in late seventeenth- and early eighteenth-century natural philosophy,[7] but also for those having social, religious and political agendas in mind: with these inspirations everyone could find a suitable interpretation of one or another facet of Newton's philosophy, or if not, it was always possible to forge an interpretation, and to bend theoretical content so as to do the desired extra-theoretical work. In this vein, e.g. the 'General Scholium' of the *Principia* could be given Socinian and Arian readings,[8] and as its system of the world required God's intervention from time to time, it was almost natural to read it as countering Deistic tendencies.[9] Besides, the *Principia* could be exploited as a common apologetic

5 This has been a long interest of sociologically inclined historians of science. For a classic survey see Shapin S., "History of Science and Its Sociological Reconstructions", *History of Science* 20 (1982) 157–211. See also Golinski J., "The Theory of Practice and the Practice of Theory: Sociological Approaches in the History of Science", *Isis* 81 (1990) 492–505. For a more recent retrospective reflection see Shapin S. – Schaffer S., "Introduction to the 2011 Edition", in *Leviathan and the Air Pump: Hobbes, Boyle, and the Experimental Life* (Princeton: 2011) xi–l.

6 Some of these uses were certainly not against Newton's intentions. The classic piece in this respect is Shapin S., "Of Gods and Kings: Natural Philosophy and Politics in the Leibniz-Clark Disputes", *Isis* 82 (1981) 187–215. See also Cunningham A., "Getting the Game Right: Some Plain Words on the Identity and Invention of Science", *Studies in History and Philosophy of Science* 19 (1988) 365–389, and his "How the *Principia* Got Its Name; Or Taking Natural Philosophy Seriously", *History of Science* 29 (1991) 377–391; Snobelen S., "Isaac Newton, Heretic: The Strategies of a Nocodemite", *British Journal for the History of Science* 32 (1999) 133–153.

7 See Schofield R., *Mechanism and Materialism. British Natural Philosophy in an Age of Reason* (Princeton: 1970), and more recently Reill P.H., *Vitalizing Nature in the Enlightenment* (Berkeley: 2005).

8 See e.g. Stewart L., "Seeing through the Scholium: Religion and the Reading of Newton in the Eighteenth Century", *History of Science* 36 (1996) 123–165.

9 On Newton and Deism see e.g. Force J.E., "Biblical Interpretation, Newton and English Deism" in Popkin R. – van der Jagt A. (eds.), *Skepticism and Irreligion in the Seventeenth and Eighteenth Centuries* (Leiden: 1993) 282–305.

source of Court Whigs and Tories while forging their alliance in support of the Glorious Revolution: the *Principia*'s image of nature was interpreted as commenting on the scope of monarchical will and could be taken to serve as its model.[10] Just like Newton's own theories, Newtonian medicine and physiology, as well as the 'science of man', could be read with an affinity to their religious and political implications.

It is in this sense that the medical theories of Archibald Pitcairne and other Scottish medical men like George Cheyne and John Arbuthnot seem to be supportive of certain aspects of Jacobite ideology with respect to the social and religious order.[11] In their works written around the turn of the century they advocated a mathematical *cum* mechanical approach to physiology, or animal oeconomy as it was called then. This approach was partly motivated by their conviction that only mathematical learning can produce certain, i.e. demonstrative knowledge, and thereby it facilitates overcoming endless and infertile disputes and prevents sectarian violence. A mathematical ideal of knowledge seemed to be a safe heaven in the midst of religious and political turbulences in Scotland after the Glorious Revolution: it provided a desirable model of order in political and philosophical scenes.

The image of man developed both in mid-century Scottish physiological theories, mainly by Robert Whytt and William Cullen, and by those working on a 'science of man', mainly David Hume, Adam Smith and Adam Ferguson, was informed by a different vision. This image is much less congruent with the ideals of monarchical than liberal forms governmentality: the drive for a mathematical, demonstrative and mechanical representation of human functioning had been abandoned by them in favour of a qualitatively oriented approach centred on the sensitivity of self-governing individuals. These theories posed a challenge to various aspects of the Jacobite ideology from its inclination to mysticism and "Highlandization" to its social imagery.

10 See Shapin, "Of Gods and Kings".

11 See most importantly Guerrini A., "The Tory Newtonians: Gregory, Pitcairne, and Their Circle", *Journal of British Studies* 25 (1986) 288–311; Schaffer S., "The Glorious Revolution and Medicine in Britain and the Netherlands", *Notes and Records of the Royal Society of London* 43 (1989) 167–190; Friesen J., "Archibald Pitcairne, David Gregory and the Scottish Origins of English Tory Newtonianism, 1688–1715", *History of Science* 41 (2003) 163–191.

Physiology in Turbulent Times

In between the publication of the first edition of *Principia* in 1687 and the *Opticks* in 1704, or rather the publication of the latter's Queries in the second edition in 1717,[12] 'Newtonianism' had had a narrower meaning than it had afterwards.[13] In the interim decades it was naturally identified with a mathematical way of studying phenomena, with what I.B. Cohen later happily termed as 'the Newtonian style': primarily, it meant taking *mathematics as the model of reality*: constructing 'the mathematical analogue of a natural situation' and then to advance from this idealized case by the addition of further conditions toward more accurate mathematical analogues of actual situations.[14]

During the last decade of the seventeenth and the first decades of the eighteenth century the dominant language of nature in Scotland was that of Newtonian mechanism.[15] This meant the adoption of what Peter Hanns Reill called 'the strong program of mechanical natural philosophy',[16] which had a double commitment to an ontology of qualitatively homogenous, corpuscular, inert matter on the one hand, and to mathematics as the model of reality on the other. Accordingly, explanations must have been couched in terms of particles different only in shape and size, properties to which mass and interparticulate forces were added by Newton.[17] This program reached well beyond the boundaries of physics: mechanical-mathematical explanations were sought for chemical reactions, and for the activities of living matter as well, and it had its influence on moral philosophy too. The program was first undermined by the ether hypothesis put forward in the passages of the *Opticks* that provided the main inspiration for the idea of a natural world populated by active

12 On their influence influence in Britain see Schofield, *Mechanism and Materialism*, and particularly in Scotland more recently see Wilson D.B., *Seeking Nature's Logic: Natural Philosophy in the Scottish Enlightenment* (University Park, Pennsylvania: 2009).

13 See e.g. Guerrini, "Tory Newtonians" 70.

14 See Cohen I.B., *The Newtonian Revolution* (Cambridge: 1980) 51, 151.

15 See Shepherd C.M., "Newtonianism in Scottish Universities in the Seventeenth Century", in Campbell R.H. – Skinner A. (eds.), *The Origins and Nature of the Scottish Enlightenment* (Edinburgh: 1982) 65–85; Barfoot M., "Hume and the Culture of Science" in Stewart M.A. (ed.), *Studies in the Philosophy of Scottish Enlightenment* (Oxford: 1990); Wilson, *Seeking Nature's Logic* 1–32.

16 Reill P.H., "The Legacy of the 'Scientific Revolution'", in Porter R. (ed.), *Cambridge History of Science. Volume 4. Eighteenth-Century Science* (Cambridge: 2003) 23–43.

17 On the central importance of mass in Newton see Janiak A., *Newton as Philosopher* (Cambridge: 2008) 81–86.

principles. Although initially 'ether' was interpreted as a mechanistic concept, and it was ascribed the role of transmitting forces between bodies, its re-interpretation first as a materialistic concept and then as a vitalistic active principle was widespread and increasingly popular among eighteenth-century naturalists.[18]

Fairly early in the 1690s the Edinburgh professor of medicine Archibald Pitcairne took up the torch of the strong program and began working on placing medicine on a mathematical *cum* mechanical footing. As his *Principia*-style epistemological warning has it, 'our Knowledge of Things is confined to the Relations they bear to one another, the Laws and their Properties of Power, which enable them to produce Changes in some Things, and to become altered by other Things'.[19] It is thus only functional and dispositional properties through which knowledge of things is possible, not through their intrinsic natures: the study of structure and function are the key to understanding the human body. Aspiring to knowledge of natures can only lead to speculation and endless disputes resulting in philosophical sectarianism. Instead of forcing experience and observation into the Procrustean bed of some sectarian metaphysics, they should be processed in a mathematical, and therefore demonstrative, disinterested manner.

Pitcairne's medical theory centred exclusively upon the circulation of blood and humours secreted from the blood, and he understood illness in terms of decreased circulatory hydraulics. He adopted the general laws of Newtonian mechanics in order to explain the functioning of body on an analogy drawn between gravity and the heartbeat. Although it seems he implicitly acknowledged short-range attractions and saw the limits of mechanical explanation in chemistry, his physiology remained within the boundaries of mechanical philosophy. However, he fiercely opposed its Cartesian variant e.g. by denying the explanatory significance of the particles' shape and size in his hydraulic account of bodily functions, instead he invoked primary particles building up various molecules by mechanically inexplicable bonds.[20] As Anita Guerrini puts it, Pitcairne's account was an exercise in demonstrating "mechanical necessity" by a "mathematical method": the tendency of Newton's natural

18 See Schoefield, *Mechanism and Materialism*; Reill, *Vitalizing the Enlightenment*; Wilson, *Seeking Nature's Logic*.

19 Pitcairne Archibald, "An Oration proving the Profession of Physic free from the Tyranny of any Sect of Philosophers", in *The Whole Works of Dr. Archibald Pitcairn* (London: 1727) 6–22, 9.

20 Guerrini A., "James Keill, George Cheyne, and Newtonian Physiology, 1690–1740", *Journal of the History of Biology* 18 (1985) 247–266, 250–251.

philosophy to reify mathematics in a general account of motion was thus applied in the particular context of human physiology.[21] The main epistemic virtue of a mathematical method in Pitcairne's eyes was its *certainty*: unlike Paracelsian and Cartesian speculations, the Newtonian style was capable of producing demonstrative and indubitable conclusions. So the proper way of medical learning did not lead through the search for physical causes but in a deductive theory of medicine: instead of empirical hypotheses on the nature of causes he urged mathematical demonstrations of relations.[22]

It has already been suggested that Pitcairne's vision of a mechanical *cum* mathematical physiology could be read as having political and religious connotations. Andrew Cunningham has pointed out that the ideological significance of Pitcairne's theory can be interpreted in the aftermath of William Harvey's *De motu cordis*. Published in 1628, its preface proclaims that the role of heartbeat is analogous to that of the king in an absolute monarchy: absolutely everything depends on it.[23] Certainly, this image suited well with Pitcairne's Jacobite politics. In a similar vein Anita Guerrini has also suggested that Pitcairne's analogy drawn between gravity and the heartbeat finds its place in Jacobite political imagery in which the monarch around whom the political life of the absolute monarchy revolves is frequently compared to the sun. Besides having analogous function in different structures, the source of heartbeat, gravity, and hereditary right have a common source, that is God.[24] This image of human functioning could be conveniently extended to and projected onto higher levels of human organization, i.e. society and the political community.

And indeed, as Guerrini, Simon Schaffer and John Friesen have explored,[25] Pitcairne's high esteem for mathematical learning did not arise exclusively in the context of his aversion to theoretical sectarianism in medicine, but also in that of religious and political sectarianism. The Glorious Revolution brought significant changes to Scottish society and academic life as a part of it. These included the ejection of Episcopal ministers because of their Jacobite sentiments and the restoration of Presbytarianism, and setting up a visitation committee in 1690 that was responsible for ensuring the allegiance of Scottish

---

21 Guerrini A., *Obesity and Depression in the Enlightenment: The Life and Times of George Cheyne* (Norman: 2000) 41.

22 Friesen J., *The reading of Newton in the early eighteenth century: Tories and Newtonianism*, PhD Thesis (University of Leeds: 2004) 51.

23 Cunningham A., "Sydenham versus Newton: the Edinburgh fever dispute of the 1690s between Andrew Brown and Archibald Pitcairne", *Medical History* 25 (1981) 71–98, 90–91.

24 Guerrini, *Obesity and Depression* 42–43.

25 See footnote no. 11.

universities to the new government. As a consequence, Alexander Monro, the president of the Edinburgh University who labelled the process as 'Presbyterian Inquisition', and several other professors had been expelled, but other Scottish universities fared even worse.

These experiences inspired Pitcairne, an Episcopalian, to write satirical plays, *The Assembly* and *Babel*, both written in 1692,[26] in which Presbyterians are represented as dogmatists taking the word of the scripture at face value, and religious enthusiasts undermining the significance of reason. Presbyterianism is thus portrayed in these plays as a form of anti-intellectualism that opposes mathematical medicine and natural philosophy in the name of a narrow-minded scholasticism. Besides, Pitcairne's Jacobite sympathies are also transparent in these plays: the stubborn, sectarian Presbyterians are also enemies of the monarchy and hereditary right. In his eyes these views were liable to cause social turbulence and sectarian violence, and he proposed mathematical learning as the only useful way of fighting them. The language of the *Principia*, as opposed to the speculative tone of competing natural philosophies, was thus exploited in an extra-theoretical context, and it was presented as the ideal model for avoiding religious fanaticism, dissent and faction.

As a consequence of 'Presbyterian Inquisition', Pitcairne and his friend David Gregory, professor of mathematics, also left Edinburgh: Pitcairne took up a professorship of medicine in Leiden in 1692. Gregory went to Oxford where was appointed as Savilian Professor of astronomy in 1691 on the recommendation of Newton. In Leiden a "Tory Newtonian" circle consisting of immigrant Scottish students formed around Pitcairne which included George Cheyne, George Hepburn, and William Cockburn.[27] Gregory also had an influence on Scottish students interested in medicine, most notably on James Keill, still in Edinburgh, and on John Arbuthnot in Oxford. In the 1690s all these medical men shared a common commitment to the extension of Newtonian natural philosophy, and especially its mathematizing tendencies to medicine, and with varying degree of commitment and publicity to the Jacobite cause.

26 On Pitcairne's Jacobitism as manifested in his other poems and its context see McQueen J., "From Rome to Ruddiman: The Scoto-Latin Tradition", in Clancy T.O. – Pittock, M. (eds.), *The Edinburgh History of Scottish Literature* (Edinburgh: 2007) 184–208, 203–206.

27 Guerrini, "Tory Newtonians" 303. It is an interesting question for the history of iatromechanism whether Pitcairne also taught Boerhaave. According to Guerrini and others (like e.g. Wolfe C.T., "On the Role of Newtonian Analogies in Eighteenth-Century Life Science: Vitalism and Provisionally Inexplicable Explicative Devices", in Biener Z. – Schliesser E. (eds.), *Newton and Empiricism* (Oxford: forthcoming)) believe that Pitcairne did so, but this is challenged by others e.g. Knoeff R., *Herman Boerhaave (1668–1738): Calvinist Chemist and Physician* (Amsterdam: 2002) 186.

They shared Pitcairne's convictions: in their eyes mathematics, beyond its capacity to produce theoretical certainty, also served as a model of ensuring social hierarchy and stability, and the image of human functioning inspired by the Newtonian style provided useful theoretical and metaphorical analogies in the context of their Jacobite politics.

It was not unprecedented in early modern Britain to turn to the certainties of mathematics in politically turbulent times. In the mid-seventeenth century Hobbes, for example, invoked the authority of geometry to secure universal assent in questions of philosophy, both 'civil' and 'natural', in a period of political and religious sectarianism.[28] In this Hobbesian manner William Petty developed his 'political arithmetick' in response to the theoretical controversies generated by the English civil war: the representation of individuals and their social relations by mathematical means generated an air of disinterestedness that was very much strived for in a time of heated and sometimes violent discussions.[29] Hobbes' and Petty's theories and methods were not only motivated by religious and political considerations, but they could be exploited for apologetic purposes. The mathematization of society was congruent with the social imagery centred upon social stability and rule by coercion: knowledge was to be collected and processed so as to ensure the interests of and conformity to the central government. In the turbulent years around 1700, Pitcairne and his circle developed physiological theories that conformed to this image: a mathematically represented mechanical image of human functioning could do good apologetic service to the Tory-Jacobite cause.[30]

Transforming Medical Theory and Its Social Background

By the 1720s and 1730s both the theoretical and political climate had been changed. From the theoretical angle the two editions of Newton's *Opticks*, the second edition of the *Principia* (1713), and the publication of his 'De Natura Acidorum' (1710) are the milestones for the future development of Scottish medicine. In this context the most important tenets of these writings are the idea of short-range attractions between particles of matters that contributed

28 See Shapin – Schaffer, *Leviathan and the Air Pump* 100–103.

29 See especially Poovey M., *A History of the Modern Fact: Problems of Knowledge in the Sciences of Wealth and Society* (Chicago: 1998) 120–137.

30 See especially Cunningham, "Sydenham versus Newton".

to the dissolution of Pitcairne's rigidly mechanistic framework,[31] the idea of aether in the form of a 'certain most subtle spirit', and a general tendency that drives towards experimentalism and away from mathematical representation. From the political angle the Union of 1707 gradually brought political stability and after two decades of economic depression and social fragmentation the economy and cultural life of Scotland started to revive, and by midcentury Jacobitism also ceased to pose a real threat to the settlement of 1689.[32]

By the time of writing his most popular book, *The English Malady* published in 1733, George Cheyne, as well as other members of the Pitcairne-Gregory circle, had already distanced himself from his earlier commitment to a mechanical and mathematical treatment of animal oeconomy, and adopted a more vitalistic stance, inspired by Newton's 'aether', toward the mind and body as united. Although he had been criticised for abandoning his earlier iatromechanical views already in his *Essay on Health and Long Life* (1724), he had not given up his self-definition nor his reputation as a Newtonian,[33] but by this time he had indeed given up the idea of writing the *Principia Medicinae* in which he intended to follow the *Principia*'s 'Newtonian style'.[34] Having dismissed the idea of explaining physiological phenomena 'from Matter and Motion alone, and all the powers of our Numbers and Geometry join'd to them' as 'mere *Jargon* and *Ignorance*',[35] he now emphasized instead the unity of matter and mind, because 'the Works of *Imagination* and *Memory*, of *Study*, *Thinking*, and *Reflecting*, from whatever Source the Principle on which they depend springs, must necessarily require bodily Organs'.[36] In his new approach Cheyne exploited Newton's aether hypothesis in the *Opticks*:[37] as 'MECHANISM takes Place and operates in it self only, on dead *Matter*', Cheyne proposes to study its concurrence and homology with the '*Self-active* Principle' to which mechanism is subordinated in 'ORGANIZED bodies fit for *Animation* and living

31 Pitcairne knew Newton's *De natura acidorum* since 1692, and discussed it with friends and students. See Guerrini A., "Archibald Pitcairne and Newtonian Medicine", *Medical History* 31 (1987) 70–83.

32 For a nice survey relevant in the present context see Lawrence C., "The Nervous System ad Society in the Scottish Enlightenment", in Barnes B. – Shapin S. (eds.), *Natural Order: Historical Studies of Scientific Culture* (London: 1979) 19–40.

33 Guerrini, *Obesity and Depression* 153.

34 On Cheyne's intellectual development see Guerrini A., "Isaac Newton, George Cheyne and the '*Principia Medicinae*'" in French R. – Wear A. (eds.), *The Medical Revolution of the Seventeenth Century* (Cambridge: 1989).

35 George Cheyne, *The English Malady* (London – Bath: 1733) 91.

36 Cheyne, *The English Malady* 53.

37 See Cheyne, *The English Malady* 43 and 75.

Functions'.[38] In this enterprise he now advertised, in place of numbers and geometry, an analogical approach to the 'Whole of *Animal Nature*' based on the insight that 'we find always *similar* Effects have *similar* Causes', a principle without which 'many Appearances in *Generation, Nutrition,* and *Animation*' would 'otherwise appear unaccountable'.[39]

The tendency from mechanistic and mathematical to vitalistic schemes of explanation was not restricted to the Pitcairne-Gregory circle; it was much more pervasive in contemporary medicine in Britain.[40] For example, in Bernard Mandeville's *A Treatise of the Hypochondriack and Hysterick Passions*, first published in 1711, and a second revised edition followed in 1730.[41] As John P. Wright points out, there is a significant difference between the two editions.[42] In the first, Mandeville denies the possibility of thinking matter, but in the second edition he claims material implementation necessary for thinking. Beside the apparent turn from iatromechanism, Mandeville had grown critical about the usefulness of mathematics in medicine, and instead of elaborating detailed theories of diseases he recommended careful observation as the key to successful treatment of diseases.[43] He was highly critical of vain theoretical, for him quasi-religious debates between Aristotelians and Cartesians, and recommended cognitive humility in medical matters, because e.g. 'Our shallow Understandings will never penetrate into the Structure of Parts of that amazing as well as mysterious Composition, the Mass of Blood' therefore we should not 'assert any more of it than what Observation will allows us'.[44]

By the middle decades of the eighteenth century, the scholarly community had grown much less enthusiastic about the prospects of the general applicability mathematics to various fields of learning. Pitcairne, and some of his Jacobite and Newtonian allies like John Friend,[45] had been accused of making mathematics subservient to sectarian interest, and in connection with a

38 Cheyne, *The English Malady* 94f.

39 Cheyne, *The English Malady* 76 and 99.

40 See e.g. Brown T.M., "From Mechanism to Vitalism in Eighteenth-Century Physiology", *Journal of the History of Biology* 7 (1974) 179–216.

41 Bernard Mandeville, *A Treatise of the Hypochondriack and Hysterick Passions* (London: 1711). The second edition appeared with a slightly different title: *A Treatise of the Hypochondriack and Hysterick Diseases*.

42 See Wright J.P., *Hume's* Treaties of Human Nature*: An Introduction* (Cambridge: 2009) 9.

43 Guerrini, *Obesity and Depression*, 99

44 Mandeville, *A Treatise* 115–116 and 137.

45 On Friend's role in this context see Rowlinson J.S., "John Freind: Physician, Chemist, Jacobite, and Friend of Voltaire's", *Notes and Records of the Royal Society of London* 61 (2007) 109–127.

controversy with Leibniz,[46] Pitcairne himself had been forced to realize that even Newton's *Principia* could be 'orangically & Hanoverianlie abus'd'.[47] Similar, but perhaps less dramatic tendencies took place in moral philosophy at the same time. Francis Hutcheson's *An Inquiry into the Original of Our Ideas of Beauty and Virtue*, first published in 1725, was originally subtitled as 'an Attempt to introduce a Mathematical Calculation in Subjects of Morality'. In later editions the subtitle has been removed,[48] and the role of mathematical formulae started to diminish.[49] Besides, his appeal to the crucial role of our moral sense in his theory of moral evaluation paved the way for philosophical theories emphasizing various forms of sensitivity in explaining human cognitive and social functioning—most importantly exemplified by David Hume's philosophy.

The flight from mathematics as a safe heaven in the human sciences can be related to the specific social and political context of early eighteenth-century Scotland. The weakening of the mathematical ideal took place in an atmosphere of general, i.e. political, economic and cultural improvement—a context in which sensibility, refinement and civility provided the fundamental categories of understanding human functioning. The problems of social stability, sectarian violence and the legitimate scope of monarchical power gradually faded away and gave way to the issues associated with economic, social and cultural backwardness.[50] In this context the ideal of liberal government replaced that of the monarchical one: the image of self governing agents cooperating and being bond together due to their natural sociability rather than coercion becomes central.[51] Against this background a new image of man started to emerge with an increased emphasis on the sensitivity and affections of humans rather than on their calculable mechanical functioning subsumed under mathematical formulae.

---

46 See Hall R.A., *Philosophers at War: The Quarell between Newton and Leibniz* (Cambridge: 1980) 132–134.

47 Schaffer S., "The Glorious Revolution and Medicine in Britain and the Netherlands", *Notes and Records of the Royal Society of London* 43 (1989) 167–190, 175–176.

48 Hutcheson F., *An Inquiry into the Original of Our Ideas of Beauty and Virtue* (Indianapolis: 2008) 199.

49 Four versions of the fourth edition have been identified, two of which lacks the mathematical formulation, and speak about 'Rule' instead of 'Canon', although the calculating spirit remains in the text. See Hutcheson, *An Inquiry* 240–241.

50 See Lawrence, "The Nervous System".

51 For a useful discussion see Poovey, *A History of the Modern Fact* 157–175.

Vitalistic Image of Man and Liberal Governmentality

By the mid-eighteenth century, in plausible connection with the changing social, political and economic situation, the intellectual climate in Scotland had changed, and anti-Jacobite ideologies could also find its expression from theories of nature to society. A new image of man started to take shape which conformed much better to the new social order emerging in the aftermath of the Glorious Revolution and Union of 1707 with England. This new image had been elaborated in terms of physiological, moral, social and economic theories that can be understood as commenting on and justifying the new, liberal form of government that replaced the monarchical ideals and practices admired by those with Jacobite sympathies.

Politically, the Union eventually offered the stability that was very much strived for in Scotland. After two decades of economic depression the economy started to revive, and the benefits of the Union could be gradually felt. From the 1720s economic development opened up more optimistic perspectives in various fields of social life. The emerging new political and economic elites—especially Lowland Whigs, most notably the Argylls—were both supportive of and actively engaged in initiatives to reform universities, establish scholarly societies or improve agriculture. Their devotion to philosophical and institutional issues was motivated by the need of catching up with England intellectually and economically, and in order to promote development in these fields they instituted a conscious policy of patronage.[52] This policy was focused on the *improvement* of various aspects of Scottish life and their active engagement in this enterprise lent them political legitimacy and served their political and social ambitions, particularly their struggle against religious fundamentalists and Highland clans, and it also contributed to their economic influence by the implementation of more refined technologies in agriculture, forestry and coal-mining. In this process the University of Edinburgh had been restructured and a medical school was founded in 1726. Within a few decades the newly founded medical faculty developed into a leading centre of medical learning in Europe comparable only to Montpellier, Halle and Leiden.[53]

---

52 For a detailed discussion of the history of patronage in the Scottish Enlightenment see Emerson R.L., *Academic Patronage in the Scottish Enlightenment* (Edinburgh: 2008).

53 See e.g. Emerson, R.L., "The Founding of the Edinburgh Medical School", *Journal of the History of Medicine and the Allied Sciences* 59 (2004) 183–218, and Dingwall, H., "The Importance of Being Edinburgh: The Rise and Fall of the Edinburgh Medical School in the Eighteenth Century", in Grell O.P. – Cunningham A. – Arrizbalaga J. (eds.), *Centres of Medical Excellence? Medical Travel and Education in Europe 1500–1789* (Aldershot: 2010) 305–324.

It is not to say that the theories to be discussed here were unique in the European context and as such they are to be ascribed to the specific social-political context in which they developed. Enlightenment vitalism is a pan-European phenomenon, and not specific to Scotland or Britain. It is rather that in the Scottish context a special meaning can be attached to these theories as they provide support to a specific social imagery from outside political discourse. But the most important feature that distinguishes the developments in Scotland is that in a period of some four decades there emerged in the Glasgow-Edinburgh axis a continuum of theories ranging from the phenomena of nature to society that bear the traces of a vitalistic worldview. These theories, taken as a whole, provide an exposition of a *Weltanschauung* that took shape in this temporally and spatially local context.

It is in the context of large-scale improvement that physiology and the 'science of man' in Scotland took a decisively vitalistic turn: a turn away from the "strong program" of Newton's *Principia* to a different one largely inspired by the experimentalism and the Queries of the *Opticks*. Although the first professors of the newly founded Edinburgh medical school were educated at Leiden, and they imported to Scotland a Boerhaavean mechanistic approach, Edinburgh quickly turned into a centre of vitalistic physiology and, from the second third of the century, offered alternative theories to Boerhaavean medical orthodoxy. Edinburgh professors like William Porterfield and Robert Whytt emphasized the *active* influence the mind exerts on physiological processes.[54] Porterfield developed a vitalistic account of binocular motions that enable us to judge the distance of objects, and then extended it to other bodily motions too. In his view it is custom and habit, arising from a rational and voluntary decision, that stabilize the processes as a result of which we cannot but constantly focus our eyes. This habit thus becomes a law that the mind imposes on itself because of the intrinsic utility it has in judging distance.

As his student notes testify, Whytt attended the classes of George Young, an adjunct teacher at Edinburgh medical school, who taught him to be sceptical about mechanical explanations of animal oeconomy because, as he saw it, presupposing a hidden mechanism behind muscular motion is empirically ungrounded. Whytt, similarly explained bodily responses as arising from 'an active sentient principle' of which we may lack sufficient theoretical knowl-

54 See Wright J.P., "Metaphysics and Physiology: Mind, Body, and the Animal Economy in Eighteenth-Century Scotland", in Stewart M.A. (ed.), *Studies in the Philosophy of the Scottish Enlightenment* (Oxford: 1990); and his "Materialism and the Life Soul in Eighteenth-Century Scottish Physiology", in Wood P. (ed.), *The Scottish Enlightenment: Essays in Reinterpretation* (Rochester: 2000).

edge, but we can know its workings from the direct experience of how it *feels*. Although its workings are frequently unconscious, it is due to us being habituated to them and to them being gentle themselves.

Gradually distancing himself from Porterfield's theory, with the introduction of a sentient principle Whytt offered the core idea of a potentially unified account of bodily processes replacing mechanism (and rationality) with feeling as its guiding principle. Whytt's was a picture of various parts of the body communicating via the nervous system and responding to stimuli involuntarily and unconsciously. Although they disagreed in several respects, Porterfield and Whytt agreed on at least one point crucial in the present context which may be called their common vitalistic stance: namely that living organisms are active in the sense that they respond with more energy than contained in the stimuli, so they cannot be studied along the same lines as dead matter, and particularly, they cannot be subject to study on mechanical principles. In the explanation of living matter the perspective of mechanical aggregation must therefore give way to that of organization.

William Cullen's chemical and medical investigations from the 1740s, first in Glasgow then in Edinburgh, were at the forefront in the process of reinterpreting Newton's aether as a vitalistic concept and along with it the refocusing of natural and medical inquiry. Cullen was dissatisfied with the extension of the strong program to chemical and medical investigation because it was in these fields in which the limitations of this program were clearly perceived. As his early nineteenth-century biographer, John Thomson noted, Cullen perceived that while earlier investigations "showed what might be achieved by mechanical principles and mathematical reasoning to physiology, indicated also what they were unable to accomplish."[55] Cullen appreciated their explanatory potential for 'phenomena depending on the general properties of matter',[56] in chemistry he thought the mechanical hypothesis ill-founded because he saw the reducibility of chemical phenomena, i.e. the particular properties of substances, to mechanical phenomena far from being warranted, and he was unsure about its possibility—even in principle. There are phenomena seemingly unexplainable by reference to those general properties: in order, for example, to explain how ice turns into water mechanical accounts frequently return to the supposition that heat changes the angular particles of solid ice

55 Thomson J., *An Account of the Life, Lectures and Writings of William Cullen*, 2 vols. (Edinburgh: 1832), vol. II, 677. See also William Cullen, "Lectures Introductory to the Course on the Practice of Physic", in William Cullen, *The Works of William Cullen*, vol. I, London & Edinburgh: 1827) 398. See also Schofield, *Mechanism and Materialism* 191.

56 Quoted in Thomson, *An Account*, vol. I, 665.

into spherical ones of fluid water, the latter being more 'easily moved, which is fluidity'. But as we cannot deduce fluidity from spherical parts, nor *vice versa*, we have to appeal to some other cause,[57] which may not fit the ideals of mechanists. Therefore, seeking explanations in terms of mechanical philosophy for chemical phenomena is neither possible nor desirable. Should we be able to find a mechanical basis for chemical phenomena, it would still fall short of an explanation in terms of proximal causes. And these, not the fundamental mechanisms, are the causes we are mostly interested in when pursuing philosophical chemistry whose chief use is to improve existing practices from agriculture to medicine.

Instead of dubious mechanical hypotheses, most of the explanatory work for Cullen's chemical enterprise is done by *elective attractions* that are to be described and classified on the phenomenal level, because their underlying causes are proclaimed to be unknown. Elective attractions thus become the cement of the chemical universe, but not in a sense modelled on Newtonian gravity: while gravity is a universal attraction, Cullen's elective attractions are selective depending on the particular properties of substances and their relative attractions, and not on their density. The business of chemistry is thus to describe and arrange elective attractions systematically, and to account for various combinations and separations of substances in terms of general principles established by such classifications.

On similar grounds, Cullen considered relying on exclusively mechanical principles equally problematic in physiology, not only because we do not perceive the mechanical means of our internal functioning, but also because a mechanical outlook cannot lead to satisfactory explanation in too many cases: the stomach, for instance, 'does not seem by any mechanical powers to contribute' to the food's 'division' while digesting; nor can the workings of the lungs be fully described in terms of the 'mechanical powers of pressure, commonly spoken of'.[58] Harvey's discovery of the circulation of blood gave an impetus toward understanding animal economy as a '*hydraulic system*', and thereby it contributed one aspect to its understanding as an '*organic system*', but this approach could not supply the mathematical means with which to study physiological phenomena.[59] And not only that: it also overshadowed the adequate complex outlook from which 'the human system can only be viewed [...] that is, as a

57 See Donovan A., *Philosophical Chemistry in the Scottish Enlightenment* (Edinburgh: 1975) 100.

58 William Cullen, "Institutions of Medicine", in William Cullen, *The Works of William Cullen*, vol. I, (London & Edinburgh: 1827) 111 and 191–193.

59 Cullen, "Lectures Introductory" 396 and 400–402.

chemical mixt, as a *hydraulic machine*, and as an *animated nervous frame*'.[60] The combination of these three points of view can make 'the system of physic' complete, but two of which, i.e. the chemical mixt and the animated nervous system, are hardly susceptible of understanding in mechanical terms.

The language of mechanism, for Cullen, is thus not the universal language of nature. This insight inspired a closer understanding of what specific forms the internal activity of a living body may take, i.e. 'the state and affections of the primary moving powers in it'.[61] Most of the crucial bodily functions Cullen ascribes to the 'mechanism of the brain' which could not fulfil its various functions 'without being united with a sentient principle or mind that is constantly present in the living system'.[62] Without there being such a principle not even the mechanical functions of the body could be adequately explained: how can the heart keep pumping blood without running down? For Cullen the explanation came from the brain and its close connection to the mind, the sentient principle responsible for the effects greater than the stimuli.

However, Cullen did not see this internal active force as centralized exclusively in the mind/brain, he distributed some of it throughout the various parts of the body: activity for him partly resides in the 'inherent power' of the muscles. And thus, while the mind/brain is the central unit, some of the bodily activities depend on various local forces that together form an organic whole. Cullen's physiological outlook is well represented in this telling passage:

> opium, alcohol, mephitic air, applied to our bodies, induce a state of sleep; they are known to diminish the motions in general, and have got the appellation of sedatives. With regard to the chief of them all, opium, the question has been often put, *quomodo opium facit dormiret* and the variety of theories offered by the mechanical physicians has amounted to little more than that of the Galenists, *quia habet in se facultatem dormifaciendi*. It has been alleged by some, that opium coagulates, and by others, that it rarifies the blood; but we say, that opium produces its effect independently of the fluids and of their circulation. Whatever difficulties Dr. Haller has raised upon this subject, I say that the experiments of Alston, Whytt, and Monro, our colleagues, upon animals, after the circulation of the blood had ceased, are quite conclusive; that though opium acts slower, it most certainly does act, after all motion of the fluids have ceased; nay, that it acts upon every separate and detached part, even

60 Ibid. 409.

61 Ibid.

62 Ibid. 114.

> when the communication with the brain is destroyed, that it acts upon the inherent power, so that we need not discuss the matter whether it coagulates or rarifies the blood, as its direct operation is upon the nervous power, the mobility, sensibility, and irritability of which, it destroys in every particular part to which it is applied.[63]

Beyond the diagnosis that mechanists do not fare better than Galenists as far as the intelligibility of their explanations go, Cullen here seems convinced that the effect of the opium is local and does not presuppose circulations in the body. Its effect is diminishing the characteristic activity of some part of the body by influencing its local 'nervous power' which Cullen considered to reside in the relevant muscles themselves.[64] As such they belong to the 'animated nervous frame', partly decentralized, which is itself part of a harmonious mechanical, chemical and physiological whole.

The vitalistic reorientation of natural inquiry and medical theory was not a unique Scottish phenomenon. As Peter Hanns Reill has shown, there was a vitalistic movement in the Enlightenment which responded to problems, particularly those of living matter, that mathematized mechanical theories could not solve. This led to a revival of natural history, most importantly represented by Buffon—and in the domain of studying moral phenomena, by the moral philosophers of the Scottish Enlightenment. Its methods were not based on mathematics, but 'on the principles of comparison, resemblance, affinity, analogical reasoning', and its explanations in terms of 'inner, active forces as central agents in nature' replaced the mechanists' view of external forces acting on inert matter.[65]

Andrew Cunningham suggested that a vitalistic outlook is characteristic of Hume's theory of the mind whose essential feature is self-activity,[66] and I have argued elsewhere that Hume's vitalism is in concert with his qualitatively oriented experimental method that aims at revealing the distinctive contribution various faculties make.[67] These faculties exert *active* influence by trans-

63 Ibid. 124.

64 Bynum W.F., "Cullen and the Nervous System", in Doig A. et al. (eds.), *William Cullen and the Eighteenth-Century Medical World* (Edinburgh: 1993) 152–162, 159–160.

65 Reill, *Vitalizing Nature* 69.

66 Cunningham A., "Hume's Vitalism and Its Implications", *British Journal for the History of Philosophy* 15 (2007) 59–73.

67 Demeter T., "Hume's Experimental Method", *British Journal for the History of Philosophy* 20 (2012) 577–599, and see also Demeter T., "Liberty, Necessity and the Foundations of Hume's Science of Man", *History of the Human Sciences* 25 (2012) 15–31.

forming the passive material of impressions and ideas: sympathy turns ideas into impressions thus enabling us to feel what others feel; upon experiencing one event regularly following another habit provides us with a secondary impression as the basis of our idea of necessary connection, and so on.[68] Human nature thus becomes a compound entity consisting of functional components characterised by their active contribution in terms of predominantly non-mechanical principles. The science of man is essentially the enterprise of charting them through the study of their interactions.

By deploying a vitalistic language in the study of moral beings, David Hume identifies 'elective attractions' in human cognitive architecture and a similar 'sentient principle' in human society. Hume's famous principles of association can be understood on the analogy of chemical processes directed by *elective attractions* between ideas that depend on their qualitatively different representational content. The principles of association are elective: they do not hold universally between all ideas, only between some, and there are, of course, pairs of ideas that do not stand in associative relations at all. The possible associative links between any two ideas depend on their *content*: the principle of cause-effect, for example, can connect two ideas that may not be connected by resemblance. If seen through a mechanical lens, different chains of association cannot be adequately distinguished: focusing on the solidity, number and structure of ideas does not give a fine enough resolution for that purpose: one needs to take into account the representationally heterogeneous nature of ideas. Beside this qualitative heterogeneity, one should also consider the particular *principle*, 'some associating quality',[69] by which ideas are linked in a chain of association. On any occasion indefinitely many ideas can be associated to an idea actually given to the mind. Only by pointing out the mind's internal self-activity can one explain why a particular principle of association is applicable in any given case, and thus also why a particular idea (and not some other) is associated to the actually given one.

Similarly, Hume's concept of sympathy can be read as an active sentient principle, which transforms ideas into impressions, facilitating communication of opinions and affections. As it makes us sensitive to the feelings of others, this principle can aptly be called the basis of sociability. Sympathy is responsible for the bonds in the social world, and as such it is analogous with the cohesive force in the world of living organisms:

---

68 Hume David, *Treatise of Human Nature* (Oxford: 2007) 2.1.11.3 and 1.3.14.16–18.

69 Ibid. 1.1.4.1.

> this is still more remarkable, when we add a *sympathy* of parts to their *common end*, and suppose that they bear to each other, the reciprocal relation of cause and effect in all their actions and operations. This is the case with all animals and vegetables; where not only the several parts have a reference to some general purpose, but also a mutual dependance on, and connexion with each other.[70]

Sympathy establishes similar reciprocal relations in human interaction, as it is due to its functioning that 'the minds of men are mirrors to one another'.[71] It is thus by the concept of sympathy where the ideas of an organic nature and human nature, the language of chemical reactions and human interactions are contiguous: living things, human minds and society are all organized by the peculiar principles of their own economy into an organic whole.

It is arguably through Hume's work that a vitalistic vocabulary extended its influence beyond the disciplinary boundaries of chemical and medical investigations into the wider realm of the moral sciences, and by the second half of the eighteenth century it had become the received language of discussing human phenomena. Adam Ferguson emphasizes the explanatory deficiencies of analogies drawn between the inanimate material world and society on the basis that the latter is composed of 'living and active members'.[72] For him, adopting the perspective of mechanical theories of inert matter in moral philosophy can only result in overlooking the dynamic nature of social phenomena.

In a similar vein, there are traces of a profound influence of a vitalistic outlook and language in Adam Smith's economic theory. The terms and examples Smith relied on in describing the economical processes are frequently borrowed from the contemporary vocabulary of discussing organized living bodies. For him the functioning of the 'body politick' is governed and preserved in a healthy state by the principles of a 'political oeconomy' described in terms similar to those of the 'animal oeconomy' of living creatures, and more specifically humans.[73] His economic theory is centred on the idea of a natural balance in the economic body maintaining itself with its own internal active forces that tend to restore its 'health' when some of its 'vital parts are overgrown'.[74] Smith depicts this body as a living organism whose activities are conceived as

70 Ibid. 1.4.6.12.

71 Ibid. 2.2.5.21.

72 Ferguson Adam, *Essay on the History of Civil Society* (Edinburgh: 1966) 268–269.

73 See Reill P.H., "Eighteenth-century Uses of Vitalism in Constructing the Human Sciences" in Alexander D.R. – Numbers R.L. (eds.), *Biology and Ideology: From Descartes to Dawkins* (Chicago: 2005) 61–87.

74 Smith Adam, *The Wealth of Nations* (Oxford: 1976) iv.vii.c.

interconnected parts of a larger whole whose balance is maintained by 'some unknown principle of preservation' explicitly compared to the unknown, vital 'principle of animal life'.[75]

Arguably, in mid-eighteenth-century Scotland a continuum of vitalistic theories from physiology through the science of man to the study of society has been developed. In connection with its focus on qualitative differences, feelings and sensitive capacities, this thrust of inquiry was much more sceptical about the use of mathematics in describing human phenomena. A mathematical representation is more suited to the study of phenomena represented as consisting of qualitatively homogeneous constituents interacting in mechanically respectable ways rather than of those described in terms of qualitative differences, affinities and sensitivity. Not surprisingly therefore, the turn to a vitalistic language of description brought along a decline in the reputation of mathematics in inquiry.

Hume draws a sharp epistemological distinction between two kinds of reasoning.[76] Demonstrative reasoning is *a priori*, it is concerned with relations of ideas, and mathematics is one of its exemplary fields. Probable reasoning is based on the relation of cause and effect, it is *a posteriori*, and it provides the foundations of theorizing concerning all matters of fact. This means that *a priori* mathematical constructions cannot be taken as representations of reality as "the only objects of the abstract science or of demonstration are quantity and number, and that all attempts to extend this more perfect species of knowledge beyond these bounds are mere sophistry and illusion."[77]

This does not entail, however, that mathematics is altogether useless in natural inquiry, but it could no longer serve as a model of reality, only as a handmaiden of experiment and observation while exploring the operations of nature. As Hume says:

> Mathematics, indeed, are useful in all mechanical operations, and arithmetic in almost every art and profession: But 'tis not of themselves they have any influence. Mechanics are the art of regulating the motions of bodies to some design'd end or purpose; and the reason why we employ arithmetic in fixing the proportions of numbers, is only that we may discover the proportions of their influence and operation.[78]

---

75 Ibid. ii.iii.31. See also Packham C., "The Physiology of Political Economy: Vitalism and Adam Smith's Wealth of Nations", *Journal of the History of Ideas* 63 (2002) 465–481, 468.

76 Hume, *Treatise* 2.3.3.2.

77 Hume David, *Enquiry concerning Human Understanding* (Oxford: 2000) 12.27.

78 Hume, *Treatise* 2.3.3.2.

Quantification is thus all right in Hume's eyes if it is about measuring proportions, or the magnitude of causes and effects. We can rely on mathematics as a useful tool in natural philosophy, and especially in its application, but we cannot proceed on *a priori* mathematical principles in our inquiries concerning matters of fact. Natural philosophy, being concerned with matters of fact, cannot be based on mathematical axioms, and we cannot have it as *essentially* mathematical. Given human cognitive capacities, mathematics just cannot be a meaningful language in which the book of nature is written, and mathematics cannot be taken meaningfully as the model of reality.

Cullen, whose knowledge of mathematics was admittedly limited,[79] was not in disfavour of quantification and measurement either. As John Christie points out, it was Cullen who initiated a research program of thermometrical quantification, which characterised Scottish chemistry in the second half of the eighteenth century through the work of Joseph Black, John Robison and William Irvine.[80] However, Cullen shared Hume's aversion from proceeding on mathematical principles concerning matters of fact. For example, he reluctantly acknowledges that the combination of Galileo's mathematical and Bacon's experimental approach had some role to play in the history of medicine: 'we must observe, that whether it was with advantage or not, many improvements have been derived from mathematics to the system of physic: they have certainly contributed to put physic in the good condition in which it is at present'.[81] But he also emphasizes that this contribution had been limited and mathematical principles in medical matters cannot have a bright future either: 'it neither could, nor ever can be, applied to any great extent; in explaining the animal economy'.[82] Just like its mechanization, the mathematization of animal economy could not deliver the complete system it had promised, and for very much the same reason: only some parcels of medicine could be effectively treated this way.

For Hume, Cullen, Smith and many of their contemporaries a qualitatively oriented and vitalistic outlook provided a more natural framework for discussing the phenomena of nature and human nature than a mechanistically oriented mathematical one. The vitalistic language and theories developed by various Scottish natural and moral philosophers reflect a common outlook on systems of natural and social phenomena in which active parts cooper-

79 See Thompson, *An Account*, vol. I, 138.

80 Christie J.R.R., "William Cullen and the Practice of Chemistry" in Doig A. et al. (eds.), *William Cullen and the Eighteenth-Century Medical World* (Edinburgh: 1993) 98–109, 106.

81 Cullen, "Lectures Introductory" 401.

82 Ibid. 398.

ate in order to achieve a common end. This vitalistic outlook on nature and human nature was more congruent with the ideology of improvement than the *Principia*'s language of nature with its emphasis on inert matter moved and organized by external forces. In the same way as the Edinburgh elite perceived itself, so nature, society, human physiology and cognitive architecture were seen as structures of interacting sensibilities, binding together and controlling the whole. There was no room for sovereign functions in the description of the body and society. Rather than looking for the influences of philosophy on medicine or the reverse, both can be interpreted by referring them to the common social context. To do this is not to suggest the models of the body and human nature was a celebration of the social order. They arose concurrently with that social order and as part of the whole ideology of improvement and its associated *Weltanschauung*.

Selective Bibliography

Cullen William, *The Works of William Cullen*, vol. 1 (London & Edinburgh: 1827).

Donovan A., *Philosophical Chemistry in the Scottish Enlightenment* (Edinburgh: 1975).

French R. – Wear A. (eds.), *The Medical Revolution of the Seventeenth Century* (Cambridge: 1989).

Guerrini A., *Obesity and Depression in the Enlightenment: The Life and Times of George Cheyne* (Norman: 2000).

Lawrence C., "The Nervous System ad Society in the Scottish Enlightenment", in Barnes B. – Shapin S. (eds.), *Natural Order: Historical Studies of Scientific Culture* (London: 1979) 19–40.

McLynn F., *The Jacobites* (London: 1985).

Pitcairne Archibald, *The Whole Works of Dr. Archibald Pitcairn* (London: 1727).

Poovey M., *A History of the Modern Fact: Problems of Knowledge in the Sciences of Wealth and Society* (Chicago: 1998).

Reill P.H., *Vitalizing Nature in the Enlightenment* (Berkeley: 2005).

Schofield R., *Mechanism and Materialism. British Natural Philosophy in an Age of Reason* (Princeton: 1970).

Wilson D.B., *Seeking Nature's Logic: Natural Philosophy in the Scottish Enlightenment* (University Park, Pennsylvania: 2009).

Index Nominum

Index Animalium—Index of Animals

This Index contains the names of both existing and imaginary animals. As far as they can be specified, animal species (and occasionally subspecies) are indicated by their English and scientific names (in italic), according to the *EOL* (*Encyclopedia* of Life, http://eol.org/). Ancient and early modern Latin and Greek names are also included in the list. Names of class, order, family, or genus are generally not included, but general and common names ("Butterfly", "Chameleon", "Chicken", etc.) are. Imaginary hybrids between man and animal are doubly indicated: alphabetically, and thematically under "Man, Imaginary hybrids".

Note on "Leopard" and "Panther": In early modern natural history, leopard and panther are often seen as distinctive species, the leopard being a hybrid from a lion and a panther, as is indicated by its etymology.

www.ingramcontent.com/pod-product-compliance
Lightning Source LLC
Chambersburg PA
CBHW071328230926
51849CB00002B/343
9789004268234